R & D,
Patents, and
Productivity

A National Bureau
of Economic Research
Conference Report

R & D,
Patents, and
Productivity

Edited by   Zvi Griliches

 The University of Chicago Press

Chicago and London

ZVI GRILICHES is professor of economics at Harvard University, and program director, Productivity and Technical Change, at the National Bureau of Economic Research.

The University of Chicago Press, Chicago 60637
The University of Chicago Press, Ltd., London

LIBRARY OF CONGRESS CATALOGING IN PUBLICATION DATA

Main entry under title:

R & D, patents, and productivity.

Papers presented at a conference held in Lenox, Mass., in the fall of 1981, and organized by the National Bureau of Economic Research.
Includes indexes.
1. Research, Industrial—United States—Congresses.
2. Patents—United States—Congresses. 3. Industrial productivity—United States—Congresses. I. Griliches, Zvi, 1930–   . II. National Bureau of Economic Research. III. Title: R and D, patents, and productivity.
HD30.42.U5R2   1984   338′.06   83-18121
ISBN 0-226-30883-9

Relation of the Directors to the
Work and Publications of the
National Bureau of Economic Research

1. The object of the National Bureau of Economic Research is to ascertain and to present to the public important economic facts and their interpretation in a scientific and impartial manner. The Board of Directors is charged with the responsibility of ensuring that the work of the National Bureau is carried on in strict conformity with this object.

2. The President of the National Bureau shall submit to the Board of Directors, or to its Executive Committee, for their formal adoption all specific proposals for research to be instituted.

3. No research report shall be published by the National Bureau until the President has sent each member of the Board a notice that a manuscript is recommended for publication and that in the President's opinion it is suitable for publication in accordance with the principles of the National Bureau. Such notification will include an abstract or summary of the manuscript's content and a response form for use by those Directors who desire a copy of the manuscript for review. Each manuscript shall contain a summary drawing attention to the nature and treatment of the problem studied, the character of the data and their utilization in the report, and the main conclusions reached.

4. For each manuscript so submitted, a special committee of the Directors (including Directors Emeriti) shall be appointed by majority agreement of the President and Vice Presidents (or by the Executive Committee in case of inability to decide on the part of the President and Vice Presidents), consisting of three Directors selected as nearly as may be one from each general division of the Board. The names of the special manuscript committee shall be stated to each Director when notice of the proposed publication is submitted to him. It shall be the duty of each member of the special manuscript committee to read the manuscript. If each member of the manuscript committee signifies his approval within thirty days of the transmittal of the manuscript, the report may be published. If at the end of that period any member of the manuscript committee withholds his approval, the President shall then notify each member of the Board, requesting approval or disapproval of publication, and thirty days additional shall be granted for this purpose. The manuscript shall then not be published unless at least a majority of the entire Board who shall have voted on the proposal within the time fixed for the receipt of votes shall have approved.

5. No manuscript may be published, though approved by each member of the special manuscript committee, until forty-five days have elapsed from the transmittal of the report in manuscript form. The interval is allowed for the receipt of any memorandum of dissent or reservation, together with a brief statement of his reasons, that any member may wish to express; and such memorandum of dissent or reservation shall be published with the manuscript if he so desires. Publication does not, however, imply that each member of the Board has read the manuscript, or that either members of the Board in general or the special committee have passed on its validity in every detail.

6. Publications of the National Bureau issued for informational purposes concerning the work of the Bureau and its staff, or issued to inform the public of activities of Bureau staff, and volumes issued as a result of various conferences involving the National Bureau shall contain a specific disclaimer noting that such publication has not passed through the normal review procedures required in this resolution. The Executive Committee of the Board is charged with review of all such publications from time to time to ensure that they do not take on the character of formal research reports of the National Bureau, requiring formal Board approval.

7. Unless otherwise determined by the Board or exempted by the terms of paragraph 6, a copy of this resolution shall be printed in each National Bureau publication.

(Resolution adopted 25 October 1926, as revised through 30 September 1974)

Contents

Acknowledgments

This volume consists of papers and discussions presented at a conference held in Lenox, Massachusetts, and of related papers distributed as background papers at the conference. The conference was organized by the Productivity and Technical Change Studies Program of the National Bureau of Economic Research and was financed primarily by grants from the Policy Research Analysis Division of the National Science Foundation (PRA79-13740; PRA81-08635) and a National Science Foundation grant to Harvard University (SOC78-04279), which also supported the research program on which many of the papers presented at the conference were based. In addition the conference benefited directly and indirectly from funds provided by the National Bureau of Economic Research Capital Formation Project. We are grateful to the sponsors of that project.

We are indebted to many people who helped make the conference a success and who made the production of this volume possible, but especially to Jeanette DeHaan, Kirsten Foss, Rochelle Furman, Mark Fitz-Patrick, and Annie Zeumer.

1 Introduction

Zvi Griliches

In the spring of 1960 a conference was held at the University of Minnesota under the auspices of the Universities-NBER Committee for Economic Research, resulting in the publication of *The Rate and Direction of Inventive Activity* (Nelson 1962), a volume that still serves as a major statement and source book of economic ideas in this field. In the fall of 1981 the Productivity and Technical Change Studies Program of the National Bureau of Economic Research (NBER) organized a conference on R & D, Patents, and Productivity at Lenox, Massachusetts. This was the first NBER conference devoted entirely to R & D related topics since 1960. This volume which is an outgrowth of that conference, contains revised versions of papers presented at the conference plus a number of additional related papers which were distributed at the conference as background papers. Most of the latter papers report on ongoing research projects at NBER.

The major themes of research in this field were already clear at the Minnesota conference: The belief that invention and technical change are the major driving forces of economic growth; that economists have to try to understand these forces, to devise frameworks and measures which would help to comprehend them and perhaps also to affect them; that much of technical change is the product of relatively deliberate economic investment activity which has come to be labeled "research and development" and that one of the few direct reflections of this activity is the number and kind of patents granted to different firms in different years. At the earlier conference there had already been much discussion of the validity and utility of patent statistics (cf. the papers by Kuznets, Sanders, and Schmookler and the associated discussion in Nelson 1962) and

Zvi Griliches is professor of economics at Harvard University, and project director, Productivity and Technical Change, at the National Bureau of Economic Research.

attempts to relate R & D investments to their subsequent effects on the growth of total factor productivity (cf. Minasian 1962).

In the two decades that passed, the field of economics expanded tremendously with a concomitant increase in specialization in this and related fields. The Lenox conference, therefore, had a much narrower focus. While the Minnesota conference included papers by historians, social psychologists, and economic theorists, the Lenox conference concentrated primarily on applied econometric papers describing data building and data analysis efforts which focus on the role of R & D investments as generators of economic progress at the firm and industry level and on the role of patent statistics in helping to illuminate these issues.

The volume before us does not cover the whole field of R & D and innovation studies. It contains few purely methodological or taxonomic papers, and it ignores much of the interesting literature on the success and failure of individual R & D projects (such as the work of Freeman 1974; Teubal 1981; and von Hippel 1982), the work of economic historians and historians of science and technology about individual inventions or industries (cf. Nelson 1962 or Rosenberg 1976), the work of sociologists of science about the operation of big and little science (Cole and Cole 1973; Price 1963) and the construction of science indicators (Elkana et al. 1978; National Science Board 1981), and studies of how optimally to plan, organize, and evaluate R & D projects. It also connects only very loosely with the growing theoretical literature in economics on the interaction of R & D and market structure, the impact of R & D investment on the relative market positions of firms, and the impact of the market structure on the firm's incentives to innovate (see Dasgupta and Stiglitz 1980; Kamien and Schwartz 1982; and Spence 1982 for examples).

The focus of this volume is much narrower. Most of the papers deal with one of the following issues: What is the relationship of R & D investments at the firm and industry level to subsequent performance indicators such as patents, productivity, and market value? How does one formulate and estimate such relationships? What makes them vary across different contexts and time periods? To what extent can one use patent counts as indicators of R & D output? Can one detect the output of R & D in the market's valuation of the firm as a whole? What determines how much R & D is done and how many patents are received? A large number of papers in this volume grew out of several National Science Foundation (NSF) sponsored NBER projects which have been motivated in part by two data developments: (1) the increasing propensity of U.S. firms to report their R & D expenditures publicly (especially since 1972) and the availability of such data in machine readable form (Standard and Poor's Compustat tapes); and (2) the recent computerization of the U.S. Patent Office records and their public availability at reasonable cost. Much of the effort in these projects has been devoted to the acquisition, cleaning, merging, and preliminary analysis of these two data bases. The first two papers (Bound et al.; Pakes and Griliches) report the results of such preliminary analyses, as do several other papers in this volume (Ben-Zion; Griliches; Griliches and Mairesse; Mairesse and Siu; Pakes).

Bound et al. describe the construction and merging of the various data sets, present the basic statistics for the resulting samples (essentially most of the universe of large- and medium-size, publicly traded firms in U.S. manufacturing), and illustrate their potential use by presenting relatively simple analyses of the R & D and firm size and the patents and R & D relationships. They find that the elasticity of R & D expenditures with respect to firm size (measured by sales and gross plant) is close to unity with some indication of slightly higher R & D intensities for both very small and very large firms in the sample. The estimated R & D to patents relationship implies a decreasing propensity to patent with growth in the size of the R & D program throughout the range of their sample, though the conclusion is clouded somewhat by the way the issue of a large number of zero patents observations (for the smaller firms) is treated econometrically. The question of selection bias also remains in their sample. To be publicly traded, small firms must be above averagely successful, which may explain their higher patents to R & D ratios.

The Pakes-Griliches paper is based on an earlier, smaller (but longer) sample and tries to estimate something like a patent production function, focusing especially on the degree of correlation between patent applications and past R & D expenditures and on the lag structure of this relationship. Their main finding is a statistically significant relationship between R & D expenditures and patent applications. This relationship is very strong in the cross-section dimension. It is weaker but still significant in the within-firm time-series dimension. Not only do firms that spend more on R & D receive more patents, but also when a firm changes its R & D expenditures, parallel changes occur in its level of patenting. The bulk of the relationship in the within-firm dimension between R & D and patent *applications* appears to be close to contemporaneous. The lag effects are significant but relatively small and not well estimated. They interpret their estimates as implying that patents are a good indicator of differences in inventive activity across firms, but that short-term fluctuations in their numbers within firms have a large noise component in them. They also find that, except for drug firms, there has been a consistent, negative trend in the number of patents applied for and granted relative to R & D expenditures during their time period of observation (1968–75). Some of the econometric issues raised by their work have been pursued further by Hausman, Hall, and Griliches (1984) and Pakes and Griliches (1982).

Pakes and Schankerman use older European data on patent renewal fees and rates of renewal to infer the rate of depreciation in the private value of patents. This is an important question both for the computation of net rates of return to R & D and for the construction of the various R & D "capital" measures. Since the value of a patent may decline both because of the inability of its owner to maintain complete proprietary rights over the information contained in it and because of the appearance of new information which may supplant it in part or entirely, the resulting decay rate need not be similar to the deterioration rates observed or assumed for physical capital. In fact, they find it to be quite a bit higher, on the order of 25 percent per year, and argue that this may explain, in part, the conflict between relatively high estimates of *gross* private rates of return to R & D investments by Griliches and Mansfield and the rather sluggish growth in R & D during the late 1960s and 1970. The actual net rate of return may not have been all that high.

Evenson examines a variety of international data on trends in patenting, R & D, the number of scientists and engineers, and the ratio of patents granted to nationals versus foreigners by country of origin and country of patent grant. From the many interesting numbers reported in his tables, two major facts emerge: (1) Most of the foreign patents granted in all countries originate in the industrial countries. This is consistent with the hypothesis that developing countries have the advantage of being able to draw on and adopt the technologies of the developed world and that it is in their comparative advantage to do so. Moreover, the data show patenting patterns in the developing countries that are similar to the production patterns in these countries. (2) Patenting per scientist and engineer (and per R & D dollar) declined from the late 1960s to the late 1970s in almost all of the fifty countries for which data are available. Whether this indicates a significant decline in the real productivity of invention was one of the more debated topics at the conference. I shall return to this below.

In his paper, Mansfield reviews a wide-ranging research program on R & D, innovation, and technological change. The studies summarized deal with the composition of R & D, price indexes of R & D, international and "reverse" technology transfer and its impact on U.S. productivity growth, imitation costs, the role of patents and their effects on market structure, and other topics. The first study reviewed indicates that the relationship of productivity increases in an industry to R & D depends on the extent to which it is long-term and basic. Larger firms (but not in more concentrated industries) did devote more of their research funding to basic research but did less than their share of risky and long-term R & D projects. In the second study, data were collected and price indexes of R & D were constructed for a number of industries for 1979 (1969 = 100), showing that the price of R & D rose by more than the GNP deflator during this period. The third study dealt with the effects of federally financed R & D on private R & D investments. The findings could be interpreted as indicating that federally supported R & D expenditures substituted for about 3 to 20 percent of private expenditures and induced an additional 12 to 25 percent increase in private R & D investments. While the direct returns from federally financed R & D projects may be lower, the projects do seem to expand the opportunities faced by firms and induce additional R & D investments by them. Several studies of international technology flows showed significant "reverse flows." Overseas R & D by U.S. corporations resulted (in over 40 percent of the cases examined) in technologies that were transferred "back" to the United States. Moreover, for a small sample of chemical and petroleum firms, the estimated impact of overseas R & D investments on their own overall productivity is quite large and statistically significant. The last set of studies reviewed by Mansfield found that imitation costs were close to two-thirds of the original innovation costs; that they were higher for patented innovations but not by much (except in the drug industry); that patented innovations were imitated surprisingly often (60 percent within four years of their initial introduction); that less than one-fourth of the patented innovations outside the drug industry would not have been introduced without patent protection; and that, in some major industries, concentration-decreasing innovations are a very substantial proportion of all product innovations.

Beggs examines historical data on patenting for twenty selected industries and thirteen census years between 1850 and 1939. By looking at the relationship of movements in relative patenting to movements in relative value added across industries, he arrives at conclusions opposite to those of Schmookler: industries that do relatively better can afford to and do slacken off their inventive efforts. (This interpretation is problematic and is inconsistent with the Griliches-Schmookler 1963 results which also can be interpreted as "relative" comparisons.) Beggs also examines the long annual series on total patents issued from 1790 to date with the help of spectral analysis techniques and finds a five-year cycle and a secondary eight-year cycle, which he interprets as several rounds of "follow-up" inventions which follow major technological breakthroughs. His results are very interesting and suggestive but not very firm both because of the high level of aggregation in this analysis and because they do not appear to be reproducible on the shorter but possibly more relevant series on patents by date of application (rather than issue).

The next set of three papers deals with different aspects of the interrelationship between R & D and the market structure of industries that firms find themselves in and have an influence on. Levin and Reiss present a detailed industry equilibrium model where concentration, R & D intensity, and advertising intensity are all jointly determined by the profit-maximizing actions of individual firms. While emphasis is put on both differences in technological opportunity across industries and differences in the conditions of appropriability of the results of technological innovations, very little directly relevant data are available on these topics. Levin and Reisss use a variety of proxies and estimate their model using simultaneous equations techniques and data for twenty manufacturing industries and three years (1963, 1967, and 1972). They get a strong positive effect of R & D on industry concentration and a negative effect of concentration on R & D intensity which becomes positive for industries with a high share of product rather than process R & D. They also find that government-supported R & D does increase private R & D investments, though the estimated effect is not very large (on the order of .1). The main difficulty with their findings (discussed also in the Tandon comment) is that while the model is fundamentally dynamic, it is estimated on industry *levels* without leaning more heavily on the changes that occurred in these industries over time.

The second Pakes-Schankerman paper also deals with the determinants of R & D intensity, primarily at the firm level but with some attention to the issue of aggregating to the industry level. They set up a model in which the optimal (from a firm's point of view) R & D level depends on the expected market size for the firm's products, the conditions of appropriability of the firm's technological innovations, and the technological opportunities facing the firm. The latter two, which are the determinants of market structure and R & D spillovers and the driving forces in the Levin-Reiss model, are largely subsumed here in an unobservable parameter which varies across firms. They reexamine some of the earlier firm R & D data reported in Grilliches (1980) and conclude that, though the coefficient of variation of research intensity is much larger than that of more traditional factors of production, very little of the observed differences in R & D intensity across firms can be explained by either past or even expected rates of growth in sales or by transitory fluctuations (errors) in the various variables. They attribute the remaining differences to interfirm variance in appropriability environments and technological opportunities. An interesting empirical fact emerges in the second part of their paper, when they expand their data base and add industry sales growth rates to their analysis. These turn out to be more important in explaining firm differences in R & D intensity than the firm's own sales growth rates, which helps explain the empirical fact that at the industry level of aggregation the variance in growth rates does account for much of the variance in R & D intensity. One interpretation of the finding, which they emphasize and which is also consistent with the Levin-Reiss results, is that industry differences in technological and market opportunities predominate in a firm's decision processes. The industry potential matters much more than the firm's own specific past history.

Scott uses the newly collected FTC line of business level data to investigate several interesting hypotheses below the firm level. Given the fact that many of the major R & D performing firms in the United States are large, diversified, and conglomerate, it is interesting to ask: Is their R & D behavior primarily determined by the industrial location of their "lines of business" (division or establishment) or does a common "company" R & D policy exist? Without an affirmative answer to the last part of this question there would be grave doubts about the applicability of various R & D optimizing models which relate to such firmwide variables as the cost of capital or their managerial style. Luckily Scott does provide an affirmative answer. In his data (473 companies, 259 different four-digit level FTC lines of business, and a total N of 3387) he can observe the variation in the R & D to sales ratio (R/S) within firms across their various lines of business. He finds that approximately half of the overall variance in R/S can be accounted for by common company effects, common industry effects, and their interaction, in roughly equal parts. Thus, there appear to be significant differences in company R & D policy above and beyond what would have been predicted just from their differential location within the industrial spectrum. Scott also finds that government-supported R & D encourages company-financed R & D. The effect here is statistically significant but small (on the order of .1). Unfortunately, no clear behavioral model is developed here (or for Mansfield's similar finding) to explain these results.

The next five papers (Griliches, Pakes, Abel, Mairesse and Siu, and Ben-Zion) are connected by their use of the stock market value of the firm either as an indicator of the success of R & D programs or as a measure of expectations and a driving force of subsequent R & D investments. In a brief note (that is republished here), Griliches sets the stage for some of this research by using the market value of a firm as an indicator of the market's valuation of the firm's intangible capital, especially its R & D program and accumulated patent experience. Using combined cross-section time-series data, he finds significant effects for both R & D and patent variables in a market value equation. These effects persist when both individual firm constants and lags in market response are allowed for, but the interpretation of the coefficients become obscure. In particular, holding previous market values and current dividends constant, there should be no additional effects from past or current R & D expenditures on subsequent market valuation unless something unanticipated happens. That is, only the news component in the R & D and patent series should affect changes in market value, and this is, in fact, what some of the results seem to imply.

In a theoretically and econometrically much more ambitious study, only whose abstract is included in this volume, Pakes tries to unravel such effects using modern time-series analysis methods. He uses the reduced form of an intertemporal stochastic optimizing model to interpret the time-series relationships between patents, R & D, and the stock market rate of return. In this interpretation, events occur which affect the market value of a firm's R & D program and what one estimates are the reduced form relationships between percentage increases in this value and current and subsequent changes in the firm's R & D expenditures, its patent applications, and the market rate of return on its stock. His empirical results indicate that though most of the variance in the stock market rate of return is noise (in the sense that it is not related to either the firm's R & D expenditures or its patent applications) there is still a significant correlation between movements in the stock market rate of return and unpredictable changes in both patents and R & D expenditures (changes which could not be predicted from past values of patents and R & D). Moreover, the parameter values indicate that these changes in patents and R & D are associated with large movements in stock market values. The R & D expenditure series appear to be almost error free in this context. Patents, however, contain a significant noise component (a component whose variance is not related to either the R & D or the stock market rate of return series). This noise component accounts for only a small fraction of the large differences in the number of patent applications of different firms, but plays a much larger role among the smaller fluctuations that occur in the patent applications of a given firm over time. The timing of the response of patents and R & D to events which change the value of a firm's R & D effort is quite similar. One gets the impression from the estimates that such events cause a chain reaction, inducing an increase in R & D expenditures far into the future, and that firms patent around the links of this chain almost as quickly as they are completed, resulting in a rather close relationship between R & D expenditures and the number of patents applied for. Perhaps surprisingly, he finds no evidence that independent changes in the number of patents applied for (independent of current and earlier R & D expenditures) produce significant effects on subsequent market valuations of the firm. The data cannot differentiate between different kinds of events that change a firm's R & D level. If his model were expanded by adding sales and investment data, it may prove possible to differentiate between pure technological shocks and demand-induced shifts in the R & D and patenting variables. Even without this distinction and without a precise structural interpretation of the estimated relationships, the current model does yield a useful description of the relationships between the various variables and their timing and a very suggestive interpretation of them.

In his note, which developed out of his comment on the Pakes paper at the conference, Abel shows how to construct a structural model in which there is an explicit connection between the market value of the firm and its current and past expenditures on R & D. This model, which is rather simple and primarily illustrative of how one might go about constructing such structural models, yields two interesting conclusions: (1) It is possible, in his one-capital world, to write the value of the firm as a linear function of the stock of R & D capital (which provides some comfort to the empirical approach outlined in the Griliches note). (2) The value of the firm depends directly on demand shocks (the innovations in the output price process) and their square and so does R & D activity (and presumably also on technological shocks which are not contained in his model), but R & D *expenditures* are a function only of the square of these shocks, of the variance in the output price process which measures the degree of uncertainty in the firm's environment. The last result may help to explain the relatively low observed correlation between the two (R & D expenditures and changes in market value).

Mairesse and Siu analyze the time-series interrelationships between changes in the market value of the firm, sales, R & D, and physical investment using what they call the extended accelerator model. This paper follows the Pakes paper both in approach and in the use of essentially the same data. It differs by not focusing on patents, adding instead sales and investment to the list of series whose interrelationship is to be examined. Two additional differences between these papers should be noted: (1) Pakes defines q (the stock market rate of return) as of the year preceding the R & D expenditures, while Mairesse and Siu define it as concurrent. (2) Parkes relates q to the (logarithmic) level of R & D and patents, while Mairesse and Siu use first differences of the logarithms of levels for their R & D, investment, and sales variables. They find that a relatively simple "causal" model fits their data: "innovations" in both market value and sales "cause" subsequent R & D and investment changes without further feedback from R & D or investment to either the stock market rate of return or sales. There is little evidence of a strong feedback relationship between physical and R & D investment, though there is some evidence of contemporaneous interaction. An interesting conclusion of their paper is that independent changes in sales explain a significant fraction of the changes in R & D (and physical investment) above and beyond what is already explained by changes in the market value of the firm and by lagged movements in R & D itself, implying that using different variables one might be able to separate out the effects of different kinds of shocks in the R & D process. This finding could, of course, be just a reflection of a substantial noise (error) level in the observed fluctuations of the stock market rate of return pointed out earlier.

Ben-Zion examines the cross-sectional determinants of market value, following an approach similar to that outlined in the Griliches note. It differs by not allowing for specific firm constants (except in the last table) and by including other variables, such as earnings and physical investment, in the same equation. He also finds that R & D and patents are significant in explaining the variability of market value (relative to the book value of its assets), in addition to such other variables as earnings. His most interesting finding, from our point of view, is the relative importance of total patents taken out in the industry on the firm's own market value. In his interpretation, patents applied for indicate new technological opportunities in the industry, and these overall opportunities may be more important than a firm's own recent accomplishments, though here again this could arise just from the high error rate in the firm's own patent counts. In the last table, Ben-Zion comes close to reproducing the market value change or stock market rate of return equations in the Griliches, Pakes, and Mairesse-Siu papers. He finds that "unexpected" changes in R & D affect market value significantly, in addition to the major impact of unexpected changes in the firm's earnings.

This set of papers clearly opens up an interesting research area but still leaves many issues unresolved. Like the proverbial research on the characteristics of an elephant, different papers approach this topic from slightly different points of view. Pakes analyzes movements in patents, R & D, and market value; Mairesse and Siu investigate the relationship between R & D, investment, sales, and market value; while Ben-Zion (in his change regressions) looks at R & D, earnings, and market value. We should be able to do a more inclusive analysis in the future, incorporating the various variables in one overall model (or at least description) of these processes.

The Schankerman and Nadiri paper is motivated by the availability of R & D investment anticipations in the McGraw-Hill Surveys data base. It sets up an optimal R & D investment model and derives the equations of motion for actual R & D, anticipated R & D, and their difference—the realization error. Given the anticipations data, the paper shows a way to formulate cost of adjustment models which permit the testing of various expectational schemes: rational versus adaptive versus static expectations. The results are somewhat inconclusive. The pure rational expectations model is rejected by the data. As is true in many such endeavors, it is not clear whether the specific hypothesis is rejected or whether the model is failing for other reasons (errors in variables or wrong functional form). The adaptive expectations version fares better. The main driving variables in their model are current and past sales and the price index of R & D. Since there are no data on specific firm prices, an aggregate and rather smooth price index is used as an approximation, resulting, unfortunately, in mostly nonsignificant estimates of the various price coefficients. The paper shows how difficult it is to formulate a rigorous theory of R & D investment and to derive explicit functional forms for its estimation. It is an important first step on a rather long road with at least three major tasks still ahead of us: (1) developing a rigorous and effective model of two types of investments with two capital stocks (R & D and physical); (2)

treating output (or sales) as endogenous, being planned simultaneously with the R & D program (after all, it is the R & D that is supposed to generate new sales in the future); and (3) finding more relevant proxies for the "price of R & D" than different versions of aggregate wage and price indexes.

The last set of papers considers the impact of R & D on total factor productivity. The first three consider the issue primarily at the micro firm level, while the last two analyze the R & D-productivity relationship at the more aggregated industrial level. The relationship of R & D to productivity growth is of interest for at least two reasons: (1) Productivity growth is a major source of overall economic growth and the understanding of its sources has been a major research goal of economists over the last three decades. (2) Many of the effects of R & D may be social rather than private, in the sense that they are not appropriated by the unit that produced the particular R & D results and may not show up in its profits or output measures. To the extent that they are of benefit to other firms and industries, they may show up in the more aggregate industry and economywide productivity numbers. Unfortunately, the productivity measures themselves are subject to much error and may not reflect many of the technological changes that are not ultimately embodied in easily measurable products. For example, the gains in space exploration and medical research do not show up directly in the productivity figures as currently constructed, nor do most of the gains in complex commodities, such as consumer electronics, for which no good price indexes are available to use in the construction of real product measures. (See Griliches 1979 for further discussion of some of these issues.)

The Griliches and Mairesse paper uses a production function framework to analyze the impact of past cumulated R & D expenditures on the output (deflated sales) of over a hundred large U.S. firms, covering a twelve-year time period (1966–77). They find that there is a strong relationship between firm productivity and the level of its past R & D investments in the cross-sectional dimension, but that in the within-firm time-series dimension of the data, this relationship almost vanishes. This may be the result of a higher degree of collinearity between the time trend which is used as a proxy for more general outside sources of technical change and the growth in physical and R & D capital stocks in the within-firm dimension, and to the greater importance of measurement errors and other transitory fluctuations in these data. When they constrain the other coefficients to reasonable values, the R & D coefficients are sizeable and significant even in the within-firm dimension of the data. They also develop a simultaneous equations interpretation of their model and estimate what they call "semireduced form" equations, which again yield rather high estimates of the contribution of R & D capital to productivity growth relative to that of physical capital. Two important topics are raised but not pursued very far: (1) In these data some of the return to R & D appears to come in the form of mergers. (2) The conventional model used in this type of research does not allow for a noncompetitive firm environment. They do reinterpret their numbers by an individual firm monopoly model but do not pursue its implications for estimation.

In their companion paper, Cuneo and Mairesse reran the Griliches-Mairesse models on similarly constructed French data with largely similar results. While their sample covers a shorter time period, they had access to better data and this seems to matter. In particular, they show that having data on materials use (or value added) does reduce the difference between the estimated coefficients in the total and the within-firm dimension. They can also take out the R & D related components of labor, materials, and capital from the conventional input measures and thereby avoid the usual double counting problem: treating R & D expenditures as a separate variable while the actual inputs bought with these expenditures are already counted once in standard measures of labor, etc. They show that this type of double counting biases the estimated R & D coefficients downward, more so in the total than in the within-firm dimension. They also confirm a not so surprising fact already discovered by Griliches (1980) and Griliches-Mairesse: the R & D intensive industries, the "scientific" industries, are also the industries with higher estimated R & D coefficients. They do more R & D and therefore, on the margin, have rates of return similar to those found in the other less R & D intensive industries. Thus there is evidence for differences in R & D elasticities (across industries) and for the view that R & D is pushed far enough in the more "scientific" industries to come close to equalizing the private rates of return to R & D across industries.

This assumption of equalization of private rates of return (rather than equality of elasticities) across firms is the motivation behind the approach taken by Clark and Griliches in their productivity study based on the PIMS (profit impact of market strategies) data base. They use a measure of R & D intensity (R/S) instead of the change in the R & D capital stock as their basic variable, postulating that private gross rates of return to R & D are more nearly equalized across firms in different industries than would be the case for production function elasticities. Their paper is also interesting for the use of data on "businesses" below the level of a company. These data are based on concepts that are similar to the FTC "line of business" classification used in the Scott and Scherer papers. This is a much more relevant level for data analysis when companies are large and conglomerate. They also had access to data on the composition of R & D expenditures (product vs. process R & D), on the importance of patents and other proprietary processes for the firm, and on the prevalence of technological change in the firm's particular line of business. Their main finding is a statistically significant relationship between R & D intensity and the growth in total factor productivity, implying a gross excess rate of return to R & D of about 20 percent. This return is bigger for process R & D than for product R & D. The effects of the latter are not all caught, presumably, by the firm's own deflated sales measures but are passed on in some part to the other product-using industries. They find no significant decline in the productivity of R & D during the 1970s, but they do find that these returns depend crucially on the presence of previous major technological changes in the respective industries, implying (without being able to measure it) a major role for spillovers from the previous R & D efforts of other firms and industries.

This is also the topic that motivates Scherer's important contribution. His paper describes in detail a major and valuable data construction effort whose basic purpose was to reallocate R & D expenditures from an industrial "origin" classification (where they are done) to a classification of ultimate "use" (where they will have their major productivity-enhancing impact). This was accomplished by examining over 15,000 patents in detail and assigning them to both industrial origin and industrial use categories and categorizing them into product and process patent categories. The detailed R & D by line of business data collected by the FTC were then reallocated from industry of origin to industries of use in proportion to the "use" distribution of their patents, thereby generating a kind of technological flow table. The many conceptual and practical difficulties in such an enterprise are discussed by Scherer in some detail. The appendix to his paper presents the most detailed data on R & D by three- and four-digit Standard Industrial Classification (SIC), by origin, and by use ever made available. These data will prove invaluable in future studies of productivity growth and differential industry R & D activity. Scherer reports briefly on an analysis of productivity growth in which, once the quality of the output growth data is controlled for, the newly generated R & D by industry of use data prove superior to the industry of origin data in the explanation of interindustry productivity growth differences.

In the final paper in this volume, Griliches and Lichtenberg examine the relationship between R & D and productivity growth in U.S. manufacturing industries at the two- and three-digit SIC levels. They use the NSF classification of R & D by product group rather than by industry of origin to approximate better the ultimate industrial location of the effects of these R & D expenditures and the Census–Penn–Stanford Research Institute (SRI) data base to construct detailed total factor productivity indexes. They look at the question of a possible secular decline in the fecundity of R & D and find no evidence for it. While there has been an overall decline in productivity growth, including R & D intensive industries, the statistical relationship between productivity growth and R & D intensity did not disappear. If anything, it grew stronger in the 1970s.

The conference concluded with a discussion session in which different speakers expressed their perception of the state of research in this field. That discussion and the discussions following the presentations of the various papers (parts of which are reproduced in this volume) ranged over a variety of topics with the following receiving the most attention: (1) the ambiguities of the patent data; (2) the aggregation level at which the R & D process should be studied: project, establishment, firm, industry, or economywide; (3) the absence of data on what really drives R & D—the changing state of technological and market opportunities; (4) the low quality and the dubious relevance of the available productivity data and the absence of alternative indicators of social returns to R & D; and (5) the difficulties in and the importance of modeling the spillover of knowledge and technology from one firm or industry to another.

The critics emphasized the fact that patents differ greatly in their economic significance and play different roles in different industries. The authors of some of the papers found it difficult to reconcile their basic reliance on the law of large numbers with their desire to analyze microdata. Even though the meaning of any individual patent may be highly variable, one hopes that large differences in the number of patents applied for across firms or over time do convey relevant information about the underlying trends and fluctuations in inventive output. This reasoning is not very helpful, however, when applied to data on individual firms, many of whom take out only an occasional patent or two, especially if patent counts are to be used as an independent variable, helping to explain some further measure of the consequences of inventive activity. It is doubtful whether small fluctuations in patent counts convey much information. Even though they should have known better, the authors were surprised, I think, by the large amount of randomness and by the low fits that they encountered in trying to analyze such data.

Nevertheless, the main conclusion that did emerge (though not unanimously) was that something is there, something worth working on and analyzing. Patents and patent counts are, after all, one of the few direct quantitative glimpses into the innovation process available to us. The studies do show a strong cross-sectional relationship between R & D and patents and a weaker, but still statistically significant one between their fluctuations over time. Thus, to a first approximation, one can use patent data as indicators of technological activity in parallel with or in lieu of R & D data. This is of significant practical import since in many contexts detailed patent data are more readily accessible than R & D data.

The work reported to date has yet to establish that there is net information added in patent counts, that patents as a measure of output of the

R & D process provide superior explanatory power in modeling productivity change or other performance indicators. Some scattered results implied an independent contribution of patents to the explanation of differences in the market value of firms (above and beyond what was already accounted for by R & D variables), but no study had connected it yet to productivity growth. Perhaps the greatest promise of the patent data is in the level of detail contained in them and in the potential for using this detail to reclassify and illuminate other data. Scherer's paper is a prime example of such work where the information contained in the patent documents was used to reclassify and reallocate R & D expenditures into more relevant industrial boundaries. Another use discussed at the conference and currently being pursued at NBER is to study the overlap in the patenting of different firms across different patent classes in an effort to develop measures of technological similarity or distance between pairs and groups of firms. The notion here would be to use such distance measures in the analysis of spillovers of R & D effects between firms and industries, assuming that they are more prevalent at shorter "distances," and to derive better, technologically more homogeneous, industry groupings for the various firm samples.

There was much debate about the appropriate level and detail for the study of the R & D process. Many of the interesting questions and decisions are taken at the "project" level but little data are available on this level. Nor is it clear how generalizable some of the cases and smaller survey studies really are. It would be valuable, however, as is illustrated by Mansfield's work reported in this volume, to have greater detail on the composition of R & D itself. It matters to the analysis whether most of R & D is spent on basic and long-term research or almost entirely on adaptive and short-term research. Most econometric research uses, however, the available data, even if they are not entirely relevant. Clearly we would like to have better and different data, but it is my belief that we have not yet digested and understood much of what is already available. Even so, the currently available data have already produced some interesting findings, confirming, for example, the relationship between R & D and productivity growth and indicating no apparent deterioration in it recently.

One of the issues discussed at the conference was the apparent worldwide decline in patenting in the 1970s. This emerged from the negative trend coefficients in the Pakes-Griliches paper, a perusal of Evenson's tables, and also a look at the raw overall data (as summarized, for example, in *Science Indicators 1980* [National Science Board 1981]. Since the resources of the various patent offices are largely fixed and since worldwide cross-patenting has increased, some of the apparent decline may be an artifact of the "crowding out" of applications, but patent office data on applications filed (not just granted) do suggest that there may have been a real decline in the 1970s in patents applied for by corporations. Patent applications by all corporations (which were ultimately granted) peaked around 1969, roughly coincidentally with peaks in real R & D expenditures in industry and in the employment of scientists and engineers. Since 1969 the level of corporate patenting has somewhat declined on the order of 10 percent. Whether this should be interpreted as reflecting an exhaustion of technological opportunities is doubtful. It is more likely a reflection of the deteriorating macro conditions around the world and of a possible decline in the value of patenting, because of rising costs of litigation to keep them in force and faster rates of disclosure and subsequent imitation following patent granting and publication, because of improved communication systems and more continuous surveillance of the patents being granted by computer accessible literature search services. Unfortunately, detailed data on applications filed but not granted are inaccessible, and the currently available patent data sets do not go back far enough (before 1965) to allow us to distinguish a trend from a cycle.

Several papers set up models of the determinants of R & D investment and others talk about the necessity of considering the R & D decision within some wider, multi equational, simultaneous framework. Unfortunately, the standard variables that are brought in to "explain" movements in R & D and the other factors of production do not appear to be all that relevant to the R & D story. First, little good data are available on the "price" of R & D, but even if we had them, they would move largely in parallel with the major cost component of R & D—the cost of labor. Second, the price of capital story is likely to be similar for physical and R & D capital except for some differences in the tax treatment of depreciation, the effects of which one is unlikely to be able to detect well in the kind of data available to us. Thus, I am somewhat pessimistic about the promise of approaches which can be caricatured as defining and treating R & D as just another "capital stock" and reducing the analysis to the previous case. This misses whatever it is that makes R & D a different endeavor from just buying another plant or a new set of machines.

Unfortunately, when one starts thinking about what is special about R & D—the importance of technological opportunities, scientific know-how levels, and expectations about eventual market size for particular products—it is difficult to see how these characteristics can be quantified and forced into the Procrustean bed of our standard models. The most that one can do at the moment is to model them as unobservable "shocks," along the lines of Pakes's paper and the earlier "unobservables" literature, and trace out their effects on and interaction with the other variables of interest, such as patents, physical investment, and market value. While it may prove possible to distinguish between demand and supply (technological opportunity) shocks in such models and provide insight into and an interpretation of the interdependence between these variables, this line of research is unlikely to lead to models with clear policy handles.

The productivity slowdown and possible reasons for it created much interest at the conference, but there were expressions of pessimism about our ability to detect the major effects of R & D in such data. Several problems are evident: (1) the poor quality of output price indexes in the R & D intensive industries, such as computers and electronic components; (2) the long and variable lags in the impact of particular technological developments on subsequent productivity growth; (3) the diffuse nature of such effects and the arbitrariness of many of the industrial boundaries in our data; and (4) the absence of good measures of real product in some of the final demand sectors with important R & D effects such as health, defense, and space exploration. In spite of the uncertainties introduced by measurement issues and the fact that different total factor productivity measures (by different researchers) do not agree closely when it comes to an examination of individual industry trends, there was general concurrence in the notion that the R & D slowdown is not implicated directly in the sharp and worldwide productivity slowdown which started in 1974–75. Less clear-cut evidence and less agreement are found for the milder and longer term total factor productivity growth slowdown which may have started in the late 1960s. Here the slowdown in R & D growth may have been a contributing factor. Whether it reflects an exhaustion of technological opportunities is not clear, but it is likely to contribute to a slower growth rate in the underlying potential of the world economy in the future.

From a methodological point of view, in spite of all the talk about technology flows and externalities, the main unsolved research problem is still how to handle the interdependence between projects, firms, and industries. Almost all our methodology is based on the individual unit of observation, be it firm or industry. We do have models which allow for a one- or two-dimensional interdependence (such as serial correlation or variance components), but we have little experience and skill in modeling the clustering of and interaction between a relatively large number of actors. Even the work based on transaction flow tables does not do the job since it is fundamentally unidirectional. Our existing methodological tools cannot handle this and predispose us to ignore such problems. I believe that the development of methods for the analysis of large group interactions will be the major task and challenge during the next decade of research in this area.

What have we learned since the last NBER R & D conference more than twenty years ago? What were the substantive findings reported at this conference? I do believe that we have made progress in understanding the questions, in developing more rigorous models and better tools of analysis, and in accumulating and analyzing much larger data sets. We have described and documented a significant relationship between R & D investments and subsequent productivity growth. This relationship remains even in these trying times, though shrouded by data and measurement uncertainties. We have also concluded that federal R & D expenditures do not have much of a direct effect on productivity as it is conventionally measured but do stimulate private R & D spending and may thereby have a nonnegligible indirect effect. We are more aware of both the conceptual and the measurement difficulties involved in productivity measurement and less sure about the relevance of the existing measures to the issues at hand. We do have evidence that there may be something interesting in the patent data after all. They do appear to be useful indicators of innovative activity (though less so in the small and over short time periods), and there may be fruitful ways of using them in further analysis. We have also learned that the relationship between R & D, firm size, concentration, and all the rest is much looser and more obscure than is implied by the usual statements of the Schumpeterian hypothesis. While much of the R & D effect is concentrated in large firms, it is more likely that they have become large because of their R & D successes rather than that they do more and more fruitful R & D because they are large.

However, we have not provided, except indirectly, many policy handles. Nor is it likely that we will do so in the future. This is not because we do not want to be helpful to the National Science Foundation or the rest of the policymaking establishment, but because what we are studying is not really amenable to short-run policy intervention or manipulation. R & D investment and performance are largely determined by the evolution of scientific opportunities in a field and by peoples perceptions and expectations of the future economic climate within which new products or processes are to be sold or used. These can only be affected indirectly and imperfectly by supporting science in general and basic research in particular and by pursuing wise macroeconomic policies. All else, I believe, is of secondary importance.

References

Cole, J. R., and S. Cole. 1973. *Social stratification in science*. Chicago: University of Chicago Press.

Dasgupta, P., and J. Stiglitz. 1980. Industrial structure and the nature of innovative activity. *Economic Journal* 90:266–93

Elkana, Y., J. Lederberg, R. K. Merton, A. Thackray, and H. Zuckerman, eds. 1978. *Toward a metric of science*. New York: Wiley.

Freeman, C. 1974. *The economics of industrial innovation*. Baltimore: Penguin.

Griliches, Z. 1979. Issues in assessing the contribution of research and development to producivity growth. *Bell Journal of Economics* 10, no. 1 (Spring): 92–116

————. 1980. Returns to research and development expenditures in the private sector. In *New developments in productivity and analysis measurement and analysis*, ed. J. W. Kendrick and B. N. Vaccara, 419–54. Conference on Research in Income and Wealth: Studies in Income and Wealth, vol. 44. Chicago: University of Chicago Press for the National Bureau of Economic Research.

Griliches, Z., and J. Schmookler. 1963. Inventing and maximizing. *American Economic Review*, September 53(4):725–29

Hausman, J., B. H. Hall, and Z. Griliches. Econometric models for count data with an application to the patents R & D relationship. 1984. *Econometrica* (in press).

Kamien, M. I., and N. L. Schwartz. 1982. *Market structure and innovation*. New York: Cambridge University Press.

Minasian, J. P. 1962. The economics of research and development. In *The rate and direction of inventive activity*, ed. R. R. Nelson, 93–142. Princeton: Princeton University Press for the National Bureau of Economic Research.

National Science Board. 1981. *Science indicators 1980*. Washington, D.C.: National Science Foundation.

Nelson, R. R., ed. 1962. *The rate and direction of inventive activity: Economic and social factors*. Universities-NBER Conference Series no. 13. Princeton: Princeton University Press for the National Bureau of Economic Research.

Pakes, A., and Z. Griliches. 1982. Estimating distributed lags in short panels. NBER Working Paper no. 933. Cambridge, Mass.: National Bureau of Economic Research.

Price, D. J. 1963. *Little science, big science*. New York: Columbia University Press.

Rosenberg, N. 1976. *Perspectives on technology*. New York: Columbia University Press.

Spence, A. M. 1982. Cost reduction, competition and industry performance. Harvard Institute of Economic Research Discussion Paper no. 897, Cambridge, Mass.

Teubal, M. 1981. The R & D performance through time of young, high technology firms. Maurice Falk Institute for Economic Research in Israel Discussion Paper 814, Jerusalem.

von Hippel, E. 1982. Appropriability of innovation benefit as a predictor of the source of innovation. *Research Policy* 11(2):95–115.

2

Who Does R & D and Who Patents?

John Bound, Clint Cummins, Zvi Griliches,
Bronwyn H. Hall, and Adam Jaffe

2.1 Introduction

As part of an ongoing study of R & D, inventive output, and productivity change, the authors are assembling a large data set for a panel of U.S. firms with annual data from 1972 (or earlier) through 1978. This file will include financial variables, research and development expenditures, and data on patents. The goal is to have as complete a cross section as possible of U.S. firms in the manufacturing sector which existed in 1976, with time-series information on the same firms for the years before and after 1976. This paper presents a preliminary analysis of these data in the cross-sectional dimension, laying some groundwork for the future by exploring the characteristics of this sample and by describing the R & D and patenting behavior of the firms in it. This paper follows previous work on a smaller sample of 157 firms (see Pakes and Griliches 1980 and Pakes 1981).

We first describe the construction of our sample from the several data sources available to us. Then we discuss the reporting of our key variable,

John Bound, Clint Cummins, and Adam Jaffe are graduate students in the Department of Economics at Harvard University. Zvi Griliches is professor of economics at Harvard University, and program director, Productivity and Technical Change, at the National Bureau of Economic Research Research. Bronwyn H. Hall is a research analyst for the National Bureau of Economic Research.

This paper is a revision of an earlier draft presented at the NBER conference on R & D, Patents, and Productivity in Lenox, Massachusetts, October 1981. That version contained preliminary results on patenting in the drug and computer industries which have been replaced in this revision by a section on patenting in all manufacturing industries.

This work has been supported by the NBER Productivity and Technical Change Studies Program and NSF grants PR79–13740 and SOC79–04279. We are indebted to Sumanth Addanki and Elizabeth Stromberg for research assistance. The research reported here is part of NBER's research program in productivity. Any opinions expressed are those of the authors and not those of NBER.

research and development expenditures, and relate this variable to firm characteristics, such as industry, size, and capital intensity. An important issue is whether the fact that many firms do not report R & D expenditures will bias results based only on firms which do. We attempt to correct for this bias using the well-known Heckman (1976) procedure.

Section 2.4 describes the patenting behavior of the same large sample of firms. We attempt to quantify the relationship between patenting, R & D spending, and firm size, and to explore the interindustry differences in patenting in a preliminary way. Because of the many small firms in this data set, we pay considerable attention to the problem of estimation when our dependent variable, patents, takes on small integer values. The paper concludes with some suggestions for future work using this large and fairly rich data set.

2.2 Sample Description

The basic universe of the sample is the set of firms in the U.S. manufacturing sector which existed in 1976 on Standard and Poor's Compustat Annual Industrial Files. The sources of data for these tapes are company reports to the Securities and Exchange Commission (SEC), primarily the 10-K report, supplemented by market data from such sources as National Association of Securities Dealers Automated Quotations (NASDAQ) and occasionally by personal communication with the company involved. The manufacturing sector is defined to be firms in the Compustat SIC groups 2000–3999 and conglomerates (SIC 9997).[1]

Company data were taken from four Compustat tapes. The Industrial file includes the Standard and Poor 400 companies, plus all other companies traded on the New York and American Stock Exchanges. The Over the Counter (OTC) tape includes companies traded over the counter that command significant investor interest. The Research tape includes companies deleted from other files because of acquisition, merger, bankruptcy, and the like. Finally, the Full Coverage tape includes other companies which file 10-K's, including companies traded on regional exchanges, wholly owned subsidiaries, and privately held companies. From these tapes we obtained data on the capital stock, balance sheets, income statements including such expense items as research and development expenditures, stock valuation and dividends, and a few miscellaneous variables such as employment.

Unfortunately, our patent data do not come in a form which can be matched easily at the firm level. Owing to the computerization of the

1. This limitation is primarily for convenience; about 97 percent of company-sponsored R & D was performed in the manufacturing sector in 1976 (NSF 1979). It does, however, exclude a few large performers of R & D in the communications and computer service industries.

U.S. Patent Office in the late 1960s, we are able to obtain a file with data on each individual patent granted by the Patent Office from 1969 through 1979. For each such patent we have the year it was applied for, the Patent Office number of the organization to which it was granted, an assignment code telling whether the organization is foreign or domestic, corporate or individual, and some information on the product field and SIC of the patent. We also have a file listing the Patent Office organization numbers and the correspondent names of these organizations. The difficulty is that these patenting organizations, although frequently corporations in our sample, may also be subsidiaries of our firms or have a slightly different name from that given on the Compustat files ("Co." instead of "Inc." or "Incorporated" and other such changes or abbreviations).[2] Thus, the matching of the Patent Office file with the Compustat data is a major task in our sample creation.

To do the matching, we proceeded as follows: All firms in the final sample (about 2700) were looked up in the *Dictionary of Corporate Affiliations* (National Register 1976). Their names as well as the names of their subsidiaries were entered in a data file to be matched by a computer program to the names on the Patent Office organization file. This program had various techniques for accommodating differences in spelling and abbreviations. The matched list of names which it produced was checked for incorrect matches manually, and a final file was produced which related the Compustat identifying Committee on Uniform Securities Identification Procedures (CUSIP) number of each firm to one or more (in some cases, none) Patent Office organization numbers. Using this file, we aggregated the file with individual patent records to the firm level. As this paper is being written, we are engaged in a reverse check of the matching process which involves looking at the large patenting organizations which are recorded as domestic U.S. corporations, but which our matching program missed. The results of this check may further increase some of our patent totals.

In assembling this data set we have attempted to confine the sample to domestic corporations, since the focus of our research program is the interaction between research and development, technological innovation, and productivity growth within the United States. Inspection of the Compustat files reveals that at least a few large foreign firms, mostly Japanese, are traded on the New York Stock Exchange, and they consequently file 10-K's with the SEC and would be included in our sample, although presumably their R & D is primarily done abroad and their U.S. patents are recorded as foreign owned. To clean our sample of these firms we did several things: First, we were able to identify and delete all firms which Compustat records as traded on the Canadian Stock Exchange.

---

2. The vast majority of patents are owned by principal companies. In our earlier sample about 10 percent of total patents were accounted for by patents of subsidiaries.

Then we formed a ratio of foreign-held U.S. patents to total number of U.S. patents for each firm in our sample. For most of our sample, this ratio is less than 15 percent; the list of firms for which it is larger includes most of the American Deposit Receipts (ADR) firms on the New York Stock Exchange and several other firms clearly identifiable as foreign. After deleting these firms from the sample, as a final check we printed a list of the remaining firms with "ADR" or "LTD" in their names. There were eighteen such firms remaining, which we deleted from the sample.

The firms which were left still had a few foreign-owned patents (about 2 percent of the total number of patents in 1976) from joint ventures or foreign subsidiaries. Since their Compustat data are consolidated and include R & D done by these subsidiaries in the R & D figure, we added those patents to the domestic patents to produce a total successful patent application figure for the firm.

Our final 1976 cross section consists of data on sales, employment, book value in various forms, pre-tax income, market vaue, R & D expenditures, and patents applied for in 1976 for approximately 2600 firms in the manufacturing sector. The selection of these firms is summarized in table 2.1. Except for a few cases, firms without reported gross plant value in 1976 are firms which did not exist in 1976. Seventy-seven firms were deleted because they were either wholly owned subsidiaries of another company in our sample or duplicates in the Compustat files; another thirty-one had zero or missing sales or gross plant value. The final sample consists of 2595 firms, of which 1492 reported positive R & D in 1976. In section 2.3 we present some results on the R & D characteristics of these firms.

2.3 The Reporting of Research and Development Expenditures

In 1972 the SEC issued new requirements for reporting R & D expenditures on Form 10-K. These requirements mandate the disclosure of the

Table 2.1   **Creation of the 1976 Cross Section**

Compustat File	Manufacturing Firms on Compustat Tape	Gross Plant Reported in 1976	Positive Gross Plant & Sales in 1976[a]	Positive R & D
Industrial	1299	1294	1248	770
OTC	489	472	458	292
Research	414	138	132	83
Full coverage	1019	867	757	347
Total number of firms	3221	2771	2595	1492

[a]Duplicates, subsidiaries, or foreign not included.

estimated amount of R & D expenditures when (a) it was "material," (b) it exceeded 1 percent of sales, *or* (c) a policy of deferral or amortization of R & D expenses was pursued. Acting on these new requirements, the Financial Accounting Standards Board issued a new standard for reporting R & D expeditures in June 1974. Until this time, accepted accounting practices appear to have allowed the amortizing of R & D expenditures over a short time period as an alternative to simple expensing, but the new standard allows only expensing (San Miguel and Ansari 1975). Accordingly, we believe that by 1976 most of our firms were reporting R & D expense when it was "material" and that the expense reported had been incurred that year.

For the purpose of this paper, we make no distinction among firms whose R & D is reported by Compustat as "not available," "zero," or "not significant."[3] All such firms are treated as not reporting positive R & D because of both the nature of the SEC reporting requirements for R & D and the way Compustat handles company responses. As noted above, companies are supposed to report "material" R & D expenditures. If the companies and their accountants conclude that R & D expenditures were "not material" (possibly zero but not necessarily), they sometimes say this in the 10-K report, in which case Compustat records "zero."[4] Alternatively, a company may say nothing about R & D, in which case Compustat records "not available." It is also likely that companies reported as "not available" include some which are "randomly" missing, that is, a company performs "material" R & D but for some reason Compustat could not get the number for that year.[5]

Another source of data on aggregate R & D spending by U.S. industry is the National Science Foundation which reports total R & D spending in the United States every year, broken down into approximately thirty industry groupings. These data are obtained from a comprehensive survey of U.S. enterprises by the Industry Division of the U.S. Bureau of the Census, which covers larger firms completely and samples smaller firms. Although there are several important differences between these data and those reported by Compustat, it is interesting to compare the aggregate figures, which we show in table 2.2. The company R & D figures are the most directly comparable to our Compustat numbers, but we also show the figures for total R & D since NSF does not provide a breakdown between company-sponsored and federal-sponsored R & D expenditures for many of the industries (to avoid disclosing individual company data). There are several reasons for the discrepancies between the Compustat

---

3. The "not significant" code is a 1977 Compustat innovation which appears in 1976 data only for the Full Coverage tape companies.

4. Or, more recently, "not significant." See note 3.

5. Also included in "missing" are companies that reported R & D but Compustat concluded that their definition of R & D did not conform.

and NSF totals. First, the industry assignment of a company is not necessarily the same across the two sets of data: the most striking difference is in the communications industry, which includes AT & T in the NSF/Census sample, while AT & T is assigned to SIC 4800 on the

Table 2.2   Comparison of Aggregate R & D Spending Reported to Compustat and NSF for 1976 (dollars in millions)

Industry	NSF[a] Total	Federal	Company	Compustat
Food & kindred products	329	—	—	336
Textiles & apparel	82	—	—	92
Lumber, wood products & furniture	107	0	106	53
Paper & allied products	313	—	—	128
Chemicals & allied products	3017	266	2751	3173
Industrial chemicals	1323	249	1074	1604
Drugs & medicines	1091	—	—	1053
Other chemicals	602	—	—	516
Petroleum refining & extraction	767	52	715	908
Rubber products	502	—	—	346
Stone, clay & glass products	263	—	—	218
Primary metals	506	26	481	302
Ferrous metals & products	256	4	252	151
Nonferrous metals & products	250	22	229	151
Fabricated metal products	358	36	322	186
Machinery	3487	532	2955	2898
Office, computing, & accounting machines	2402	509	1893	2035
Electrical equipment & communication	5636	2555	3081	2543
Radio & TV receiving equipment	52	0	52	119
Electronic components	691	—	—	327
Communication equipment & communication	2511	1093	1418	231
Other electrical equipment	2382	—	—	866
Motor vehicles & motor vehicles equipment	2778	383	2395	2847
Other transportation equipment	94	—	—	54
Aircraft & missiles	6339	4930	1409	851
Professional & scientific instruments	1298	155	1144	1195
Scientific & mechanical measuring instruments	325	6	318	315
Optical, surgical, photographic & other instruments	974	148	826	880
Other manufacturing	217	5	212	93
Conglomerates	—	—	—	563
Total manufacturing	26093	9186	16906	15470

Note: Columns do not add up due to NSF suppression of cells with small numbers of firms.

[a]Source: *Research and Development in Industry, 1977*. Surveys of Science Resources Series, National Science Foundation, Publication no. 79–313.

Compustat files and is therefore not in our sample. Adding the 1976 R & D for AT & T and its subsidiary, Western Electric, to the Compustat communications total would raise it to about $1 billion, not enough to account for the difference.

There are also definitional differences between the Form 10-K R & D and that in the Census survey. The 10-K includes international and contracted out R & D, while these are entered on a separate line of the Census survey.[6] The total amount involved is about $1.7 billion in 1976. This is likely to explain why our industrial chemicals figure is too high, for example. Some firms include engineering or product testing on one survey but exclude it on the other, apparently because the Census survey is quite explicit about the definition of research and development, while the 10-K allows considerably more flexibility. Finally, the coverage of firms in the U.S. manufacturing sector by Compustat is less complete than by the Census for two reasons: (1) privately held firms are not required to file Form 10-K, and (2) some large firms which do file a 10-K record their R & D as not "material" even though a positive figure is reported to the Census Bureau. In spite of all these caveats, the Compustat and NSF numbers do seem to match fairly well across industries, and the total is within 15 percent after correcting for AT & T and the international and contracted out R & D.

Table 2.3 presents some summary statistics for the firms in the sample, broken down into twenty-one industry categories. The categories are based approximately on the NSF applied R & D categories shown in table 2.2, with some aggregation, and the separation of the lumber, wood, and paper, and consumer goods categories from miscellaneous manufacturing. The exact industry category assignment scheme which we used throughout this paper, based on SIC codes, is presented in the appendix. A few firms with exceptionally large or small R & D-to-sales ratios have been "trimmed" from the sample in this table (see below for an exact definition of the criterion used). As the table shows, the population of the industry categories and the fraction of firms reporting R & D varies greatly, from 20 percent for the miscellaneous category to above 80 percent for drugs and computers.

Table 2.4 shows the size distribution of firms in the sample. A large number of small firms are included; there are about seventy firms with less than $1 million in sales, and over six hundred with less than $10 million. These firms, however, account for less than 1 percent of total sales of firms in the sample. As might be expected, larger firms tend to report R & D more often even though they do about the same amount as

6. This comparison of the definitions in the two surveys is drawn from a letter detailing the differences, from Milton Eisen, Chief, Industry Division, U.S. Bureau of the Census, to Mr. William L. Stewart, R & D Economic Studies Section, Division of Science Resources Studies, National Science Foundation, in April 1978.

Table 2.3   **Statistics for the 1976 Cross Section: Trimmed Data**

Industry	NFIRMS	AVEPLANT	AVESALES
Food & kindred products	182	178.7	585.7
Textile & apparel	188	55.2	137.8
Chemicals, excl. drugs	121	503.2	693.6
Drugs & medical inst.	112	116.6	301.7
Petroleum refining & ex.	54	3200.1	4622.8
Rubber & misc. plastics	98	122.4	214.8
Stone, clay & glass	81	186.1	243.6
Primary metals	103	499.6	488.5
Fabric. metal products	196	57.8	131.0
Engines, farm & const. equip.	64	186.9	457.3
Office, comp. & acctg. eq.	106	288.2	352.9
Other machinery, not elec.	199	40.8	116.1
Elec. equip. & supplies	105	155.0	405.5
Communication equipment	258	31.8	89.9
Motor veh. & transport eq.	105	464.2	1233.6
Aircraft and aerospace	37	237.4	754.1
Professional & sci. equip.	139	73.4	130.5
Lumber, wood, and paper	163	204.2	260.4
Misc. consumer goods	100	81.6	232.5
Conglomerates	23	1174.3	2202.3
Misc. manuf., n.e.c.	148	36.3	89.3
All firms	2582	230.9	417.2

Note:
    NFIRMS = Total number of firms in industry.
    AVEPLANT = Average gross plant in millions of dollars.
    AVESALES = Average sales in millions of dollars.
    AVEEMP = Average employment in thousands.

a fraction of sales. This is shown graphically in figure 2.1. Up until about $100 million in sales, only about half the companies report R & D, but above $10 billion almost 90 percent do. Previous analysts have suggested that this may be because big companies are able to do their accounting more carefully (San Miguel and Ansari 1975), but it is surprising *how* big a company must be before it has a 75 percent probability of reporting R & D.

As we indicated above, the nature of SEC reporting rules results in ambiguity in the interpretation of firms' reporting zero R & D or not reporting R & D. This ambiguity has implications for the analysis of the subsample of firms that do report R & D ("the R & D sample"). Although we do not believe that the non–R & D sample firms all do zero R & D, it is likely that they do less than the firms that report it. Also, they possibly do less R & D than would be expected, given their other characteristics such as industry, size, and capital intensity. If so, then their exclusion from regressions of R & D on firm characteristics will

AVEEMP	NRNDFIRM	AVERND	AVERATIO	NPATFIRM	AVEPAT
8.9	62	5.4	0.005	46	5.8
4.3	49	1.9	0.018	33	5.9
9.1	92	18.6	0.021	67	39.0
6.8	96	14.4	0.045	64	28.2
20.0	26	34.9	0.005	25	72.0
5.3	59	5.9	0.016	35	12.2
5.3	31	7.0	0.019	26	22.4
8.6	39	7.7	0.013	44	14.6
2.6	102	1.8	0.011	77	5.4
8.8	51	10.2	0.016	42	25.7
8.3	94	21.6	0.061	42	39.0
2.8	149	2.3	0.021	111	5.8
10.7	77	11.2	0.023	56	34.3
2.5	199	3.4	0.040	110	13.3
22.2	59	49.2	0.012	48	25.0
15.6	26	32.7	0.042	17	39.0
3.3	118	8.0	0.051	65	16.0
4.7	64	2.8	0.007	49	6.9
5.2	44	1.8	0.013	41	5.2
50.1	13	43.3	0.014	20	37.3
2.1	29	0.7	0.027	16	2.1
6.8	1479	10.5	0.027	1034	19.1

NRNDFIRM = Number of firms with nonzero R & D.
AVERND = Average R & D expenditure in millions of dollars for firms with nonzero R & D.
AVERATIO = Average R & D to sales ratio for firms with nonzero R & D.
NPATFIRM = Number of firms with nonzero patents.
AVEPAT = Average number of patents for firms with nonzero patents.

result in biased estimates of the association of these characteristics with the firms' propensity to do R & D.

To shed light on this problem, the distribution of reported R & D was examined in several ways. First, if firms consider R & D expenditures to be immaterial if they fall below some absolute amount, then the distribution of R & D would be truncated from below. We find no evidence of such truncation in the R & D distribution. R & D may also be considered immaterial if it is small *relative* to firm size. This seems particularly likely because, in addition to the requirement to report material R & D expenditures in item 1(b)(6) of the 10-K, the SEC requires firms to report *all* expense categories that exceed 1 percent of sales. Figure 2.2 is a histogram of R & D as percent of sales; once again, no truncation is apparent. In fact, the mode of the distribution occurs at about .3 percent of sales.

Although no obvious truncation was visible, either in absolute magnitude or as a percent of sales, we cannot rule out the likelihood that a combination of cutoffs, both absolute and relative (as interpreted by a

Table 2.4   Size Distribution of Firms

Size Class (sales in 1976 dollars)	Number of Firms	Number of Firms Reporting R & D	Percent of Firms Reporting R & D	Percent of Total Sales	Percent of Total R & D
Less than 1 million	72	33	46	0.003	0.019
1 to 10 million	545	293	54	0.23	0.42
10 to 100 million	1097	575	53	4.1	3.4
100 million to 1 billion	663	412	62	19.1	14.8
1 to 10 billion	205	167	81	48.3	50.6
Over 10 billion	13	12	92	28.2	30.7

firm's accountants), are in effect, implying an indeterminate bias in the relationship of observed R & D to a firm's characteristics. Therefore, we attempt to quantify the reporting and not reporting of R & D with a probit equation after presenting results for the firms which do report R & D.

In figure 2.3 we show a plot of log R & D vesus log sales for the R & D sample, which summarizes the basic relationship between R & D and firm size in our data. It is apparent from this plot that the slope and degree of curvature of this relationship are likely to be influenced strongly by a few outlying points; some very small firms do large amounts of R & D, and a few firms in the intermediate size range do very little R & D. To test for the sensitivity of the results to these few points, the sample was trimmed by eliminating seven firms (.5 percent) with the lowest R & D/sales ratios, and seven firms with the highest. The firms removed are those outside the diagonal lines drawn on the plot. This reduces the mean ratio of R & D to sales from 4.1 percent to 2.7 percent and the standard deviation from 35 percent to 3.8 percent. The effects on the log distribution are much less dramatic. The smallest ratio that was deleted from the upper tail was .716; the largest from the lower was .0002. These are beyond three standard deviations of even the untrimmed distribution, whether it is viewed as normal or (more plausibly) lognormal. Since the results with trimmed data were not strikingly different from those with untrimmed data, we present only one set of results for our regressions, using the trimmed data throughout.

The first question we investigated in this sample was the nature of industry variation in R & D performance and the relationship between R & D and firm size. Equations of the form

(1) $\log R = \alpha + \beta \log S + \epsilon,$

where R is R & D and S is sales, were estimated separately for the twenty-one industries in table 2.3. Except for the textile industry and miscellaneous manufacturing, the estimated betas were not significantly

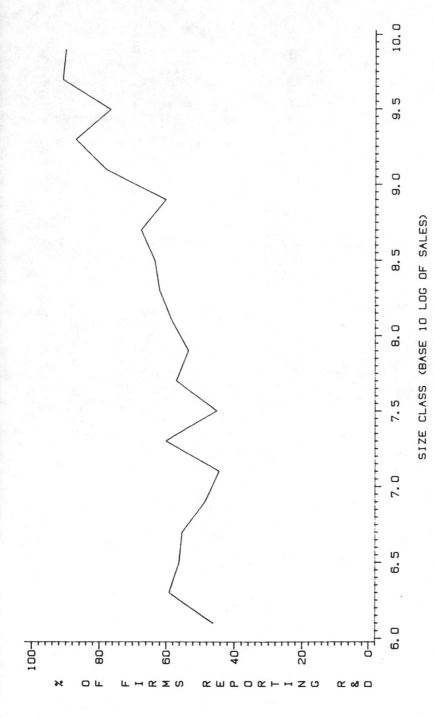

Fig. 2.1 Fraction of firms reporting R & D by size class. Firms with less than $1 million in sales were added to the smallest size class, and those with more than $10 billion were added to the largest.

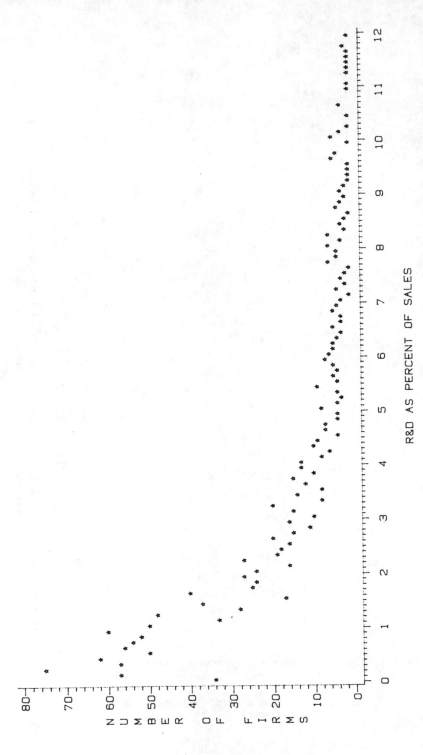

Fig. 2.2   Distribution of R & D as percent of sales for firms reporting R & D. Observations with R & D/sales percentage greater than 12 are not shown.

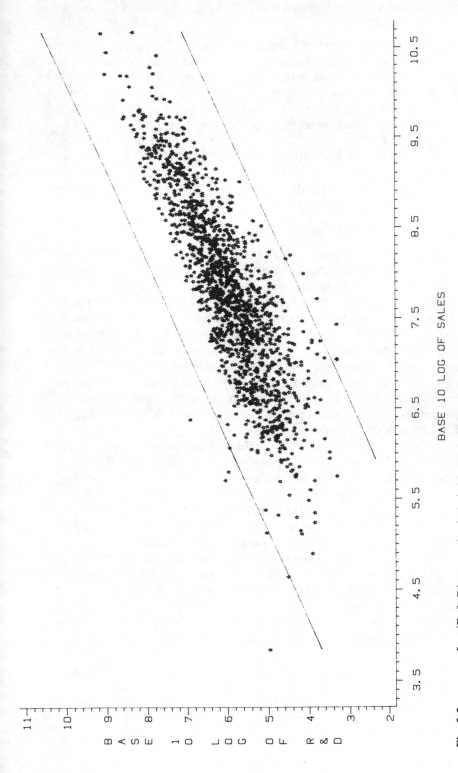

Fig. 2.3   Log(R & D) versus log(sales) for 1976 cross section.

different from one another statistically, and the R-squares were above
.65. The remainder of the analysis was performed using uniform slope
coefficients, while allowing for different industry intercepts by using
industry dummies. This was done primarily for convenience, but it is not
inconsistent with the individual industry results. While such aggregation
is rejected by a conventional F-test for the simple regression of log R & D
on log sales ($F[20, 1437] = 3.34$), given the size of our sample one should
really use a much higher critical value (about 8), in which case one need
not reject it.[7]

After accepting the hypothesis of equality of the slope coefficients, we
estimated equations of the form

(2) $\log R = \beta_1 \log S + \beta_2 \log A + \beta_3 (\log S)^2 + \gamma_i + \epsilon,$

where R and S are as previously defined, A is gross plant, and γ_i is a set of
industry intercepts. Simple statistics on the regression variables are
shown in table 2.5 and basic regression results in table 2.6.

The first column in table 2.6 gives the results of the simplest regression.
Although we know that this story is incomplete, this equation indicates
almost no fall in R & D intensity with increasing firm size. An analysis of
variance using this equation and restrictions on it is also interesting. Log
sales explains 73 percent of the total variance in log R & D and 79 percent
of the variance remaining after we control for the variations in industry
means. Looked at the other way, the industry dummies explain 10
percent of the total variance and 30 percent of the variance remaining
after we control for log sales.

The second column shows the effect of capital intensity on R & D
intensity. If we interpret this equation in terms of the equivalent regression of log R & D on log sales and log of the capital-sales ratio, we find it
implies a sales coefficient of .95, almost identical to that of the first
column, and a complementarity between capital intensity and R & D
intensity (coefficient of .24 for log [gross plant/sales]). While this effect is
highly significant, its additional contribution to the fit is small.

The third and fourth columns in table 2.6 indicate significant nonlinearity in the relationship between log R & D and log sales. These estimates
imply that the elasticity of R & D with respect to sales varies from .7 at
sales of $1 million to 1.2 at sales of $1 billion. This nonlinearity is also
apparant in the scatter plot of log R & D and log sales presented in figure
2.3. While a fairly linear relationship may exist for large firms, it clearly
breaks down for smaller firms. This may be a result, at least in part, of the
selection bias discussed above; more will be said about this below.

7. Leamer (1978) suggests using critical values for this F-test based on Bayesian analysis
with a diffuse prior as a solution to the old problem of almost certain rejection of the null
hypothesis with a sufficiently large sample. Using his formula (p. 114), the 5 percent level for
this F-test is 7.8, implying that we would accept the hypothesis of equal slopes in these data.

Table 2.5   Key Variables for the R & D Sample
(number of observations = 1479)

Variable	Mean	Standard Deviation	Minimum[a] (thousands)	Maximum[a] (billions)
Log R & D	−0.15	2.19	$30	$1.3
Log sales	4.10	2.19	$79	$49
Log gross plant	2.99	2.43	$37	$30
R & D/sales	0.026	0.038	0.00024	0.57

[a]The antilogs of the extreme are shown for the first three variables.

Table 2.6   Log R & D Regression Estimates (observations = 1479)

				4		
Variable	1	2	3	All Firms	Small Firms	Large Firms
Log sales	.965	.713	.684	.519	.576	.641
	(.013)	(.043)	(.036)	(.050)	(.105)	(.101)
Log gross plant	—	.240	—	.187	.113	.187
		(.039)		(.039)	(.074)	(.046)
(Log sales)^2	—	—	.035	.031	.044	.020
			(.004)	(.004)	(.052)	(.008)
Standard error	.954	.942	.932	.925	.910	
R^2	.813	.818	.821	.824	.832	

Note: All regressions include twenty-one industry dummies, except that for small firms, in which the primary metals and conglomerate dummies were dropped because of lack of firms. There are 319 small firms (less than $10 million in sales) and 1160 large firms.

In the last two columns of table 2.6 we present the results for the fourth regression estimated separately for small firms (up to $10 million in sales) and large firms (all others). The fit is improved slightly; the F ratio for aggregation of the two subsamples is 3.29 (22, 1433). Allowing for differences in the slopes of log sales and log gross plant together diminishes the significance of the log sales squared term, particularly for the small firms.

Our measurement of the contemporary relationship between R & D and sales may be a biased estimate of the true long-run relationship because of the transitory component and measurement error in this year's sales, particularly if we are interpreting sales as a measure of firm size. To correct for these errors in variables bias, we obtained instrumental variable estimates of a regression of log R & D on log sales, log sales squared, and the industry dummies using log gross plant and its square as instruments for the sale variables. The estimated coefficients were .755 (.042) and .028 (.005) for log sales and its square, implying an elasticity of R & D with respect to sales of .985 at the sample mean. This compares to an elasticity of .972 for equation (3) in table 2.6 and suggests that the errors in variables bias, although probably present, are not very large in magnitude.

As a first step in our attempts to correct for possible bias from nonreporting of R & D, we estimated a probit equation whose dependent variable was one when R & D was reported and zero otherwise. The model underlying this equation is the following: The true regression model for R & D is

(3) $$\log R_i = X_i\beta + \epsilon_{i1},$$

where X_i is a vector of firm characteristics such as industry and size, and ϵ_{i1} is a disturbance. We observe R when it is larger than some (noisy) threshold value C_i, different for each firm. This model is a variation of the generalized Tobit model, described by many authors; this particular version is in Nelson (1977) and is equivalent to a model described by Griliches, Hall, and Hausman (1978). C_i contains the 1 percent of sales rule and anything else the firm uses to decide whether R & D is "material," plus a stochastic piece, ϵ_2, which describes our inability to predict exactly when a firm will report:

(4) $$C_i = Z_i\delta + \epsilon_{i2}.$$

In this framework, the probability of observing R & D may be expressed as Prob $(\epsilon_1 - \epsilon_2 > Z_i\delta - X_i\beta \mid Z_i, X_i)$. If we assume ϵ_1 and ϵ_2 are distributed jointly as multivariate normal, we get the standard probit model

(5) $$\text{Prob } (R_i \text{ observed}) = 1 - F[(Z_i\delta - X_i\beta)/\sigma],$$

where σ is the variance of $\epsilon_1 - \epsilon_2$, and $F(\cdot)$ is the cumulative normal probability function. Since the probit model is only identified up to a scale factor, we can only estimate δ/σ and β/σ. Deriving the model in this way also reveals what it is we are estimating when we run a probit on this data: presumably Z_i and X_i include many, if not all, of the same variables. For example, if the Z_i were only log sales and the 1 percent rule was being followed, the coefficient δ would be unity, and if the true elasticity of R & D with respect to sales were also unity, the probit equation would yield a sales coefficient of zero. However, if reporting depended only on the absolute amount of R & D performed, then C_i would be a constant, and predicting large R & D would be equivalent to predicting high reporting probability; this hypothesis implies that the coefficients in the probit should be the same as those in the R & D regression (up to a scale factor). Finally, if reporting depends in a more complex way on industry and size of the firm, then no obvious relationship is needed between the coefficients of the probit model and those of the regression.

Table 2.7   **Log R & D Regression Corrected for Selectivity Bias**

Variable	Probit Estimates[a]		Log R & D Regression			
			Uncorrected		Corrected	
Log sales	.016	(.051)	.519	(.050)	.536	(.050)
(Log sales)^2	.0018	(.0050)	.031	(.004)	.032	(.004)
Log gross plant	.140	(.039)	.186	(.039)	.246	(.044)
Mills ratio	—		—		.933	(.326)
Standard error	—		.925		.923	
R^2	—		.824		.825	

Note: All models contain industry dummies.

[a]These are the maximum likelihood estimates of the coefficients in equation (5), the probability of R & D reporting. There are 2582 observations and 1479 report R & D. The χ^2 for the three variables besides the industry dummies is 233.

The results of the probit estimation are presented in the first column of table 2.7. The coefficient on log sales is .016 (.05) compared to .52 (.05) in the comparable ordinary least squares (OLS) equation for log R & D. At the mean of log sales for the whole sample, the coefficient is .077. The coefficient on log gross plant is reduced somewhat from OLS estimates. These results suggest that the first of our two hypotheses above is closer to the truth: R & D reporting depends primarily on R & D intensity and not on the absolute level of R & D spending, with perhaps a smaller effect from firm size.

If it is true that the nonreporting firms are characterized only by lower than average R & D as percent of sales, the OLS estimates of elasticities presented earlier are not necessarily biased, although the constant term and industry dummy coefficients could be. Since it is also true, however, that the nonreporting firms are smaller on average,[8] the OLS elasticity estimates may be biased downward. This possibility was investigated using the procedure popularized by Heckman (1976). For each observation with R & D reported, the "inverse Mills ratio" was calculated as:

$$(6) \qquad M = \frac{f(\hat{u})}{F(\hat{u})},$$

where $\hat{u}$ is the argument of the probit equation $(Z_i\delta - X_i\beta)/\sigma$ evaluated for this observation's data and the estimated probit coefficients, and $f(\cdot)$ and $F(\cdot)$ are the standard normal density and cumulative distribution functions, respectively. When M is added to the OLS estimations, it "corrects" for selectivity bias.

A regression including the Mills ratio variable is presented in the third

8. Average sales for reporting firms is $620 million, for nonreporting firms, $240 million.

column of table 2.7, together with the "uncorrected" estimates for comparison. The coefficients on the Mills ratio is positive and significant, indicating the presence of selectivity bias. There is only a slight rise in the sales coefficients, however, and the nonlinearity is about the same. The largest increase is in the log gross plant coefficient, which was also the best predictor of R & D reporting. Thus we would underestimate the complementarity of capital intensity and R & D intensity if we did not take into account the fact that non–capital-intensive firms also tend to be those which do not report R & D expenditures.

It should be emphasized that in this application of the Heckman technique the Mills ratios are nonlinear functions of all the other independent variables in the equation, because we have no variables that predict reporting but not quantity of R & D. For this reason, the incremental explanatory power of the M variable is caused solely by the nonlinearity of its relationship to the other variables in the model. We know, however, that the dependence of R & D on these variables is likely to be nonlinear to begin with. In the absence of a reporting predictor that is excluded from the quantity equation, it is impossible to distinguish selectivity bias and "true" nonlinearity in the R & D-size relationship. This makes it impossible to draw a definitive conclusion regarding the possibility of bias in the OLS estimates.

2.4   Patenting

The matching project described in the section 2.1 yielded 4,553 patenting entities which were matched to the companies in our sample. Of our 2582 companies, 1754 were granted at least one patent during the 1965–79 period, but only about 60 percent of that number applied for a patent in 1976. Firms with R & D programs are far more likely to apply for patents: about 20 percent of the firms with zero or missing R & D have at least one patent in 1976, but this fraction rises rapidly with size of R & D program until well over 90 percent of firms with R & D larger than $10 million have patents in 1976.

If we look at the size of the firm rather than the R & D program, 28 percent of the small firms (less than $10 million in sales) applied for a patent in contrast to the 53 percent which reported R & D, but this difference results primarily from the integer nature of the patents data: When we consider all years rather than just 1976, the percentage who patent rises to sixty. These same small firms account for 4.3 percent of sales, 3.8 percent of R & D, but 5.7 percent of patent in our sample. However, the latter number may be an overestimate since we know that approximately one-third of all domestic corporate patents remain unmatched in 1976 in our sample, and it is likely that some of these belong to subsidiaries of our larger companies which we have overlooked. Further checking of these patents is being done.

In table 2.3 we show the mean number of patents and number of firms which have one or more patents for each of our twenty-one industry classes. As we expect, patenting is higher in the science-based or technological industries in terms of both the fraction of firms which patent and the average number of patents taken out by the patenting firms. The industries with more than twenty-five patents per firm are chemicals, drugs, petroleum, engines, computers, electrical equipment, motor vehicles, aircraft and aerospace, and conglomerates. Presumably petroleum, motor vehicles, and conglomerates appear on this list partly because of the average size of the firms in those industries. On the other hand, the scientific instrument and the machinery industries have a large number of patents per R & D dollar but are composed of relatively small firms.

Earlier studies by Pakes and Griliches (1980) on a sample of 157 large U.S. manufacturing firms show a strong contemporaneous relationship between patent applications and R & D expenditures across firms in several industries, and they suggest that patents are a fairly good indicator of the inventive output of the research department of a firm. We consider the relationship again in figure 2.4. Because of the large size range of our firms, the patents–R & D relationship will be obscured by the simple correlation between number of patents and size of firm. Therefore, we plot the log of patents normalized by gross plant versus the log of R & D normalized by the same quantity for the firms which both do R & D and patent. The plot shows a strong correlation between patenting and R & D for those firms with a slope slightly greater than one and a hint of nonlinearity in the relationship (increasing slope for higher R & D). There is considerable variance: the range of patents per million dollars of R & D for the firms which patent is from about one-seventh of a patent to ninety patents. The typical firm has a ratio of about two, that is, half a million dollars of R & D per patent.[9]

This picture is slightly misleading, however, since it covers only one-third of our sample. Accordingly, when we turn to modeling the relationship, we want to include the zero observations on both patents and R & D in our estimation. We attempt to solve this problem in two ways: First we set log patents to zero for all zero patent observations and allow those firms to have a separate intercept (PATDUM) in our regressions, as suggested by Pakes and Griliches (1980). It should be emphasized that there are about 1700 such observations, which suggest that the significance level of our estimates needs to be interpreted with caution. The

9. Scherer (1981), using data on 443 large industrial corporations comprising 59 percent of corporate patenting activity in the United States, found an R & D cost per patent of $588,000 for the period of June 1976 through March 1977 (adjusted to annual basis).

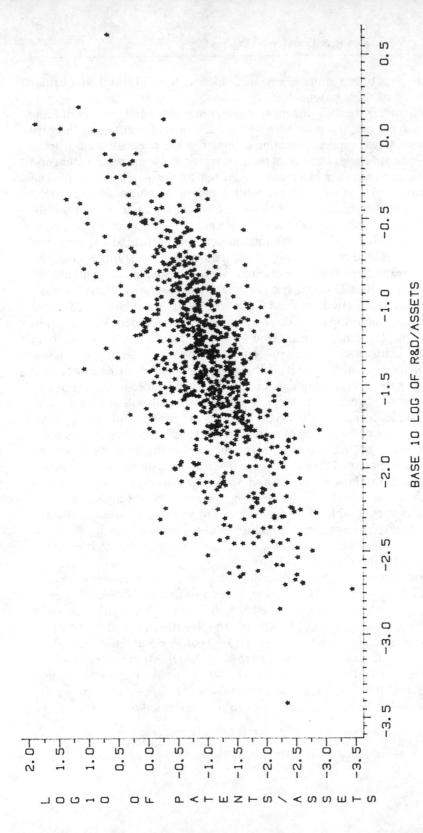

Fig. 2.4   Plots of log(patents/assets) versus log(R & D/assets) for 1976 cross section.

estimates we obtain imply that the observations with no patents have an expected value of about one-half of a patent. Second, we model the patents properly as a counts (Poisson) variable, taking on values 0, 1, 2, etc., as suggested by Hausman, Hall, and Griliches (1984). In this case, with our many small and few very large observations, the Poisson model turns out to give quite different results from the logarithmic OLS model.

The first column of table 2.8 displays the results of a regression of log patents on log R & D expenditures, dummies for zero or missing R & D and patents, and our twenty-one industry dummies. The estimate of the log R & D coefficient is considerably lower than the comparable estimates by Pakes and Griliches (1980), .61 (.08), or by Hausman, Hall, and Griliches (1984), .81 (.02). The difference could be attributed to the size range of firms in our sample which is far greater than in the earlier work and also to the large number of zeroes in our variables. For comparison, the coefficient of log R & D is .59 (.02) when we use only firms with nonzero patents and R & D. The overlap of this last sample of firms with the Pakes and Griliches sample is about 100 firms out of 831, consisting primarily of the larger firms from the complete sample. We will return to the question of how to handle the enormous size range of our complete sample after we discuss the Poisson and negative binomial results for this model.

The industry dummies from the regression in the first column of table 2.8 are a measure of the average propensity to patent in the particular industry, holding R & D expenditures constant. Relative to the overall mean, the industries with significantly higher than average patenting propensity are chemicals, drugs, petroleum, engines, farm and construction machinery, electrical equipment, aircraft, and the conglomerates. Several industries which are highly technology based, such as communications equipment and computers, do not seem to patent any more

Table 2.8   **Log Patents Regressions (number of observations = 2582)**

Variable	1		2		3		4	
Log R & D	.38	(.01)	.37	(.008)	.37	(.008)	.32	(.010)
Log gross plant	—		—		—		.064	(.008)
Log R & D squared	—		.083	(.002)	.084	(.002)	.081	(.002)
PATDUM	−.79	(.04)	−.82	(.03)	−.85	(.03)	−.76	(.03)
Other variables included	R&DDUM, industry dummies		R&DDUM, industry dummies		R&DDUM, intercept		R&DDUM industry dummies	
Standard error	.713		.589		.595		.583	
R^2	.653		.763		.756		.768	
Test for industry dummies			$F_{20,2557} = 3.5$					

than the average: in fact, a firm in the computer industry has 85 percent of the patents of an average firm doing the same amount of R & D.

To allow for possible nonlinearity in the patenting-R & D relationship, we add the log of R & D expenditures squared to the regression in column two of table 2.8. This coefficient is highly significant and implies a substantially higher propensity to patent for firms with larger R & D programs, with an elasticity of .25 at R & D of half a million, rising to over unity at R & D expenditures of $100 million. The F-test for the industry dummies is now $F(20, 2557) = 3.5$, implying very little difference in the average propensity to patent across industries once we allow R & D to have a variable coefficient. This is a bit surprising and probably reflects the nonhomogeneity of the firms in our industry classes and the problems associated with assigning each firm to one and only one industry. The industries which have coefficients significantly different from the average are the petroleum industry (patenting 30 percent higher on average), engines, farm and construction machinery (28 percent), conglomerates (76 percent), and computers (20 percent less on average). We reestimated the equation with no industry dummies (column three of table 2.8) and found that the slopes hardly changed; this result held true for several different specifications of the model, including one with only the log of R & D in the equation.[10] Although we believe that there are significant differences in the relationship of R & D and patenting at the detailed industry level from inspection of the distribution of the two variables by industry, these differences do not affect the basic results of this aggregate study. We have therefore omitted the industry dummies for the sake of simplicity in what follows.

In the fourth column of table 2.8 we add the log of gross plant value to the regression to control for firm size independently of R & D expenditures. Larger firms may patent more often simply because they are bigger and employ patent lawyers and other personnel solely for this purpose. The coefficient estimate for log gross plant lends some support to this hypothesis. However, one should be careful in interpreting the estimated size (assets) effects. To a significant extent they may be just compensating for transitory and timing errors in our R & D measure. The equation estimated assumes that this year's patents applied for depends only on this year's R & D expenditures. We know that this is not exactly correct (see Pakes and Griliches, this volume). Some of the patents applied for are the result of R & D expenditures in years past, while not all of the R & D expenditures in this year will result in patents, even in subsequent

---

10. We also looked at this question for two different size classes of firms: above and below $100 million in gross plant. We found that the smaller firms had a lower R & D coefficient (.26 in contrast to .36) and slightly less curvature. For the smaller firms, the industry dummies were completely insignificant, whereas they remained at about the same level for the large firms.

years. In this sense, the R & D variable is subject to significant error which will be exacerbated once we control for size, thereby reducing the signal-to-noise ratio. This may explain both the reduction of the R & D coefficient when assets are introduced as a separate variable and the rather large estimated pure size effect. We cannot do much about this in this paper, but we shall return to this topic when we turn to the panel aspects of this data set in later work.

We now turn to the Poisson formulation of the patents model. This model treats the patents for each firm as arising from a Poisson distribution whose underlying mean is given by $\exp(X\beta)$, where $X\beta$ is a regression function of the independent variables in our model. Coefficients estimated for this model are directly comparable to those from a log patents regression; we have merely taken account of the fact that the dependent variable is nonnegative counts rather than continuous. However, for our data we might expect the Poisson formulation of the model to give quite different answers from a simple log patents regression for two reasons: First, over half of our observations on patents are zero, and many are quite small. Second, the Poisson objective function tends to give the largest observations more weight than least squares on log patents, therefore these observations will have more influence on the results. This is what we find in our results, which are shown in the second column of table 2.9, together with the OLS estimates for comparison. The OLS estimates imply an elasticity of patenting with respect to R & D which rises from zero at $100,000 of R & D to well above one at $1 billion. For the Poisson model, on the other hand, the elasticity is one at $4 million of R & D and falls to one-half at $1 billion. This is because the very largest firms do less patenting per R & D dollar than would be predicted by a linear regression of log patents on log R & D, and they are having more influence on the Poisson estimates than the OLS. We show this graphically in figure 2.5: What is plotted is the predicted logarithm of patents versus the logarithm of R & D expenditures, superposed on the actual data. Clearly the differences in fit of the models are most pronounced in the tails of the distribution.

As was pointed out by Hausman, Hall, and Griliches (1984), the Poisson model is highly restrictive, since it imposes a distribution on the data whose mean is equal to its variance. This property arises from the independence assumed for the Poisson arrival of "events" (patent applications) and is unlikely to be true, even approximately, of our data. One way out of this problem is the negative binomial model in which the Poisson parameter is drawn from a gama distribution with parameters $\exp(X\beta)$ and δ. We estimated such a model in the third column of table 2.9 and found that the results, although qualitatively closer to the OLS estimates than to the Poisson, produce quite different predictions over the range of the data and imply a lower and less varying elasticity of

Table 2.9   Comparison of Patents Models (number of observations = 2582)

Variable	OLS $\log P = X\beta + u$		Poisson		Negative Binomial		Nonlinear Least Squares $P = \exp(X\beta) + \epsilon$	
Log R & D	.37	(.008)	1.13	(.010)	.58	(.018)	2.18	(.10)
Log R & D squared	.084	(.002)	-.047	(.002)	.012	(.003)	-.16	(.009)
Dummy (R & D = 0)	-.11	(.03)	-.43	(.01)	-1.37	(.08)	1.85	(.58)
Constant	.97	(.02)	.61	(.02)	-1.33	(.04)	-1.67	(.26)
D(patents = 0)	-.85	(.03)	—		—		—	
$\hat{\delta}$	—		—		.059	(.0016)	—	
Standard error	.595		—		—		20.57	
Log likelihood	—		56,171.		63,558.		—	

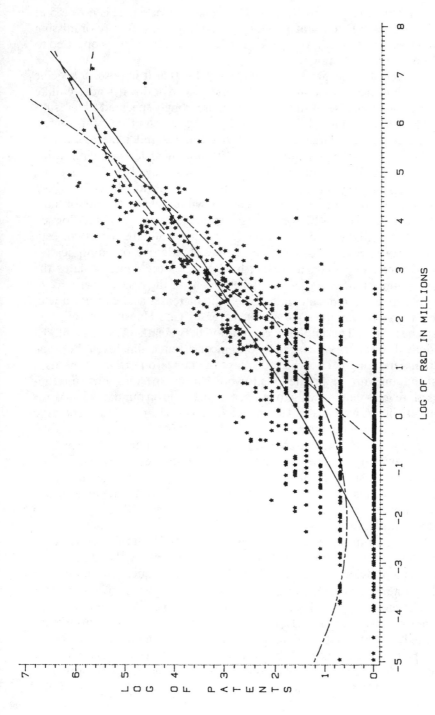

Fig. 2.5   Predictions for models with log R, $(\log R)^2$, $D(R = 0)$, no industry dummies, 831 observations plotted (total = 2.582)(*** = data, — = NB, --- = NLLS, ——— = OLS, —— = Pois).

patenting with respect to R & D. The range of elasticities is now .55 at $100,000 in R & D to .66 at $1 billion. A typical firm with zero or missing R & D is predicted to have applied for 1.3 patents in 1976, as opposed to 2.4 under the OLS model.

A defect of the negative binomial model is that it imposes a specific distribution, namely gamma, on the multiplicative disturbance. Unlike the least squares case, if this distribution is wrongly specified, the resulting maximum likelihood estimates may be inconsistent. For this reason and because of the large swings in our estimates under the models we tried, we also estimated our model with nonlinear least squares using patents as the dependent variable, which was proved by Gourieroux, Monfort, and Trognon (1981) to be consistent for a wide class of Poisson-type models. This produced the result shown in the last column of table 2.9. The discrepancies between these estimates and those of the Poisson model are a kind of "specification" test, since both are consistent estimates of a large class of count models with additive or multiplicative disturbances. Our data, however, have one feature which violates the assumptions of most of these models: not only is the residual variance of patents larger than the mean, but the ratio increases as the magnitude of the exogenous variables (R & D) increases (see Hausman, Hall, and Griliches 1984). This implies a correlation between the X's in the model and the disturbance which can lead to inconsistent estimates of the slope parameters. Figure 2.5, which displays the nonzero portion of the data distribution with the predictions for our various specifications superimposed, reveals that in trying to impose a quadratic on our data to look for scale effects we may mislead ourselves seriously because of the very large range of the data and the peculiar distribution of the dependent variable. It appears that the form we choose for the error distribution of the patents variable will have a considerable effect on the results. It should be emphasized that this result does not depend only on the large number of zero observations in the data: we obtained qualitatively the same results when we reestimated, including only those firms with both nonzero patents and R & D.

Because of the increasing variance with R & D and the difficulty of choosing a proper functional form for both tails of the distribution simultaneously, we chose to look at the interesting questions in this data (the existence of a patenting threshold and the measurement of returns to scale at the upper end of the R & D distribution) by dividing the sample into two parts, using R & D as the selection variable. To do this we first plotted the patents–R & D ratio for firms with both patents and R & D grouped by R & D size class, as shown in figure 2.6. This plot is consistent with a patenting elasticity of considerably less than one up to about $1 or $2 million of R & D and an elasticity of about one after that, with a hint of

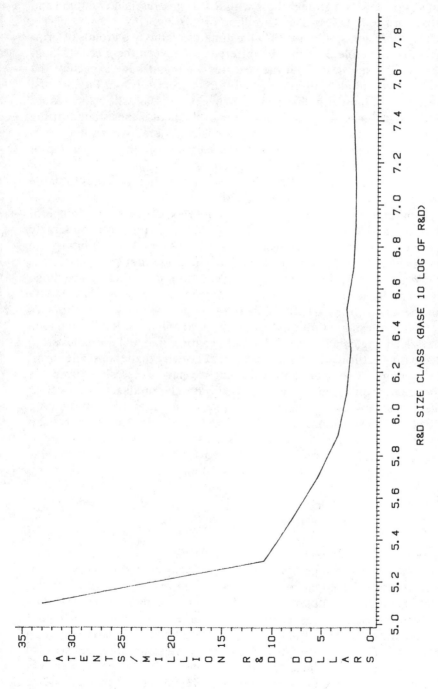

Fig. 2.6   Patents per million R & D dollars by R & D size class for firms with both R & D and patents.

downturn at the upper end (above $100 million). Accordingly, we divided our sample into two groups: those with R & D greater than $2 million and those with R & D less than $2 million or missing.

The coefficients of interest from estimates on the two groups of firms are shown in table 2.10 and the differences between them are striking. The small firms show both the features we might have expected: the Poisson-type models all are quite different from OLS on log patents, since most of these firms have less than five patents, and the estimates are all much closer to each other, since the problem of inconsistency arising from the increasing variance of patents is considerably mitigated. Substantively, there is no real evidence of curvature in the relationship of R & D and patents at this end of the distribution, and the elasticity of patenting with respect to R & D is close to the earlier estimates for large firms, albeit not very well determined.

Turning to the larger firms, as we might expect, since the range of R & D is about ten times that of the smaller firms, there is considerably more variation in the estimates. The log patents regression estimates are much closer to the others, since the integer nature of the patents data is not much of a problem here. However, there does seem to be some evidence of a decrease in the elasticity of patenting with respect to R & D for the largest firms. The Poisson and nonlinear least squares estimates exhibit increasing returns up to about $20–40 million of R & D and then start declining, whereas the OLS and negative binomial estimates show decreasing returns with a slightly higher elasticity than the smaller firms throughout. It is clear, however, that we have not really solved the specification problem for these large firms. The predicted values from these estimates exhibit nearly the same sensitivity to exactly how we weight the observations as did those from the whole sample. Our tentative conclusion is that there are nearly constant returns to scale in patenting throughout the range of R & D above $2 million, with decreasing returns setting in some place above $100 million.

2.5 Conclusion

We began this paper with a question: Who does R & D and who patents? We can now provide at least a partial answer. We have seen that research and development is done across all manufacturing industries with much higher intensities in such technologically progressive industries as chemicals, drugs, computing equipment, communication equipment, and professional and scientific instruments. We have found an elasticity of R & D with respect to sales of close to unity, but we also found significant nonlinearity in the relationship, implying that both very small and very large firms are more R & D intensive than average-size

Table 2.10   Estimates for Two R & D Size Classes

Variable	OLS $\log P = X\beta + u$	Poisson	Negative Binomial	Nonlinear Least Squares $P = \exp(X\beta) + \epsilon$
Small Firms ($N = 2102$)				
Log R & D	.10 (.03)	.62 (.06)	.49 (.10)	.58 (.32)
Log R & D squared	.017 (.006)	.014 (.020)	.026 (.030)	-.004 (.193)
Elasticity (R & D = $100k)	.02	.56	.37	.60
Elasticity (R & D = ($2M)	.12	.64	.53	.57
Large Firms ($N = 480$)				
Log R & D	.90 (.12)	1.57 (.02)	.90 (.08)	2.19 (.22)
Log R & D squared	-.003 (.02)	-.098 (.003)	-.034 (.009)	-.16 (.02)
Elasticity (R & D = $2M)	.90	1.43	.85	1.97
Elasticity (R & D = $100M)	.65	.67	.59	.72

firms.[11] This effect remained after an attempt to account for the (possibly) nonrandom selection of the dependent variable, although the lack of an exclusion restriction in this procedure casts some doubt on the completeness of this correction. We also found evidence of complementarity between capital intensity and R & D intensity, which was increased when we corrected for the selectivity of R & D.

These results are contrary to the preponderance of previous work on the size–R & D intensity relationship.[12] Hamberg (1964) and Comanor (1967) found a weakly decreasing relationship between R & D intensity and firm size. Scherer (1965a) found that R & D intensity increased with firm size up to an intermediate level, and then decreased (except in the chemical and petroleum industries, in which it increased throughout). This has been interpreted to imply, for most industries, a threshold size necessary before R & D is performed, presumably because of fixed costs in performing R & D (Kamien and Schwartz 1975). As noted above, our results suggest the opposite, though the selectivity issue precludes a definitive conclusion. In any case, these data cast strong doubt on the existence of any significant R & D threshold.[13]

There are several possible reasons for these conflicting results. First, earlier studies were based on small samples of larger companies of the *Fortune* 500 variety. An attempt was made to approximate these samples by estimating equation (3) of table 2.6 on those firms with sales of $500 million or more (256 observations). This regression indicates that this sample difference is not the source of the discrepancy; the relationship was close to linear with an implied elasticity of R & D with respect to sales of 1.23 at sales of $1 billion.[14]

In addition, our R & D variable is an expenditure variable, whereas much of the previous work used the number of R & D employees. If R & D expenditures per research employee rise fast enough with increasing firm size, perhaps because of greater capital intensity of R & D, we would expect the observed difference in the results. It is not possible, with these data, to test this hypothesis.

Finally, it is possible that the size–R & D intensity relationship has changed since the earlier work was done.[15] Because that work did not look at small firms at all, it would be sufficient to postulate increased

11. It should be emphasized, however, that our finding of increasing R & D intensity as firm size rises does not necessarily imply returns to scale in R & D unless one assumes homogeneity of some degree in the R & D production function (see Fisher and Temin 1973, 1979).

12. See Kamien and Schwartz (1975) for a summary.

13. These data also do not support the existence of a peculiar size–R & D intensity relationship in the chemical or petroleum industries.

14. The coefficients (standard errors) were: log sales: 1.29 (.61); log sales squared: $-.008$ (.038).

15. Hamberg used 1960 data; Comanor used 1955 and 1960 data; Scherer used 1955 data.

relative R & D intensity by the largest firms to reconcile their results with ours. We hope that our examination of the time-series component of this data set will shed some light on this question.

Turning to the second question in our title, we have found that some, but not all, of the firms which do R & D also patent, and that there is a strong relationship between the two activities throughout our sample. The small firms which do R & D tend to patent more per R & D dollar than larger firms, and firms with R & D programs larger than about $1 or $2 million have a nearly constant ratio of patenting to R & D throughout the sample, except for the firms with the very largest R & D programs.

Previous research on the relationship of R & D and patenting, in particular Scherer (1965b), has tended to focus on the largest U.S. corporations. Scherer found an elasticity of patenting with respect to R & D *employment* of unity with a hint of diminishing returns at the highest R & D input intensity. Our data do not contradict this result, but they do suggest that for these larger firms the elasticity of patenting with respect to R & D may have fallen slightly between 1955 and 1976. However, measurement issues cloud this conclusion since we are relating contemporary R & D expenditures and successful patent applications, while Scherer looks at patents granted and the number of R & D employees (lagged by four years). It is not easy to say a priori which relationship will be most free of noise, and we must wait for time-series studies to give us a better reading on the precise relationship of the two variables. Work thus far (Pakes and Griliches 1980; Pakes 1981) has shown a strong contemporaneous relationship of R & D and patent applications, but it has also found a total elasticity closer to one when lagged R & D is included.

These data also confirm and extend what others, including Scherer, have observed: a higher output of patents per R & D dollar for smaller firms. However, our results are for many more smaller firms than previously, and they show much sharper decreasing returns both in the measured elasticity and in the basic patents-to-R & D ratio. We also found that for this sample it mattered very much whether we used a model and estimation method which allowed for zero-valued observations.

In looking at these results on smaller firms, however, it is important to emphasize that although we include all manufacturing firms in our sample, whether or not they do R & D or patent, another kind of selectivity is at work: for a smaller firm, whether or not it appears on the Compustat file in the first place is a sign of success of some sort, or of a need for capital. The basic definition which gets a firm into the sample (if it is not automatically included as a result of being traded on a major stock exchange) is that it "commands sufficient investor interest." One of the likely causes of interest is a successful R & D program, and hence some patent applications. Thus we tend to observe small firms only when they have become "successful," whereas almost all large firms are publicly traded and will appear in our sample whether or not they have been particularly successful recently in research or innovation. We find it difficult to argue purely from this data that small firms have a higher return to R & D when we have reason to believe that only those which are successful at R & D are likely to be in our sample in the first place.

This is our first exploration of this rather large and rich data set. We hope to focus in the near future on the time-series characteristics of these data. We expect to be able to construct a consistent set of data for at least seven years (1972–78) for over a thousand firms. This should allow us to investigate more thoroughly some of these same questions and also many other aspects of R & D and patenting behavior.

Appendix

Composition of Industry Classes

Industry	Included SIC Groups
Food and kindred products	20
Textiles & apparel	22, 23
Chemicals, excluding drugs	28, excluding 2830, 2844
Drugs & medical instruments	2830, 2844, 3841, 3843
Petroleum refining & extraction	29
Rubber & misc. plastics	30
Stone, clay, and glass	32
Primary metals	33
Fabricated metal products	34, excluding 3480
Engines, farm & construction equipment	3510–3536
Office, computers, & accounting equipment	3570, 3573
Other machinery, not electric	35, excluding 3510–3536, 357
Electric equipment & supplies	36, excluding 3650–3679
Communication equipment	3650–3679
Motor vehicles & transportation equipment	37, excluding 3720–3729, 3760
Aircraft & aerospace	3720–3729, 3760
Professional & scientific equipment	38, excluding 3841, 3843
Lumber, wood & paper	24, 25, 26
Miscellaneous consumer goods	21, 31, 3480, 3900–3989
Miscellaneous manufacturers, n.e.c.	27, 3990

References

Comanor, W. S. 1967. Market structure, product differentiation, and industrial research. *Quarterly Journal of Economics* 81, no. 4:639–57.

Fisher, Franklin M., and Peter Temin. 1973. Returns to scale in research and development: What does the Schumpeterian hypothesis imply? *Journal of Political Economy* 81, no. 1:56–70.

———. 1979. The Schumpeterian hypothesis: Reply. *Journal of Political Economy* 87, no. 2:386–89.

Gourieroux, C., A. Monfort, and A. Trognon. 1981. Pseudo maximum likelihood methods: Applications to Poisson models. Université Paris IX, CEPREMAP, and ENSAE, December.

Griliches, Zvi, Bronwyn H. Hall, and Jerry A. Hausman. 1978. Missing data and self-selection in large panels. *Annales de l'INSEE* 30–31 (April–September): 137–76.

Hamberg, D. 1964. Size of firm, oligopoly, and research: the evidence. *Canadian Journal of Economics* 30:62–75.

———. 1967. Size of enterprise and technical change. *Antitrust Law and Economics* 1, no. 1:43–51.

Hausman, Jerry A., Bronwyn H. Hall, and Zvi Griliches. 1984. Econometric models for count data with an application to the patents–R & D relationship. *Econometrica*, forthcoming.

Heckman, James J. 1976. The common structure of statistical models of truncation, sample selection, and limited dependent variables and a sample estimator for such models. *Annals of Economic and Social Measurement* 5: 475–92.

Kamien, Morton I., and Nancy L. Schwartz. 1975. Market structure and innovation: A survey. *Journal of Economic Literature* 13:1–37.

Leamer, Edward E. 1978. *Specification searches: Ad hoc inferences with nonexperimental data.* New York: Wiley.

National Register Publishing Company. 1972. 1976. *Dictionary of corporate affiliations.* Skokie, Illinois.

National Science Foundation. 1979. *Research and Development in Industry, 1977.* Surveys of Science Resources Series, Publication no. 79–313.

Nelson, Forrest D. 1977. Censored regression models with unobserved stochastic censoring thresholds. *Journal of Econometrics* 6:309–22.

Pakes, Ariel. 1981. Patents, R & D, and the one period rate of return. NBER Discussion Paper no. 786.

Pakes, Ariel, and Zvi Griliches. 1980. Patents and R & D at the firm level: A first report. *Economics Letters* 5:377–81.

San Miguel, Joseph G., and Shahid L. Ansari. 1975. Accounting by business firms for investment in R & D. New York University, August.

Scherer, F. M. 1965a. Size of firm, oligopoly, and research: A comment. *Canadian Journal of Economics* 31, no. 2:256–66.

———. 1965b. Firm size, market structure, opportunity, and the output of patented inventions. *American Economic Review* 55, no. 5:1097–1125.

———. 1981. Research and development, patenting, and the microstructure of productivity growth. Final report, National Science Foundation, grant no. PRA-7826526.

Standard and Poor's Compustat Service, Inc. 1980. *Compustat II*. Englewood, Colorado.

3   Patents and R & D at the Firm Level: A First Look

Ariel Pakes and Zvi Griliches

3.1 Introduction

This paper is the first report from a more extensive study of knowledge-producing activities in American industry initiated by the National Bureau of Economic Research. Perhaps the most serious task facing empirical work in the area of "technological change" and "invention and innovation" is the construction and interpretation of measures (indices) of advances in knowledge.[1] If one defines K as the level of economically valuable technological knowledge, and $\dot{K} = dK/dt$ as the net accretion to it per unit of time, then the first task of our research program is to evaluate the usefulness of several indicators of $\dot{K}$, focusing particularly on patents and the value of the firm, variables which have yet to receive the attention that we think might be warranted in this context.[2]

The basic structure of our project is illustrated succinctly by the path analysis diagram in figure 3.1. In that diagram $\dot{K}$ is a central unobservable which, together with the observables, the X's, and the disturbances, the v's, determines the magnitude of several interrelated indicators of invention and innovation, the Z's. The latter include the stock market value of the firm, the productivity of traditional factors of production, and invest-

Ariel Pakes is a lecturer in the Department of Economics in the Hebrew University of Jerusalem, and a faculty research fellow of the National Bureau of Economic Research. Zvi Griliches is professor of economics at Harvard University, and program director, Productivity and Technical Change, at the National Bureau of Economic Research.

1. For a thoughtful discussion of this point, see Kuznets (1962).
2. Most of the previous work on patents is either quite ancient or inconclusive. Professional opinion has not really progressed much past the disagreement about the utility of patent statistics reflected in the discussions between Kuznets, Sanders, and Schmookler (Nelson 1962). The most recent review of the literature and independent contribution is found in Taylor and Silberston (1978). The papers that come closest to the topics treated here are Scherer (1965) and Comanor and Scherer (1969).

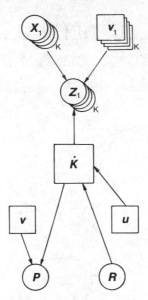

Fig. 3.1   A simplified path analysis diagram of the overall model.

ment expenditures on traditional capital goods. We report on an investigation of the lower half of this diagram in this paper. We see in figure 3.1 that $\dot{K}$ is produced by a knowledge production function (KPF) which translates past research expenditures, R, and a disturbance term, u, into inventions. The disturbance term reflects the combined effect of other nonformal R & D inputs and the inherent randomness in the production of inventions. Patents, P, are an imperfect indicator of the number of new inventions, with v_o representing the noise in the relationship between P and $\dot{K}$. It is clear from the figure that the patent equation, the equation connecting patents to past research expenditures, combines the properties of both the KPF and the indicator function relating to P and $\dot{K}$. Without additional indicators of $\dot{K}$ one cannot separate the two types of effects. For example, both u and v_o enter the relationship between R and P, but only u affects the Z's. In the context of a larger model, one could separate out the effects of u from v_o by calculating the effect of the residual in the patent equation on the Z's, but this cannot be done from the patent equation alone.

We have made several simplifications in drawing and discussing this diagram. For example, the relationship between K and $\dot{K}$ should be defined explicitly to allow for the possibility of decay in the private value of knowledge. $\dot{K}$ may be determined by the absolute level of K as well as by past investments in research resources. If, as in likely, the u's are correlated over time, then one would expect any realization of u to feed back into the demand for research resources. Moreover, conditions (economic, technological, and legal) should be specified under which the benefits from applying for a patent outweigh the costs of the patenting process, adding thereby more structure to the relationship between P and $\dot{K}$.[3] Figure 3.1 does, however, provide an overview of our project and is sufficiently precise for the discussion of the two issues on which this paper will concentrate: (1) the "quality" of patent counts as an indicator of knowledge increments, and (2) the time shape of the lag between research expenditures and patentable results.

The recent computerization of the U. S. Patent Office's data base has made it possible for the first time to follow the patenting behavior of a large cross section of firms over a significant time interval. This makes patent counts an easily accessible, perhaps the most easily accessible, indicator of the number of inventions made by a firm. Moreover, patents are a quantitative and rather direct indicator of invention; an indicator not contaminated by many of the X's which also affect the Z's. However, the patent measure does have several problems, the major ones being that not all new innovations are patented and that patents differ in their economic impact. These considerations have led to doubts about the "quality" of patent counts as an indicator of knowledge increments (see the literature cited in note 2). We attempt to respond to such concerns by first presenting a more precise description of the patent equation in section 3.2 and then reporting in section 3.3 on one particular measure of the "quality" of patent statistics.

Patent counts have another advantage over other indicators of knowledge production. Patents are applied for at an intermediate stage in the process of transforming research input into benefits from knowledge output. They can be used, therefore, to separate the lags that occur in that process into two parts: one which produces patents from current and past research investments, and another which transforms patents, with the possible addition of more research expenditures, into benefits. Such a breakdown should allow us to estimate more precisely the overall lag

3. Such a theory, we think, would be based on the underlying notion of a research project whose success depends stochastically on both the amount of resources devoted to it and the amount of time that such resources have been deployed. Each technical success is associated with an expectation of the ultimate economic value of a patent to the inventor or the employer. If this expectation exceeds a certain minimum, the cost of patenting, a patent will be applied for. That is, the number of patents applied for is a count of the number of successful projects (inventions) with the economic value of a patent exceeding a minimal threshold level. If the distribution of the expected value of patenting successful projects remains stable, and if the level of current and past R & D expenditures shifts the probability that projects will be technically successful, an increase in the number of patents can be taken as an indicator of an upward shift in the distribution of $\dot{K}$. Whether the relationship is proportional will depend on the shape of the assumed distributions and the nature of the underlying shifts in them. What we are dealing with here is at best a very crude reduced-form-type equation whose theoretical underpinnings still remain to be worked out. But one has to start someplace.

structure, a structure which has confounded and confused previous empirical work in this area.[4] Section 3.4 presents our first-round estimates of the distributed lag between research expenditures and patentable results.

The data used in this study are at the firm level and are based on a merger of the information provided in the Standard & Poor's (1980) Compustat file (based on the 10-K firm reports to the SEC) and patent data tabulated by the Office of Technology Assessments and Forecasts of the U.S. Patent Office. These data and the particular sample chosen are described in greater detail in Appendix A. Most of the work reported here is based on the patenting behavior of 121 firms during 1968–75.

3.2 The Model

We report in Appendix B a preliminary investigation into the functional form of the relationship between patents and past R & D expenditures. That analysis supports a rather simple patent equation: the logarithm of patents (p) as a function of a time trend (t), current and five lagged values of the logarithm of research expenditures (r), and a set of firm-specific dummy variables. In this section we provide an interpretation of this patent equation in terms of a simple model relating past r to the logarithm of current knowledge increments ($\dot{k}$), and $\dot{k}$ to p.

Consider first the transformation function from r to $\dot{k}$ or the KPF. Assuming it to be of the Cobb-Douglas form but allowing for firm constants and a time trend, we have:

$$(1) \qquad \dot{k}_{i,t} = a_i + bt + \sum_{\tau=0}^{5} \theta_\tau r_{i,t-\tau} + u_{i,t},$$

where $u_{i,t}$ is an independent and identically distributed disturbance which is not correlated with r and represents randomness in the KPF. The a_i represent firm-specific differences in the private productivity of research effort caused by either variation in appropriability environments, opportunities, or differences in managerial ability. Such differences will, in general, be transmitted to differences in research expenditures; firms with more productive research departments investing more in research. Thus, the a_i have two roles in the subsequent analysis. First, they cause differences in $\dot{k}$, and this should be considered in an analysis of the determinants of the variance in p. Second, their correlation with the $r_{t-\tau}$ must be accounted for in any attempt to estimate the θ_τ or else the coefficient estimates will be a combination of the effect of the $r_{t-\tau}$ on $\dot{k}$ (the θ_τ) and the effect of a_i or r. To be more explicit about the latter point,

---

4. See, for example, how two different assumptions about the lag structure lead to very different calculations of the private rate of return to research expenditures from the NSF-Griliches data: Griliches (1980b) versus Pakes and Schankerman (this volume).

we simply project the a_i on all in-sample research expenditures. Since the a_i are constant over time they should only be correlated with the means of the research variables. We can write, therefore,

$$(2) \qquad a_i = \sum_{\tau=0}^{5} \psi_\tau r_{i,\cdot-\tau} + u_{i\cdot},$$

where

$$r_{i\cdot 0} = T^{-1} \sum_{t=1}^{T} r_{it}, \ r_{i\cdot-1} = T^{-1} \sum_{t=0}^{T-1} r_{it-1}, \text{ etc.,}$$

and u_i is by construction uncorrelated with all in-sample research variables.[5]

Patents are our indicator of knowledge increments. If one allows for a time trend in the relationship between p and $\dot{k}$, that relationship is written as:

$$(3) \qquad p_{i,t} = dt + \beta \dot{k}_{i,t} + v^*_{i,t},$$

where v^* is uncorrelated with $\dot{k}$ and t by construction.

Equation (3) should be interpreted as a reduced form from the appropriate patenting model. In that reduced form, β is the elasticity of patents with respect to knowledge increments, and d is a measure of the trend in factors determining the propensity to patent. On the other hand, $v^*_{i,t}$ is that part of the (detrended) variance in patents which cannot be accounted for by (detrended) movements in knowledge increments; that is, variance in $v^*_{i,t}$ is "noise" in the patent measure. To facilitate interpretation we will make two assumptions on $v^*_{i,t}$. First, we let $v^*_{i,t}$ be composed of a firm-specific component, v_i, which reflects differences among firms in their *average* propensity to patent, and a second, independent, identically distributed disturbance, $v_{i,t}$, reflecting the *variations* (around a trend) in the propensity to patent of a given firm over time. Thus, $v^*_{i,t} = v_i + v_{i,t}$. Second, since $v^*_{i,t}$ is uncorrelated with $\dot{k}_{i,t}$ (by choice of β), we shall also assume that its determinants, v_i and $v_{i,t}$, are each uncorrelated with the determinants of $\dot{k}$ (the r's and u's) given by equations (1) and (2).[6]

---

5. In econometric terminology, the model we are working with is a variant of the partial transmission model of Mundlak and Hoch (1965). The unobservable portion of the KPF, which is transmitted to the research demand equation is assumed to remain constant over time. This assumption, plus the nature of the panel, will allow us to use single equation estimation techniques to estimate parameters of the patent production function. A more precise discussion of the econometric techniques underlying the estimation procedures to be used in this paper is found in Mundlak (1978) and Pakes (1978, chap. 3).

6. The first assumption allows us to provide standard errors for our estimates of the regression coefficients. The second is a rather strong assumption. We are assuming that randomness in the KPF, above or below average success in converting research expenditures into knowledge increments, does not influence the patenting decision, that the two sources of randomness are distinct and independent. We need this assumption to make the interpretations that follow.

Substituting (1) and (2) into (3), we can now provide an interpretation to the equation preferred in our analysis of functional form, that is to the equation:

$$(4) \qquad p_{i,t} = \alpha + \gamma t + \sum_{\tau=0}^{5} w_\tau r_{i,t-\tau} + \sum_{\tau=0}^{5} \phi_\tau r_{i\cdot,-\tau} + \eta_i + \epsilon_{i,t},$$

where

$$w_\tau = \beta\theta_\tau, \quad \gamma = \beta b + d, \quad \phi = \beta\psi,$$

$$\eta_i = v_i + \beta u_i, \text{ and } \epsilon_{i,t} = v_{i,t} + \beta u_{i,t}$$

for

$$i = 1, \ldots, N \text{ and } t = 1, \ldots, T.$$

The first point to note from equation (4) is that though one cannot estimate the elasticities of knowledge increments with respect to research resources, the θ_τ, one can investigate the form of the distributed lag connecting k and r, since $w_\tau / \Sigma w = \theta_\tau / \Sigma\theta$. The sum of the estimated lag coefficients, $w^* = \Sigma_{\tau=0}^{5} w_\tau$, estimates the product of the degree of economies of scale in the KPF, $\Sigma_{\tau=0}^{5} \theta_\tau$, and the elasticity of patents with respect to knowledge increments (β). These two parameters can be identified separately only in a larger model which includes additional indicators of the benefits from knowledge-producing activities (see section 3.1).

Recall that the various variance components which combine to form the disturbance term is (4) are mutually uncorrelated. It follows that $\text{Var}(\eta_i + \epsilon_{i,t}) = \sigma^2$, the variance of the total disturbance in the patent equation, is greater than $\text{Var}(v_i + v_{i,t})$, the variance of the noise in patents as an indicator of k. It also implies that, temporarily ignoring the time trend in the patent indicator equation (assuming $d = 0$), the ratio of σ^2 to the total variance in the logarithm of patents $(1 - \bar{R}^2)$ provides an upper bound for the noise-to-total-variance ratio in the patent measure. The upper bound will be called λ^{uT}, and its complement, the relevant $\bar{R}^2$ measure, is a measure of the quality of patents as an indicator of knowledge increments. If, instead of assuming $d = 0$, we assume $b = 0$, that is, the entire trend effect is caused by differences in the average propensity to patent over time, then one can derive an analogous measure of λ^{uT} for detrended patents by filtering out time from both the patent and the R & D variables. In practice, the two measures of λ^{uT} were always almost identical. In section 3.3 we also present the comparable information on the noise-to-total-variance ratio in the between firm variance in patents (i.e., in the variance of $p_{i\cdot} - p \ldots$), labeled λ^{uB}, and in the within firm variance in patents (the variance in $p_{it} - p_{i\cdot}$), λ^{uW}. The latter two statistics provide some indication of the usefulness of patent counts as an indicator of knowledge increments for studies of invention and innovation that focus either on cross-section differences in the production of knowledge between firms or on the within firm fluctuations over time.

3.3 Measures of the Quality of the Patent Variable

Table 3.1 presents estimates of $1 - \lambda^{uT}, 1 - \lambda^{uW}, 1 - \lambda^{uB}$, the lower bounds to the systematic-to-total-variance ratios, σ_ϵ^2 and σ_η^2, and some relevant sample moments for each of the seven industries in our data (rows 0 through 6), all firms in our sample (row 7) and firms in the industries defined by rows 1 through 6 (row 8). The latter sample concentrates on firms in research-intensive industries.

Starting with the measures of $1 - \lambda^{uT}$ in the separate industries, it is clear, even from our simplistic model, that much of the patent variance is systematic, providing a good indicator of the underlying variance in $\dot{k}$. For the seven industries in our sample, about 85 percent of the variance in p is associated with variance in r, and in some industries, notably scientific instruments and office, computing, and accounting machinery, the lower bound of the systematic to total variance in patents is closer to .95.

These estimates hide, however, some relevant information. Moving to column (2), we are clearly far less certain of whether changes over time in p within any given firm reflect systematic changes in knowledge production by that firm. In the within firm calculations it mattered whether or not we first filtered out time trends from p and r. Therefore, the numbers in parentheses beside column (2) refer to systematic-to-total-variance ratios in detrended patents. Averaging over the seven industries, we find that the lower bound $(1 - \lambda^{uW})$ is only around 20–25 percent, though it does reach 50 percent in office, computing, and accounting machinery. Without the larger model alluded to in section 3.1, one cannot really tell whether the smaller systematic-to-total-variance ratios in the "within" data reflect true randomness in the knowledge production function (small differences in research expenditures over time within a given firm having very sporadic effects on the production of inventions in particular years) or whether they arise because firms decide to patent different proportions of their inventions in different years.

Two more points should be noted about the results for the separate industries. Column (6) shows that over 90 percent of the total variance in p is between firm variance. As a result $1 - \lambda^{uB}$ is very close to $1 - \lambda^{uT}$. Second, though σ_ϵ^2 does not vary too much between the sample industries, σ_η^2 varies a lot, being much larger in the less homogeneous industries (rows 0, 1, and 3). This is likely to reflect greater differences in the average propensities to patent in those industries.

Looking at the samples which aggregate the various industries (rows 7 and 8), we find that λ^{uW} actually decreases after pooling different industry samples. This implies that, at least in our sample, the elasticity of

Table 3.1 Lower Bounds to the Systematic Variance Ratio in p and Some Sample Moments for the Seven Industries

| Industry Description[a] | Lower Bound to the Systematic Variance ratio[b] | | | | | Variance in p | Ratio of Within to Total Variance in p | Variance in r | Ratio of Within to Total Variance in r | Firms N | Observations NT |
	$1-\lambda^{uT}$ total	$1-\lambda^{uW}$ within	$1-\lambda^{uB}$ between	σ_ϵ^{2c}	σ_η^{2d}						
	(1)	(2)	(3)	(4)	(5)	(6)	(7)	(8)	(9)	(10)	(11)
0 Other manufacturing	.74	.19 (.16)	.77	.17	.050	2.71	.08	1.83	0.04	41	328
1 Industry 28 except 283 (chemicals & allied products except drugs & medicines)	.82	.13 (.11)	.86	.10	.26	2.02	.06	1.16	0.02	19	152
2 Industry 283 (drugs & medicines)	.80	.33 (.22)	.85	.07	.14	.85	.10	.56	0.03	19	152
3 Industry 35 except 357 (machinery except office, computing & accounting)	.82	.11 (.06)	.87	.16	.32	2.97	.07	2.17	0.04	13	104
4 Industry 357 (office, computing & accounting)	.84	.52 (.50)	.96	.16	.09	3.78	.11	2.06	0.11	10	80
5 Industry 366/367 (electronic components & communications)	.51	.46 (.42)	.89	.07	.07	.83	.13	1.57	0.03	8	64
6 Industry 38 (professional & scientific instruments)	.95	.28 (.06)	.97	.08	.05	2.55	.04	2.21	0.04	11	88
7 Total sample	.66	.33 (.23)	.69	.14	.66	2.41	.07	1.72	0.04	121	968
8 Firms in research-intensive industries	.73	.33 (.22)	.76	.12	.48	2.20	.07	1.61	0.04	80	640

[a]Two- and three-digit industry identification numbers refer to SIC codes.

[b]The numbers in parentheses refer to lower bounds to the systematic variance ratio in patents after a time trend has been filtered out of both p and r: p − log patents; r − log R & D.

[c]Calculated as the variance of estimate from an OLS regression of $p_{i,t}$ on current and five consecutive lagged years of $r_{i,t}$, and firm-specific constants.

[d]Calculated as the variance of estimate from an OLS regression of equation (4) in the text minus the estimate of σ_ϵ^2.

patents with respect to knowledge increments (β) and the response of k to current and past $r(\theta_\tau)$ do not vary much between the industries aggregated; a result which will be confirmed in section 3.4.

3.4 Coefficient Estimates

Table 3.2 presents the estimates of the w_τ and the coefficient of the trend term based on data from all of the 121 firms and estimates based on two subsamples: firms in research-intensive industries and "other manufacturing" firms. Row 10 presents the estimates value of the F statistic for the null hypothesis that these coefficients do not differ between the industries aggregated. The test statistics indicate that, after we allow for a separate trend and intercept for the drug industry (row 9), our sample cannot really pick up any additional interindustry differences in coefficients.[7]

Turning to the coefficient of the trend term, that coefficient was negative, and significantly so, for all industries except for the drug industry. This result has two alternative explanations, and they cannot be separated out without the larger model alluded to in section 3.1. First, the negative trend is consistent with impressionistic evidence on the declining propensity to patent in U.S. manufacturing. The drug industry is indeed an exception since, during the period concerned, there occurred both a relaxation in the Patent Office's acceptance procedures regarding patents on natural substances and significant changes in regulatory conditions facing that industry.[8] The same result, however, could have been caused by a secular decline in the private productivity of research resources, a hypothesis which is consistent with the observed negative growth rate of employment of R & D scientists and engineers during the period considered.[9]

The individual coefficients are not estimated very precisely. The sum of the lags, w^*, is estimated with a fair amount of precision and equals about .60 with a standard error of 0.08. If one ignores the fact that some of the estimated lag coefficients are negative and computes a "mean lag," it equals about 1.6 years for the all-firm sample. Unless substantial R & D is done on projects after patents are applied for, this should approximately equal the mean R & D project gestation lag, the lag between project

7. The possible exception here is the drug industry. When that industry was dropped from the first two samples, the observed values of the F test dropped significantly to 1.37 and 1.67, respectively. Still the estimated coefficients for the drug industry were not very different from those of the other industries in the sample, except for the trend coefficient.
8. For a description of the effect of these events, see Temin (1979).
9. See Griliches (1980a) for a similar finding on aggregate data.

Table 3.2   Distributed Lag Estimates[a]

Variables	All Firms (1)	Firms in Research-Intensive Industries (2)	Other Manufacturing Firms (3)
1. r_0	.56	.52	.62
	(.07)	(.10)	(.14)
2. r_{-1}	−.10	−.01	−.22
	(.09)	(.12)	(.16)
3. r_{-2}	.05	.08	−.02
	(.09)	(.12)	(.16)
4. r_{-3}	−.04	−.21	.13
	(.09)	(.13)	(.15)
5. r_{-4}	−.05	−.01	−.08
	(.10)	(.15)	(.16)
6. r_{-5}	.19	.25	.13
	(.08)	(.11)	(.14)
7. Sum (w^*)	.61	.62	.61
	(.08)	(.09)	(.04)
8. t	−.04	−.05	−.03
	(.007)	(.008)	(.012)
9. t_{drugs}	.07	.07	—
	(.10)	(.01)	
10. F aggregation (critical	1.54	2.08	—
values, 1%, 5%)	(1.39, 1.58)	(1.45, 1.69)	
11. Degrees of freedom	837	550	279

[a]Standard errors are in parentheses below coefficient estimates.

inception and project completion. The scattered empirical evidence on gestation lags indicates that this is indeed the case.[10]

Still, the estimated form of the lag is rather disturbing. There are large, significant, postive coefficients in the first and last years and very little effect of interim R & D on patent applications. Though the current year's coefficient could indicate the presence of simultaneous equations bias, that is not really a necessary implication of the results. The R & D project level data cited above do point to a gestation lag highly skewed with large early year coefficients, and any minor misspecification in the model could push all this effect into the coefficient of r_0. The coefficient estimate which is perhaps more disturbing is that of the last year since it could be indicating the presence of a "truncation" problem in our distributed lag

10. Sources of project level data are Wagner (1968) and Rapoport (1971). This evidence is summarized in terms of mean gestation lags in Pakes and Schankerman (this volume). The average of the mean gestation lags presented in the latter paper was 1.34 years.

estimates. That is, the coefficient of the fifth year could be proxying for a series of small effects of the more basic research done six years ago or earlier.[11] These estimates of the form of the lag should be treated with caution, both because of the possible truncation problem and because they are not really consistent with our prior beliefs about the form of this lag structure.

3.5 Conclusions and Extensions

Our first look at the patent equation suggests the following conclusions. First, the data were quite clear on the form of that equation; log-log with (correlated) firm effects and a time trend being preferred over alternatives. Second, our major positive finding is given by the $1 - \lambda^{uB}$ estimates presented in table 3.1. They show that patents are a good indicator of between firm differences in advances of knowledge. Since the between firm component dominates the total variance in patents, a similar comment also applies to the total variance. If this result changes at all in the more sophisticated models we are beginning to estimate, it is only likely to improve. Use of a longer series of past R & D expenditures can only increase the fit of the patent equation, and adding another indicator of benefits will separate out the effect of randomness in the KPF, the u, from the effect of noise in the patent measure, the v^*, allowing us to narrow the bound further.

The rest of our results are not as heartening. While a part of the within firm variance in patents is related to the variance in R & D expenditures, a significant portion (about 75 percent) is not. At this stage we cannot tell whether the fault lies in the patent measure (the variance in v^*), in randomness in the KPF (the variance in u), or in simple errors of measurement in both p and r. Most of the coefficients, except for trend, were not estimated very precisely. This is a result of two factors: First, only the within firm variance in p and r can be used to estimate w_τ, and this variance is a small part of the total variance in these variables (see table 3.1). The second factor leading to imprecise estimates is the small sample size (maximum $T = 8$; $N = 121$). We can and will increase our sample significantly in the future by not insisting that firms had to have reported R & D expenditures before 1972 (see appendix A). Including such firms will force us, however, to use only a few lagged terms of r or assume a specific functional form for the distributed lag between patents and R & D expenditures, even though we have yet to acquire much information on the shape of this distribution. Because our estimates indicate that even with five lagged R & D terms we still may have a truncation problem, we have been developing a technique for estimating

11. See Griliches and Pakes (1980) for further discussion of such problems.

distributed lags in panel data when the time series on the independent variable is short. We are also investigating the impact of other sources of bias in the estimated coefficients, in particular the effect of measurement errors in the R & D variables. Finally, once an appropriate specification for the patent equation has been determined, we will combine it with the other equations in our model in the hope of providing a fuller understanding of the process of invention and innovation in American industry.

In short, a great deal of work remains to be done, but we have made a start. It is already clear that something systematic and related to knowledge-producing activities is being measured by patents and that they are, therefore, very much worthy of further study.

Appendix A

Data Sources, Sample and Variable Definitions, and Sample Characteristics

The data base used in this preliminary round is neither complete nor representative. We have tried to gather from published sources as large a sample of firms as possible covering 1963–77. The main selection variable is R & D. Until recently (1972 and later) most firms did not report their R & D expenditures publicly. The firms that did report R & D expenditures reported company-financed R & D expenditures, and those numbers are recorded on the Standard and Poor's Compustat tape, which served as a major source of our data.[12] An earlier study by Nadiri and Bitros (1980) had used both the Compustat tape and a mail survey to fill in some of the gaps on this tape to construct time series of R & D for 114 firms during 1963–72. Starting with a later edition of the tape, we found 146 firms with no more than three years of R & D data missing during 1963–75. Combining it with the Nadiri and Bitros sample yielded an unduplicated total of 172 firms. Fifteen firms were eliminated from this total either because they were foreign, had undergone large mergers, or had other unreconcilable jumps in their data. This left a total of 157 firms which constitute the data on which a number of recent NBER studies have been based.[13] Based on preliminary experimentation (see appendix B), the sample for this paper was further restricted to firms that had data (did not undergo any major reorganization) throughout the whole period

---

12. Only company-financed R & D ought to lead to patents since government R & D contracts most often include clauses which put the output of government-funded projects in the public domain.

13. For further description of this data base, see Bound and Hall, 1980. A much larger sample is possible to construct if one is willing to restrict oneself to post-1972 data. See Bound et al. (this volume).

($N = 144$) and had an R & D program of more than minimal size (R & D $\geq$ \$0.5 million) in any one year ($N = 121$). What we have done, then, is to expand the Nadiri and Bitros sample slightly, update it to 1977, *and* add patent data to it.[14]

The patent data were supplied to us by the Office of Technology Assessment and Forecasts (OTAF) of the U.S. Patent Office. They are based on a tape of all patents granted during 1969–78. These data are then reclassified by year of application rather than by year of grant. One of our tasks was to be sure that we had all the subsidiaries and names used by a particular corporation. For this purpose we scrutinized the alphabetical index of patenting organizations provided by OTAF and checked it against the list of firms' subsidiaries given in the *Dictionary of Corporate Affiliations* (National Register 1972, 1976) and a list of past mergers given in *Mergers and Acquisitions* (1974–77). If a firm had acquired another firm during this period, we added in the patents of the acquired organization (and its R & D expenditures, when known). In a few cases, where the mergers were large and occurred toward the end of the period, we left the two firms unmerged and instead declared the recent (postmerger) years as missing.

Because the patent data are based on patents *granted* during 1969–78, patents by year applied for cannot really be used before 1968. While only less than 1 percent of all patents granted is granted within the year of application, about 10 percent are granted in the following year. Thus, only about 89 percent of the patents applied for in 1967 would appear among the patents counted by us. Similarly, one probably cannot use the patent data by year of application after 1975, since it takes about four years after the application before more than 96 percent of the patents applied for in that year will be eventually granted are actually granted.[15] Thus, at best, we have about eight or nine years of usable patent data. In most of the analyses we used the eight years, 1968–75. Eight years and 121 firms give us an effective sample size of 968.

Table 3.A.1 gives means and standard deviations for a few of the major variables in the various samples and industries represented in this study. The industrial classification was chosen to approximate the industrial breakdown used by the NSF in its reports. It is clear from this table that these firms are rather large, that the exclusion of firms with R & D budgets of less than half a million dollars makes them even larger, that the size distribution of the firms is quite skewed (standard deviations are

---

14. Some of the missing years have been interpolated by us. Also, the definition of expenditures reported as R & D by different firms may change over time. Where such changes were obvious or stated in the 10-K forms, we tried to adjust for them. Where we could not and the discrepancies were large, we eliminated the firm from our sample

15. These estimates are based on an unpublished tabulation of patents granted by date applied for, for 1965–77, made available to us by OTAF.

Table 3.A.1   **Characteristics of Sample Firms by Industry: Averages 1963–75 and Standard Deviations**

Variable		Entire Sample			Firms with Complete Data and Min. R & D ≥ 500K	
	N	Mean	Standard Deviation	N	Mean	Standard Deviation
--- Ind = 0 ---						
DEFRND	13	5.678	7.291			
GROPLA72	13	211.994	272.923			
PATS	13	15.788	21.781			
--- Ind = 28 ---				--- Ind = 28 ---		
DEFRND	21	28.353	32.953	19	31.258	33.362
GROPLA72	21	1053.298	1248.221	19	1162.378	1264.633
PATS	21	92.804	104.502	19	102.520	105.298
--- Ind = 28.3 ---				--- Ind = 28.3 ---		
DEFRND	20	26.665	20.009	19	28.013	19.603
GROPLA72	20	264.486	206.233	19	277.698	203.002
PATS	20	54.531	40.089	19	57.316	39.151
--- Ind = 35 ---				--- Ind = 35 ---		
DEFRND	14	17.143	25.603	13	18.407	26.191
GROPLA72	14	327.631	429.287	13	352.300	436.366
PATS	14	47.464	64.881	13	50.990	66.119
--- Ind = 35.7 ---				--- Ind = 35.7 ---		
DEFRND	13	25.422	30.165	10	32.169	31.521
GROPLA72	13	544.321	767.480	10	665.861	843.293
PATS	13	62.490	98.328	10	79.912	106.895
--- Ind = 36 ---				--- Ind = 36 ---		
DEFRND	15	15.457	34.068	8	28.427	43.694
GROPLA72	15	393.137	1248.032	8	731.354	1683.749
PATS	15	37.975	68.848	8	70.031	83.442
--- Ind = 38 ---				--- Ind = 38 ---		
DEFRND	15	25.507	46.550	11	34.452	51.996
GROPLA72	15	352.326	815.241	11	477.772	930.342
PATS	15	63.592	99.897	11	85.920	109.149
--- Ind = 99 ---				--- Ind = 99 ---		
DEFRND	46	20.489	29.581	41	22.884	30.500
GROPLA72	46	3074.459	10476.016	41	3445.557	11052.898
PATS	46	56.068	90.813	41	61.808	94.687
--- Combined ---				--- Combined ---		
DEFRND	144	22.612	30.994	121	26.709	32.228
GROPLA72	144	1331.104	6044.733	121	1578.299	6569.299
PATS	144	59.854	84.891	121	70.565	88.661

Note: DEFRND = deflated R & D expenditures, in million dollars.
GROPLA72 = book value of gross plant in 1972, in million dollars.
PATS = number of patents, by year applied for.
Ind = 0 = firms with incomplete data for the whole period.
Ind = 28 = chemicals and allied products, except drugs and medicines.
Ind = 28.3 = drugs and medicines.
Ind = 35 = machinery, except office, computing, and accounting.
Ind = 35.7 = office, computing, and accounting machinery.
Ind = 36 = electronic components and communications.
Ind = 38 = professional and scientific instruments.
Ind = 99 = other manufacturing.

on the order of the means or larger), and that the industrial distribution is quite uneven. The firms represented in the sample are those who reported their R & D expenditures publicly in the 1960s, with drug and chemical firms overrepresented.

The R & D expenditures have been deflated by an R & D "deflator" index constructed along the lines suggested by Jaffe (1972): a weighted average of the index of hourly labor compensation and the implicit deflator in the nonfinancial corporations sector, with .49 and .51 as relative weights.

The main problem with our sample is its peculiar nature. It is based on those companies that reported R & D expenditures in the mid-1960s. Since it is selected on the "independent" variable in this study, one need not anticipate much of a selectivity bias in equations where patents or the market value of the firm are the dependent variables. Also, since much of our analysis will be "within" firms, any fixed selectivity adjustment would be incorporated in the constant term and would not affect our inferences.

Appendix B
The Form of the Patent Equation

Because there was little prior empirical or theoretical research on the R & D-to-patents relation, we began our analysis with an investigation of the functional form of the equation that might connect these two variables in our data.

Functional form questions were examined, allowing the parameters of all estimated equations to differ in each of our seven industries and between firms with large and small R & D departments within each industry.[16] That is, fourteen sets of parameters were estimated. The independent variables included in the estimating equations were a set of time dummies, the current and five consecutive lagged values of both the logarithm of R & D expenditures and R & D expenditures per se, and a set of firm-specific dummy variables (constants). To simplify matters we assumed that the appropriate form of the dependent variable was either $\log(P) = p$ or P itself. Hence log-log, semilog, and linear functions, each with firm and time effects, were all special cases of the model with which we started.

A variant of the Box and Cox (1964) procedure was used to choose the

---

16. Small firms were defined, quite arbitrarily, as firms whose R & D expenditures over the sample period (1963–1975) fell below half a million dollars in at least one year. The size breakdown had the effect of separating out the recently born science-based firms from the others in the sample and allowed for the possibility that the characteristics of the KPF differed in the firms with smaller, less established, research departments.

form of the dependent variable. It indicated that the logarithm of patents was clearly preferred over the absolute number of patents by the data for each separate grouping and for the sample as a whole. We then asked whether the parameters of the relationship between p and the independent variables within each industry differed between firms with large and small R & D departments. The test statistic was significant at any reasonable level of significance, indicating that the form of the relationship between patents and research expenditures was different for firms with small R & D departments. The twenty-six firms in the small group were dropped from all the subsequent computations reported in this paper. Next, we wanted to know whether the model could be simplified by assuming either that the coefficients of current and all lagged values of R & D, or that the coefficients of the logarithmic forms of these variables, were all zero. The $F^{36, 734}$ statistic for the joint significance of the R & D variables in their natural form was a rather small 1.18, whereas that test statistic for the logarithmic form of the R & D variables was a highly significant 3.30. We, therefore, accepted the former hypothesis and rejected the latter and went on to test another simplification: whether or not the seven time dummies could be approximated by a linear time trend. The observed value of the $F^{30, 770}$ deviate for this hypothesis was .95, which is below the expected value of that test statistic, given that the time dummies were in fact representing a simple trend. Two other hypotheses were tested but both were clearly rejected by the data. The first was that the distribution of the firm-specific constants was degenerate, that there were no "firm effects." After rejecting this hypothesis we went on to test whether it was reasonable to assume that the firm effects were uncorrelated with research expenditures. It was not. Thus the form of the equation we settled on was rather simple: the logarithm of patents as a function of a time trend, current and five consecutive lagged values of the logarithm of R & D expenditures, and (correlated) firm-specific constant terms.[17]

17. There is one issue which we have not dealt with here because it is not very important in our sample. For observations where $P = 0$, log (P) is undefined. This exposes an underlying truncation problem in our model. That problem, however, is of minor importance for our sample since only 8 percent of the observations are at $P = 0$. This is less than the percentage of observations at $P = 1$ (14 percent), indicating that the truncation problem is not large. It is even smaller for the larger R & D firm sample ($N = 121$) where the zero patents percentage is only three. As a result we treated the whole problem as one of finding a point on the logarithmic scale for $P = 0$. This was accomplished by adding a dummy variable to the independent variables for observations where $P = 0$. The estimated coefficients of this dummy variable are stable across models, implying roughly the value of 0.1–0.7 for the $P = 0$ observations. It does raise the issue, though, of whether our functional form (log-log) is appropriate for low patenting level observations. We intend to investigate explicitly probabilistic models of the patenting process in subsequent work. These issues are discussed in more detail in Hausman, Hall, and Griliches (1984).

References

Bound, J., and B. H. Hall. 1980. The R & D and patents data master file. Mimeo.

Box, G. E. P, and D. R. Cox. 1964. An analysis of transformations. *Journal of the Royal Statistical Society*, series B, 26:211–243.

Comanor, W. S., and F. M. Scherer. 1969. Patent statistics as a measure of technical change. *Journal of Political Economy* 77, no. 3 (May–June):392–98.

Griliches, Z. 1980a. R & D and the productivity slowdown. *American Economic Review* 70, no. 2:343–48.

———. 1980b. Returns to research and development expenditures in the private sector. In *New developments in productivity measurement and analysis*, ed. J. W. Kendrick and B. N. Vaccara, 419–54. Conference on Research in Income and Wealth: Studies in Income and Wealth, vol. 44. Chicago: University of Chicago Press for the National Bureau of Economic Research.

Griliches, Z., and A. Pakes. 1980. The estimation of distributed lags in short panels. NBER Working Paper no. 4, October. Cambridge, Mass: National Bureau of Economic Research.

Hausman, J., B. H. Hall, and Z. Griliches. 1984. Econometric models for count data and with application to the patents–R & D relationship. *Econometrica*, in press.

Jaffe, S. A. 1972. A price index for deflation of academic R & D expenditures. Washington, D.C.: NSF 72–310.

Kuznets, S. 1962. Inventive activity: Problems of definition and measurement. In *The rate and direction of inventive activity: Economic and social factors*, ed. R. R. Nelson, 19–51. Princeton: Princeton University Press for the National Bureau of Economic Research.

Mergers and acquisitions, vols. 8–11. 1974–77. *Journal of Corporate Venture*. McLean, Va.: Information for Industry Inc.

Mundlak, Y. 1978. On the pooling of time series and cross section data. *Econometrica* 46, no. 1:69–86.

Mundlak, Y., and I. Hoch. 1965. Consequences of alternative specifications in estimation of Cobb-Douglas production functions. *Econometrica* 33, no. 4:814–28.

Nadiri, M. I., and G. C. Bitros. 1980. Research and development expenditures and labor productivity at the firm level: a dynamic model. In *New developments in productivity measurement and analysis*, ed. J. W. Kendrick and B. N. Vaccara, 387–412. Conference on Research in Income and Wealth: Studies in Income and Wealth, vol. 44. Chicago: University of Chicago Press for the National Bureau of Economic Research.

National Register Publishing Company. 1972. 1976. *Dictionary of corporate affiliations*. Skokie, Illinois.

Nelson, R. R., ed. 1962. *The rate and direction of inventive activity: Economic and social factors*. Universities–NBER Conference Series no. 13. Princeton: Princeton University Press for the National Bureau of Economic Research.

Pakes, A. 1978. Economic incentives in the production and transmission of knowledge: An empirical analysis. Ph.D. diss., Harvard University.

Rapoport, J. 1971. The anatomy of the product-innovation process: Cost and time. In *Research and innovation in the modern corporation*, ed. E. Mansfield, J. Rapoport, J. Schnee, S. Wagner, and M. Hamburger, 110–35 New York: Norton.

Scherer, F. M. 1965. Firm size, market structure, opportunity, and the output of patented inventions. *American Economic Review* 55, no. 5:1097–1125.

Standard and Poor's Compustat Services, Inc. 1980. *Compustat II*. Englewood, Colorado.

Taylor, C. T., and Z. A. Silberston. 1978. *The economic impact of the patent system: A study of the British experiment*. Cambridge: Cambridge University Press.

Temin, P. 1979. Technology, regulation, and market structure in the modern pharmaceutical industry. *Bell Journal of Economics* 10:429–46.

Wagner, L. U. 1968. Problems in estimating research and development investment and stock. In *Proceedings of the business and economic statistics section*, 189–98. American Statistical Association. Washington, D.C

4 The Rate of Obsolescence of Patents, Research Gestation Lags, and the Private Rate of Return to Research Resources

Ariel Pakes and Mark Schankerman

The recent interest of economists in knowledge-producing activities has two main strands. The first is an attempt to explain the growth in the measured productivity of traditional factors of production by incorporating research resources in production function and social accounting frameworks (for a review see Griliches 1973, 1980). The second derives from two fundamental characteristics of knowledge as an economic commodity, its low or zero cost of reproduction and the difficulty of excluding others from its use. These features give knowledge the character of a public good and suggest that the structure of market incentives may not elicit the socially desirable level (or pattern) of research and development expenditures. In particular, it has been argued that market incentives may create either underinvestment or overinvestment in knowledge-producing activities (see Arrow 1962; Dasgupta and Stiglitz 1980). To investigate this possibility, economists have applied the techniques of productivity analysis and estimated the private (and social) rate of return to research from production functions incorporating research resources as a factor of production. The estimates of the *private* return for the late 1950s and early 1960s fall in the range of 30–45 percent.[1] Despite these

Ariel Pakes is a lecturer in the Department of Economics at the Hebrew University of Jerusalem, and a faculty research fellow of the National Bureau of Economic Research. Mark Schankerman is an assistant professor in the Department of Economics at New York University, and a faculty research fellow of the National Bureau of Economic Research.

This paper was originally circulated as Harvard Institute of Economic Research, Discussion Paper 659 (October 1978). The authors have benefited from discussions in an informal seminar on technological change at Harvard, chaired by Zvi Griliches, and from comments by Gary Chamberlain. All remaining errors are the authors. Financial assistance from the National Science Foundation, grant GS-2726X, is gratefully acknowledged.

1. For example, see Griliches (1980) and Mansfield (1968). Note that both these studies are based on firm or microdata. More aggregative data bases are not directly relevant to estimates of the private rate of return to research.

high estimated private rates of return, the share of industrial resources allocated to research expenditures did not increase over the succeeding decade.[2] This suggests a paradox: Why has research effort not been receiving more attention from industrial firms if the private rate of return to research is so attractive?

Two important parameters in these calculations of the private return to research are the rate of decay of the private revenues accruing to the industrially produced knowledge and the mean lag between the deployment of research resources and the beginning of that stream of revenues. These parameters, of course, are necessary ingredients in *any* study involving a measurement of the stock of privately marketable knowledge. The rate of decay in the returns of research has not previously been estimated. In this paper we present a method of explicitly estimating that parameter. We also use information provided by others to calculate the approximate mean R & D gestation lags. Since previous research has not included the latter and seems to have seriously understated the rate of decay of appropriable revenues in calculations of the private rate of return to research expenditures, we then use our estimates to improve on previous results on this rate of return.

Of course, all previous work in this area has been forced to make some assumption, either implicit or explicit, about the value of the decay rate. The problem arises because it has been assumed to be similar to the rate of decay in the *physical* productivity of traditional capital goods. The fact that the rate of deterioration of traditional capital and the rate of decay in appropriable revenues from knowledge arise from two different sets of circumstances seems to have been ignored.[3]

The employment of research resources by a private firm produces new knowledge, with some gestation lag. The new knowledge or innovation may be a cost-reducing process, a product, or some combination of the two. The knowledge-producing firm earns a return either through net revenues from the sale of its own output embodying the new knowledge, or by license and nonmonetary returns collected from other firms which lease the innovation. Since the private rate of return to research depends on the present value of the revenues accruing to the sale of the knowledge produced, the conceptually appropriate rate of depreciation is the rate at which the appropriable revenues decline for the innovating firm. However, as Boulding (1966) noted, knowledge, unlike traditional capital,

2. The share of net sales of manufacturing firms devoted to total R & D (publicly and company funded) actually declined from 0.046 in 1963 to 0.029 in 1974, or by about 40 percent, though the share devoted to company-funded R & D remained constant at 0.019. See National Science Foundation, (1976, table B-36).
3. The commonly assumed rate of decay of the knowledge produced by firms is between 0.04 and 0.07 (Mansfield 1968). Griliches (1980), noting some of the conceptual distinctions between the rates of decay in traditional capital and in research, assumes an upper bound of 0.10 for the latter.

does not obey the laws of (physical) conservation. The rate of decay in the revenues accruing to the producer of the innovation derives not from any decay in the productivity of knowledge but rather from two related points regarding its market *valuation*, namely, that it is difficult to maintain the ability to appropriate the benefits from knowledge and that new innovations are developed which partly or entirely displace the original innovation. Indeed, the very use of the knew knowledge in any productive way will tend to spread and reveal it to other economic agents, as will the mobility of scientific personnel. One might expect then that the rate of decay of appropriable revenues would be quite high, and certainly considerably greater than the rate of deterioration in the physical productivity of traditional capital.[4]

In section 4.1 we examine two independent pieces of evidence bearing on the rate of decay of appropriable revenues. The information from various sources on the mean lag between R & D expenditures and the beginning of the associated revenue stream is summarized in section 4.2. In section 4.3 we attempt to get a rough idea of how seriously the existing estimates of the private rate of return to research overstate the true private rate of return. Brief concluding remarks follow.

4.1 The Rate of Decay of Appropriable Revenues

The first piece of evidence on the rate of decay of appropriable revenues (hereafter, the rate of decay) is based on data presented in Federico (1958). Federico provides observations on the percentage of patents of various ages which were renewed by payment of mandatory annual renewal fees during 1930–39 in the United Kingdom, Germany, France, the Netherlands, and Switzerland. A theoretical model of patent renewal will lead directly to a procedure for estimating the rate of decay from these data.

Consider a patented innovation whose annual renewal requires payment of a stipulated fee. Letting $r(t)$ and $c(t)$ denote the appropriable revenues and the renewal fee in year t, the discounted value of net revenues accruing to the innovation over its life span, $V(T)$, is

(1) $$V(T) = \int_0^T [r(t) - c(t)]e^{-it}dt,$$

where i is the discount rate and T is the expiration date of the patent.

---

4. The models used in this paper do not assume that the rate of decay in appropriable revenues is exogenous to the firm's decision-making process. In a dynamic context, a firm processing an innovation has to choose between increasing present revenues and inducing entry, and charging smaller royalties to forestall entry. This choice is the basis of Gaskins's (1971) dynamic limit pricing analysis of situations involving temporary monopoly power. The Gaskins model can be used to show that the optimal revenue stream will decline over time and that the rate of decline will depend on certain appropriability parameters.

Differentiating (1) with respect to T, the optimum expiration date, T^*, is written implicitly as

(2) $$r(T^*) = c(T^*),$$

provided that $r'(t) < c'(t)$ for all t. Equivalently, the condition for renewal of the patent in year t is that the annual revenue at least covers the cost of the renewal fee

(3) $$r(t) \geq c(t).$$

Let the annual renewal fee grow at rate g, and the appropriable revenues decline at rate δ. Then condition (3) can be written as

(4) $$r(0) \geq c(0)e^{(g+\delta)t}.$$

Allowing for differences in the initial appropriable revenues among patents, and letting $f(r)$ represent the density function of the distribution of their values, the percentage of patents renewed in each year, $P(t)$, is[5]

(5) $$P(t) = \int_{C(t)}^{\infty} f(r)dr,$$

where $C(t) = c_0 e^{(g+\delta)t}$. It follows that

(6a) $$P'(t) = -C'(t)f(C),$$

and

(6b) $$P''(t) = -f(C)C''(t)\left[1 + C\frac{f'(C)}{f(C)}\right],$$

where the primes denote derivatives. That is, as long as $(g + \delta) > 0$, the percentage of patents renewed will decline with their age. The curvature of $P(t)$, however, will depend on the distribution of the values of the innovations patented.[6] For example, if $f(r)$ is lognormal, $P(t)$ will have one point of inflection, being concave before it and convex thereafter (see curve 1, fig. 4.1). Alternatively, Scherer (1965) cites evidence presented in Sanders, Rossman, and Harris (1958) which indicates that the value of patents tends to follow a Pareto-Levy distribution. If $f(r)$ is Pareto-Levy, $P(t)$ will be a strictly decreasing convex function of patent age, as shown by figure 4.1, curve 2. Figure 4.2 presents the actual time paths of $P(t)$

5. We are implicitly assuming that the rate of decay of appropriable revenues, δ, does not differ among patented innovations. This assumption allows us to compare directly our estimate of δ to those assumed in previous empirical work (since the same implicit assumption is prevalent in that work) and to consider the economic implications of the different values of δ (see section 4.3). The model could be generalized to allow for differences in decay rates and, if sufficient data were available, one could estimate the parameters of the joint distribution of the values of the initial revenues and the decay rates.

6. Since the value of patents (eq. [1]) is a monotonic transformation of the adjusted initial revenues, we shall use the two terms interchangeably.

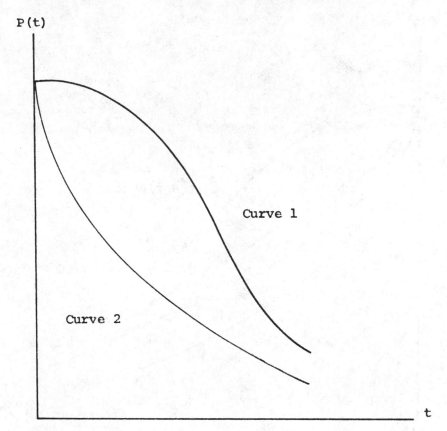

Fig. 4.1

from Federico (1958). Four of the five curves tend to support Sanders et al.'s data and are consistent with an underlying Pareto-Levy distribution of patent values. The time path for Germany, however, indicates that the underlying distribution for that country has at least one mode. Since it is futile to estimate both the parameters of the underlying lognormal (for example) and the decay rate in appropriable revenues from only eighteen observations available on Germany, we shall disregard the German data in the remainder of this empirical work.[7]

We now use this simplified model to obtain rough estimates of the decay rate of appropriable revenues δ. Consistent with the evidence in figure 4.2, the relative density function of initial revenue is taken to be of the Pareto-Levy type:

(7) $$f(r) = \beta r_m^\beta r^{-(\beta+1)}, \ r_m > 0, \beta > 0,$$

---

7. The United Kingdom patent system requires no renewal payments until the fifth year. Hence, the underlying distribution of patent values may have an inflection point, but it cannot be ascertained from the data.

P(t)

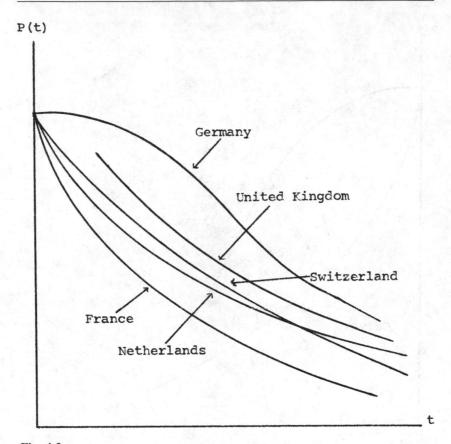

Fig. 4.2

where r_m is the minimum value of r in the population. Using equations (5) and (7), the percentage of patents renewed in year t can be expressed as

$$(8) \qquad P(t) = (r_m/c_0)^\beta e^{-\beta(g+\delta)t}.$$

Two error terms differentiate the observed value of the logarithm of P_t, log P_t^m, from the value predicted by equation (8).[8] The first, v_1, is a sampling or measurement error, while the second, v_2, is a structural error in the model. Assuming that P_t^m is derived from a binomial sampling process around the actual value, that is, $P_t^m \sim b(P_t, N)$, it follows that

$$\sigma_{v_1}^2 = V(\log P_t^m) \approx \frac{1 - P_t}{P_t N},$$

where N is the (unobserved) number of patents sampled. The structural

8. Since P_t is bounded by zero and unity, the composite error cannot be independently and identically distributed. The analysis which follows corresponds closely to the treatment of similar problems in logit regressions. See Berkson (1953) and Amemiya and Nold (1975).

error, v_2, will be assumed to be an independent, identically distributed normal deviate with variance σ^2.

Letting j index a country, for sufficiently large N_j the logarithmic transform of P_{tj}^m can be written as

$$(9) \qquad \log P_{tj}^m = \alpha_{0j} + \alpha_{1j}t_j + \mu_{tj},$$

where $\mu_{tj} \sim N[0, \sigma^2 + (1 - P_{tj})/P_{tj}N_j]$, $\alpha_{1j} = -\beta_j(g_j + \delta_j)$, and $\alpha_{0j} = \beta_j\log(r_m/c_0)_j$. The estimating question (9) embodies the basic prediction of the model, namely a negative relationship between α_{1j} and g_j, where the slope coefficient is the parameter of the underlying distribution of patent values.

Consistent estimates of α_{0j}, α_{1j} and their standard errors can be derived from the following two-stage procedure. First, estimate (9) by ordinary least squares. Next, define e^2 as the squared residuals from (9) and regress

$$(10) \qquad e_{tj}^2 = \sigma^2 + \frac{1}{N_j}\frac{1 - P_{tj}^m}{P_{tj}^m}.$$

Letting F be the fitted value from (10), use $F^{-\frac{1}{2}}$ to weight and perform weighted least squares on (9).

If our model is correct, and if β and δ do not vary between countries, then

$$(11) \qquad \alpha_{1j} = -\beta\delta - \beta g_j.$$

Since g_j, the rate of growth of renewal fees, is available from Federico's data,[9] (11) can be tested by using $F^{-\frac{1}{2}}$ to weight and by performing weighted least squares on the equation

$$(12) \qquad \log P_{tj}^m = \alpha_{0j} - \beta\delta t_j - \beta(g_j t_j) + \mu_{tj},$$

where all symbols are as defined above.

If (11) is the true specification, then minus twice the logarithm of the likelihood ratio from (12) and the weighted least-squares version of (9) will distribute asymptotically as a χ_2^2 deviate. Moreover, equation (12) will provide estimates of both the rate of decay of appropriable revenues (δ) and the underlying distribution of patent values (β).

Table 4.1 summarizes the empirical results. The observed value of the χ^2 test statistic is 5.4, while the 5 percent critical value is 5.99. Though a little high, the test statistic does indicate acceptance of the hypothesis in (11). The estimates of $1/N_j$ and σ^2 are all positive thereby lending support to the weighting procedure described above.

Turning to the parameters of interest, the point estimate of β is 0.57

---

9. These growth rates were calculated from a semilog linear regression of costs against time for each country. The growth rates (and their standard errors) for the Netherlands, the United Kingdom, France, and Switzerland were 0.085, (0.002), 0.129 (0.015), 0.089 (0.006), and 0.143 (0.008), respectively.

Table 4.1   Estimates from the Patent Renewal Model[a,b]

Common Parameters (57 observations)		Country-Specific Parameters		
			α_{0j}	$1/N_j$
$\beta\delta$	0.14 (0.01)	France	0.04 (0.02)	0.00014
β	0.57 (0.07)	United Kingdom	0.55 (0.02)	0.00016
δ		Netherlands	0.09 (0.02)	0.00040
Point estimate[c]	0.25	Switzerland	0.32 (0.02)	0.00032
Confidence interval[d]	0.18–0.36			
σ^2	0.0002			
R^2	0.996			

[a]The data on patent renewal are taken from Federico (1958). These data cover the percentage of patents of different ages in force during 1930–39 which were renewed by payment of a mandatory annual renewal fee. For example, if a patent was granted in 1925, it would appear in the data as five years old in 1930, six years old in 1931, and so on. Therefore, the percentage of patents renewed after five years is based on the total number of patents issued five years earlier in a particular country.

[b]Standard errors are in parentheses.

[c]$\hat{\delta} = \widehat{\beta\delta}/\hat{\beta}$.

[d]The confidence interval corresponds to the 95 percent Fieller bounds on δ.

with a standard error of 0.07. One can check this estimate against an independent source of information: as mentioned earlier, Sanders, Rossman, and Harris (1958) provide evidence on the distribution of the value of patents in the United States. Fitting a Pareto-Levy distribution to these data, we obtain a point estimate for β of 0.63 with a standard error of 0.06. That is, the estimate of β from Sanders et al.'s data is very close to that obtained using our model and Federico's data.[10]

Our primary interest is in δ, the (average) decay rate in appropriable revenues. The point estimate of δ is 0.25, while a 95 percent confidence interval places the true value of δ between 0.18 and 0.36.[11] An estimated δ

10. These data correspond to appropriable revenues minus costs associated with the patent, but since cost data were not available we were forced to use net value data. A cumulative Pareto-Levy distribution was fitted to the five positive net value observations on expired patents which therefore have observable net values. The R^2 from this regression was 0.97. Note that Pareto-Levy distributions with $\beta < 1$ do not have either a finite mean or variance and hence do not behave according to the law of large numbers. Therefore, if the distribution of patent values approximates the distribution of project values, diversification into many independent projects will not reduce risk. This point was originally made by Nordhaus (1969). Of course, if the returns to different projects are negatively correlated, diversification may still reduce risk.

11. Note that since the estimate of δ is obtained as the ratio of two coefficients, its confidence interval (obtained by using Fieller bounds) is not symmetric around its point estimate. Two remarks on the robustness of these results are also in order. First, the assumption that the revenue stream can be described by an exponential rate of decay can be

of 0.25, though consistent with theoretical arguments concerning the unique characteristics of knowledge as an economic commodity, implies that earlier researchers have assumed values of δ which are far too small. In particular, the *lower* bound of the 95 percent confidence interval for δ is nearly twice the *maximum* value of the rate of decay of private returns used in previous research.

Of course, our estimate of δ may reflect some sample selection bias. The rate of decay of patented innovations may differ from that of all innovations. The direction of the bias is indeterminate since it depends on the correlation between the patent selection process and the rates of obsolescence in the universe of all innovations. However, the estimates of δ may be biased downward for two reasons: First, the fact that patents create property rights in the embodied knowledge may result in a lower rate of obsolescence for those patentable innovations. Second, given a patentable innovation, it is easy to show that the innovator will actually take out a patent only if patenting lowers the rate of decay. As we show presently, however, evidence of a completely different nature suggests that whatever bias exists is negligible.

The second source of evidence on the magnitude of the decay rate of appropriable revenues is derived from data presented in Wagner (1968) on the life span of applied research and development expenditures. Survey data on applied research and development were collected from about thirty-five firms with long R & D experience in thirty-three product fields, using the product field description employed by the National Science Foundation (NSF) in its annual industry reports. Included in the survey was a question on the life span of R & D defined as the period after which the product of the R & D was "virtually obsolete."

This definition does not correspond directly to the decline in the appropriable revenues accruing to research and development. However, a rough correspondence can be established by assuming that R & D is virtually obsolete when the appropriable revenues reach some small fraction of the initial value, and then by experimenting with different fractions to examine the sensitivity of the implied decay rate to the assumption. Table 4.2 presents the average life span of R & D for durable and nondurable product field categories, product- and process-oriented R & D, and the implied decay rates based on various reasonable definitions of virtual obsolescence. While the implied decay rates do vary with

---

viewed as a first-order (logarithmic) approximation to a more general stream of revenue. We also experimented with a second-order approximation, namely, $r(t) = r(0)\exp(At + Bt^2)$. The estimates of B and its standard error were both zero to two decimal places and the rest of the results were almost identical to those reported here. Apparently market-induced obsolescence is well approximated by an exponential pattern. Griliches (1963) reaches the same conclusion with respect to the obsolescence component of the deterioration in the value of traditional capital goods. Second, the results from the unweighted version of (12) yielded a point estimate of $\delta = 0.22$, with Fieller bounds of 0.16 to 0.33, and an estimate of $\beta = 0.62$.

Table 4.2   **Estimates of δ from Average Life Span of R & D[a,b]**

	Ratio of Revenue in Year T to Initial Revenue ("virtual obsolescence")		
	0.15	0.10	0.05
Durable goods R & D			
Product ($T = 9$)	0.21	0.26	0.33
Process ($T = 11$)	0.17	0.21	0.27
Nondurable goods R & D			
Product ($T = 9$)	0.21	0.26	0.33
Process ($T = 8$)	0.24	0.29	0.38

[a]Taken from Wagner (1968, p. 196, table 5), which refers to "applied research and development" (AR & D). These life span figures (denoted by T in parentheses) are averages of survey responses, weighted by 1965 product-field expenditures and by frequencies of the response distribution.

[b]Calculated as $\delta = -(\log x)/T$, where x is the assumed ratio of revenue in year T to initial revenue accruing to the R & D.

the definition of virtual obsolescence, the rate of values is nearly identical to the Fieller bounds on δ in table 4.1.[12]

The responses of firms to Wagner's question can also be used to check the reasonableness of the rates of obsolescence commonly assumed in the literature. If in fact $\delta = .05$ ($.10$), that would imply (using $T = 9$ from table 4.2) that firms consider the product of their R & D virtually obsolete even though the annual revenue flow is still 64 (41) percent of its initial value. This seems highly implausible and casts additional doubt on the conventionally assumed values of δ.

4.2 Mean R & D Lags

Two independent sources of information are used to estimate the mean R & D lags, defined as the average time between the outlay of an R & D dollar and the beginning of the associated revenue stream. This lag consists of a mean lag between project inception and completion (the gestation lag), and the time from project completion to commercial application (the application lag).

Rapoport (1971) presents detailed data on the distribution of costs and time for forty-nine commercialized innovations and the total innovation time for a subset of sixteen of them in three product groups—chemicals, machinery, and electronics. The innovation process is decomposed into

12. The only other estimate of the decay rate is produced knowledge of which we are aware is reported in a footnote in Griliches (1980). A regression of productivity growth against R & D flow and stock intensity variables in his microdata set yielded an estimated δ of 0.31. Griliches points out the discrepancy between this result and the rest of his analysis but offers no reconciliation.

Table 4.3   Estimates of the Mean R & D Lag (years)

	R & D Gestation Lag	Application Lag	Total Lag
Rapoport			
Chemicals	1.48	0.24	1.72
Machinery	2.09	0.31	2.40
Electronics	0.82	0.35	1.17
Wagner			
Durables	1.15	1.47	2.62
Nondurables	1.14	1.03	2.17

Source: Calculated from data contained in Rapoport (1971) and Wagner (1968).

five stages: applied research, specification, prototype or pilot plant, tooling and manufacturing facilities, and manufacturing and marketing start-up. Since the expenditures on manufacturing and marketing start-up are not included in the NSF definition of R & D expenditures, the time involved in that stage is treated here as the application lag. The remaining data are used to calculate the gestation lag. The first part of table 4.3 summarizes the R & D gestation and application lags for the three product groups.[13]

Additional information on the average R & D and application lags is provided in Wagner (1968). Survey data on process- and product-oriented R & D were gathered from about thirty-six firms with long R & D experience in a variety of durable and nondurable goods industries. Included was information on the duration of applied research and development, project duration for projects successfully completed in 1966, the distribution of R & D expenditures for successfully completed projects classified by project duration, the percentage of total funds accounted for by projects abandoned before completion together with the time of abandonment, and the interval between the completion of R & D and commercial application of the innovations. These data are used to calculate both an application lag and a mean gestation lag which, unlike those based on Rapoport's data, take into account expenditures on both technically successful and unsuccessful projects. The results are given in the second part of table 4.3.

The gestation lags based on Wagner's data are broadly similar to those

13. The details of the calculations are omitted here for the sake of brevity but are available on request. However, the limitations of these estimates should be noted. First, all the projects analyzed by Rapoport resulted in significant innovations, and as Scherer (1965) and Mansfield (1968) have noted, mean lags tend to be longer for more significant technical advances. Second, we have not taken into account the time overlap between stages, which, according to Rapoport, is considerable. Both of these factors would tend to cause upward biases in our estimates of the mean lag, θ. On the other hand, the R & D costs of technically unsuccessful projects should be taken into account, which would tend to raise the estimates of θ.

derived from Rapoport, but the application lags are considerably longer,[14] causing some discrepancy between the two sets of results. Mansfield (1968), using data gathered from extensive personal interviews with R & D project evaluation staff, concluded that the mean application lag was about 0.53 years. Substitution of this number for Wagner's would bring the two sets of results closer together and put the total lag at about 1.75 years. For present purposes, however, a range of values between 1.2 and 2.5 years is good enough.[15]

4.3 Implications for Measuring the Private Rate of Return to Investment in Research

The preceding sections of this paper provide estimates of the decay rate and the mean R & D lag whose values are substantially higher than those assumed in previous research. These estimates are now used to get a rough indication of the implications for production function estimates of the private rate of return to research expenditures.

Let Q_R denote the increment in value added (or sales) generated by a unit increase in research resources θ years earlier. Then the equation for the private (internal) rate of return to investment in research is

$$(13) \qquad e^{\theta r} = \int_0^\infty Q_R e^{-(r+\delta)\tau} d\tau,$$

where θ and δ were defined earlier, and r is the private rate of return or the implicit discount rate that would make investment in research marginally profitable. Integrating (13) yields the following nonlinear equation for r:

$$(14) \qquad e^{\theta r}(r+\delta) - Q_R = 0.$$

In the special case where $\theta = 0$, this reduces to $r = Q_R - \delta$, corresponding to the equation used by previous researchers.

Given estimates of Q_R, δ, and θ, we can compute the private rate of return from equation (14). Two points should be noted. First, since research expenditures are usually included in the measures of traditional capital and labor expenditures in the production functions used to estimate Q_R, the private rates of return to research reported in the literature

14. Since Wagner does not precisely define the "end of AR & D" or the "application of innovations," some caution should be exercised in interpretating the application lags. Wagner does indicate that the longer application interval in durables reflects in large part the defense-space-atomic-energy-oriented fields, so that the application lag for other industries is probably closer to the nondurable estimate. On the other hand, Rapoport's "manufacturing and marketing start-up" stage may understate the actual application lag.

15. The maximum of this range is considerably shorter than the midpoint of the interval between project inception and marketing, reflecting the fact that the distribution of research expenditures on projects is considerably skewed to the left.

represent excess returns above and beyond the normal remuneration to traditional factors (see Griliches 1973). To avoid this problem, we base the calculations on estimates of Q_R corrected for this double-counting (Schankerman 1981). Second, these estimates of Q_R are calculated by multiplying the estimated sales (value added) elasticity of the stock of knowledge times the ratio of sales to the stock of knowledge. The stock of knowledge is taken as the undepreciated sum of research expenditures over the period of observation. For the calculations in (14) to be consistent, however, the stock of knowledge must be calculated according to declining balance depreciation. We therefore calculate the depreciated sum of research expenditures with a decay rate of δ and then use this stock of knowledge to convert the estimated sales elasticity into a value of Q_R.[16] This has been done for three values of δ, corresponding to the point estimate and Fieller bounds obtained earlier (0.18, 0.25, and 0.36), and for three different values of Q_R (0.30, 0.35, and 0.40), corresponding to the pooled (across industries) point estimate plus or minus one standard deviation computed from Schankerman (1981). The results are presented in table 4.4 for $\theta = 2$.

Turning to the results, it is apparent that the net private rates of return to investment in research are greatly reduced by our adjustments. The net private rate of return varies between .075 and .174. If the normal (net) rate of return to traditional capital is about 0.08 (Griliches 1980), this implies risk premiums for investment in research of between zero and about 9 percent. In view of the abnormal riskiness associated with research expenditures, these risk premiums appear modest.

In short, table 4.4 suggests that the private rates of return to investment in research and traditional capital are roughly equated at the margin. Another way of checking this possibility is to ask: What is the decay rate of appropriable revenues implied by the assumption that firms equate, at the margin, the private rates of return to investment in research and traditional capital? With a mean R & D lag of θ, the return to a dollar of research is $(r + \delta + 1)(1 + r)^{-\theta}$, while for traditional capital, with depreciation rate δ_c, it is $(r + \delta_c + 1)$. Using $\delta_c = 0.06$ and $r = 0.08$ from

16. We thank Zvi Griliches for pointing out this problem and suggesting a solution. The data for the calculations are taken from the National Science Foundation (1976), table B-1. Three additional points should be noted. First, Schankerman's estimates of Q_R are based on the large microdata set used by Griliches (1980). Griliches's (uncorrected) pooled estimate of Q_R was about 0.30, while his estimates for research intensive and nonintensive industries were 0.40 and 0.20, respectively. Mansfield's (uncorrected) estimates, averaged over the ten firms he used, range from about 0.20 to 0.30, depending on the specific assumptions made. Second, if private returns to knowledge do in fact decay, there is an error in the measured stock of knowledge used as an independent variable in the regressions to estimate Q_R. However, it can be shown that the ratio of the variance in measurement error to the variance in the true stock is very small (less than 0.0026), so the sales elasticity of the stock of knowledge can be taken directly from Schankerman's regressions. Finally, the private rates of return are much less sensitive to θ than to δ.

Table 4.4   Estimates of the Net Private Rate of Return to Research ($\theta = 2$)

	Q_r		
δ	.30	.35	.40
.18	.090	.118	.144
.25	.111	.144	.174
.36	.075	.107	.139

Griliches (1980), the value of δ which equates these two terms is $\delta = 0.25$ if $\theta = 2.0$. This value is identical to the point estimate of δ in table 4.1.

4.4 Concluding Remarks

In this paper we stress the conceptual distinction between the rates of decay in the physical productivity of traditional capital goods and that of the appropriable revenues which accrue to knowledge-producing activities. An estimate of the private rate of obsolescence of knowledge is necessary in any study which requires constructing a stock of privately marketable knowledge. We estimate this parameter from a simple patent renewal model and find the estimate comparable to evidence provided by firms on the life span of the output of their R & D activities. The empirical results indicate that the rate of obsolescence is considerably greater than the rates typically assumed in the literature. The estimated decay rates, together with mean R & D gestation lags, are used to calculate the net private rate of return to investment in research. Our results suggest that the *private* rate of return to research expenditures, at least in the early 1960s, was *not* unreasonably high. It is important to emphasize, however, that to draw conclusions regarding the divergence between the private and social rates of return to knowledge-producing activities, information on the social rate of return must be added to the information contained in this paper.[17] Nonetheless, if our calculations of the private rate of return are even approximately correct, they do suggest a partial resolution to the paradox presented in the introduction to this paper: Why did private firms not increase the share of their resources devoted to R & D if their previous research efforts were so highly profitable? Part of the answer may be that research was not as privately profitable as has been thought.

17. In this connection, the social rate of decay may well be smaller than the rate of decay of appropriable revenues. See Hirshleifer's (1971) distinction between real and distributive effects in the production of knowledge.

References

Amemiya, Takeshi, and F. Nold. 1975. A modified logit model. *Review of Economics and Statistics* 57:255–57.

Arrow, Kenneth J. 1962. Economic welfare and the allocation of resources for invention. In *The rate and direction of inventive activity: Economic and social factors*, ed. R. R. Nelson. Universities-NBER Conference Series no. 13. Princeton: Princeton University Press for the National Bureau of Economic Research.

Berkson, J. 1953. A statistically precise and relatively simple method of estimating the bioassay with quantal response, based on the logistic function. *Journal of the American Statistical Association* 48:565–99.

Boulding, Kenneth E. 1966. The economics of knowledge and the knowledge of economics. *American Economic Review: Proceedings* 56:1–13.

Dasgupta, P., and J. Stiglitz. 1980. Uncertainty, industrial structure, and the speed of R & D. *Bell Journal of Economics* 11:1–28.

Federico, P. J. 1958. *Renewal fees and other patent fees in foreign countries*. Subcommittee of Patents, Trademarks, and Copyrights of the Committee of the Judiciary, United States Senate, Study No. 17. Washington, D.C.: Government Printing Office.

Gaskins, Darius W., Jr. 1971. Dynamic limit pricing: Optimal pricing under threat of entry. *Journal of Economic Theory* 3:306–23.

Griliches, Zvi. 1963. Capital stock in investment functions: Some problems of concept and measurement. In *Measurement in economics: Studies in mathematical economics and econometrics in memory of Yehuda Grunfeld*, ed. Carl F. Christ. Stanford, Cal.: Stanford University Press.

———. 1973. Research expenditures and growth accounting. In *Science and technology in economic growth*, ed. B. R. Williams, 59–83. London: Macmillan.

———. 1980. Returns to research and development expenditures in the private sector. In *New developments in productivity measurement and analysis*, ed. J. W. Kendrick and B. N. Vaccara, 419–54. Conference on Research in Income and Wealth: Studies in Income and Wealth, vol. 44. Chicago: University of Chicago Press for the National Bureau of Economic Research.

Hirshleifer, Jack. 1971. The private and social value of information and the reward to inventive activity. *American Economic Review* 61:561–74.

Mansfield, Edwin. 1968. *Industrial research and technological innovation: An econometric analysis*. New York: Norton for the Cowles Foundation for Research in Economics, Yale University.

National Science Foundation. 1976. *Research and development in industry*. Washington, D.C.: Government Printing Office.

Nordhaus, William D. 1969. *Invention, growth, and welfare: A theoretical treatment of technological change*. Cambridge: MIT Press.

Rapoport, John. 1971. The anatomy of the product-innovation process: Cost and time. In *Research and innovation in the modern corporation*, ed. E. Mansfield, 110–35. New York: Norton.

Sanders, J., J. Rossman and L. Harris. 1958. The economic impact of patents. *Patent, Trademark and Copyright Journal of Research and Education* 2:340–62.

Schankerman, Mark. 1981. The effects of double-counting and expensing on the measured returns to R & D. *Review of Economics and Statistics* 63:454–58.

Scherer, Frederic M. 1965. Firm size, market structure, opportunity, and the output of patented inventions. *American Economic Review* 55:1097–1125.

Wagner, Leonore U. 1968. Problems in estimating research and development investment and stock. In *Proceedings of the business and economic statistics section*, 189–98. Washington, D.C.: American Statistical Association.

5 International Invention: Implications for Technology Market Analysis

Robert E. Evenson

This paper examines international data on patented inventions, R & D expenditures, and scientists and engineers engaged in inventive activity. It reaches two principal conclusions which have some bearing on the modeling of firm behavior and possibly on policy actions which might be taken toward the stimulation of invention. First, the data show comparative advantage patterns in invention similar to patterns observed in products. The production of pioneering invention is concentrated in certain firms located in countries with the best economic laboratories for invention. Large parts of industry in most countries import inventions and concentrate on adaptive invention rather than investing heavily in R & D. Second, the data show that inventions per scientist and engineer have declined from the late 1960s to the late 1970s in almost all of the fifty countries for which data are available.

These conclusions are based on data on patented inventions from many countries. To defend them one must argue not only that patented inventions are a reasonable proxy for inventions in general but also that this proxy relationship has a reasonable degree of international comparability. Further, to support the second conclusion one must argue that no major changes in the proxy relationship have taken place over the past ten to fifteen years.

These conclusions have a threefold defense. First, because of international conventions regarding patenting and the requirements for patentability and the high degree of international patenting (i.e., patents granted to foreigners), a general standardized legal basis for patenting exists. This is further standardized by the widespread adoption of the International Patent Classification system. Second, patent data show regular patterns

Robert E. Evenson is a professor at the Economic Growth Center at Yale University.

and consistency. Patenting is highly correlated with R & D spending in the United States and other countries where reasonably good data exist. Most patents granted are subsequently cited as "next best art" in the United States and other countries with citation requirements.[1] Patent infringement cases are important enough in most countries to indicate that patents are not trivial or irrelevant. Finally, there is little evidence that standards of patentability have changed drastically in recent years in most major centers of invention. Nor is there evidence to suggest that firms in almost all industries in almost every country of the world have changed their policy toward obtaining patents to a degree sufficient to explain the data.

Section 5.1 of this paper presents a descriptive summary of patent data and discusses different types of patent systems and standards for patentability. Section 5.2 shows the trade patterns of the data. Section 5.3 provides data supporting the conclusion that inventions per scientist and engineer have declined and argues the case for interpreting this phenomenon as the result of exhaustion of "invention potential." Section 5.4 discusses implications for technology market analysis.

5.1   International Invention: A Descriptive Summary

To interpret data on patenting it is first useful to summarize the options open to a firm to alter the technology it uses:

(1) It can engage in fundamental or basic research to obtain findings that will improve the efficiency of its more applied research.

(2) It can engage in applied research designed to invent a new product or process and bring it to the development stage.

(3) It can engage in the testing, pilot production, and plant design work required to bring inventions developed by its own applied research into use.

(4) It can purchase inventions (or in the case of unprotected inventions, imitate them) and engage in strictly adaptive research and development bringing them into use.

(5) It can purchase semi- or fully developed inventions "embodied" in machines, chemicals, or "turnkey" plants, making only minor modifications of other inventions.

Of these activities, (1) produces few patentable inventions; (2) produces most conventional or utility patents; (3) produces a number of utility patents (especially of process inventions) and a number of "petty" patents (utility models); (4) produces most petty patents; and (5) generally does not produce patented inventions.

1. Wright and Evenson (1980) reported that approximately 75 percent of the patents granted in specialized chemical fields (oils and food chemicals) are subsequently cited as next best art in other patents.

Legal systems and industrial organization policies in different countries influence the types of inventive activities undertaken by firms and the patentability of inventions. Some countries pursue policies which encourage the holding of inventions in trade secrecy. When industrial organization structures effectively discourage competition in an industry, firms may have little incentive to sell new technology in direct form and will attempt to capture rents through the sale of new technology embodied in products. This tendency is reinforced by trade secrecy laws which provide penalties for the pirating of trade secrets.

The traditional "invention patent" is designed to provide an alternative form of protection by granting the inventor legal means to prevent others from copying or using the invention without permission for a limited period of time (usually fifteen years). Invention patent documents are required to provide an "enabling disclosure" which sufficiently describes the invention to enable one skilled in the technology field to replicate or make the invention.[2]

Three fundamental requirements must be met by an invention to qualify for the *standard invention patent*:

(1) The invention must be "novel."

(2) The invention must be "useful."

(3) The invention must exhibit an "inventive step" (i.e., it must be unobvious to practitioners skilled in the technology field).

These requirements are important in understanding international patent data when considered in conjunction with international patent "conventions," chiefly the Paris Convention. Membership in these conventions generally requires: (1) that the three requirements for patentability be judged by international standards and (2) that member countries grant patent protection to inventors from other countries provided these standards are met.[3]

An important alternative to the invention patent used in some countries is a *"petty" patent or utility model.* Petty patents generally have a very weak inventive step requirement and in practical terms do not always require novelty against the world's inventions but only against national or regional inventions. In addition, *design patents,* which do not require inventive steps and have relatively weak usefulness require-

2. The legal literature sees this enabling disclosure, which enables or induces further inventions by others, as an important part of the bargain in which monopoly rights are granted in return for disclosure. (Economists, by contrast, see invention incentives as the principal benefit obtained in the bargain.)

3. I will argue in the final section of this paper that membership in international conventions has been very costly (and unwise) for many countries. The cost of searching the world's patent literature to establish novelty is high, and many small countries cannot adequately undertake this task. Furthermore, adhering to a strong international standard of the inventive step requirement effectively removes patent protection from "adaptive" inventions which are about the only types of inventions many developing countries can produce.

ments, are granted by most countries. *Trademarks*, which require only novelty, are likewise granted by most countries. In addition, a number of countries also grant *plant patents*, primarily for asexually reproduced plants.

Table 5.1 provides data for forty-nine countries on numbers of invention patents granted during four periods: 1967, 1971, 1976, and 1980. The countries have been grouped into six classes: (1) industrialized market economies with moderate to rapid growth rates over the past twenty years; (2) industrialized market economies with slow growth rates; (3) semi-industrialized economies with rapid growth rates; (4) semi-industrialized economies with slow to moderate growth rates; (5) middle-to-low-income developing economies; and (6) industrialized planned economies.[4]

Reference to the table will reveal a few anomalies, particularly for the developing countries where some data are missing. Table 5.9 provides a summary by type of economy and a number of generalizations are best drawn at that level. In discussing this and the next several tables attention will be given to individual country data. Table 5.1 shows that the relative ranking of patenting by national inventors has changed appreciably over the period. The United States was the clear leader in 1967 with more than twice as many patents granted as the Soviet Union in second place. France, Japan, East Germany and the United Kingdom followed. By 1980 both the Soviet Union and Japan had surpassed the United States. West Germany had moved into fourth place, with both France and the United Kingdom experiencing substantial declines in patents granted to nationals.

Patents granted to nationals in the United States were only 72 percent of the 1967 level in 1980 (only 60 percent in 1979). For all other industrialized market economies, patents granted to nationals actually increased slightly (2 percent) from 1967–1980. Patents granted to foreigners in the United States rose by 71 percent over the period. For other industrialized nations, patents granted to foreigners declined to only 66 percent of the 1967 level (about 43 percent of this decline was attributable to the decline in patenting abroad by U.S. inventors). In consequence the share of foreigner's patenting in the United States rose from 22 percent in 1967 to 40 percent in 1980.

Of the industrialized economies both Japan and West Germany markedly expanded patenting activity at home. Only Japan, among large industrialized nations, realized a significant expansion of patenting abroad. The United States continued to be the dominant country in patenting abroad with West Germany, Japan, and France following.[5]

---

4. These classifications are based on World Bank (1980).

5. Patenting abroad is influenced by cost considerations. The European countries have recently introduced the Europatent which provides low-cost patent protection in a group of

Table 5.1   Invention Patents Granted by Country: Selected Years

	Patents Granted to Nationals				Patents Granted to Foreigners				Patents Granted to Nationals in Foreign Countries			
	1967	1971	1976	1980	1967	1971	1976	1980	1967	1971	1976	1980
Industrialized Market Economies												
Moderate to Rapid Growth												
Japan	13,877	24,759	32,465	38,032	6,896	11,652	7,582	8,074	6,843	15,832	20,246	20,663
Austria	1,188	1,230	1,177	1,227	6,896	7,460	5,235	4,745	1,913	2,399	1,065	1,669
France	15,246	13,696	8,420	8,433	31,749	37,760	21,334	19,622	14,393	17,150	12,677	12,511
Denmark	338	252	208	192	2,002	2,212	2,068	1,453	1,165	1,650	1,217	1,103
W. Germany	5,126	8,295	10,395	9,826	8,300	9,854	10,570	10,362	41,775	44,862	37,316	33,708
Belgium	1,586	1,345	1,034	837	15,041	15,004	12,110	5,081	2,701	2,894	1,905	1,720
Norway	225	386	210	276	1,831	2,343	1,883	1,843	618	658	617	549
Netherlands	322	318	370	417	1,913	2,396	3,219	2,907	7,283	8,745	5,901	5,964
Slow Growth												
Canada	1,263	1,587	1,301	1,503	24,573	27,655	20,449	22,392	2,789	3,201	2,661	2,200
Italy	9,076	4,320	—	1,810	26,180	13,180	—	6,190	5,621	6,749	5,416	5,877
Ireland	28	16	27	24	635	788	1,055	1,407	113	151	146	106
Switzerland	5,388	4,165	3,482	1,475	16,462	11,914	8,818	4,486	12,452	15,409	10,954	9,827
Sweden	1,776	2,245	1,888	1,394	7,532	7,748	6,956	3,604	5,031	6,327	5,719	4,769
United States	51,274	55,988	44,162	37,152	14,378	22,328	26,074	24,675	73,960	87,589	90,273	54,360
Australia	752	979	910	620	10,371	9,662	10,074	7,805	905	986	1,065	2,690
United Kingdom	9,807	10,376	8,855	5,158	28,983	31,178	30,942	18,646	17,579	21,179	14,072	11,140
Finland	231	350	291	439	739	1,312	912	1,467	345	559	650	928
New Zealand	—	—	211	137	—	—	1,314	1,122	135	1,420	91	235

Table 5.1 (cont.)

	Patents Granted to Nationals				Patents Granted to Foreigners				Patents Granted to Nationals in Foreign Countries			
	1967	1971	1976	1980	1967	1971	1976	1980	1967	1971	1976	1980
Semi-industrialized Market Economies												
Rapid Growth												
Spain	2,758	2,042	2,000	1,485	6,827	7,764	7,500	7,739	627	933	766	1,180
Israel	178	202	200	305	935	1,225	1,200	1,419	219	231	146	316
Greece	975	1,227	1,343	1,114	2,302	698	1,285	942	61	70	81	691
Singapore	5	2	—	1	26	334	—	548	—	—	5	5
Portugal	84	214	46	95	1,045	3,238	1,319	2,200	53	57	50	50
Brazil	262	429	450	349	684	1,543	1,500	3,494	63	85	88	113
S. Korea	207	200	1,593	258	152	117	1,727	1,161	20	20	50	50
Slow to Moderate Growth												
Chile	80	58	60	60	1,237	1,115	514	514	—	—	—	—
Venezuela	41	237	50	55	954	1,599	514	408	—	—	—	—
Argentina	1,244	1,346	1,300	1,264	4,488	3,484	2,800	2,843	81	152	102	133
Mexico	1,981	412	300	174	7,922	5,199	3,000	1,831	149	148	181	171
Turkey	30	52	35	34	438	357	588	424	—	—	—	—
Uruguay	165	88	46	41	351	161	110	236	—	—	—	—

Developing Economies												
Ecuador	5	8	7	7	126	180	103	103	—	—	—	—
Iraq	22	5	12	14	146	67	150	24	—	—	—	—
Morocco	28	24	23	21	391	313	334	330	—	—	—	—
United Arab Republic (Egypt)	48	13	16	10	873	236	511	317	—	—	—	—
Colombia	49	62	30	36	851	651	600	808	—	—	—	—
Philippines	16	46	108	82	498	946	767	755	—	—	—	—
Kenya	0	1	5	—	104	121	98	97	—	—	—	—
India	428	661	433	500	3,343	3,256	2,062	2,000	72	70	73	57
Sri Lanka	1	10	4	5	4	148	156	36	—	—	—	—
African Intellectual Property Organization	1	15	3	26	573	455	545	545	—	—	—	—
Planned Economies												
E. Germany	11,520	8,295	3,755	4,455	8,351	9,854	2,735	1,371	976	2,240	1,652	992
Czechoslovakia	3,613	2,824	4,880	6,763	787	1,276	2,220	1,854	1,718	1,735	927	515
Soviet Union	24,008	33,534	40,259	92,897	662	2,098	1,883	7,852	1,379	2,973	3,309	2,601
Hungary	414	559	594	760	663	1,054	1,155	1,081	596	1,020	1,116	1,294
Poland	1,564	2,331	5,619	5,736	485	543	2,380	1,962	447	538	347	629
Bulgaria	423	674	750	1,271	90	240	393	102	78	164	167	242
Yugoslavia	173	143	58	58	650	706	355	355	95	90	87	110
Rumania	2,955	1,075	1,123	1,194	1,283	1,246	572	814	224	313	106	103

Source: *Industrial Property Statistical Report*, annual issues, World Intellectual Property Organization, Geneva.

The semi-industrialized nations have a varied experience in patenting. Most of the rapid-growth countries show expansion in patents granted to nationals (or have relatively high levels of patenting, e.g., Spain). The slower growing semi-industrialized countries in general have experienced some decline in national patenting. Patents granted to foreigners have tended to increase in the fast growing semi-industrialized countries and to decrease quite drastically in the slow-growth countries (for the group, patenting by foreigners was only 40 percent of its 1967 level in 1980). This decline reflects policy changes by this group of countries and other developing countries toward multinational firms. In general, through adminstrative procedures and through exclusion of certain technology areas from patentability (chiefly food and drugs), patenting by foreigners has been cut back. Unfortunately, as will be discussed later, these policies have not produced significant expansion in patenting by nationals.

The developing countries on the whole have relatively low levels of national patenting and high ratios of patenting by foreigners (policies in India have curtailed the latter). While data on patenting abroad are incomplete, available data for both semi-industrialized and developing countries indicate that the ratio of patenting abroad to patenting at home is much lower than is the case for industrialized countries.

The industrialized planned economies in general have relatively high levels of patenting by nationals and low levels of foreign patenting and patenting abroad. With the exception of East Germany and Rumania, the planned economies have expanded patenting activity over the period. This and the low levels of patenting by planned economy inventors in industrialized market economies suggests that patentability standards may differ considerably between industrialized market and planned economies.[6]

Table 5.2 provides a summary of data for nine countries operating utility model or petty patent systems. All of these countries are relatively successful in invention given their levels of development (Brazil introduced its utility model in 1970 and we have only recent data; Italy has not reported recent data). Petty patents are granted primarily to nationals (although West Germany has granted a significant number to foreigners from countries without petty patent systems). They are also granted primarily to individuals rather than to large corporate firms. Most are granted in mechanical technology rather than in chemical or biogentic technology.

---

member countries. This legal instrument will have important implications for future data interpretation but has had little impact on the data reported here. Proximity of markets is also a factor in patenting abroad—particularly in the case of Canada and the United States.

6. Table 5.4 indicates that a considerable part of patenting abroad by the planned economies is in other planned eonomies. The planned economies also have ratios of patents granted to scientists and engineers comparable to those in industrialized market economies, see table 5.9.

Table 5.2   Utility Models (Petty Patents) Granted 1967

| | Applications | | | | | | Utility Models Granted | | | | | |
| | Nationals | | | Foreigners | | | Nationals | | | Foreigners | | |
	1967	1975	1980	1967	1975	1980	1967	1975	1980	1967	1975	1980
W. Germany	42,214	30,114	26,094	11,344	11,938	8,153	20,948	12,099	10,252	2,400	2,181	1,879
Italy	4,418	—	—	778	—	1,397	3,935	—	—	702	—	—
Japan	109,154	178,992	190,388	1,906	1,668		20,601	47,449	49,468	721	957	533
Philippines	141	565	762	2	7	24	94	331	465	—	9	3
Poland	1,647	1,896	2,523	22	31	36	411	1,775	1,680	4	25	20
Portugal	139	78	118	25	13	15	77	153	159	9	25	6
Spain	7,601	7,650	5,380	710	1,353	1,162	6,177	4,128	3,845	600	2,041	11,131
Brazil	—	—	1,657	—	—	89	—	—	131	—	—	13
S. Korea	—	7,052	7,936	—	238	622	—	1,032	1,315	—	14	438

Source: *Industrial Property Statistical Report*, annual issues, World Intellectual Property Organization, Geneva.

The advantage of the petty patent is that it broadens the invention base by providing incentives to encourage individuals and small firms to develop inventions. Some semi-industrial countries, notably South Korea and now Brazil, are using this legal system effectively. Japan and West Germany have used it effectively in the past.

Table 5.3 provides data for two weaker legal instruments: the industrial design patent and the trademark. In a sense, a design patent is a petty patent and may serve a similar purpose. Those countries with petty patent systems also have relatively active design patent systems. Design patents have generally not experienced the same pattern of decline observed in invention patents. Except for Canada and the smaller European countries, design patenting by foreigners is a relatively small fraction of total patenting. This is particularly true for semi-industrialized and developing countries where multinational firms have not utilized this instrument for protection (in contrast to the use of invention patents).

The data on trademarks, on the other hand, show that foreign firms are using trademark protection in most markets, including the semi-industrialized and developing countries. An expansion of trademark registration to nationals and foreigners is observed in the majority of economies of all types except the planned economies. This is consistent with the general pattern of industrial trade expansion.

5.2 Comparative Advantage Patterns

Table 5.1 provides data on patents granted to nationals at home, on patents granted to nationals abroad, and on patents granted to foreign inventors. The ratio of patents granted to nationals to total patents granted varied from a high of .76 in the planned economies (and the United States) to a low of .11 for all developing economies in the late 1960s (see table 5.9 for a summary). The ratio of patents granted to nationals to patents granted to nationals abroad ranged from over 2.0 for many developing countries to around .1 for developing and slow-growth semi-industrialized countries.

The first ratio is related to the level of development of the country in question and to its size and degree of economic integration with other countries (particularly for the European Economic Community members and Canada and the United States). It is relevant to this discussion, however, because it indexes technology trade between countries. A firm has an incentive to obtain patent protection in a second country either because it is exporting products protected by the patent to the second country, producing such products in the second country, or selling technology directly through a licensing or technical agreement. The cost of obtaining patents abroad will be a factor in the firm's decision to patent abroad as will the expected market for the protected invention.

Table 5.3   Industrial Design Patents and Trademarks Granted (1975, 1980)

| | Industrial Designs Granted | | | | Trademarks Granted | | | |
| | Nationals | | Foreigners | | Nationals | | Foreigners | |
	1975	1980	1975	1980	1975	1980	1975	1980
Industrialized Market Economies								
Moderate to Rapid Growth								
Japan	34,129	30,696	700	593	104,156	41,577	5,010	5,290
Austria	3,987	4,260	1,517	1,744	1,458	3,333	1,247	2,148
France	11,320	13,209	857	1,560	12,645	37,332	4,312	9,784
Denmark	390	314	486	630	1,520	1,324	3,704	3,339
W. Germany	54,231	70,701	2,609	4,844	9,396	13,006	3,432	3,838
Benelux	1,671	1,691	1,376	1,262	5,529	4,418	3,571	3,082
Norway	243	252	364	434	522	464	2,531	2,675
Slow Growth								
Canada	337	337	1,168	978	3,507	8,779	3,391	6,755
Ireland	34	46	176	284	107	162	893	2,098
Switzerland	465	351	213	325	2,552	2,462	1,508	1,507
Sweden	1,283	1,558	364	588	1,397	1,577	2,591	2,608
United States	3,428	3,056	854	892	28,353	17,319	2,578	1,566
Australia	1,165	1,377	568	580	2,835	1,860	4,252	2,715
United Kingdom	1,665	2,166	1,354	2,799	5,878	3,356	5,562	3,352
Finland	165	371	222	350	276	703	1,126	3,542
New Zealand	157	170	167	173	845	524	2,015	1,318

Table 5.3 (cont.)

	Industrial Designs Granted				Trademarks Granted			
	Nationals		Foreigners		Nationals		Foreigners	
	1975	1980	1975	1980	1975	1980	1975	1980
Semiindustrialized Market Economies								
Rapid Growth								
Spain	3,234	2,239	224	407	—	11,119	—	12,822
Israel	115	266	42	56	224	255	1,064	868
Greece	—	—	—	—	1,546	1,260	1,469	1,800
Singapore	266	335	216	228	—	784	—	2,499
Portugal	—	136	—	81	770	1,035	481	581
Brazil	1,583	3,917	6	154	—	136,808	—	42,821
S. Korea	—	—	—	—	348	603	1,182	1,647
Hong King								
Slow to Moderate Growth								
Chile	—	—	—	16	2,883	1,986	2,810	1,735
Venezuela	59	77	34	na	635	2,360	1,452	1,961
Argentina	2,426	na	159		12,428	—	2,032	—
Costa Rica	—				521		974*	
Mexico	—				3,352	8,637	3,117	8,292*
Turkey	—				557*	1,129**	1,171*	1,181**
Uruguay	—				1,293	6,414	1,152	541

Developing Economies

Ecuador	—	—	—	210	513	612	1,077
Iraq	19	9	—	68	184	236	885
Morocco	82	116	15	40	428	541	309	443
United Arab Republic (Egypt)	127	166	8	27	234	145	396	408
Colombia	11	na	5	na	702	584**	1,542	672**
Philippines	151	na	19	na	539	1,225	341	1,013
Kenya	—	—	—	—	153	443	585	747
Ghana	—	—	—	—	27	8	263	167
India	723	na	29	na	3,019	na	640	na
Sri Lanka	8	na	—	na	43	160	130	376
Indonesia	—	—	—	—	1,160	6,479	697	2,741
Pakistan	74	93	14	36	283	494**	640	780**
Zambia	—	—	3	—	22	4	441	215
African Intellectual Property Organization	26	—	57	—	62	na	954	na

Planned Economies

E. Germany	—	—	—	—	299	150	325	265
Czechoslovakia	577	1,304	8	20	182	134	302	258
Soviet Union	—	—	—	—	48	1,627	5	559
Hungary	165	120	11	28	107	149	290	194
Poland	139	124	16	28	288	116	640	544
Bulgaria	27	38	5	6	15	73	434	492
Yugoslavia	102	na	30	na	156	na	154	na
Rumania	—	—	—	—	205	418	334	53

Source: *Industrial Property Statistical Report*, annual issues, World Intellectual Property Organization, Geneva.

*1976.

**1979.

The ratio of patents granted to nationals to total patents has risen in most rapidly growing economies and declined in most slow growing economies. For example, in the United States the ratio fell from .78 in 1967 to .60 in 1980. In Japan it rose from .66 to .82 over the same period. This can be taken as an index of changing comparative advantage. Table 5.4 presents patent (trade) "balance" data for 1967 and 1980. These data are organized to present the persepective of the granting country (i.e., row proportions sum to one). These data show that the great bulk of foreign patents granted in all countries, whether industrial, semi-industrial, developing, or planned, originate in industrial countries. Even the

Table 5.4   **Patent Balance Data, 1967 and 1980: Perspective of Granting Country**

| Granting Country | Patents Granted to Foreigners | | Percent Originating in | | | | | |
| | | | United States | | United Kingdom | | Germany | |
	1967	1980	1967	1980	1967	1980	1967	1980
Industrial								
Japan	6,896	8,074	.49	.49	.09	.06	.16	.05
Austria	3,920	4,481	.21	.13	.07	.04	.09	.46
France	31,749	19.622	.34	.28	.11	.07	.24	.26
Denmark	1,997	1,453	.23	.22	.11	.09	.22	.23
W. Germany	8,300	10,362	.41	.31	.12	.06	—	—
Belgium	na	5,081	na	.32	na	.04	na	.17
Norway	1,817	1,843	.26	.23	.11	.08	.19	.15
Netherlands	1,913	2,907	.31	.30	.10	.05	.22	.21
Canada	24,753	22,392	.55	.60	—	.05	.28	.08
Italy	na	6,190	na	.01	na	.01	na	.29
Ireland	635	1,407	.29	.32	.24	.17	.11	.16
Switzerland	16,462	4,486	.22	.22	.08	.05	.38	.32
Sweden	7,532	3,604	.32	.29	.10	.06	.25	.22
United States	14,378	24,675	—	—	.19	.09	.26	.14
United Kingdom	28,893	18,646	.47	.36	—	—	.24	.21
Finland	739	1,464	.20	.18	.05	.07	.17	.20
Semi-industrial								
Spain	6,827	7,739	.27	.25	.08	.07	.17	.20
Israel	935	1,419	.39	.46	.10	.09	.13	.16
Greece	1,319	942	.29	.21	.06	.08	.12	.22
Portugal	1,038	2,200	.18	.23	.11	.08	.19	.17
S. Korea	152	1,446	.45	.26	.02	.04	.28	.08
Brazil	679	6,228	.42	.36	.08	.04	.13	.22
Chile	1,224	na	.46	na	.07	na	.12	na
Venezuela	961	408	.59	.47	.04	.04	.05	.11
Argentina	4,479	na	.50	na	.08	na	.08	na
Mexico	5,817	2,389	.44	.62	—	.03	.05	.02
Turkey	427	—	.29	—	.12	—	.21	—
Uruguay	350	236	.41	.24	.08	.12	.13	.13

Eastern European planned economies grant the bulk of their foreign patents to Western European inventors.

The dominance of industrial countries in origination of patents granted abroad reflects their general comparative advantage in invention. The ratio of patents granted to nationals abroad to patents granted to nationals at home is a rough index of the degree of "pioneeringness" or "adaptiveness" of invention. This index is affected by size of country and proximity of similar countries (as in the EEC) and thus is not ideal. It varies so markedly between industrial countries (around 2) and semi-industrial countries (.15–.25) and developing countries (.1) that no rea-

Percent Originating in									
Japan		Other Industrial		Planned		Semi-industrial		Developing	
1967	1980	1967	1980	1967	1980	1967	1980	1967	1980
—	—	.25	.31	.01	.04	—	.002	—	—
.02	.04	.66	.27	.09	.05	—	.02	.01	.002
.04	.10	.21	.21	.05	.05	.01	.03	—	.001
.02	.06	.39	.37	.02	.03	.01	.08	—	.002
.04	.23	.37	.32	.05	.05	.01	.03	—	.001
na	.06	na	.38	na	.03	na	.03	na	.001
.02	.05	.50	.47	.005	.02	.005	.004	.006	.001
.02	.15	.32	.26	.02	.02	—	—	.01	.006
.08	.09	.03	.15	.04	.01	.012	.02	.008	.001
na	—	na	.55	na	.05	na	.09	na	.005
.01	.01	.33	.32	—	—	.01	.02	.005	.005
.02	.09	.27	.27	.02	.04	.01	.006	—	.001
.01	.07	.28	.32	.03	.04	—	.003	.01	.001
.10	.25	.40	.53	.02	.03	.02	.02	.01	.001
.07	.12	.16	.28	.01	.01	.04	.02	.01	—
.01	.04	.53	.39	.03	.09	.005	.02	.003	.01
.01	.05	.45	.40	.01	.02	.004	.007	.01	.001
.01	.02	.35	.26	.01	.01	.01	.001	.001	.001
.02	.03	.46	.38	.03	.05	.02	.02	.001	.01
.01	.03	.45	.42	.005	.01	.05	.05	.01	.01
—	.50	.24	.09	—	.02	—	.002	.007	.01
.01	.06	.32	.30	.001	.01	.03	.01	.01	.001
.03	na	.26	na	.01	na	.04	na	.007	—
.03	.03	.28	.23	—	.01	.01	.09	.001	.02
.01	na	.31	na	.005	na	.01	na	.008	—
.10	.03	.31	.26	.06	.01	.04	.02	—	.07
.002	—	.33	—	.04	—	.007	—	—	—
.02	.02	.22	.31	.01	.01	.12	.15	.009	.01

Table 5.4 (cont.)

Granting Country	Patents Granted to Foreigners		Percent Originating in					
			United States		United Kingdom		Germany	
	1967	1980	1967	1980	1967	1980	1967	1980
Developing								
Ecuador	126	103	.41	.56	.04	.01	.21	.04
Iraq	161	24	.34	.08	.10	.20	.17	.20
Morocco	387	330	.21	.22	.04	.03	.10	.12
U.A.R. (Egypt)	867	317	.27	.31	.06	.07	.14	.19
Colombia	848	808	.57	.58	.08	.04	.10	.05
Philippines	496	775	.67	.54	.05	.03	.03	.08
Kenya	104	97	.24	.28	.40	.21	.05	.14
India	3,329	na	.33	na	.24	na	.11	na
Sri Lanka	53	174	.21	.42	.15	.17	.06	.08
O.A.P.I.	513	136	.12	.10	.03	.06	.06	.61
Planned								
E. Germany	1,553	1,371	.03	.20	.09	.04	.40	.27
Czechoslovakia	787	1,546	.05	.18	.05	.07	.17	.27
Soviet Union	503	1,572	.07	.22	.17	.05	.17	.20
Hungary	512	1,018	.06	.17	.08	.06	.18	.26
Poland	484	1,962	.07	.22	.08	.07	.14	.03
Bulgaria	166	425	.03	.06	.02	.08	.28	.29
Yugoslavia	644	na	.07	na	.07	na	.18	na
Rumania	1,283	814	.06	.24	.06	.04	.21	.24

Patents Originating Summary

	Patents Granted		Patents Originated	
	1967	1980	1967	1980
Industrial	149,994	146,714	181,243	172,651
Semi-industrial	24,208	23,716	1,796	1,913
Developing	6,884	6,221	805	750
Planned	5,932	8,708	6,225	6,426

Source: *Industrial Property Statistical Report*, annual issues, World Intellectual Property Organization, Geneva.

sonable adjustment for these factors would alter the picture. Invention in developing countries is almost entirely adaptive in nature. Some of the more advanced semi-industrial economies (Spain, Israel, Brazil) appear to have significant pioneering invention, but they are still predominantly adaptive. The data from the planned economies are more difficult to interpret as they may be subject to considerable domestic policy effects.

The overall picture that emerges from these data supports both the notion of technology trade based on the differentiation of invention along a pioneering-adaptive continuum. Developing and semi-industrialized

				Percent Originating in					
Japan		Other Industrial		Planned		Semi-industrial		Developing	
1967	1980	1967	1980	1967	1980	1967	1980	1967	1980
—	.05	.31	.28	—	—	.02	.04	.008	.02
.06	—	.23	.44	.09	.08	.006	—	.006	—
.003	.01	.57	.52	.03	.03	.04	.04	.003	.03
.018	.05	.32	.32	.19	.02	.01	.04	.002	.003
.01	.02	.21	.26	.006	.01	.02	.03	.006	.01
.09	.12	.15	.18	—	.02	.002	.02	.008	.01
.02	.01	.26	.35	—	.01	—	—	.03	—
.04	na	.22	—	.05	na	.008	na	.004	—
.02	.04	.54	.25	.02	.02	—	.01	—	.01
.002	—	.78	.20	—	.01	.004	.02	.004	—
.01	.02	.32	.23	.13	.22	.003	.02	.02	.002
.01	.07	.35	.10	.37	.28	.005	.03	—	—
.07	.10	.48	.30	.04	.13	.004	.002	—	.001
.01	.04	.42	.21	.24	.24	.004	.02	.01	.003
.004	.03	.35	.51	.35	.14	.004	.001	.002	—
.02	.02	.45	.19	.19	.36	.006	.002	.006	—
.02	na	.46	na	.20	na	—	na	—	na
.02	.02	.39	.29	.25	.16	.01	.01	.005	.005

countries are overwhelmingly importers of technology and they special-ize in adaptive invention at home. The jockeying for position among industrialized-country exporters has changed somewhat in the past four years with Japan moving into a strong competitive position. The U.S. share of patent exports fell from .37 in 1967 to .30 in 1979. Japan's share rose from .03 to .11.

The notion that developing countries engage in mostly adaptive inven-tion suggests that inventions made in more highly developed countries "disclose" possibilities for modifications of these inventions in the "downstream" developing countries. Clearly their invention is different in character than that of developed countries. It is apparently of low value "upstream" in industrialized, developed countries with high wages. Since these countries do obtain some patents abroad it will be useful to look at the patterns of this invention.

Table 5.5 provides data organized from the perspective of the originat-ing country for 1980. Predictably, most patenting abroad originates in developed countries. This is where the large markets are. Argentina, Brazil, and Mexico, however, do appear to be patenting "downstream" in developing Latin American countries to a significant extent. This

Table 5.5   Patent Balance Data, 1980: Perspective of Originating Country

		Percent Granted in				
Origin Country (Patents Originated)		United States	Japan	Ger- many	Other Indus- trial	Planned
Japan	(20,663)	.35	—	.11	.422	.022
Austria	(1,669)	.158	.022	.015	.70	.013
France	(12,511)	.167	.035	.073	.479	.055
Denmark	(1,103)	.148	.027	.061	.63	.057
W. Germany	(33,708)	.171	.040	—	.584	.006
Belgium	(1,720)	.144	.028	.038	.604	.039
Norway	(549)	.144	.044	.064	.562	.086
Netherlands	(5,964)	.109	.059	.090	.571	.031
Canada	(2,200)	.503	.036	.037	.326	.015
Italy	(5,877)	.137	.025	.052	.506	.061
Ireland	(107)	.206	.009	.047	.599	—
Switzerland	(9,827)	.127	.041	.096	.50	.059
Sweden	(4,769)	.173	.042	.058	.594	.044
United States	(54,360)	—	.073	.059	.678	.033
United Kingdom	(11,140)	.219	.041	.054	.478	.045
Finland	(928)	.133	.022	.033	.599	.100
Spain	(1,180)	.028	.006	.009	.85	.007
Israel	(316)	.377	—	.044	.474	—
Greece	(691)	.006	—	.003	.986	.003
Brazil	(113)	.204	.018	.009	.397	.009
Argentina	(133)	.211	.015	.008	.187	.015
Mexico	(171)	.275	.029	.018	.326	.029
India	(57)	.175	.053	.018	.489	.018
Panama	(233)	.009	—	.021	.562	.069
Bahamas	(103)	.058	—	.049	.524	.058

provides some support for the adaptiveness hypothesis, but a full treatment would require detailed industry data.[7]

Data on receipt of royalties and fees to U.S. residents for the use of intangible property such as patents, techniques, process designs, trademarks, and other technology-related activities show that the export of technology is not a trivial activity. Total receipts of royalties and fees were $5.5 billion in 1978.[8] Product trade data also show that exports of R & D intensive products have been important to the U.S. economy. In 1964 the trade balance in R & D intensive manufactured products showed

7. Such data are now becoming available from the International Patent Documentation Center (INPADOC 1981). Patents can be classified by International Patent Class (IPC). A concordance between IPC and Standard Industrial Classes (SIC) has been made by INPADOC. F. M. Sherer (this volume) has questioned the value of such concordances, but for reasonably broad industrial classes they may be adequate.

8. See table 5.8 for data from several OECD countries on receipts and payments of royalties and fees.

	Percent Granted in					
	Semi-industrial				Developing	
Latin America	Africa	Asia	Europe	Latin America	Africa	Asia
.021	—	.05	.022	.001	.001	.001
.022	—	.017	.033	.002	.015	.003
.016	—	.016	.128	.002	.023	.006
.014	—	.012	.044	.001	.006	—
.049	—	.014	.064	.002	.005	.005
.024	—	.020	.075	.005	.015	.008
.024	—	.016	.056	—	.004	—
.045	—	.008	.078	.002	.003	.004
.033	—	.012	.022	.009	.006	.001
.066	—	.015	.118	.003	.011	.006
.065	—	.019	.037	.009	—	.009
.044	—	.030	.080	.003	.013	.007
.030	—	.006	.048	.0002	.004	.001
.066	—	.027	.048	.005	.007	.004
.027	—	.038	.074	.003	.013	.008
.055	—	.002	.044	.009	.001	.002
.042	—	.003	.039	.006	.008	.002
.070	—	.003	.032	—	—	—
.001	—	—	.001	—	—	—
.150	—	—	.186	.009	.009	.009
.406	—	—	.090	.068	—	—
.088	—	.006	.053	.123	.006	.047
.053	—	—	.123	—	.053	.018
.039	—	.034	.150	.004	.026	.086
.039	—	.029	.126	.049	.049	.019

net exports of $8.8 billion and net imports of $3.7 billion in non-R & D intensive manufactured products. In 1979 net exports of R & D intensive manufactured products had grown to $39.3 billion, but net imports of non-R & D intensive products had grown to $34.8 billion. (These data do not include agricultural and mineral products, also important in trade.)

5.3 Evidence of Declining Patent/Inventive Input Ratios

I now turn to four data sets on patents and inventive inputs (scientists and engineers engaged in R & D and R & D spending) to examine the question of "inventor productivity." All four data sets show that the ratio of patents granted per unit of inventive input has fallen from 1964 to 1979–80. This decline shows up for almost all of the industries in the two data sets (United States and Japan) where industry-specific data are available. The decline shows up in each of the five countries for which

OECD data are available (United States, United Kingdom, France, West Germany, and Japan), and it shows up in most of the forty-four countries for which UNESCO data are available.

A decline in the ratio of patents granted to inventive inputs need not imply that *real* invention per unit of inventive input has declined. A change in the "propensity to patent" (i.e., patents granted per unit of real invention) could have produced the results reported here. A rise in the cost of obtaining and enforcing patents, changes in legal systems, and changes in company policies could produce changes in the propensity to patent. We know that some changes have occurred particularly because of rising patent enforcement costs. A few countries have changed their legal systems as well. However, many countries have not experienced rises in patent enforcement costs and have actively encouraged invention through subsidies and favorable tax treatment. Changes in the propensity to patent are unlikely to explain the universal decline in patenting per inventive input unit shown by the data.[9]

Consider first the data by industry for the United States. Table 5.6 provides the most detailed industry data readily available on patents granted to nationals and foreigners, R & D expenditures (in 1972 constant dollars), and scientists and engineers engaged in R & D. Data on the proportion funded by government and on the proportion considered "basic" and "development" are also provided by the National Science Foundation.

Table 5.6 shows that R & D spending per scientist and engineer, while varying somewhat by industry, has changed little from 1964 to 1978. It increased at an annual rate of only .0047. Patenting per scientist and engineer fell at an annual rate of $-.0126$ from 1964–66 to 1971–72 and $-.0439$ from 1971–72 to 1976–78 ($-.0283$ over the entire period.) Regression (1) provides a statistical description of this decline, controlling for industry effects. The annual rates are unchanged by the correction for industry effects, and the decline from 1964–66 to 1976–78 is highly significant from a statistical perspective. (These data do not include patenting for 1979 and 1980.) At the national level, patents granted to national inventors declined by 10.2 percent from 1976–78 to 1980 (this excludes the extraordinarily low patenting in 1979). Numbers of scientists and engineers and R & D expenditures rose by roughly 10 percent during this period. (In table 5.6, R & D and scientists and engineers data are lagged behind patents granted by two years).

Since R & D spending rose relative to scientists and engineers only slightly, patents granted per dollar spent on R & D declined only slightly more than is the case for scientists and engineers. The table also shows that the ratio of national to foreign patenting fell in every industry in both

9. The data utilized in table 5.6 are summarized in National Science Board (1981).

periods. The change in this ratio is positively correlated with the change in labor productivity across industries over the 1966–78 period ($r = .583$). There is also a positive correlation between the change in the national to foreign patent ratio and the change in national patenting per scientist and engineer (.346 over the entire period; .556 in the second half). Changes in national patenting and foreign patenting are positively correlated for the 1966–78 period ($r = .701$), but changes in national patenting in the second half of the period are negatively correlated with changes in foreign patenting in the first period ($r = -.631$). The converse is also true ($r = .342$).[10]

Regressions (2) and (3) in table 5.6 report a simple effort to control for some characteristics of the research system on patenting per unit of inventive input. They provide some evidence that government funding increases inventive output while emphasis on basic research decreases it.

Our second set of data reported in table 5.7 includes industry level data for Japan. These data show that patent applications per scientist and engineer were lower in 1975–76 than in 1967–68 in all industries except textiles and foods, where they were unchanged. Patent applications per dollar expended on R & D also declined because R & D per scientist rose. (R & D spending is expressed in millions of 1970 constant yen.) A positive correlation between the changes in patents per scientists and engineers by industry between the United States and Japan exists with the transport (motor vehicles) and nonelectric machinery industries in both countries experiencing the largest declines.

Table 5.8 reports our third data set. These data were collected by the OECD and are somewhat more reliable than the fourth data set collected by UNESCO (summarized in table 5.9). The five countries included in the table undertake the bulk of the world's R & D. The decline in patents granted to national inventors per scientist and engineer (the scientist and engineer and R & D data are lagged two years prior to the patenting data, i.e., for the 1967 column S & E and R & D data are for 1965) shows up in each country. In the cases of Japan and Germany, patents per scientist and engineer peaked in 1971. In the United States, United Kingdom, and France, this ratio has declined since 1967. In terms of the date of investment in R & D, these declines set in two years earlier. The decline was therefore not directly associated with the energy price increases of the early 1970s.[11]

---

10. Some variation in patents granted is from changes in the "backlog" of patents applied for but not examined. A decline in patents granted in period T due to an increase in the backlog will produce an increase in patents granted in a later period. In the U.S. Patent Office, 1979 was a particularly bad year in this regard, and patenting was low because of an increase in the backlog. The 1979 data for the United States, United Kingdom, and France are not used in any of the calculations made in this paper because of this problem.

11. The data on R & D and Industrial Product (IDP) have first been converted to 1971 constant currency using national GDP deflators. They were then converted to U.S. dollars

Table 5.6   Industry Patenting (National and Foreign), R & D Expenditures, and Scientists and Engineers, United States, 1964–78

Industry	Ratios 1976–78			Annual Percent Changes 1964–66 to 1971–72				Annual Percent Changes 1971–72 to 1976–78			
	R & D / S & E	PN / S & E	PN / PF	PN / S & E	PN / R & D	R & D / S & E	PN / PF	PN / S & E	PN / R & D	R & D / S & E	PN / PF
Food	40.2	86.3	2.10	.057	.034	.023	−.065	−.039	.039	.001	−.027
Textiles	33.7	241.1	1.35	.021	.036	−.015	−.084	−.014	−.094	.040	−.066
All chemicals	50.3	159.1	1.38	−.018	−.016	−.002	−.068	−.030	−.038	.008	−.047
Industrial chemicals	56.9	173.1	1.35	.017	.014	.002	−.055	−.031	−.081	.011	−.039
Drugs	47.1	71.0	1.06	−.093	−.078	−.016	−.034	.010	.005	.005	−.057
Petroleum products	69.2	84.2	4.14	−.031	−.099	−.010	−.036	−.007	−.036	.033	−.070
Rubber products	42.7	276.7	1.89	−.029	−.072	.043	−.098	−.006	−.081	−.020	−.054
Stone and glass	42.5	214.2	1.84	−.018	.001	−.019	−.123	−.076	−.078	.002	−.054
Primary metals	46.2	55.7	1.11	−.040	−.009	−.030	−.088	−.089	−.094	.005	−.061
Fabricated metals	37.9	766.2	2.17	−.002	−.049	.038	−.086	−.064	−.063	−.002	−.046
Nonelectric machines	49.9	206.5	1.54	−.074	−.108	−.015	−.077	−.100	−.110	.010	−.057

Electric machines	52.2	100.1	1.78	.009	.002	.006	−.082	−.046	−.051	.005	−.062
Motor vehicles	79.4	95.5	1.91	−.029	−.021	−.005	−.061	−.034	−.060	.025	−.025
Aircraft	66.6	13.0	1.21	.041	.054	−.013	−.061	.023	.021	.003	−.062
Scientific and professional instruments	52.4	253.2	1.67	.001	−.030	.029	−.094	−.097	−.048	−.007	−.056
All industries	51.2	186.4	1.77	−.0126	−.0138	.0011	−.0742	−.0439	−.0529	.0082	−.0523

Regressions:

(1) $LN(PN/S \& E) = 5.0986 + ind. dummies - .0756 \ T2 - .3392 \ T3 \ (R^2 = .947; F = 38.27).$
$\qquad\qquad\qquad\qquad\qquad\qquad\qquad\quad (.105) \qquad\quad (.105)$

(2) $LN(PN) = 5.0147 + 1.18 \ LN(S \& E) + 2.526 \ proportion \ govt. \ funded - 13.91 \ proportion \ basic$
$\qquad\qquad\qquad\qquad (.166) \qquad\qquad\quad (1.058) \qquad\qquad\qquad\qquad\qquad\qquad\qquad (10.97)$
$\qquad + 1.419 \ proportion \ development + ind. \ dummies - .132 \ T2 - .433 \ T3 \ (R^2 = .97; F = .512).$
$\qquad\qquad (.98) \qquad\qquad\qquad\qquad\qquad\qquad\qquad\qquad\qquad (.106) \qquad (.130)$

(3) $LN(PN) = .107 + 1.23 \ LN(R \& D) + 1.93 \ proportion \ govt. \ funded - 13.77 \ proportion \ basic$
$\qquad\qquad\qquad\qquad (.15) \qquad\qquad\qquad (1.00) \qquad\qquad\qquad\qquad\qquad\qquad\qquad\quad (10.15)$
$\qquad + 1.38 \ proportion \ development + ind. \ dummies - .155 \ T2 - .526 \ T3.$
$\qquad\qquad (.55) \qquad\qquad\qquad\qquad\qquad\qquad\qquad\qquad\qquad (.099) \qquad (.126)$

Source: *Science Indicators 1980*.

Note:
$T2 = 1$ if year equals 1971.
$T3 = 1$ if year equals 1978.
Standard errors in parentheses.

Table 5.7   Patent Applications, Scientists and Engineers, and R & D Expenditures in Japan, 1967–76

Industry	Patents/S & E			Patents/R & D			R & D/S & E		
	1967–68	1971–72	1975–76	1967–68	1971–72	1975–76	1967–68	1971–72	1975–76
Chemicals	1.60	1.30	1.42	.425	.187	.218	3,762	6,969	6,528
Nonelectrical machinery	3.28	2.66	2.33	.948	.422	.273	2,914	11,402	4,714
Electrical machinery	1.23	.93	1.12	.378	.132	.197	3,251	7,042	5,667
Transport and construction equipment	2.08	1.49	1.32	.328	.136	.110	6,332	10,959	11,832
Textile and household goods	4.82	5.54	4.89	1.413	1.071	.988	3,412	5,169	4,954
Foods	1.37	.84	1.38	.494	.169	.270	2,778	4,949	5,108
All industries	1.80	1.43	1.54	.733	.200	.236	3,610	7,150	6,540

Source: *Statistical Yearbook of Japan*, Ministry of Trade and Industry, Tokyo.

Table 5.8   OECD Data: Patents, R & D, and Scientists and Engineers

	Patents to Nationals (PN)	Scientists and Engineers (S & E)	PN / S & E	PN / R & D	R & D / IDP	Royalties and Fees			
						Received	Paid	Received PN	Paid PF
United States									
1967	51,274	494.5	103.8	2.85	2.49	23.84	2.28	.32	.16
1971	55,988	555.2	100.8	3.145	2.12	31.91	3.02	.36	.14
1975	44,162	525.1	84.1	2.484	1.98	28.96	3.19	.32	.12
1980	37,652	573.9	65.6	2.341	1.91	26.89	2.55	.49	.14
United Kingdom									
1967	9,807	49.9	196.5	7.36	2.00	2.51	2.38	.14	.08
1971	10,376	52.8	196.5	8.45	1.85	3.61	3.28	.17	.11
1975	8,855	80.7	109.7	8.75	1.75	3.32	2.73	.24	.09
1980	5,158	80.7	63.8	5.23	1.82	na	na	na	na
West Germany									
1967	5,126	61.0	84.0	2.30	1.28	1.24	2.64	.03	.31
1971	8,295	74.9	110.7	2.23	1.54	1.53	3.94	.04	.40
1975	10,395	102.5	101.4	2.69	1.59	1.73	3.85	.05	.36
1980	9,826	111.0	88.5	2.29	1.64	1.91	4.64	.06	.40
France									
1967	15,246	42.8	356.2	10.44	1.42	2.68	3.16	.19	.10
1971	13,696	57.2	239.4	7.90	1.29	4.42	4.59	.26	.12
1975	8,420	64.1	131.4	4.07	1.39	6.54	5.44	.51	.25
1980	8,433	68.0	124.0	3.61	1.35	9.19	7.42	.73	.42
Japan									
1967	13,877	117.6	118.0	9.67	.84	.37	3.28	.06	.47
1971	24,795	157.1	157.8	7.70	1.11	.75	6.12	.05	.52
1975	32,465	238.2	136.3	8.49	1.19	1.01	4.79	.05	.63
1980	38,032	272.0	139.8	8.50	1.29	1.33	5.85	.07	.63

Source: *Science Indicators 1980.*

The data on the ratio of expenditures on R & D to industrial product show this ratio to be declining sharply in the United States and rising significantly in Japan and West Germany. In 1980 these five countries did not differ greatly on this measure.

We also have data on royalties and fees paid and received for these five countries (expressed in "real" U.S. dollars as was R & D). The ratios of royalties and fees received per patent granted to nationals abroad are relatively low for West Germany and Japan, while the reverse is true for the ratio of payments made per patent granted to foreigners. These two countries are "aggressive" about expanding their R & D investments and patenting in foreign countries. Their strategy has generally been to borrow or import technology to build their own capacity. This is reflected in the fact that they pay substantial fees for imported technology and receive relatively low payments for their patenting abroad (although it should be noted that the payments data are for patents granted in prior years and do not match up with the patents data in the table.)[12]

Table 5.9 provides a summary of international data for forty-four countries (see table 5.1 for classification by region). Rows (1)–(5) provide means of the patent and trademark data reported in tables 5.1 and 5.3 by region. These means highlight the major features of the patent data. They show the decline in the importance of the United States in world patenting and the rise in importance of Japan, West Germany, and the planned economies. They also show the marked differences in patents granted abroad to patents granted at home—a measure of adaptiveness of invention—between the industrialized nations and the semi-industrialized and developing economies. They further show the high degree of foreign patenting in most of the world's economies.

Rows (6)–(9) provide data on ratios of patenting to numbers of scientists and engineers in the productive sector and on patenting to R & D expenditures in the productive sector. It should be noted that both the S & E data and the R & D data are subject to considerable errors. The UNESCO data provide a breakdown of both for the "productive" (i.e., industry, transport, commerce), education, and service sectors. I have used the data from the productive sector. A further problem with these data is that they are not available for all years and some interpolation was

using the purchasing power parity exchange rate of Kravis, Summers, and Heston (1980). This exchange rate is designed to enable better comparability of incomes between countries. The real costs of undertaking research may not be closely related to the real costs of producing goods generally. We do not have an ideal deflator for R & D spending in any single country and obviously no ideal deflator exists to achieve cross-country comparability. This paper does not attempt to draw strong conclusions from cross-country comparisons as a consequence. They are reported as a matter of convenience (but see note 12).

12. This inference requires comparability in the real dollar conversions. While expressing skepticism about conversion rates (see note 11), one can probably say that the problems are less serious for this group of countries than for most others.

required.[13] The most serious problem, however, is the exchange rate conversion. This conversion is relevant only if one wishes to make cross-section comparisons. Comparisons over time require only an appropriate deflator to convert expenditures to a constant currency unit. Row (8) provides R & D data in constant 1972 U.S. dollars where standard exchange rate conversions to dollars were made and where the U.S. deflator was used. Row (7) utilizes the purchasing power parity exchange rates developed by Kravis, Summers, and Heston (1980). This deflator modifies both the time series and cross-section aspect of the conversion and, while imperfect for the task at hand, is probably the best available.

Examination of the data in row (6) shows that patents per scientist and engineer have declined in the industrialized and slow growing semi-industrialized countries (which account for most of the world's patents—see row [1]). The numbers in parentheses are regression estimates of the decline in the ratios within countries (i.e., country dummies were included in the regressions). Statistically significant declines are shown for the last two periods relative to the first for these groups. In addition, virtually all individual countries in each group showed declines in the ratios. These ratios do vary considerably by type of economy, with the semi-industrialized countries (notably those with rapid growth) showing ratios far above the industrialized country standard. Developing countries are generally far below the industrialized countries in this regard.

In developing countries a relatively high proportion of time is devoted to adaptive invention, much of which is not patentable. Many of these countries have vented frustration over the terms on which technology is purchased in international forums. Few have shown imagination in developing legal systems suited to their competitive position in international invention. Most invention from these countries is adaptive. Yet they have generally not modified their patent systems to encourage adaptive invention. They have instead opted to weaken the scope of patent coverage in an attempt to discourage foreign patenting. In this the slow-growth industrialized economies and the developing economies have been successful. Unfortunately, they have also discouraged national invention in the process.

The data on patents per dollar expended on R & D are somewhat less regular in showing declines in patenting per unit of inventive activity than are the data on patents per scientist and engineer. Part of this is because of the problem of deflating these data appropriately. The data show

---

13. UNESCO statistical yearbooks provide data for available years and it is not possible to match up the data for all relevant years. Simple interpolation was used to fill in missing years. The classification of R & D and S & E data by type of performing organization is also subject to some differences between countries. Personnel data are classified as scientists and engineers, technicians, and other personnel. The inclusion of technicians in the data reported in table 5.9 would not have altered the results.

Table 5.9   International Data: Patenting and Invention Input Means by Region and by Period

	Industrialized Economies			Semi-industrialized Economies		Developing Economies	Planned Economies
	Rapid Growth	USA	Other	Rapid–Moderate	Slow Growth		
(1) Share of world's invention patents							
1967–71	.251	.316	.110	.026	.017	.004	.277
1976–79	.310	.233	.075	.033	.010	.004	.336
(2) Share of world's design patents							
1975–80	.852	.025	.041	.046	.017	.010	.029
(3) Share of world's trademarks							
1975–80	.412	.080	.064	.309	.092	.036	.007
(4) Ratio: patents granted to nationals abroad to patents (N)							
1967–71	1.94	1.51	2.28	.28	.092	.10	.155
1976–79	1.31	1.69	2.65	.20	.165	.09	.109
(5) Ratio: patents (N) to total patents							
1967–71	.39	.75	.19	.25	.17	.11	.76
1976–79	.51	.62	.18	.27	.20	.12	.84
(6) Patents per scientist and engineer (PN/S & E)[c]							
1967	.238	.248		.998	.380	.053	.269
1971	.258 (.08)	.214 (−.11)		.876 (.12)	.337 (−.02)	.066 (.08)	.218 (−.07)
1976	.201 (−.14)*	.152 (−.36)**		.494 (−.12)	.185 (−.69)**	.055 (.13)	.187 (.28)*
1979	.200 (−.09)	.108 (−.69)**		.550 (.05)	.154 (−.98)**	.052 (.02)	.243 (−.08)

(7) Patents per dollar expended R & D (PN/R & D)[a,b,c]						
1967–71	1.007	1.660	4.054	3.429	.340	1.092
1975–79	1.276	1.463	6.803	2.181	.337	1.297
(8) Patents per dollar expended R & D (PN/R & D)[a,c]						
1967–71	1.276	1.775	6.799	6.858	.777	1.092
1975–79	1.119	1.202	9.733	3.621	.822	1.297
(9) R & D/GDP[a,c]						
1967	.0325	.0230	.0094	.0040	.0056	.0357
1971	.0227 (−.38)**	.0168 (−.29)**	.0056 (−.53)**	.0043 (.13)	.0053 (−.05)	.0261 (−.26)**
1976	.0206 (−.48)**	.0159 (−.36)**	.0043 (−.77)**	.0041 (−.16)	.0041 (−.26)**	.0337 (−.08)*
1979	.0196 (−.52)**	.0171 (−.30)**	.0043 (−.78)**	.0041 (−.16)	.0039 (−.27)**	.0329 (−.11)*
(10) Regression of LN(R & D) on:[c]						
LN(MFG)	1.032**	1.25**	.627*	−.124	.301*	.767*
Industrial growth	−.0025	−.040	−.014	−.133*	.050**	.035
Trade intensity	2.722**	−1.46*	−1.46*	−3.36*	−1.45*	—
R^2	.99	.99	.98	.96	.97	.99
F	238	769	61	25	55	160

Source: UNESCO Yearbooks.

Note: On time dummy variables. Country and time dummies included in regressions. Numbers in parentheses are coefficients.

*Coefficient > 1.5 times standard error.

**Coefficient > 2 times standard error.

[a]Numbers in parentheses are time dummy coefficients in a regression LN( ) on country and time dummies.

[b]These R & D data are deflated by the Kravis, Summers, and Heston (1980) purchasing power parity exchange rate.

[c]U.S. and other industrialized countries are combined.

general declines in patenting per dollar expended in R & D except in the rapidly growing semi-industrialized economies, in the planned economies, and for the data deflated by the purchasing power parity exchange rate for the rapidly growing industrialized economies.[14]

Row (9) of the table corroborates the pattern observed in the OECD countries of a decline in the share of industrial product devoted to R & D. This share has declined in virtually all countries in the data set, including the planned economies. This is consistent with the proposition that invention has become more costly (i.e., that the probability of discovery has declined). The magnitude of this deline in investment is highly significant and has important policy implications for growth when considered along with the evidence for declining productivity of invention.

Table 5.9 also reports an investment regression for each of these regions that is more "descriptive" in nature than analytical. In each region the log of R & D is regressed on the log of industrial Gross Domestic Product (GDP), the industrial growth rate in the previous ten years, and the trade intensity of the country (i.e., the ratio of the value of trade to GDP). Country and time dummy variables are included to pick up constant country effects. These regressions, while not particularly remarkable given the data and the problem of international comparability, suggest that investment decisions are reasonably systematic. Except for the slow growing semi-industrialized economies (a rather mixed bag of countries), investment in R & D is related to industry size. There is little evidence that past industry growth affects investment decisions (although this might differ if we had detailed industry data). Openness to trade appears to have a positive effect on R & D spending in the fast growing economies and a negative effect in the slow growing economies. The interpretation of this result is not readily obvious since openness to trade and willingness to invest in R & D may be jointly determined by a set of political factors. It is tempting to suggest that aggressive growth strategies, as by Japan and West Germany, produce this positive correlation, while the reverse is true for those countries pursuing less aggressive growth policies. This type of data, however, is not really suited to test that proposition, and these regressions are accordingly presented here in a data table and labeled descriptive.[15]

To conclude that a significant decline in real productivity of invention has taken place does not require proof that the propensity to patent has not changed. The magnitude of the declines in patenting per scientist and

14. A simple regression: $LN(PN/S \& E) = bLN(R \& D/S \& E)$ plus country and time dummies was run for each country group. The b coefficient was positive in all cases and greater than its standard error in all but the slow growing industrialized countries. This indicates that R & D data are measuring real scientific resources rather than scientists' and engineers' time. The time dummy coefficients were similar to those reported in row (6).

15. The problem of shifts between industries is particularly problematic for such comparisons.

engineer and per unit of R & D in the OECD countries for which we have good data is large. Furthermore, these countries have reduced spending on R & D relative to the market for inventions by significant amounts, presumably in response to a decline in invention potential. Broad cycles in growth potential have marked our history before. The 1970s may well have been more normal in this regard than the 1950s and 1960s.

Our data force us to deal with broad aggregates. If we had more detailed data by technology field, we would probably find that even prior to 1965 many technology fields were exhibiting declines in invention potential. This was true in the 1970s as well. It is just that the declines are outweighing the increases. One need only look at detailed patenting data by subclass to note cycles. Patenting activity may be sporadic for a period, then increase to a peak, and then decline. Of the patent subclasses utilized in the U.S. Patent Office today, the majority are considered "dead art" (i.e., patenting activity has ceased).

The natural model underlying these data is a search model in which a pool of potential invention is determined by existing technical and scientific knowledge. The pool is depleted by inventive activity and recharged in various ways. Other related inventive activity can recharge pools through disclosure effects. More basic scientific and technical research can produce findings which recharge as well.[16]

5.4 Implications for Studies of Technology

Most studies of the economic determinants of R & D spending by firms or of the economic outcomes attendant to that spending have not taken trade effects into account. Many studies have implicitly, if not explicitly, supposed that firms do not have the option to purchase new technology directly, except in "embodied" form in capital goods. In addition, many studies presume that the probability of discovery from a firm's R & D is constant over significant periods of time. Most studies recognize that industry-specific effects may be present in this probability, but few make any attempt to take into account the degree of adaptability of the R & D and its dependence on discoveries made by other firms, including international firms.

The data summarized in this paper, as well as the evidence on "overseas R & D" undertaken by U.S. multinational firms, suggest that for many problems the international dimension cannot be easily set aside.[17] Many firms have international R & D strategies with laboratories located in different markets and economic environments. Virtually all R & D activities have some elements of adaptiveness, and the probability of

16. Kislev and Evenson (1975) apply a simple search model to R & D processes. Such models require some enrichment but may be a useful starting point for further study.
17. See the paper by Edwin Mansfield in this volume.

discovery will depend on other firms making closely related discoveries. Studies based on a sample of only large firms in one industry (or industries), such as many in this volume, do not provide a realistic picture of industry equilibria regarding R & D strategy. Even in these large firms the variation in R & D spending has been noted to be much higher than for normal factors of production.[18] Had the entire industry been sampled, we would find that some firms in some industries engage in practically no formal R & D. Yet they exist in competitive equilibrium with other firms engaging in significant levels of R & D.

International data show this pattern of high variation in formal spending on R & D across countries. They also show that the degree of adaptiveness of R & D is highly correlated with the level of R & D spending. A further piece of evidence suggesting that significant trade in technology takes place is that many patented inventions are granted to individuals either not associated with firms or associated with very small firms. This would not occur to a very significant degree if it were not possible to sell invented technology in forms disembodied from a product.

The patent system is often seen primarily as a means for a firm to prevent infringement on the technology it has discovered and is using in production. A well-functioning patent system has two further important aspects: First, patent systems enable the exchange of technology by providing the basis for legal transactions. Second, patent systems require an "enabling disclosure" which legal scholars regard to be of great importance. Removing invention from secrecy is considered the main social benefit offsetting the cost of the limited monopoly granted. (Economists tend to stress incentives for invention as the major benefits.) We observe that when patent systems are functioning efficiently (i.e., the cost of obtaining and enforcing patent production is low) it encourages technology trade. When patent systems are not efficient, technology trade becomes closely integrated with product trade.

For certain types of studies, the fact that technology can be purchased and sold, that R & D activities can vary in adaptiveness, and that R & D productivity may be influenced by other firms' discoveries requires that we develop better "price" data for technology. Alternative types of technology acquisition activities also must be better specified than at present. We should, for example, be measuring a firm's investment in pioneering R & D, adaptive R & D, licensing and royalty payments, search costs for new designs, etc., if we are to understand fully the firm's investment motives. We should define more meaningful price variables facing the firm. Technology embodied in capital goods (or technical services) supplied by other firms is available at a real price and is a

18. Pakes and Schankerman (this volume).

substitute for some types of R & D activity and is quite possibly complementary to highly adaptive R & D. Other technology can be licensed for a price. A firm's own technical capacity will affect the price it pays. The supply side of these technology markets changes with new discoveries. Obviously, defining proxies for these prices will require a good deal of imagination and probably a few good case studies. A few studies are already showing progress on this score, however.[19]

The issue of changing invention productivity is of obvious importance, independent of our interest in investigating firm behavior more clearly. If invention potential pools are being depleted more rapidly than they are being recharged, economic growth will suffer. If this depletion-recharge process differs significantly by industry and by economic environment, it has important implications for comparative advantage and incomes associated with it. The data reviewed in this paper are in many ways too aggregative to investigate adequately the depletion-recharge issue. They strongly suggest that the United States and a few other developed economies may have experienced a fairly broad scale net depletion of invention potential pools. Further, the international patterns of comparative advantage appear to have changed markedly in recent years. The two phenomena are related and their net effect on the U.S. economy in the past fifteen years may have been quite significant. It is not unreasonable to suppose that the *potential* economic growth of the economy (setting aside macropolicy issues) may well have been considerably lower since 1965 or so than in the preceding fifteen to twenty years. It is also not unreasonable to suppose that some loss of international comparative advantage rents has been sustained by the economy.

As economists investigate this issue, policy attention will focus on the recharge mechanism. Progress toward measuring the effectiveness of alternative recharge strategies (investment in scientific research, etc.), however, will depend on our ability to specify the depletion mechanisms (i.e., the invention process). Patent data are now becoming available in more detailed form (IPC classes) and for more countries.[20] They provide scope for both firm level and more aggregate trade-type studies. Application of trade theory to the issue should help sort out relevant issues.

A final point can be made regarding patent system policy. International organizations have pressed strongly for the establishment of international agreements on intellectual property. These agreements are designed in part to achieve standardization of legal system treatment of intellectual

---

19. Zvi Griliches (1979) discusses a number of the relevant issues. Some of the papers in this volume, notably Mansfield and Ben-Zion, reflect concern for these points.

20. The richest data set is that provided by INPADOC (1981). Patents by IPC for some fifty countries are now available for recent years. One can trace families of patents (i.e., the same patent granted in a number of different countries), firm assignments, and data on renewals.

property rights and to lower the costs of intercountry recognition of these rights. Implicitly, these international conventions seek to provide global (or as much of the globe as possible) property rights to inventors in a particular country. This may be a perfectly reasonable trade agreement between certain countries (e.g., EEC countries). We have observed in this paper, however, that trade in intellectual property is a very unequal trade, with developing countries having a strong competitive disadvantage in supplying intellectual property to developed country markets. Their inventors do not have the economic laboratories and other resources to enable them to be competitive.

Ironically, nations do not recognize global property rights in nonintellectual property and regularly intervene in commodity and capital trade markets to achieve nationalistic goals. With few exceptions, these same nations have joined international conventions freely granting intellectual property rights to citizens of other countries. By doing so they have gained some advantages in bargaining with multinational firms and in some forms of technology purchase. But unless the cost of "pirating" inventions is very high they have paid more than necessary for technology purchased from abroad.

However, the most serious impact of membership in international conventions may well be that it restricts the flexibility of many countries to design legal systems tailored to their comparative advantage, particularly regarding adaptive invention and the encouragement of indigenous secondary technology core development. Petty patent systems appear to be one alternative; there are probably others.

References

Griliches, Zvi. 1979. Issues in assessing the contribution of research and development to productivity growth. *Bell Journal of Economics* 10, no. 1:92–116.

International Patent Documentation Center (INPADOC). 1981. *International patent series*. Vienna, Austria: INPADOC.

Kislev, Y., and R. E. Evenson. 1975. *Agricultural research and productivity*. New Haven, Conn.: Yale University Press.

Kravis, I., L. Summers, and A. Heston. 1980. International comparison of real product and its composition: 1950–77. *Review of Income and Wealth*, series 26, no. 1 (March).

Mansfield, E. 1980. Basic research and productivity increase in manufacturing. *American Economic Review* 70, no. 5 (December):863–73.

National Science Board. 1981. *Science indicators 1980*. Washington, D.C.: National Science Foundation.

Statistical yearbook of Japan. Tokyo: Ministry of Trade and Industry.
United Nations economic and social council yearbook. 1980. Geneva,
    Switzerland: UNESCO.
World Bank. 1980. *World development report 1980.* Washington, D.C.:
    The World Bank.
Wright, Brian, and R. E. Evenson. 1980. An evaluation of methods for
    examining the quality of agricultural research. Economic Growth Cen-
    ter, Yale University. Mimeo.

Comment   Frederic M. Scherer

Professor Evenson has brought together a fascinating set of data on
patents and scientific and engineering employment in a broad cross
section of nations.

His most interesting finding is the apparent decline, with the notable
exceptions of Japan and West Germany, in patenting per scientist and
engineer between the late 1960s and 1970s. This supposedly occurred
because the pool of inventive possibilities became depleted. I find his
explanation consistent with my own qualitative observations on what
happened in a number of industries that experienced technological
booms and then entered a period of apparent maturity (Scherer 1978).
Nevertheless, some critical questions must be raised about what the data
mean.

If indeed the pool of inventive possibilities became "fished out," one
might suppose there would be increased effort to catch the same fish.
Given the way the patent system works, this might show up in an increased incidence of "interferences"—that is, two or more parties claim-
ing to have made the same invention. Changes in the rules governing
interference proceedings under the U.S. patent system permit a comparable time series dating back only to 1966. This, however, spans the
period analyzed by Evenson. The evidence runs contrary to the increasing interference hypothesis. From 1966 to 1970, five interferences were
declared per thousand invention patent applications. (Interferences were
lagged a year behind applications to reflect examination delays.) The
incidence of interferences *fell* during the next five years, and by 1976–80
the average number of interferences was only three per thousand applications. It would be interesting to know whether similar patterns, seemingly
at odds with the depletion theory, are observed in other industrialized
nations.

Frederic M. Scherer is a professor in the Department of Economics at Swarthmore
College.

An alternative possibility is that the propensity to patent an invention of given quality fell during the 1970s. This might be so for several reasons. A number of studies (e.g., by Taylor and Silberston, [1973], Mansfield, Schwartz, and Wagner [1981] and myself [1977]) have shown that in many industries patent protection is simply not a very important component of the incentive structure underlying R & D investment. Perhaps business decision makers came to recognize this and cut back their support of patent processing activities. The relative importance of patent protection may also have declined as marketing methods became more sophisticated and "first in" image advantages served to blunt the threat of imitative R & D. Still any "declining propensity to patent" hypothesis must also come to grips with the exceptional cases, i.e., Japan and West Germany. A long-time U.S. Patent Office official insisted in 1980 that Japanese corporations were much more aggressive in seeking patent rights than their U.S. counterparts. That West Germans place special stress on the patent system's role is suggested by the existence of a large Max-Planck-Institute in Munich studying patent matters—an operation that, to the best of my knowledge, has no peer elsewhere in the world. Yet all this is highly speculative. One would like to have firmer evidence, if possible, on whether systematic changes have occurred over time in national propensities to patent.

Before we can even begin to make such advances, we must have a clearer idea of what it is patents protect and what kinds of activities give rise to patentable "inventions." Evenson argues that applied research, as contrasted to "adaptive" research or the testing and pilot plant work of development, is the most important source of invention patents. I question whether this is so. My analysis of the Xerox Corporation's patent portfolio indicated, for example, that the vast majority of Xerox patents stemmed from activities that were clearly developmental in nature (Scherer 1977, p. 9). And as the bleary-eyed reviewer of some 15,000 patent abstracts in connection with research described elsewhere in this volume, I was struck by how narrowly incremental (adaptive?) most "inventions" are. To the extent that I am right, it becomes less clear why developing nations largely confined to adaptive work generate so few patents. Perhaps what matters is not whether inventive activity is adaptive, but whether the adaptation occurs near or far behind the frontier of what has previously been accomplished. Much of the inventive activity in developing nations may well be obvious to one having ordinary skill in the art, as U.S. patentability precedents put it. The real problem could be that few people are skilled in the art. Further research on what activities yield patentable invention (e.g., extending some of the investigations described in Professor Mansfield's paper in this volume) would be desirable.

Whether or not recent patenting trends reveal significant depletion of inventive possibilities, it seems probable to me that sooner or later such a depletion effect must be observed. Following Terleckyj (1974) and others, a standard approach in studying the contibution of R & D to productivity growth is to estimate the following cross-sectional relationship based upon production function theory:

$$(1) \qquad \text{TFP} = \frac{\dot{Q}}{Q} - \alpha\frac{\dot{K}}{K} - \beta\frac{\dot{L}}{L} = \lambda + \frac{\partial Q}{\partial R}\frac{\dot{R}}{Q},$$

where all variables are in logarithms, dots indicate time derivatives, λ is an exogenous shift parameter, Q is output, R is the stock of knowledge acquired through research and development, and the other variables are as conventionally interpreted. The last term is of central interest. $\partial Q/\partial R$ is the marginal product of additions to the R & D knowledge stock. $\dot{R}$ is the annual increase in the knowledge stock; assuming no depreciation, it is simply the annual *rate* of R & D spending. Between 1953 and 1969, real R & D effort in the United States was growing at about 6 percent per year. If one can believe the more heroic extrapolations of de Solla Price (1963, chap. 1), scientific and engineering activity has been growing at this rate for two-and-a-half centuries. Meanwhile, output per hour of work (the left-hand side of equation (1) without the $\dot{K}/K$ term) has been growing at an average annual rate of about 2 percent over the past century—the longest period for which we have tolerable data. The question arises: Is there some law of nature requiring that we increase scientific and engineering activity (undepreciated $\dot{R}$) by 6 or so percent per year to sustain labor productivity growth of 2 percent per year? That is what equation (1) and the estimated component magnitudes imply. If they are correct, we are heading for trouble, for as de Solla Price observed (1963, p. 19):

> It is clear that we cannot go up another two orders of magnitude as we have climbed the last five. If we did, we should have two scientists for every man, woman, child, and dog in the population. . . . Scientific doomsday is therefore less than a century distant.

The main way out seems to be to increase the marginal product of R & D or, given the market's tendency to equalize yields at the margin, its inframarginal counterpart (i.e., the average productivity of R & D). How can that be done? One possibility that may be less far-out than one might suppose is to utilize rapidly evolving knowledge on the biochemistry of mental processes to enhance scientific creativity. Another nearer the province of economic analysis, and consistent with some of Mansfield's recent findings, may be to secure a better allocation of resources between basic research, with its substantial externalities, and applications-oriented R & D. Or third, R & D resource allocation might be improved by effecting a better international division of labor.

Evenson's patent data provide some indication that such a division of labor is beginning to emerge. His impressions are consistent with the results of a more elaborate analysis by Slama (1981) of the Osteuropa Institute in Munich. Slama fitted a gravitational regression model to both patent and conventional trade flows data for a sizeable cross section of nations. He discovered that intercountry patenting and trade flows were greater, the larger the nations in a pair were and the smaller the geographic distance separating them. The distance slope coefficients were larger for conventional trade than for patents, the latter being much less costly than goods to "transport." Like Evenson, Slama found a significantly higher propensity to trade within the Soviet bloc planned economies than between East and West. Thus, we are beginning to find out something about the international division of labor in advancing technology. However, much more remains to be learned, especially about how market processes encourage or discourage the division of scientific and engineering labor and how much untapped potential for improving the productivity of R & D that division of labor offers.

References

de Solla Price, Derek J. 1963. *Little science, big science.* New York: Columbia University Press.

Mansfield, Edwin, M. Schwartz, and S. Wagner. 1981. Imitation costs and patents: An empirical study. *Economic Journal* 91:907–18

Scherer, F. M. 1977. The economic effects of compulsory patent licensing. Monograph Series in Finance and Eonomics, no. 1977–2. New York University Graduate School of Business Administration, Center for the Study of Financial Institutions.

———. 1978. Technological maturity and waning economic growth. *Arts & Sciences* 1 (Fall): 7–11. Evanston, Ill.: Northwestern University.

Slama, Jiri. 1981. Analysis by means of a gravitation model of international flows of patent applications in the period 1967–1978. *World Patent Information* 3 (January): 2–8.

Taylor, C. T., and Z. A. Silberston. 1973. *The economic impact of the patent system.* Cambridge: Cambridge University Press.

Terleckyj, Nestor. 1974. *Effects of R & D on the productivity growth of industries: An exploratory study.* Washington, D.C.: National Planning Association.

6   R & D and Innovation: Some Empirical Findings

Edwin Mansfield

6.1   Introduction

Until about twenty years ago, economists neglected the study of technological change, with adverse effects on both the quality and usefulness of economic analysis. During the past twenty years, a substantial corpus of knowledge has been developed in this area, and much of it is being used by policymakers in both the public and private sectors. Despite the advances that have been made, the gaps in our knowledge are great. The economics of technological change, while healthy and growing, is still at the stage where many of the basic facts, and theories are missing.

In the past two years, I have been engaged in a number of interrelated studies of R & D, innovation, and technological change. These studies have been concerned with a variety of topics, ranging from the composition of R & D expenditures to international technology transfer, from price indexes for R & D inputs to the effects of government R & D on private R & D. At this point, many of these studies have reached the point where some of the major findings are in hand, even though much more remains to be done before our understanding of the relevant topics is reasonably satisfactory.

The purpose of this paper is to bring together and discuss some of the empirical findings that have emerged. To keep the paper to a reasonable length, I shall have to be very selective and brief. Only a few findings of

Edwin Mansfield is a professor in the Department of Economics, University of Pennsylvania.

The research on which this paper is based was supported by grants from the National Science Foundation, which, of course, is not responsible for the views expressed here. The author is grateful to the more than one hundred firms that provided the data on which the work is based.

each study can be presented. In a sense, this paper provides a partial and preliminary overview of some of the recent work I have been doing in this area. Since the various studies are interrelated in many ways, such an overview should be useful.

6.2   Composition of R & D: Effects and Determinants

To begin with, let's consider the composition of R & D expenditures. In my opinion, economists have devoted too little attention to this topic. For both analytical and policy purposes, the total R & D figures are hard to interpret because they include such a heterogeneous mixture of activities. Basic research and applied research are mixed up with development. Long-term projects are mixed up with short-term projects. Projects aimed at small product and process improvements are mixed up with projects aimed at major new processes and products. Process R & D is mixed up with product R & D. To answer many important analytical and policy questions, it is essential to disaggregate R & D.

Unfortunately, little work has been done on this score. To help fill this gap, I have tried to (1) estimate the effects of the composition of an industry's or firm's R & D expenditures on its rate of productivity increase (when its total R & D expenditures are held constant), (2) investigate the relationship between the composition of a firm's R & D expenditures and its innovative output, as measured by the number of major innovations introduced, and (3) determine what factors are associated with the composition of a firm's R & D expenditures, with particular attention being directed at firm size and industrial concentration.[1]

At least four findings emerge from these studies. First, holding constant the amount spent on applied R & D and basic research, an industry's rate of productivity increase between 1948 and 1966 seems to have been directly related to the extent its R & D was long-term. Although the interpretation of this result is by no means clear-cut, it certainly is suggestive. As pointed out elsewhere,[2] many firms tend to concentrate on short-term, technically safe R & D projects. Particularly in recent years, some observers, including both public policymakers and top officials of the firms themselves, have begun to question the wisdom of this emphasis.

Second, when a firm's total R & D expenditures were held constant, its innovative output seemed to be directly related to the percentage of its R & D expenditures devoted to basic research. The data on which this result is based pertain to the chemical and petroleum industries, areas

1. Some results of these studies have been published in Mansfield (1980). Additional results appear in Mansfield (1981a). Link (1981) also has been investigating factors associated with the composition of R & D.
2. For recent evidence on this subject, see Mansfield (1981b).

where we have accumulated a considerable amount of data concerning the R & D and innovative activities of particular firms. It would be extremely useful if a similar investigation could be made of other industries. In view of the roughness of both the data and the analysis, this finding should be viewed as preliminary and tentative. In particular, it is hard to tell whether basic research is the relevant variable, or whether it is a surrogate for something else.

Third, based on data obtained from 108 firms that account for about one-half of all industrial R & D expenditures in the United States, the composition of a firm's R & D expenditures appears to be related to the firm's size. But the relationship is not as simple as one might think. Whereas the largest firms seem to carry out a disproportionately large share of the basic research (and perhaps the long-term R & D) in most industries, they do not tend consistently to carry out a disproportionately large share of the relatively risky R & D or the R & D aimed at entirely new products and processes. Instead, they generally seem to carry out a disproportionately small share of the R & D aimed at entirely new products and processes. These results are not contradictory. Basic research is by no means the same thing as R & D aimed at entirely new products and processes. Also, since both basic research and applied R & D can be relatively risky, the riskiness of a firm's R & D need not be closely correlated with the percentage of its R & D devoted to basic research.

Fourth, the more concentrated industries in our sample seem to devote a smaller, not larger, percentage of R & D expenditures to basic research. This relationship is statistically significant, but not very strong ($r^2 = .46$). Relatively concentrated industries also tend to devote a relatively small, not large, proportion of their R & D expenditures to long-term projects and to projects aimed at entirely new products and processes, but the correlation (in each case r^2 is about .09) is far from statistically significant. A positive correlation does exist ($r^2 = .15$) between an industry's concentration level and the proportion of its R & D expenditures going for relatively risky projects, but this correlation too is far from significant.

6.3   Price Indexes for R & D Inputs

Not only is relatively little known about the composition of R & D expenditures, but equally important, the available data concerning real R & D expenditures are bedeviled by the lack of a suitable price index for R & D inputs. In view of the inherent difficulties and the strong assumptions underlying the few alternative measures that have been proposed, the official government R & D statistics use the GNP deflator to deflate R & D expenditures. Many observers inside and outside the government are uncomfortable with this procedure, but little is known about the size or direction of the errors it introduces.

To help fill this gap, we constructed price indexes for R & D inputs and for inputs used in other stages of the innovative process. Detailed data were obtained from thirty-two firms in the following eight industries: chemicals; petroleum; electrical equipment; primary metals; fabricated metal products; rubber; stone, clay, and glass; and textiles. These industries account for about half of the company-financed R & D in the United States. Although our sample contains both large and small firms, it includes a substantial proportion of the R & D carried out in these industries. Indeed, the firms in our sample account for about one-ninth of all company-financed R & D in the United States.[3]

At least four findings stem from this study. First, for these industries as a whole, the Laspeyres price index for R & D inputs indicates that the price of such inputs was about 98 percent higher in 1979 than in 1969. However, the rate of inflation in R & D seems to have been higher in some industries than in others. In particular, the rate of inflation seems to have been highest in fabricated metal products, chemicals, and petroleum, and lowest in electrical equipment.

Second, turning to the innovation process as a whole, Laspeyres price indexes indicate that the price of inputs into all stages of the innovative process was about 101 percent higher in 1979 than in 1969. Thus, the rate of inflation for inputs into all stages of the innovation process seems to have been somewhat higher than for R & D alone. As in the case of R & D, the rate of inflation for inputs into all stages of the innovation process seemed to be highest in fabricated metal products, chemicals, and petroleum, and lowest in electrical equipment.

Third, if we assume that the production function for R & D in each industry is Cobb-Douglas (with constant returns to scale), an exact price index for each industry is

$$(1) \qquad I = \prod_{i=1}^{n} \left(\frac{P_{1i}}{P_{0i}} \right)^{\alpha_i} \times 100,$$

where the price of the ith input in 1979 is P_{1i}, its price in 1969 is P_{0i}, α_i is the proportion of R & D cost devoted to the ith input, and n is the number of inputs.[4] Even though there is little or no information concerning the nature of the production function for R & D, it is interesting to compare the resulting indexes with the Laspeyres indexes because, since Laspeyres indexes ignore substitution effects, they may exaggerate price increases. Table 6.1 shows the results for each industry. As you can see,

---

3. This work is being done with Anthony Romeo and Lorne Switzer. For a preliminary account of some of our findings, see Mansfield, Romeo, and Switzer (1983). For some previous work, see Goldberg (1978) and Jaffe (1972).

4. For a proof of this, see Mansfield, Romeo, and Switzer (1983).

Table 6.1   Price Indexes for R & D Inputs and for Inputs in the Innovative
                   Process, Eight Industries, 1979 (1969 = 100)[a]

| | Laspeyres Index | | |
Industry	R & D	Innovation Process	Cobb-Douglas R & D
Chemicals	222	223	217
Petroleum	222	228	218
Electrical equipment	183	186	190
Primary metals	205	210	205
Fabricated metal products	248	275	222
Rubber	209	200	206
Stone, clay, and glass	205	195	183
Textiles	200	220	220
Mean[b]	198	201	200

Source: See section 6.3.

[a]The three columns are not entirely comparable because some firms could be included in some columns but not others because of lack of data. For the innovation process, some figures have been rounded to the nearest 5 or 0 to indicate their roughness.

[b]Each industry's price index is weighted by its 1969 R & D expenditure.

those based on the Cobb-Douglas assumption are generally quite similar to those based on the Laspeyres indexes. On the average, the Cobb-Douglas indexes indicate that the price of R & D inputs was about 100 percent higher in 1979 than in 1969.

Fourth, in practically all of the industries included here, the rate of increase of the price index for R & D inputs exceeded the rate of increase of the GNP deflator. Because of the inadequacies of the GNP deflator for this purpose, the official U.S. statistics concerning deflated R & D expenditures seem to overestimate the increase during 1969–79 in industrial R & D performance. For these industries as a whole, deflated R & D expenditures increased by about 7 percent (during the period, not annually) based on the GNP deflator, but only by less than 1 percent based on our price indexes for R & D inputs. Taken at face value, this seems to indicate that the bulk of the apparent increase in real R & D in these industries was due to the inadequacies of the GNP deflator.

6.4   Effects of Federal Support on Privately Financed R & D

Just as the lack of R & D price indexes has long been recognized, the need for more information on the effects of government R & D on private R & D has also long been known. This area has been the subject of considerable controversy. Some economists argue that increases in government R & D funding are likely to reduce expenditures of the private sector because (among other reasons) firms may receive government support for some projects they would otherwise finance themselves. Other economists say that government R & D is complementary to private R & D, that increases in the former stimulate increases in the latter. This question is of great importance for both policy and analysis, but little is known about it.

To study the effects of federal support on privately financed R & D in the important area of energy, we chose a sample of twenty-five major firms in the chemical, oil, electrical equipment, and primary metals industries. Together they carry over 40 percent of all R & D in these industries. To estimate the extent to which these firms obtained government funding for energy R & D projects that they would have carried out in any event with their own funds, we obtained detailed data from each of the firms. Moreover, even more detailed data were obtained concerning a sample of forty-one individual federally funded energy R & D projects. These projects account for over 1 percent of all federally supported energy R & D performed by industries.[5]

The following are four of the conclusions stemming from this study. First, these firms apparently would have financed only a relatively small proportion of the energy R & D that they performed with government support. Based on our sample of firms, they would have financed only about 3 percent on their own. Based on our sample of individual projects, they would have financed about 20 percent. It would be useful if similar estimates could be obtained for various kinds of R & D outside the field of energy.

Second, if a 10 percent increase were to occur in federal funding for their energy R & D in 1979, the response (for all twenty-five firms taken as a whole) would be that, for each dollar increase in federal support, they would increase their own support of energy R & D by about six cents per year for the first two years after the increase in federal funds. In the third year after the increase, there would be no effect at all. This finding is based on careful estimates by senior R & D officials of each firm. However, substantial differences exist among the firms' responses. These results are quite consistent with those obtained by Levin (1981) and Terleckyj and Levy (1981) in their econometric studies of the aggregate relationship between federally funded R & D expenditures and privately funded R & D expenditures.

Third, if a 10 percent cut were to occur in federal funding for their energy R & D in 1979, the response (for all twenty-five firms taken as a whole) would be that, for each dollar cut in federal support, they would reduce their own support of energy R & D by about twenty-five cents in each of the two years following the cut. In the third year after the federal cut, they would cut about nineteen cents in their own spending. Taken at

5. This work is being conducted with Lorne Switzer.

face value, a 10 percent cut in federally funded energy R & D would apparently have a bigger effect on privately funded energy R & D than would a 10 percent increase. But until more and better data are obtained, we feel that this difference should be viewed with considerable caution.

Fourth, in modeling the effects of federally funded R & D on the economy, our results indicate that it may be more realistic to view such R & D as a factor that facilitates and expands the profitability of privately funded R & D, rather than focus solely (as most econometric studies have done) on the direct effects of federally funded R & D on the productivity of the firms and industries performing the R & D. Based on our sample of federally funded projects, such projects typically appear to make only about half as large a direct contribution to the firm's performance and productivity as would be achieved if the firm spent an equivalent amount of money on whatever R & D it chose. But in about one-third of the cases, the federally financed R & D projects suggested some further R & D into which the firm invested its own funds. (As shown in table 6.2, the likelihood of such a spin-off is enhanced if the firm helped to formulate the ideas on which the project is based, and if the project was not completely separated physically from the projects financed by the firm.)[6] If federally funded R & D is viewed in this way, econometricians may have more success in measuring its effects on productivity in the private sector.

6.5   Forecasts of Engineering Employment

Engineering manpower is one of the most important inputs required in the complex process leading to innovation and technological change. Policymakers in government, universities, and business must make decisions that depend, explicitly or implicitly, on forecasts of the number of engineers employed in various sectors of the economy at various times. For example, in evaluating the adequacy of existing engineering manpower, the National Science Foundation and the Bureau of Labor Statistics must try to forecast how many engineers will be employed in the private sector. Although such forecasts sometimes are based on a collection of forecasts made by firms of their own engineering employment, little is known about the accuracy of these forecasts.

To help fill this gap, a detailed econometric study was carried out. Data were obtained from a well-known engineering association which has collected such forecasts from firms for many years. For fifty-four firms in

---

6. Of course, we recognize the difficulty in many cases of identifying where the ideas underlying a particular project originated. But in the cases in table 6.2, this generally seemed to be a matter of agreement among all parties. Note too that, whereas the source of the project seems to have a statistically significant effect, the separation variable is not significant when both variables are included.

Table 6.2   Percentage of Federally Financed Energy R & D Projects Resulting
                   in Company-Financed R & D Done Subsequently by the Performer,
                   by Source of Idea for Project, and by Extent of Separation from
                   Company-Financed Projects, Forty Projects[a]

Characteristic of Project	Percentage
Source of idea for project:	
Firm	44
Government	15
Both firm and government	44
Separation:	
Complete	17
Not complete	38

Source: See section 6.4.

[a]One project could not be included because it was not yet clear whether it would result in company-financed R & D. The figures in this table may understate the true percentages because they pertain only to company-financed R & D resulting directly and almost immediately from these projects.

the aerospace, electronics, chemical, and petroleum industries, comparisons were made of each firm's forecasted engineering employment with its actual engineering employment during 1957–76. Since data were obtained for a number of forecasts of each firm, the accuracy of 218 such forecasts could be evaluated.[7]

At least three conclusions stem from this study. First, there appear to have been substantial differences among industries in the accuracy of the forecasts. As shown in table 6.3, the forecasting errors for individual firms in the areospace industry were much greater than in the electronics, chemical, or petroleum industries. (In chemicals and petroleum, firms' two-year forecasts were off, on the average, only by about 5 percent.) The relatively large forecasting errors in the aerospace industry may have been caused by its heavy dependence on government defense and space programs which were volatile and hard to predict.

Second, although the forecasting errors for individual firms were substantial, they tend to be smaller when we consider the total engineering employment for all firms in the sample. On the average, the six-month forecasts were in error by about 2 percent, the two-year forecasts were in error by about 1 percent, and the five-year forecasts were in error by about 3 percent. The fact that little bias was present in the forecasts is encouraging since, for many purposes, the central aim is to forecast total engineering employment in an entire sector of the economy, not the engineering employment of a particular firm.

Third, the firms' forecasts may be improved if a simple econometric model is used. Based on data from over a dozen chemical and petroleum

7. This work was done with Peter Brach. Some of the results appear in Brach and Mansfield (1982).

Table 6.3   **Frequency Distribution of Forecasts, by Ratio of Forecasted to Actual Engineering Employment, Aerospace, Electronics, Petroleum, and Chemical Industries, Six-Month and Two-Year Forecasts**[a]

Forecasted Employment ÷ Actual Employment	Aerospace	Electronics	Petroleum	Chemical
Number of 6-Month Forecasts				
0.81–0.90	0	1	0	0
0.91–1.00	8	10	12	6
1.01–1.10	7	9	13	9
1.11–1.20	0	2	0	1
1.21–1.30	2	0	0	0
1.31–1.40	2	0	0	0
Number of 2-Year Forecasts				
0.61–0.70	0	1	0	0
0.71–0.80	2	3	0	0
0.81–0.90	4	3	4	1
0.91–1.00	2	9	8	1
1.01–1.10	3	6	6	5
1.11–1.20	0	0	0	0
1.21–1.30	0	3	0	0
1.31–1.40	3	0	0	0

Source: See section 6.5.

[a]Five-year and ten-year forecasts were also included in the study but are not in this table.

firms, the proportion of the way that a firm's engineering employment moves toward the desired level is inversely related to the desired percentage increase in engineering employment[8] and is directly related to the profitability of the firm. (A similar model was used in Mansfield 1968.) Using information concerning this relationship in the past as well as the firm's desired level of engineering employment in the future, one can forecast the firm's future engineering employment. The evidence, while fragmentary and incomplete, suggests that experimentation with such an approach may be worthwhile.

6.6 International Technology Transfer

To understand a wide variety of topics, ranging from economic growth to industrial organization, economists must be concerned with international technology transfer. In my opinion, economists interested in the

8. This model assumes that desired employment exceeds actual employment, which was the typical case in these firms in the relevant time periods. Obviously, this model should be used only in cases where this assumption is true.

relationship between R & D and productivity increase have paid too little attention to this subject. In practically all econometric models designed to relate R & D to productivity increase, international technology flows are not included (explicitly at least). Yet U.S.-based firms carry out about 10 percent of their R & D overseas, and this R & D has an effect on the rate of productivity increase in the United States. In addition (and probably more important), R & D carried out by one organization in one country often has a significant effect on technological advance and productivity increase in another organization in another country. For example, productivity increase in the American chemical industry was certainly influenced by the work of Ziegler in Germany and of Natta in Italy.

To shed new light on the process of international technology transfer, we have carried out several types of studies. One study was concerned with the channels of international technology transfer and the effects of international technology transfer on U.S. R & D expenditures. Another study was concerned with the size and characteristics of overseas R & D carried out by U.S.-based firms. Still another study dealt with the transfer of technology by U.S.-based firms to their overseas subsidiaries.[9] Based on these studies, it seems that economists should reconsider some of the models that have been used most frequently to represent the process of international technology transfer.

The traditional way of viewing the process of international technology transfer has been built around the concept of the product life cycle.[10] According to the product life cycle, a fairly definite sequence exists in the relationship between technology and trade, whereby the United States tends to pioneer in the development of new products, enjoying for a time a virtual monopoly. After an innovation occurs, the innovator services foreign markets through exports, according to this model. As the technology matures and foreign markets develop, companies begin building plants overseas, and U.S. exports may be displaced by production of foreign subsidiaries. The concept of the product life cycle has had a great influence in recent decades because it has been able to explain the train of events in many industries.

At least four of our findings seem relevant in this regard. First, our data suggest that the situation may be changing, that the product life cycle may be less valid than in the past. By the mid-1970s, in the bulk of the cases we studied, the principal channel through which new technologies were exploited abroad during the first five years after their commercialization was foreign subsidiaries, not exports (see table 6.4). About 75 percent of the technologies transferred by U.S. firms to their subsidiaries in developed countries during 1969–78 were less than five years old.

9. See Mansfield, Romeo, and Wagner (1979); Mansfield, Teece, and Romeo (1979); Mansfield and Romeo (1980).
10. Vernon (1966, 1970).

Table 6.4   **Percentage Distribution of R & D Projects, by Anticipated Channel of International Technology Transfer, First Five Years after Commercialization, Twenty-Three Firms, 1974**

	Channel of Technology Transfer				
	Foreign Subsidiary	Exports	Licensing	Joint Venture	Total[a]
All R & D projects[b]	74	15	9	2	100
Projects aimed at:[c]					
Entirely new product	72	4	24	0	100
Product improvement	69	9	23	0	100
Entirely new process	17	83	0	0	100
Process improvement	45	53	2	1	100

Source: See section 6.6.

[a]Because of rounding errors, percentages mat not sum to 100.

[b]This is the mean of the percentage for sixteen industrial firms and for seven major chemical firms. The results are much the same in the two subsamples. Only projects where foreign returns were expected to be of some importance (more than 10 percent of the total for the first subsample and 25 percent of the total for the second subsample) were included.

[c]Only the chemical subsample could be included.

Based on our data, the "export stage" of the product cycle has often been truncated and sometimes eliminated. Particularly for new products, firms frequently begin overseas production within one year of first U.S. introduction. In some industries, such as pharmaceuticals, new products commonly are introduced by U.S.-based firms more quickly in foreign markets than in the United States (in part because of regulatory considerations).

Second, there seems to be a difference in this regard between products and processes. For processes, the "export stage" continues to be important (table 6.4). Firms are more hesitant to send their process technology overseas than their product technology because they feel that the diffusion of process technology, once it goes abroad, is harder to control. In their view, it is much more difficult to determine whether foreign firms are illegally imitating a process than a product.

Third, to a large extent, this change in the process of international technology transfer and trade reflects the fact that many U.S.-based (and foreign-based) firms have come to take a worldwide view of their operations. Many of them now have in place extensive overseas manufacturing facilities. As indicated above, many also have substantial R & D activities located abroad. Given the existing worldwide network of facilities and people, firms are trying to optimize their overall operations. This may mean that some of the technology developed in the United States may find its *initial* application in a Canadian subsidiary, or that an innovation developed in its Canadian subsidiary may find its *initial* application in the firm's British subsidiary, and so on.

Fourth, the product life cycle is less valid than it used to be because technology is becoming increasingly internationalized. For example, in the pharmaceutical industry it is no longer true that a new drug is discovered, tested, and commercialized all within a single country. Instead, the discovery phase often involves collaboration among laboratories and researchers located in several different countries, even when they are within the same firm. And clinical testing generally becomes a multicountry project. Even in the later phases of drug development, such as dosage formulation, work often is done in more than one country. In contrast, the product life cycle seems to assume that innovations are carried out in a single country, generally the United States, and that the technology resides exclusively within that country for a considerable period after the innovation's initial commercial introduction.[11]

6.7 "Reverse" Technology Transfer

"Reverse" technology transfer is the transfer of technology from overseas subsidiaries to their U.S. parents. Some analysts tend to dismiss technology transfer of this sort as unimportant. Yet practically nothing is known about the extent and characteristics of "reverse" technology transfer, even though such information obviously would be of relevance to public policymakers concerned with the technological and other activities of multinational firms.

To determine the extent to which overseas R & D by U.S.-based firms has resulted in technologies that have been applied in the United States, we obtained data pertaining to twenty-nine overseas R & D laboratories of U.S. firms in the chemical, petroleum, machinery, electrical equipment, instruments, glass, and rubber industries. This sample of overseas laboratories, chosen essentially at random from those of major firms in these industries in the Northeast United States, accounts for about 10 percent of all overseas R & D spending by U.S.-based firms. The industrial and geographical distribution of the sample is reasonably similar to the industrial and geographical distribution of all overseas laboratories according to the National Science Foundation and other data sources.[12]

The following four findings help to put "reverse" technology transfer into better perspective. First, over 40 percent of these laboratories' 1979 R & D expenditures resulted in technologies that were transferred to the United States. Thus, such transfer is common and by no means insignificant. However, there are vast differences among overseas laboratories in

---

11. See Mansfield et al. (1982).
12. This work is being done with Anthony Romeo.

the percentage of R & D expenditures resulting in technologies transferred to the United States. Most of this variation can be explained by three factors: (1) whether the laboratory's primary function is to produce technology for worldwide application, rather than to service or adapt technology transferred from the United States or to produce technology for foreign aplication; (2) the laboratory's total R & D expenditures; and (3) the percentage of its total R & D expenditures devoted to research rather than development.

Second, there is a very short lag (on the average) between the date when a transferred technology first is applied abroad and the date when it is first applied in the United States. Indeed, in the electrical equipment firms in our sample the average lag is negative. Because of the size and richness of the American market, firms tend to introduce new products (and processes) based on technologies developed in their overseas laboratories about as quickly in the United States as in their overseas markets. These results indicate the extent to which firms take a global view of the introduction of innovations. As pointed out in section 6.6, this is a departure from the situation years ago.

Third, based on our data, more recently developed technology tends to be transferred more quickly to the United States than technology developed years ago. Also, technologies yielding relatively large profit in the United States were transferred more quickly than those that were less profitable here.

Fourth, although much of the R & D carried out overseas is directed at the adaptation and improvement of existing technology, overseas R & D laboratories have generated technology that was the basis for new products and other innovations that contributed billions of dollars in profits to U.S. manufacturing firms in 1980, if the laboratories in our sample are representative in this respect.

6.8   Overseas R & D and Productivity Growth of U.S. Firms

As pointed out in section 6.7, "reverse" technology transfer is not included (at least explicitly) in existing models of R & D and productivity growth. Indeed, because the official R & D statistics have excluded U.S. firms' overseas R & D expenditures until recently, previous studies of the relationship between a firm's or industry's R & D expenditure and its rate of productivity increase have ignored overseas R & D. Obviously, it would be interesting and useful to include U.S. firms' overseas R & D in such models and to see how much effect it has on the productivity growth of these firms.

To do this, it is convenient to use essentially the same model as that employed by Mansfield (1968, 1980), Griliches (1980), and Terleckyj (1974), except that research and development is disaggregated into two parts: domestic R & D and overseas R & D. In a particular firm, the production function is assumed to be:

$$(2) \qquad Q = Ae^{\lambda t}R_d{}^{\beta_1}R_o{}^{\beta_2}L^{\nu}K^{1-\nu},$$

where Q is the firm's value added, R_d is the firm's stock of domestic R & D capital, R_o is its stock of overseas R & D capital, L is its labor input, and K is its stock of physical capital. Thus, the annual rate of change of total factor productivity is

$$(3) \qquad \rho = \lambda + \theta_1\frac{dR_d/dt}{Q} + \theta_2\frac{dR_o/dt}{Q},$$

where $\theta_1 = \delta Q/\delta R_d$, and $\theta_2 = \delta Q/\delta R_o$. And based on the usual assumptions,[13]

$$(4) \qquad \rho = \lambda + a_1\frac{X_d}{Q} + a_2\frac{X_o}{Q},$$

where X_d is the firm's domestic R & D expenditures, and X_o is its overseas R & D expenditures in the relevant year.

My econometric results pertain to fifteen chemical and petroleum firms, for which I have estimated ρ for 1960–76 (see Mansfield 1980). For each of these firms I obtained data concerning X_d/Q and X_o/Q. The results are shown in table 6.5.[14] Estimates of a_1 and a_2 could be obtained by least squares,[15] the results being

$$(5) \qquad \rho = 0.022 + 0.19\,X_d/Q + 1.94\,X_o/Q\cdot$$
$$\qquad (7.40)\ (2.44) \qquad (1.90)$$

These results have at least two implications. First, they indicate that overseas R & D, as well as domestic R & D, contributes to productivity growth of U.S. firms. The estimate of a_2 is positive and statistically significant. More surprisingly, the estimate of a_2 is much larger than that of a_1, indicating that a dollar's worth of overseas R & D had much more effect on productivity increase than a dollar's worth of domestic R & D. But this difference is not statistically significant. For most firms, I doubt that a_2 is this much larger than a_1, based on our other studies. But be this as it may, equation (5) certainly is consistent with our findings in section 6.7 concerning the nontrivial nature of "reverse" technology transfer.

13. These assumptions are described in detail in Mansfield (1980).

14. One firm included in Mansfield (1980) could not be included here because it is part of a foreign-based multinational firm. The data concerning X_d/Q and X_o/Q were obtained from the firms.

15. Tests were carried out to determine whether an industry dummy variable should be included in equation (5). The results provide no statistically significant evidence that this should be done.

Table 6.5   Values of X_d/Q, and X_o/Q, Fifteen Chemical and Petroleum Firms[a]

Firm	$\dfrac{X_d}{Q}$	$\dfrac{X_o}{Q}$
1	.0500	0
2	.0890	.0043
3	.0715	0
4	.0610	.0024
5	.0770	0
6	.0820	.0091
7	.0101	0
8	.0061	.0003
9	.0072	.0001
10	.0068	0
11	.0114	0
12	.0118	.0001
13	.0073	0
14	.0087	.0020
15	.0147	0

Source: See section 6.8.

[a]The data concerning X_d/Q and X_o/Q pertain to a year in the mid-1960s (1963–65). It was not possible to get data for precisely the same year, but the results should be sufficiently comparable for present purposes.

Second, these results allow a first glimpse of the nature of the bias that may have resulted from the omission of overseas R & D expenditures in some past studies. If X_o/Q had been omitted from equation (5), the result would have been

(6) $$\rho = 0.022 + 0.28 X_d/Q.$$
                         (6.61)   (3.95)

Thus, a_1 would have been higher than if both overseas and domestic R & D were included. In cases where Xo/Q has been positively correlated with Xd/Q, as in the present instance, the rate of return from domestic R & D may have been overestimated in previous studies, since a_1 has often been interpreted as such a rate of return.

6.9   Imitation Costs, Patents, and Market Structure

In the previous three sections we have been concerned with the transfer of technology from one nation to another, where the transferor and transferee are often parts of the same firm. Now let's return to technology transfer within the same nation, where the transferor and transferee are different firms, and where the transfer is involuntary from the point of view of the transferor. In particular, suppose that one firm imitates (legally) another firm's innovation. How much does it cost? How long does it take? How often does it occur? Economists have long recognized the importance of these questions. For example, they frequently have pointed out that, if firms can imitate an innovation at a cost that is substantially below the cost of developing an innovation, they may have little or no incentive to be innovative. Yet no attempts have been made to measure imitation costs, to test various hypotheses concerning the factors influencing those costs, or to estimate their effects.

To help fill this important gap, we obtained data from firms in the chemical, drug, electronics, and machinery industries concerning the cost and time of imitating (legally) forty-eight product innovations.[16] Imitation cost is defined to include all costs of developing and introducing the imitative product, including applied research, product specification, pilot plant or prototype construction, investment in plant and equipment, and manufacturing and marketing start-up. (If there was a patent on the innovation, the cost of inventing around it is included.) Imitation time is defined as the length of time elapsing from the beginning of the imitator's applied research (if there was any) on the imitative product to the date of its commercial introduction.

For present purposes, four findings of this study are of particular interest. First, innovators routinely introduce new products even though other firms can imitate these products at about two-thirds (often less) of the cost and time expended by the innovator. In our sample, imitation cost averages about 65 percent of innovation cost, and imitation time averages about 70 percent of innovation time. Considerable variation exists among products in the ratio of imitation cost to innovation cost. Much of this variation can be explained by differences in the proportion of innovation costs going for research, by whether an innovation was a drug subject to FDA regulations, and by whether an innovation consists of a new use for an existing material that is patented by another firm.

Second, the magnitude of imitation costs in a particular industry seems to have a considerable impact on the industry's market structure. How rapidly a particular innovation is imitated depends on the ratio of imitation cost to innovation cost. Also, an industry's concentration level tends to be low if its members' products and processes can be imitated easily and cheaply. The latter relationship is surprisingly close. Apparently, differences among industries in the technology transfer process (including transfers that are both voluntary and involuntary from the point of view of the innovator) may be able to explain much more of the interindustry variation in concentration levels than is generally assumed.

Third, in most cases, patents seem to have only a modest effect on imitation costs, as shown in table 6.6. However, in the drug industry, patents seem to have a bigger impact than in other industries. According to the firms, about one-half of the patented innovations in our sample

16. This work was done with Mark Schwartz and Samuel Wagner. Some of the results appear in Mansfield, Schwartz, and Wagner (1981).

Table 6.6   **Estimated Percentage Increase in Imitation Cost Due to Patents, Thirty-three New Products, Chemical, Drug, Electronics, and Machinery Industries**[a]

Percent Increase in Imitation Cost	Number of Products
Under 10	13
10–19	10
20–49	4
50–99	0
100–199	3
200 and over	3
Total	33

Source: See section 6.9.

[a]Not all innovations in our sample are included here because not all were patented or patentable.

would not have been introduced without patent protection. But the bulk of these innovations occurred in the drug industry. Excluding the drug industry, the lack of patent protection would have affected less than one-fourth of the patented innovations in the sample.

Fourth, patented innovations seem to be imitated surprisingly often and quickly. In our sample, about 60 percent were imitated within four years of their initial introduction. Reality seems to depart sharply from the commonly held belief that a patent holder is free from imitation for the life of the patent. In my view, it is very important that this fact be taken into account by the excellent economic theorists working in this area, since sometimes models of the innovation process tend to assume that the innovator receives all of the benefits from an innovation and that imitation can be ignored.

6.10   Innovation and Market Structure

In recent years, economic theorists have also begun to focus on the effects of innovation on market structure. Of course, technological change has long been recognized as one of the major forces influencing an industry's market structure. Karl Marx stressed this fact over a century ago. But the renewed interest is welcome, since traditional models of the relationship between innovation and market structure have been deficient in many respects.

Unfortunately, empirical findings on this score have also been relatively scanty. Little is known about the effects of recent major process innovations in various industries on the minimum efficient scale of plant. Almost nothing is known about the effects of recent major product innovations in various industries on the extent of concentration. To help fill this gap, I obtained information from twenty-four firms in the chemical, petroleum, steel, and drug industries about the effects of over sixty-five process and product innovations that were introduced in the past half-century.[17]

Although this study is still in a relatively early phase, several findings are emerging. First, in the chemical and petroleum industries, the bulk of the process innovations resulted in increases in miminum efficient scale of plant. In steel, only about half of the process innovations resulted in such increases, but most of the rest had little or no effect on minimum efficient scale. Thus, in all three industries,[18] scale-increasing innovations far outnumbered scale-decreasing innovations.[19] However, although relatively few major innovations in these industries have reduced minimum efficient scale, a substantial proportion have had no appreciable effect on it.

Second, the evidence of these industries does not support Blair's (1972) well-known hypothesis that, since World War II, fewer innovations tend to increase minimum efficient scale than in the past. To test this hypothesis, I compared the proportion of process innovations introduced after 1950 that resulted in such an increase with the proportion introduced before or during 1950 that did so. Contrary to Blair's hypothesis, the proportion was higher, not lower, in the later period.

Third, in all four industries combined, less than half of the product innovations in the sample seemed to increase the four-firm concentration ratio. The percentage was particularly low in drugs. The fact that only a minority of these major new products increased concentration in these industries is noteworthy, given the common tendency among economists to view technological change as a concentration-increasing force. If these industries are at all representative (and if this preliminary result holds up in my subsequent work), there should probably be more emphasis on innovation's role in reducing and limiting existing concentration.[20]

6.11   Conclusions

The findings presented here have a number of implications for public policy. With respect to government R & D policy, they suggest the following: (1) In their attempts to increase productivity, policymakers should recognize the importance of long-term R & D and basic research.

17. The lists of innovations came from Mansfield (1968), Mansfield et al. (1977), and Landau (1980).
18. The drug industry was excluded here because of its emphasis on product innovation.
19. This seems to be in accord with the observed changes in minimum efficient scale in these industries. See Scherer (1980).
20. In their paper on this subject, Nelson and Winter (1978) emphasize the concentration-increasing effects of innovation. However, they are careful to point out that their computer simulations represent a "partial view," not a "general model."

(2) Policymakers should also recognize that much of the apparent increase in real industrial R & D during 1969–79 (which was relatively modest in any event) may have been a statistical mirage, caused by the lack of better price indexes for R & D inputs. (3) Changes in federally financed R & D expenditures (at least in energy) are unlikely to be offset to any appreciable extent by changes in privately financed R & D ; on the contrary, such changes seem to induce changes in the same direction in privately financed R & D. (4)To the extent that policymakers want to increase the spillover from federally financed to privately supported R & D, the results suggest that firms should be encouraged to work with government agencies in the design of federally financed R & D projects.

With respect to patent policy, the findings seem to suggest that, except for pharmaceuticals and agricultural chemicals, patents frequently are not regarded as essential by innovators. Excluding drug innovations, more than three-fourths of the patented innovations in our sample would have been introduced without patent protection. In a minority of cases, patent protection had a very major effect on imitation costs and delayed entry significantly, but in most cases it had relatively little effect. Obviously, these findings have important implications concerning the patent system's role in stimulating technological change and innovation.

With regard to antitrust policy, our findings shed new light on the relationship between an industry's concentration level and the nature of its technological activities. Highly concentrated industries seem to devote a relatively low percentage of their R & D to basic research, and there is an inverse (but not significant) relationship between an industry's concentration ratio and the percentage of its R & D that is long-term or aimed at entirely new products and processes. Also, our results (covering the chemical, drug, petroleum, and steel industries) provide new information about the frequency with which major new products result in increases in concentration. In our sample, many new products (particularly in drugs) seem to have been introduced by firms that "invaded" the relevant market or that were not among the leaders in that market. This is not to argue that innovations do not frequently increase concentration. But it does suggest that the role of innovation in undermining existing concentration may sometimes be underestimated.

With respect to national policies concerning international technology transfer and the multinational firm, our findings underscore the extent to which technology is transferred across national boundaries, the difficulties and costs involved in trying to stem the technological outflow from U.S. firms to their foreign subsidiaries, and the benefits to the United States from the inflow of technology from these subsidiaries. "Reverse" technology flows are becoming increasingly important. Based on our econometric results, overseas R & D has a considerable effect (per dollar spent) on productivity of U.S. firms. These facts should be taken into account in the evaluation of the role of multinational firms in contributing to technological change and economic growth in the United States.

Our findings should also be of use to industrial managers. Faced with the difficult task of choosing an R & D portfolio, managers badly need evidence concerning the relationships between the composition of a firm's R & D expenditure, on the one hand, and its innovative output and rate of productivity increase, on the other. Also, they need more sophisticated and reliable indexes of the rate of inflation in R & D to budget their resources properly, and they can benefit from improved techniques for forecasting engineering employment.

Besides being of interest to policymakers, we believe that these findings have some implications for economic analysis. In my opinion, models relating R & D to productivity change should go further in disaggregating R & D, in taking account of international technology flows (and, in some cases, interindustry technology flows), and in using better R & D price indexes. For many purposes, it may also be useful to view government R & D as a factor that expands the profitability of private R & D. With regard to the role of technology in international trade, the product life cycle model should be altered or supplanted to recognize the changes that have occurred in this area. Further, students of industrial organization should devote more attention to the measurement and analysis of imitation costs (and time); this is a central concept that has been ignored entirely in econometric work.

In conclusion, the limitations of the studies described here should be noted. Although many of the samples (of firms, R & D projects, innovations, and so forth) are reasonably large, they nonetheless cover only certain industries or sectors of the economy. In many instances, the theoretical models we use are highly simplified. No pretense is made that the findings presented here are the last words on the subject. However, we believe that these findings increase our understanding of a wide variety of major topics about which relatively little (often, practically nothing) has been known.

References

Blair, John. 1972. *Economic concentration: Structure, behavior, and public policy*. New York: Harcourt Brace Jovanovich.

Brach, Peter, and Edwin Mansfield. 1982. Firms' forecasts of engineering employment. *Management Science* 28:156–60.

Goldberg, Lawrence. 1978. Federal policies affecting industrial research and development. Paper presented at the Southern Economic Association, November 9.

Griliches, Zvi. 1980. Returns to research and development expenditures in the private sector. In *New developments in productivity measurement and analysis*, ed. J. W. Kendrick and B. N. Vaccara, 419–54. Conference on Research in Income and Wealth: Studies in Income and Wealth, vol. 44. Chicago: University of Chicago Press for the National Bureau of Economic Research.

Jaffe, S. A. 1971. A price index for deflation of academic R & D expenditures. Washington, D.C.: NSF 72-310.

Landau, Ralph. 1980. Chemical industry research and development. *Innovation and U.S. research*, ed. W. N. Smith and Charles Larson. Washington, D.C.: American Chemical Society.

Levin, Richard. 1981. Toward an empirical model of schumpeterian competition. Working paper series A, no. 43. Yale School of Organization and Management.

Link, Albert. 1981. A disaggregated analysis of R and D spending. Paper presented at Middlebury College, April 17.

Mansfield, Edwin. 1968. *Industrial research and technological innovation: An econometric analysis*. New York: Norton for the Cowles Foundation for Research in Economics, Yale University.

———. 1980. Basic research and productivity increase in manufacturing. *American Economic Review* 70:863–73.

———. 1981a. Composition of R and D expenditures: Relationship to size of firm, concentration, and innovative output. *Review of Economics and Statistics* 63:610–15.

———. 1981b. How economists see R and D. *Harvard Business Review* 59: 98–106.

Mansfield, Edwin, and Anthony Romeo. 1980. Technology transfer to overseas subsidiaries by U.S.-based firms. *Quarterly Journal of Economics* 94:737–50.

Mansfield, Edwin, Anthony Romeo, and Lorne Switzer. 1981. R and D price indexes and real R and D expenditures. *Research Policy* 12:105–12.

Mansfield, Edwin, Anthony Romeo, and Samuel Wagner. 1979. Foreign trade and U.S. research and development. *Review of Economics and Statistics* 61:49–57.

Mansfield, Edwin, Mark Schwartz, and Samuel Wagner. 1981. Imitation costs and patents: An empirical study. *Economic Journal* 91:907–18.

Mansfield, Edwin, David Teece, and Anthony Romeo. 1979. Overseas research and development by U.S.-based firms. *Economica* 46:187–96.

Mansfield, Edwin, John Rapoport, Anthony Romeo, Edmund Villani, Samuel Wagner, and Frank Husic. 1977. *The production and application of new industrial technology*. New York: Norton.

Nelson, Richard, and Sidney Winter. 1978. Forces generating and limiting concentration under Schumpeterian competition. *Bell Journal of Economics* 9:524–48.

Scherer, F. M. 1980. *Industrial market structure and economic perform-ance*. 2d ed. Chicago: Rand-McNally.

Terleckyj, Nestor. 1974. *Effects of R & D on the productivity growth of industries: An exploratory study*. Washington, D.C.: National Planning Association.

Terleckyj, Nestor, and David Levy. 1981. Factors determining capital formation, R and D investment, and productivity. Mimeo.

Vernon, Raymond. 1966. International investment and international trade in the product cycle. *Quarterly Journal of Economics*.

———, ed. 1970. *The technology factor in international trade*. New York: National Bureau of Economic Research.

Comment   Zvi Griliches

I agree with Mansfield about the potential importance of good R & D deflators and I am glad to see him doing something about it. Some years ago (as part of my work with the Census-NSF data, see Griliches 1980), being in dire need of an R & D price index, I "constructed" one, patterning it on the methodology of Jaffe (1972). The resulting index is nothing but a weighted average (with almost equal weights) of the hourly compensation index and the implicit deflator in the nonfinancial corporations sector. This index has two advantages over the usual overall GNP deflator: (1) It is based on data from a more relevant subsector of the economy. (2) It gives wage rates more weight, which is as it should be since R & D is more labor intensive than the average corporate product. The resulting index is shown in table C6.1. On Mansfield's base $1969 = 1.00$; my index was at 2.01 in 1979, compared to Mansfield's mean values in table 6.1 of 1.98, 2.01, and 2.00. I do not think one could come closer if one tried. This would seem to indicate that this type of a simple approximation may be pretty good, at least in recent years. Nevertheless, it would be desirable if NSF or BLS would take on the task of constructing and keeping up-to-date an actual price index such as Mansfield's.

Two additional brief notes: (1) Other attempts to construct an R & D price index were undertaken by Goldberg (1979), Schankerman (1979), and Halstead (1977). They all come out roughly in the same place: an R & D input price index rises by more than the GNP deflator. (2) We are considering here an R & D *input* price deflator. We have no data to attempt an R & D *output* price deflator. It is not clear whether the

Zvi Griliches is professor of economics at Harvard University, and program director, Productivity and Technical Charge, at the National Bureau of Economic Research.

Table C6.1   Approximate Deflator for R & D Expenditures (1972 = 1.00)

1957	.598	1969	.855
1958	.616	1970	.906
1959	.631	1971	.956
1960	.647	1972	1.000
1961	.658	1973	1.064
1962	.670	1974	1.170
1963	.680	1975	1.285
1964	.698	1976	1.361
1965	.711	1977	1.459
1966	.737	1978	1.573
1967	.768	1979	1.7175
1968	.809	1980	1.870

Note: Index = .49 hourly compensation index + .51 implicit deflator, both for nonfinancial corporations. (Value of 1957 hourly compensation extrapolated using the hourly compensation figures for the manufacturing sector.) Underlying data from U.S. Department of Labor, *Productivity and Costs in Nonfinancial Corporations*, Washington, D.C., various issues.

"productivity" of R & D has been growing or diminishing over time. From a social point of view it could be growing. From the private point of view of a company or a university it has probably been declining, in the sense that to keep the same competitive edge, to stay in the same position in the commercial or academic market, R & D laboratories today need more expensive equipment, computers, and materials. From the point of view of a laboratory director his real "costs" of R & D are rising faster than is indicated by Mansfield's or my index.

References

Goldberg, L. 1979. The influence of federal R & D funding on the demand for and returns to industrial R & D. Alexandria, Va.: The Public Research Institute, October.

Griliches, Zvi. 1980. Returns to research and development expenditures in the private sector. In *New developments in productivity measurement and analysis*, ed. J. W. Kendrick and B. N. Vaccara, 419–54. Conference on Research in Income and Wealth: Studies in Income and Wealth, vol. 44. Chicago: University of Chicago Press for the National Bureau of Economic Research.

Halstead, Kent D. 1977. Higher education prices and price indexes, 1977 supplement. Washington, D.C.: National Institute of Education.

Jaffe, S. A. 1972. A price index for deflation of academic R & D expenditures. Washington, D.C.: NSF 72-310.

Schankerman, Mark. 1979. Essays on the economics of technical change: The determinants, rate of return, and productivity impact of research and development. Ph.D. diss., Harvard University.

Comment   George C. Eads

Ed Mansfield's paper reflects the wide range of research he has conducted on the topics covered by this conference. Hardly a single area has failed at some time or another to attract his attention. Wherever he has chosen to work, we are the richer for his contribution.

While it would be possible to evaluate each of the pieces of research reported in this paper as an individual item, I believe it instructive to group them into categories. This enables us not only to assess the items' individual worth, but also to view Mansfield's research from a broader perspective. Although others might be suggested, I propose a two-fold divsion. The first division includes the works of what I will refer to as the "neoclassical" Mansfield; the second, the works of the "pragmatic" Mansfield. The "neoclassical" Mansfield is a member of a group of researchers who have attempted to apply neoclassical production functions to the measurement of the determinants of the growth of output (in his case, concentrating on the role that technological change plays in generating that growth). The "pragmatic" Mansfield is perhaps the leading exponent of an ad hoc approach to investigating the microeconomics of technological change. While his contributions in both areas have been considerable, it is the latter body of his research that I have found to be consistently the most provocative, raising questions about the way we ought to approach the study of technical change and forcing us to reexamine our preconceptions.

Indeed, it often performs such a role in the research of the "neoclassical" Mansfield. Consider section 6.2, "Composition of R & D: Effects and Determinants." Examination of the *American Economic Review* piece from which this section is drawn reveals that the results reported early in the section stem from the estimation of a neoclassical model based on a Cobb-Douglas production function. The dependent variable is value added by a given industry during period t. The independent variables are the industry's labor input, stock of physical capital, and two kinds of R & D capital—basic and applied. After performing suitable transformations and simplifications, Mansfield ends up with a single equation model which he estimates under various specifications.

The coefficient associated with the basic research variable is found to be consistently significant. The significance of this coefficient leads Mansfield to his principal conclusion: holding constant the amount an industry spends on *applied* R & D, the higher the proportion of research that is basic (or long-term), the higher the rate of productivity over the 1948–66 period. (Mansfield's attempts to replicate these results using data from the post-1966 period have proved unsuccessful.) Mansfield is well aware

George C. Eads is a professor at the School of Public Affairs, University of Maryland.

of the pitfalls in interpreting his results. In particular, the direction of casuality is ambiguous. (These problems and their implications are discussed in more detail in the *American Economic Review* piece.) Yet the finding, even qualified, is provocative.

But there is more, for the "pragmatic" Mansfield cannot resist having a crack at the topic that the "neoclassical" Mansfield has opened. Near the end of the *American Economic Review* piece (and in the third and fourth summary points in section 6.2 of the paper), Mansfield reports the results of arraying data he has collected concerning recent changes in the composition of industry R & D expenditures. (Since he is merely looking for interesting connections, he employs simple correlation analysis and some multiple regressions.) I find his conclusions provocative: that there is a critical distinction between *basic* (i.e., "long-term") research and what I would term "breakthrough" research—research aimed at relatively new products and processes. His results suggest that large firms do indeed perform a disproportionately large share of the *basic* research, but they perform a disproportionately *small* share of the "breakthrough" research. He also finds little if any statistical relationship between changes in the proportion of a firm's R & D expenditures directed to basic research and changes in the proportion directed to "breakthrough" research. Finally, he reports the various reasons cited by his survey respondents for recent declines in the amount of "breakthrough" research they fund.

At this point the section (and the *American Economic Review* piece) ends. I wish it hadn't, for I would have liked to have seen results reported in the latter part of the section related to those reported in the earlier part. A number of interesting questions suggest themselves. For example, "breakthrough" research would seem to be more amenable than basic research to targeted incentives relating to potential risk and reward. Both this logic and Mansfield's findings suggest that the two categories are likely to be affected quite differently by changes in underlying economic conditions—inflation and regulation, for example. But is one type of research more likely to generate important advances in productivity than the other? Are government policies designed to increase basic research a substitute for other policies designed to increase risk taking of the sort that leads to more "breakthrough" research?

This brings me to a more general point. Throughout the paper, Mansfield characterizes his research as "gap filling." Yet I inevitably finish a Mansfield study, especially one of the sort whose results are reported in section 6.2, more aware of what we *don't* know about technological change and more *dissatisfied* with our traditional approaches than convinced that a "gap" has somehow been closed. I don't want to be misunderstood. I yield to no one in my admiration of Mansfield's ability to collect interesting data, to array them in provocative ways, and to spin out interesting hypotheses about what they might imply. But let's be candid. Mansfield is *not* putting the finishing touches on a well-understood edifice called "The Economics of Technical Change." Instead he—and the rest of us—are laboring at a far earlier stage; one in which the surprises produced even by simple correlations might be enough to cause us to go back to our plans to see if we are even constructing the right structure. That Mansfield's work sends us back to the drawing board more often than it produces a feeling of satisfaction that a critical piece of the structure has been completed is not a criticism, therefore, but a comment on the relatively primitive state of our knowledge.

This gap *creating* tendency of much of Mansfield's research is illustrated by the sections of the paper that report on his excursion into the world of international R & D. Just when we were despairing at our inability to explain the causes of our nation's poor productivity performance using purely domestic variables, Mansfield suggests through his work that the problem might be even more complex than we had thought. For not only must we understand the connections between domestic variables, such as basic or applied R & D and the rate of technological change and, ultimately, the rate of economic growth; we must also take into account the flows of technology to and from the United States, especially between the overseas subsidiaries of U.S. firms and their domestic parents.

In section 6.6, Mansfield provides a framework for understanding the various types of international technology transfers. In section 6.7 Mansfield begins to explore the gains to U.S. technology from research performed overseas. In section 6.8 the "neoclassical" Mansfield again ventures forth. Using a model identical in structure to the one whose results are reported in section 6.2, he attempts to measure—albeit crudely—the contribution of research performed overseas to U.S. productivity performance for fifteen firms in the petroleum and chemical industries.

This time the R & D capital variable is divided not into "basic" and "applied," but into "domestic" and "overseas." The results by Mansfield's own admission are "surprising." Not only did overseas research contribute significantly to productivity for the firms in his sample, but the "bang for the buck" is nearly ten times as great. Mansfield dismisses this result since the two coefficients, though each significantly different from zero, are not significantly different from one another. I'm surprised at the model results in view of the fact that, according to table 6.5, eight of the fifteen firms for which Mansfield has data performed *no* overseas research at all during the period under investigation. The "bang" from the work of those that did certainly must have been powerful.

Mansfield then takes his model one step further. He attempts to derive an estimate of the possible "bias" from omitting the overseas research variable by comparing the domestic coefficient in the two-factor R & D model with one in which the overseas variable is excluded. The coefficient in the latter is larger than in the former, leading him to conclude that "the rate of return from domestic R & D may have been overestimated in previous studies"—including his, I presume. However, extreme caution is required in this interpretation because, by my calculations, the two coefficients are only marginally significantly different from each other.

The paper reports on several other interesting bits of research, mostly representative of the "pragmatic" Mansfield. For example, there is an attempt to measure the costs of imitation and, hence, the value of patents. The implication of this work is that, except in limited areas, patents provide surprisingly little protection to an inventor. The exception is the drug industry. Although Mansfield briefly refers to the potential impact of FDA regulations on his results, I'd like to know more about the possible interaction of FDA drug approval procedures and the efficacy of patents.

Another interesting tidbit is contained in section 6.4 where Mansfield attempts the analytically difficult task of separating out the effects of federal support of privately financed R & D. His research seems to imply an asymmetry. Each dollar's increase in federal R & D support generates only six cents additional private R & D during the first two years and then zero thereafter. But each dollar *cut* in federal R & D support causes a fall of twenty-five cents in private support during each of the first of two years and nineteen cents after the third year.

This and results reported later in the section suggest a finding contrary to that reported by Mansfield—namely, that federal R & D support exerts a growing, not a declining, influence over time on the character of a firm's R & D spending. If, as Mansfield contends, the federal influence declines, I'd be hard pressed to explain the asymmetry he observes. But again his results are preliminary, and we must await publication to examine his detailed argument.

Taken as a whole, this paper and the articles and books it refers to, both published and unpublished, reveal a highly productive research organization, led by an extraordinary individual, investigating a remarkable variety of interesting and important topics. The research methodologies employed by this organization are "problem driven," not "tool driven," and that is indeed fortunate. The nature of the problems they are investigating requires this. Indeed, it would be too bad if the "neoclassical" Mansfield ever prevailed decisively over the "pragmatic" Mansfield and imposed a rigid theoretical structure on the work of the University of Pennsylvania team. The research would be less useful and we would all be the losers.

Reply   Edwin Mansfield

Given George Eads's generous comments concerning my paper and the research on which it is based, I feel no compulsion to take issue with him. It seems to me that our "neoclassical" and "pragmatic" work (if one accepts such a distinction) are linked together. Moreover, there clearly is a general model or theoretical framework on which all our empirical work is based. But he is quite right that this paper makes no attempt to put the pieces together. Having said this, I would like to clarify two small points. First, it is not quite true that attempts to replicate the results concerning the effects of basic research on productivity growth, using data from the post-1966 period, were unsuccessful. Although the fit is much poorer than in the pre-1966 period, the regression coefficient of the basic research variable is statistically significant in an appreciable number of specifications. Second, the apparent asymmetry in the effect of federal support on privately financed R & D may well be due to chance. The bigger apparent effect of a decrease than of an increase in federal support is attributable largely to a single firm in our sample. Although my paper pointed out that this apparent asymmetry should be viewed with considerable caution, I may have mislead Eads (and others) because my language was not stronger. Put bluntly, the apparent asymmetry may well be a fluke.

Turning to Zvi Griliches's comment on price indexes for R & D inputs, I think that the comparison he presents is interesting. Price indexes based on proxies are valuable, if they are reasonably accurate, since they are relatively cheap to construct. The Organization for Economic Cooperation and Development has also experimented with such indexes. In my view, actual price indexes should perhaps be combined with price indexes based on proxies. For example, actual price indexes might be constructed for benchmark years, and price indexes based on proxies might be used for interbenchmark years. I agree with Griliches that work should go forward to construct and compare both types of indexes. Moreover, it may well be that this work should be at the industry level, since there seem to be interindustry differences in the rate of increase of the price index for R & D inputs, a result apparently of interindustry differences in the types of R & D inputs used.

7   Long-Run Trends in Patenting

John J. Beggs

7.1   Introduction

By the beginning of the nineteenth century, three countries had firmly established patent systems. In the United States the Constitution gave Congress the power: "To promote the progress of science and useful arts, by securing for limited times to authors and inventors the exclusive right to their respective writings and discoveries" (Art. 1, Sec. 8[8]). The first patent law was passed in 1790. These laws were motivated by a concern for the justice of protecting intellectual property rights and by economic concerns, such as the need to guarantee sufficient protection from competition to allow profitable development of inventions and the need to encourage the disclosure of new ideas that could form the building blocks for future advances.

This relationship between technological change and industrial development is at the core of the economists' interest in the patent system. However, much compounding of effects makes the statistical analysis of this relationship a difficult one. Essential dynamics are present in the creative process. Single inventions suggest the follow-up direction for future research as well as create preconditions for breakthroughs in other, not obviously related, fields. Industry structure and patenting may be linked in ways that depend on more than the underlying rate of

John J. Beggs is a professor in the Department of Statistics, Faculty of Economics, Australian National University, and a faculty research fellow of the National Bureau of Economic Research.

The author has benefited from discussions with Derek de Solla Price and Gregory Dow. The author is continuing research with Gregory Dow on the issues raised in section 7.5 of this paper. Thanks are due to Dana Allen, Gary Pissano, John Berlyn, and Peter Fleisher, who sacrificed a summer of their youth to prepare the data that form the basis of this paper. Financial assistance from NSF grant PRA-8019779 is gratefully acknowledged.

technological advance in an industry. For example, firms may create patent portfolios as a direct instrument of competition by "fencing in" technologies, making new entry into their industries more difficult.

Patents are one of the few immediately applicable statistical indicators of technological change. As an itemized list of per period inventions, this statistical series contains a desirable amount of objectivity. The economic worth of individual patents varies greatly, and the interpretation of these data relies on "large-number-type" properties to help ensure that the average worth of a large number of patents is a meaningful quantity. More troublesome are the biases introduced both by changes in the laws and regulations governing patentability of inventions and by the possibility that the economy and particular industries may move through phases where a type of inventive activity is either more or less susceptible to patenting.

This paper first examines, at the industry level, the relationships between the rate of patenting and certain aggregate indicators of industry performance. Section 7.2 discusses the data set that has been prepared to investigate the question. Section 7.3 outlines certain hypotheses about the correlations between rate of patenting and industry performance variables, and section 7.4 reports statistical findings. Section 7.5 considers the dynamics of aggregate patenting and the role of inventions as preconditions for further inventions.

7.2 Data

The source of industry data for this study was the United States Census of Manufactures. The Census of Manufactures was taken as part of the Census of the United States every ten years from 1850 to 1940. The Census of Manufactures was taken separately in 1902, 1914, 1921, 1923, 1925, 1927, 1931, 1935, 1937, 1947, 1954, 1958, 1963, 1967, 1972, and 1977. In all the years when the Census of Manufactures was taken concurrently with the Census of the United States, the data on manufactures were from the year before the official Census year. The data collected included number of establishments, number of workers, average wage, capital expenditures, value added, and value of product.

The data collected from all years are generally comparable, but two changes in the Census of Manufactures could not be backdated. The Census of Manufactures data for number of wage earners include salaried employees in and before 1879 but do not include them after that date. Therefore, the data on number of wage earners and average wage include salaried employees and their salaries in 1879 and all previous years. The data for 1947 and all years thereafter use the classification "production workers" in place of "wage earners." This does not create a large ambiguity in the data, since the two classifications are similar. Both classifications exclude salaried officers, nonworking foremen, and clerical personnel. The 1947 Census of Manufactures states that the two classifications are "closely comparable." Capital data were included in the Census of Manufactures from 1850 until 1919. Data pertaining to capital were not collected after 1919 until 1933 when expenditures on plant and equipment were included.

Some small changes in industry definitions have occurred throughout the period. This generally occurred when a broadly defined industry was split into its component parts by the Census during the later years of this study. Since the earlier years often gave no breakdown of industries, the earlier definition has been used.

Data have been collected from a sample of twenty industries (listed in appendix A). The criteria for including particular industries were primarily associated with the complexity of their technologies. The industries included are chiefly those having more elementary technologies and those for which it is possible to identify the relevant patent statistics. It is important to recognize, for the purposes of later discussion, that the patents classified as belonging to a particular industry represent only a small part of the complex of technologies that must come together before a new industry can progress. For example, a patent for a new design of a sewing machine would appear in our statistics. The whole series of developments in metal alloys and machine tooling, which permitted this new sewing machine patent, would not appear in the data. As the economy has moved into the new electrical, electronics, and chemical technologies, these interdependencies have grown ever more interwoven and more difficult to unravel. For this reason, the data collecting exercise has focused primarily on "old" industries and, for the most part, on the period from 1850 through 1939.

Patent data were collected annually for each industry from published reports of the U.S. Patent Office. The data collecting procedure is described in some detail in appendix B. Patents were identified with industries by using an exhaustive alphabetical index of patents published by the Patent Office. This procedure is not entirely clean because no published (nor, apparently, unpublished) record exists of how patents were indexed. Discussions with retired patent examiners indicate that patents were indexed according to industry of predominant impact, be that either the industry of origin or the industry of use. Unfortunately, there is no entirely untained way to handle this question. Appendix B gives, for comparative purposes, a brief summary of the Schmookler procedures. Schmookler's data do not match the data collected by the Census of Manufactures as well as the new data set does.

7.3 Some Hypotheses

In his classic work *Invention and Economic Growth*, Schmookler asked the question, "Are inventions mainly knowledge-induced or demand-induced?" The up-side effect of demand-induced invention is possibly the easiest and best understood of all the mechanisms for stimulating invention. Here an expansion of the market creates the opportunity for new products, for new investment, and for the replacement of old processes by new. Schmookler (1966) demonstrated the close links at the industry level between investment in plant and equipment and successful patent applications, perhaps nowhere more so than in the well-known example of the railroad industry. An investment series for our sample of industries could not be constructed from the available data. In its place a surrogate was considered, namely wage expenditures as a percentage of value added. The wage bill would seem to fall relative to value added in times of high investment and to rise relative to value added in times of low investment levels. The surrogate suffers from the deficiency of including the effect of changes in the wage rate and changes in the price of final output, but, in the absence of an alternative, it provides a crude indicator of changes in investment. The use of this surrogate is discussed further in section 7.4.

A "down-side" effect of demand-induced invention is also possible.[1] In the event that an existing industry is challenged by the emergence of a new industry, it will likely experience a slump in sales. In the absence of any competitive response, the industry will surely be driven out of existence. The natural reaction to competition should then be an increased and more intensive search for better production processes and better products for the industry. In the time period of our study, industries such as ice making; cotton manufacturers; wool textiles; flax, hemp, and jute; turpentine and rosin; clay products; and the confectionary industry have had to face such challenges. A fall in output caused by some economy-wide decline in output would be met in a different fashion than a fall in output resulting from the encroachment of other industries. For this reason, the relevant measure of changes in output is the change in output relative to the change in, say, gross national product. Such a variable is defined in section 7.4.

The nature of the technological change in an industry will determine how wages move relative to the national average. Labor mobility and the institutional response of organizations, such as trade unions, enter into the adjustment mechanism. Proceeding by example, inventions such as power tools seem to have substantially reduced the skill levels required by the woodcraft artisan, presumably lowering the marginal product of labor and, hence, the real wage in this industry. One can think of converse examples where the initial skill levels were quite low and the introduction of inventions required higher levels of skills, such as the ability to read and write. The phenomena discussed thus far are associ-

1. The term "Indian Summer" is also sometimes used to describe this phenomenon.

ated with changes in the technical skill requirements of the work force. Technological change may also be associated with rapid expansion of the market and increased demand for certain types of skilled labor or for labor in certain geographic localities. In the event of reasonable labor mobility, these fluctuations above or below prevailing average wage levels should soon disappear. In the event of significant productivity gains in strongly unionized industries, labor will possibly be able to negotiate some share of the new surplus above what it might have earned in competitive labor markets.

7.4 Empirical Evidence

The data brought to bear on the above questions are discussed in section 7.2. The variables cover twenty industries and, after expressing the variables in rates of change, there are 363 observations. Where relevant, the variables measure rates of change relative to the national aggregate. This has the effect of purging the data of movements in the macroeconomic aggregates associated with the trade cycle. Variables are expressed in logarithms to give the coefficients an elasticity-type interpretation. The variables are then:

(1)
$$X_{it}^1 = \log\left[\frac{\text{Patents}_{it}}{\text{Patent}_t} \cdot \frac{\text{Patents}_{t-1}}{\text{Patents}_{it-1}}\right].$$

(2)
$$X_{it}^2 = \log\left[\frac{\text{Value Added}_{it}}{\text{GNP}_t} \bigg/ \frac{\text{Value Added}_{it-1}}{\text{GNP}_{t-1}}\right].$$

(3)
$$X_{it}^3 = \log\left[\frac{\text{\# Wage Earners}_{it} * \text{Av. Wage}_{it}}{\text{Value Added}_{it}} \bigg/\right.$$
$$\left.\frac{\text{\# Wage Earners}_{it-1} * \text{Av. Wage}_{it-1}}{\text{Value Added}_{it-1}}\right].$$

(4)
$$X_{it}^4 = \log\left[\frac{\text{Av. Wage}_{it}}{\text{Av. Wage}_t} \bigg/ \frac{\text{Av. Wage}_{it-1}}{\text{Av. Wage}_t}\right].$$

A subscript (i, t) indicates an observation for the ith industry in period t. Patents$_t$ is a variable for all patents issued in the United States for period t. Av. Wage$_t$ is the average wage for production workers in manufacturing and was taken from the individual Census of Manufactures. The number of patents issued in any industry in a given year has a high variance. To help eliminate chance or measurement error influences, the variable Patent $_{it}$ is the average number of patents per year in periods t, $(t-1)$, and $(t-2)$.

Examining movements in an industry series relative to movements in the national aggregate of the series is a particularly tough test of the theory. One difficulty is that the national aggregate may not be the most meaningful yardstick against which to measure performance. An industry's performance could be compared to industries of like technical characteristics (either on the product or the process side) or to industries facing similar amounts of foreign competition or located in similar geographic regions. The development of such peformance criteria is not an easy task either conceptually or as a matter of data preparation. Taken in conjunction with the difficulties in defining industry boundaries, a considerable amount of measurement error must be supposed in the data.

Interpreting the direction of causation among the above variables is difficult. The data are not particularly rich in time series, having on average only thirteen observations per industry. Furthermore, the time series data do not correspond to equally spaced time intervals. The period of time between Census of Manufactures varies from two years to ten years, and the data for each industry do not correspond to the same period of time. Some series commence earlier than others and some end earlier.

A series of two variable regressions were run and the results are reported in table 7.1. Statistical linkages appear to exist between the rate of patenting and the rate of growth of value added, and between the rate of patenting and the rate of change in the wage bill expressed as a proportion of value added. In both cases the coefficients on the regressions are negative. The wage rate variable does not appear to be correlated with the rate of change of invention in this data set. As was discussed in section 7.3, the wage bill as a percentage of value added will be taken as an inverse surrogate for the rate of investment. Schmookler (1966, pp. 151–62) used a cruder surrogate for investment, namely, value added itself. Though our variable x^3 is far from a perfect surrogate for investment, it should represent an improvement over Schmookler's use of simple value added in that it corrects for the cost of labor. The results in

Table 7.1   **Single Variable Regressions: Patenting and Industry Characteristics**

(1) $X_{it}^1 = -0.113\ X_{it}^2$ $R^2 = 0.035$
     $\quad\quad\quad(.031)$

(2) $X_{it}^1 = -0.121\ X_{it}^3$ $R^2 = 0.042$
     $\quad\quad\quad(0.041)$

(3) $X_{it}^1 = -0.165\ X_{it}^4$ $R^2 = 0.015$
     $\quad\quad\quad(0.192)$

Degrees of freedom = 362

Note: Intercept terms are insignficant as expected from the definition of the variables which effectively centers the regression around the origin. Mcasurement error will bias both the coefficients and the R^2 statistic toward zero.

equation (3) indicate that there is not an apparent link between wages and invention, giving us more confidence that movement in the variable x^3 is being driven more by investment than changes in the cost of labor. Since x^3 is an inverse surrogate for investment, equation (2) has the correct sign and supports the investment-demand-induced explanation of patenting, namely, that many new inventions are embodied in new capital equipment. While this result is in good congruence with Schmookler (1966), the theory has been put to a far more rigorous test. By defining variables in terms of rates of change *relative* to the national aggregates, one avoids the possibility of spurious relationships which might emerge as all the indicator series move together up and down the trade cycle. Since those regressions are "with-in" regressions, the relatively low value of the R^2 statistic is to be expected. Before leaving this equation, the possibility remains that the causal direction is the reverse of that discussed above. It is again useful to reflect on the nature of the patent "statistic." Patents do not measure technological change, though they are a manifestation that some change is taking place. Patents which represent *major* technological breakthroughs may well lead to growth in industrial investment.[2] But such patents are only a small percentage of total patents issued in an industry in a given year. The great bulk of patents are for inventions which represent incrementally small advances in knowledge. Such patents are for minor modifications, often of such devices as locks, switches, hinges, metal cutting devices, tools, etc. Arguably, these small inventions are less likely to explain movements in industry investment.

Of considerable interest are the results in equation (1) where there is a negative relationship between the *relative rate of patenting* and the *relative rate of growth of value added*. This result is different from the Schmookler results, which used level of value added as a surrogate for investment and found a positive relationship between the level variables of value added and patenting. The reason for the apparent differences in the results is that the equations are testing for *different effects*. Schmook-

---

2. Often these breakthroughs came very early in the sample period for the industries being studied. For example, Goodyear purchased the patent for sulphur vulcanization of rubber in 1839; most of the ideas and patents on synthetic rubber were available by 1910 (by 1939 synthetics were still less than 2 percent of the market); the ammonia absorption system for icemaking and refrigeration was patented in 1862; plate glass was first manufactured in 1852; the electric typewriter was patented by Edison in 1872; Singer patented a sewing machine in 1851 with a straight needle, stationary hanging arm, fed by roughened wheel, material held in place by presser foot beside the needle (in subsequent years, there have been as many as three hundred patents per year on sewing machines, each a small variation on an established idea); Ivory soap, special characteristics being that it was white and would float, was manufactured in 1879; first friction match was patented in 1827, and the safety match was patented in 1855; the first battery clock was patented in 1840, the self-winding watch in 1924, and the Quartz crystal clock in 1927; chocolate was invented in Switzerland in 1872, and the first packaging for national distribution of a confectionary was in 1872, when Mr. Cracker Jack (real name) launched his famous popcorn product; other technologies, such as iron, steel, and sugar refining, were well established by the 1880s.

ler's (1966, pp. 160–61) results[3] are across industry regressions with a trend variable included. Industries with large value added have larger numbers of patents per year, so there is considerable regression on the scale of the industry. Also there is possible synchronous behavior of the series through the trade cycle. The proposition being tested in equation (1) is somewhat more subtle. The question is how an industry behaves as it goes faster or slower *relative* to the other industries about it. The evidence in equation (1) is that when industries do well relative to other industries about them, they slacken their rate of patenting relative to all other industries. This would be consistent with the Kamien and Schwartz (1978) argument that, in the absence of a financial constraint, individual firms experiencing high profits will be less likely to innovate, since such innovation serves to cannibalize existing profitable market positions. Conversely, if an industry goes more slowly relative to its neighbors, it responds by quickening the rate of invention. In periods of severe competitive pressure, brought on by the encroachment of other industries onto its turf, firms may respond by quickening the tempo of their inventive efforts. Under such circumstances, there may be an undue increase in the "number" of patents if the patents are the type which attempt to modify and upgrade an existing capital stock or an existing product. Such patents will be small, low-value patents but could, given the nature of the activity, be very numerous. Inventions are made by firms and by individuals rather than by an "industry," and the extent of competitive pressures will surely change from industry to industry. However, to the extent that the fortunes of firms in an industry are tied to one another, it seems that those pressures will, in general, be greater when an industry is faring less well relative to other industries.[4]

7.5   Inventions and Further Inventions

Though invention is undoubtedly a response to market opportunities (and hence an economic phenomenon), the direction and pace of invention may well depend on previous inventions. Previous inventions may establish the necessary technological preconditions for the development of some new product or process as well as shape tastes and preferences for the developments which should follow.

The history of patenting seems to have been a complicated one, and the process of sorting out persistence effects from changing underlying trends is not easily accomplished. The longest published series of patent statis-

3. Similar results were found in the current data series; they are not reported here as they are almost an exact replication of Schmookler's findings.

4. Results similar to the above results are also reported in Beggs (1981) where the data are again industry level, but for the period 1953–78. In that paper, a short-run negative relationship is found between the growth rate of R & D expenditures and the growth rate of industry profits.

tics for the United States is for patents "issued," which runs continuously from 1790 to the present. A shorter published series is available on patent "applications," commencing some fifty years later. To study these series and their time series behavior, it is necessary to evoke types of detrending procedures. This is, at best, a hazardous undertaking (Nelson and Kang 1981), and almost all procedures attempted for these particular series result in a residual series exhibiting a long swing. While it remains possible that such long swings exist in the data, it is sufficiently easy to artificially create such cyclical behavior by incorrect detrending that this result cannot be taken seriously without much further investigation.

One detrending procedure which does not induce long swings in the data is a transformation to rate of change of patenting, that is, $(P_t - P_{t-1}/ P_{t-1})$. Some interesting results are reported below when this detrending procedure is applied to patents "issued," a series of 190 observations. A word of warning at the outset, though: The results reported here are not robust to segmentation of the data set and do not apply to the shorter time series on patent "applications." It is certainly true that the signal-to-noise ratio in these series is very high and it appears that reductions in sample size are not well accommodated. More seriously, of course, one must recognize the possibility that the results reported are merely a sampling artifact of one particular sample series. In subsequent research, when the question of detrending has been considered in greater depth, it will be necessary to reconcile any differences in the time series behavior of the patents "issued" series and the patent "applications" series. The patent "applications" series contains noise and related effects associated with changes in the general desire to patent inventions (either for economic reasons or whimsical social reasons). The patents "issued" series is a more seriously compiled series in that each patent issued has passed some rigorous technical examination of its merit. On the debit side, however, various forms of bureaucratic inertia may induce artificial cycles in this series. These questions do not arise immediately here since statistically meaningful results appear to be found only in the 1790–1980 period patents-"issued" series.

The smoothed periodogram for the rate of change of patents issued series is shown in figure 7.1. The shape of the periodogram suggests a process with a five-period lag and with a small coefficient (i.e., the periodogram is rounded rather than spiked). An autoregressive process with a five-period lag was fitted to the data and the residuals were examined. The periodogram of the residuals suggested an eight-period lag. The model finally fitted to the data was a moving average process, where y_t is the rate of growth of patents issued per year.

(5) $$y_t = \epsilon_t + 0.264\epsilon_{t-5} + 0.071\epsilon_{t-8}.$$
$$\quad\quad (0.030)\quad\quad (0.011)$$

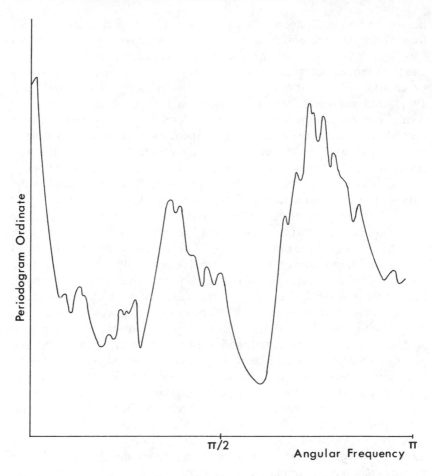

Fig. 7.1   Smoothed periodogram: annual rate of change of aggregate
U.S. patents issued, 1790–1980.

Asymptotic estimates of the standard errors are shown in parentheses.
The theoretical spectrum for the estimated moving average process is
shown in figure 7.2. Visual inspection indicates good conformity between
the periodogram and the estimated spectrum. There are two-and-one-
half waves in both (caused by the fifth-order lag term), and the peaks and
troughs are the correct relative magnitude (caused by the eighth-order lag
term).

The initial five-year lag from patent invention to patent invention is the
result of time taken to understand and develop the original patent and to
then understand and produce the appropriate follow-up invention. Since
these are national aggregate patents, one might expect longer lags than if
one simply studied a patent series within a single industry. Inventions in

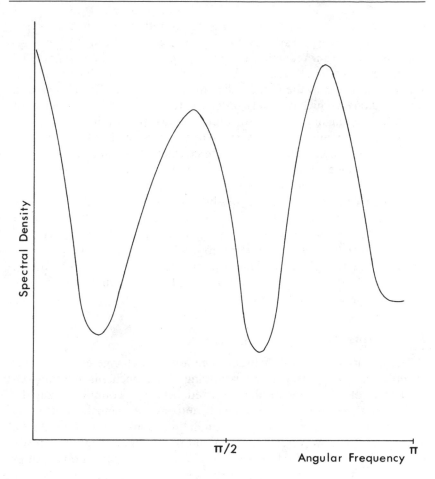

Spectral Density

π/2

Angular Frequency   π

Fig. 7.2   Theoretical spectral density function for moving average process given in equation (5).

one industry may lead to follow-up inventions in other industries, but the transmission process will be slower. For example, a patent issued for a semiconductor invention may be associated with a rapid follow-up patent in semiconductors, but the follow-up patent in, say, automated tool cutting will occur much later. Also, since the data cover the period from 1790, much of the sample is from an era when information transmission mechanisms were much less sophisticated than today, so the intuition of everyday experience in 1981 may not be particularly relevant to most of the sample.

A burst of patents in period t leads to follow-up patents in period $(t + 5)$, hence it is reasonable to expect further follow-up patents some period later. The lags associated with this second round of follow-up patents are likely to be shorter than the first round because there has been a period of growing awareness and experience of the new technology. The data indicate a reduction in the lag from five to three years. The magnitude of the coefficient on the second round should have a magnitude on the order of the first coefficient squared, $(0.264)^2$. This gives a value of 0.0696, which is remarkably close to the estimated coefficient of (0.071). The magnitude of third round follow-ups will likely be on the order of $(0.264)^3$ and, hence, too small to be estimated from the available data set. The actual magnitudes of the coefficients seem to fall within a reasonable range. A 1 percent increase in patents in period t leads to a subsequent 0.33 percent increase in patents over the next eight years (this is a rough calculation because of the nonlinearity introduced by the compounding rates of growth), which is on average 4 percent of a patent per year. This is quite close to the average rate of growth of patents issued per year over the entire sample period, which is about 5 percent. We conclude that though the model in equation (5) is not statistically robust, it is particularly rich in interpretation and, hence, of interest in guiding future research on this topic.[5]

7.6   Conclusions

The history of the links between technological change and economic progress can yield a deeper understanding of the mechanism driving our modern economy. The results reported here are conditional on the nature of the sample data employed and are very much affected by measurement errors and changes through time in institutional structures. The results are, however, amenable to interesting interpretations, and do indicate the direction for future research, both in the collection of better data and in the formulation of more exacting tests of our models.

5. I have benefited from discussions with Derek de Solla Price about the interpretation of these results.

Appendix A

Industries Included in Data

1. Pulp and paper
2. Rubber tires
3. Ice making
4. Iron and steel
5. Glass
6. Salt
7. Meat packing
8. Cotton manufacture
9. Wool
10. Flax, hemp, and jute
11. Sewing machines
12. Tobacco
13. Turpentine and rosin
14. Soap
15. Clay (including bricks)
16. Chocolate candy
17. Sugar
18. Matches
19. Watches
20. Typewriters

Appendix B
Procedures for Collecting Patent Data

Published Patent Statistics

With its founding in 1830, the U.S. Patent Office began publishing an "Annual Report of the Commissioner of Patents." This volume listed the patents issued each year under one of sixteen headings. Also included was a detailed description of each invention. By 1871 there were 145 such subheadings. In 1871 the "Official Gazette of the Patent Office" and an accompanying index replaced the annual report. The descriptions of inventions were published in a monthly magazine and the alphabetical index directed the reader to the relevant monthly volume. In 1898 the Patent Office modified the method of classification to distinguish three categories of patents: (i) method or process, (ii) function, and (iii) structure. In 1954 the Patent Office ceased publishing the alphabetical index to inventions. At this time a strictly numerical classification system was adopted. The procedure for linking patents to industries was as follows:

(a) Find desired industry in the "Index of Classification."

(b) Record headings and subheadings and obtain one-line description of headings.
(c) Check the current "Classification Bulletins" to insure that pertinent patent groups had not been reclassified during the year.
(d) Examine the technical "Definitions of the Subclasses" (a volume several thousand pages long) to determine whether subheadings are pertinent to industry.
(e) Use the "Index to the Gazette" and find patent numbers issued that year in the appropriate subheading.
(f) Finally turn to the "Official Gazette" and monthly "Volumes of Patents" to find descriptive information on the invention.

Collecting the Patent Data

The same procedure was used to obtain a patent series for each of the twenty industries. The only variation in the reports is the number of years covered. The patent series begins for each industry ten years before the Census of Manufactures commenced publishing data for that industry. The patent series continues either until Census figures were no longer available or until 1953. After 1953 the Patent Office began using a classification system which makes obtaining an accurate count difficult.

For each year the patents listed in the index under the name of industry and under related headings were counted. Each patent title was examined to determine whether it had a meaningful bearing on the industry under consideration.

Notes on Schmookler Patent Data

The patents in Schmookler's (1972) study were counted according to the date of application between 1874 and 1950. Data are given on a "when issued" basis for the years 1837–76 and 1947–57. Schmookler's study covers "*capital goods inventions* classified according to the industry expected to use them." Schmookler assigned Patent Office subclasses to standard industrial class (SIC) industries. The Patent Office classification system is based on technological-functional not industrial principles, so Schmookler had to "convert from the Patent Office classification system to the industrial classification." If an entire subclass seemed to apply to an industry, he automatically included it. Otherwise, he took a sampling, and if two-thirds of the patents seemed to belong, he included the entire subclass. Once Schmookler determined the subclasses to be included, the Patent Office counted the number of patents granted per year in each class between 1836–1957.

The interindustry features of many inventions were also addressed in the data set. If Schmookler could not determine which industry to assign a patent to, or if an invention could be used in many industries, the patent was simply disregarded. Hence, he did not include steam engines with railroad data or tractors with farm data.

Along these lines, some uncertainty arises as to whether Schmookler grouped the patents according to industry of origin or industry of use. One quote indicates that "the inventions were to be assigned to the current main producing *or* using industry." However, it was also stressed that patents be assigned to "the industry expected to use them." In some cases, patents were included twice, once in the "using" industry and once in the "manufacturing" industry.

Schmookler breaks down broad industrial classifications, like "agriculture," into activity types, like "harvesting," and finally into commodity groups, like "plows." Patent Office subclasses are assigned to commodity groups from which the data time series is constructed.

References

Beggs, J. J. 1981. The dynamic interaction between industry level profits and research and development. Cowles Foundation Discussion Paper no. 588. Yale University.

Kamien, M. I., and N. L. Schwartz. 1978. Self-financing of an R & D project. *American Economic Review* 68, no. 3:252–61.

Nelson, C. Ri, and H. Kang. 1981. Spurious periodicity in inappropriately detrended time series. *Econometrica* 48, no. 3:741–51.

Penrose, E. T. 1951. *The economics of the international patent system.* Westport, Conn.: Greenwood Press.

Schmookler, J. 1966. *Invention and economic growth.* Cambridge, Mass.: Harvard University Press.

————. 1972. *Patents, invention and economic change: Data and selected essays*, ed. Z. Griliches and L. Hurwicz. Part II. Cambridge, Mass.: Harvard University Press.

Comment   Mark Schankerman

In his seminal work, Schmookler (1966) attempted to demonstrate the importance of demand as a determinant of inventive activity. The basic idea behind demand inducement is that the monetary returns to a given

Mark Schankerman is an assistant professor in the Department of Economics at New York University, and a faculty research fellow of the National Bureau of Economic Research.

The author benefited from discussions with John Beggs on the contents and interpretation of his paper. The author retains responsibility for the views expressed here.

piece of produced knowledge vary directly with the number of output units which embody the new knowledge, or the expected size of the market. Given the cost of producing the new knowledge, it follows that the profit-maximizing level of inventive activity should vary directly with expected market size.[1] As one leading test of the hypothesis, Schmookler organized a time series of patents on capital goods inventions according to the industry in which the invention is primarily used[2] and performed log-linear regressions of patents against value added in the industry of use. As we all know, the empirical results indicated a rough proportionality between patents and value added and strongly supported the demand-inducement hypothesis.

As Schmookler emphasized, the appropriate scheme for assigning patents to industries depends on the purpose of the analysis. For example, if one were interested in relating patents as an indicator of inventive output to some measure of inventive input, such as R & D expenditures, patents should be assigned to the industry of origin where the R & D is spent to produce them. To study demand inducement, however, one should assign patents to the industry of use and then correlate them with the level of demand for the products which embody (or which are produced with the process which embodies) the patents. The proper measure of demand depends on the type of patents under study. For capital goods patents the level of investment in the industry of use is appropriate (Schmookler 1966, chaps. 6 and 7), whereas for materials-embodied patents the intermediate purchases by the industry of use is more suitable. For a mixed sample of patents, one can either employ an assignment of patents by industry of use and the level of output in the industry of use, or an assignment by industry of origin and the level of output in the industry of origin.[3]

With these principles in mind, we turn to Beggs's examination of the demand-inducement argument. He constructs patent statistics for the period 1850–1939 and assigns them to one of twenty industries according to the "industry of predominant impact, be that either the industry of origin or the industry of use." This criterion is somewhat ambiguous and may not be entirely suitable for a test of the demand-inducement hypothesis. Beggs indicates that these patent data correlate better than

1. For more detailed theoretical statements of demand inducement, see Nordhaus (1969) and Pakes and Schankerman (this volume, Chap. 9).
2. As Beggs notes, this procedure was not always followed but it does seem to have been the guiding principle. See Schmookler (1972, p. 87–91).
3. Since Schmookler worked only with capital goods inventions, most of his empirical work is based on investment goods demand. To extend the time series coverage backward, he used value added as a proxy. He explored both an industry of use (chap. 7) and an industry of origin (chap. 8) criterion, but used the criterion both for the assignment of patents and the definition of value added.

Schmookler's industry of use patents with other data collected from the Census of Manufactures. It would be useful to know more about the relationship between the two data sets. With what Census data is the comparison made? Are the trends in patenting similar? How are they correlated with each other? With what are the differences correlated?

Because investment data are not available for the entire sample period, Beggs tests the demand-inducement hypothesis by regressing the rate of growth of patents against the labor share in value added as a proxy for investment, on the heuristic argument that this proxy should vary inversely with the level of investment. (Actually, the variables are defined relative to the corresponding aggregate variables, but this is immaterial for our discussion.) I do not see how such a result can be derived from any familiar model of investment, and I am therefore skeptical of this proxy. In any event, since Beggs's sample of patents is expressly not limited to capital goods inventions, it would seem more appropriate to measure demand by the level of output or value added in the industry of "predominant impact," which is available for the entire sample period.

Beggs does use value added data, but to test the different and interesting hypothesis that an industry's patenting activity depends directly on the degree of competitive encroachment by other industries. Beggs tests this hypothesis by regressing the relative rate of growth of patents assigned to an industry (relative to the growth of total patents) against the relative rate of growth in value added, on the argument that the correlation should be positive if the hypothesis is true. He obtains a positive correlation, but I do not find this evidence very convincing because I think that the assignment of patents by "predominant impact" is inadequate to test the hypothesis. The essence of the proposition is that an industry's patenting activity in particular markets is related to the competition it faces from other industries in those markets. The hypothesis says nothing (directly at least) about the total level of patenting for use in a given industry, which would seem to depend on the strength rather than just the existence of such an effect. I think that to test this hypothesis one needs a two-way classification scheme, by industry of use (or predominant impact) and by industry of origin. For example, one might test whether the number of patents produced by industry i for use in industry j is related to the number of patents produced by all other industries for use in industry j. I do not think that any one-way classification scheme for patents is adequate to test the competitive encroachment hypothesis as presently formulated.

While Beggs correctly emphasizes that this hypothesis is different from demand inducement, his empirical finding does appear to contradict Schmookler's results on demand inducement. Beggs's regression of the relative rate of growth of patents in an industry against the relative rate of growth in value added is essentially the same as adding time dummies to Schmookler's log-linear regression of patents against value added. Yet when Schmookler included a time trend in his regressions, the results remained essentially unchanged. Does the dramatic difference in findings simply reflect time effects which are so poorly approximated by a time trend? This apparent empirical contradiction is worth exploring.

Beggs also reports some interesting findings from a long time series of patents issued, extending from 1790–1980. His remarks on the nonrobustness of the results to data segmentation and to the use of patent applications instead of patents issued should be kept in mind, but I want to focus on the interpretation of his findings. Beggs finds that the rate of patenting is well approximated by a moving average scheme with fifth- and eighth-order lags, and he interprets the result as reflecting a first and second round of information transmission, according to which a burst of patents induces subsequent patenting once the technological information has a chance to diffuse. This idea has some kinship to the innovation business cycle theory advanced by Schumpeter a long time ago. The hypothesis is worth pursuing and may explain patterns of patenting for specific classes of patents, but I find it hard to believe that this can rationalize spikes in the spectrum of the rate of patenting in aggregate data. If there is any distribution across patents in the time it takes for their technological information to diffuse (and surely there is), I would expect the spectrum of the aggregate patent series to exhibit more smoothness. I would like to suggest an alternative (perhaps complementary) explanation. If demand inducement is what moves patenting activity, then the (detrended) aggregate time series of the rate of patenting should reflect cyclical movements in aggregate output (presumably with some lag).[4] Hence, I would like to see the cross spectrum between the rate of patenting and the rate of growth of aggregate output. Using this information, one could deduce the coefficients in the (possibly two-sided) lag distribution connecting patenting and output, and bivariate exogeneity tests on the two series could be conducted. Disentangling the effects of demand inducement from information diffusion remains an interesting and important research challenge.

References

Nordhaus, William. 1969. An economic theory of technological change. *American Economic Review* 59:18–28.

4. The underlying model I have in mind consists of a behavioral link between R & D activities and expected output, an expectational link between expected output and past output, and a production function relation between patents and past research activity. This would translate into a reduced form relation between patents and past output, and possibly between output and past patents.

Schmookler, Jacob. 1966. *Invention and economic growth*. Cambridge, Mass.: Harvard University Press.

———. 1972. *Patents, inventions and economic change and selected essays*, ed. Z. Griliches and L. Hurwicz. Cambridge, Mass.: Harvard University Press.

8 Tests of a Schumpeterian Model of R & D and Market Structure

Richard C. Levin and Peter C. Reiss

8.1 Introduction

Schumpeter's (1950) ideas about the role of innovation in modern capitalist economies have inspired a substantial literature on the relationship between market structure and innovative activity. This literature has focused on Schumpeter's observation that seller concentration influences the appropriability of R & D. Unfortunately, as is apparent in the surveys of Kamien and Schwartz (1975) and Scherer (1980), it is unclear whether highly concentrated markets enhance the appropriability of the returns to R & D (e.g., Schumpeter 1950) or whether the opposite is true (Fellner 1951; Arrow 1962). In either case, industry concentration is viewed as an important determinant of R & D intensity.

With few exceptions (notably Phillips 1966, 1971), it was not until recently that economists turned their attention to the reciprocal influence of R & D on market structure. This new literature has emphasized Schumpeter's oft-cited notion of "creative destruction," where market structure is influenced by past and current innovative successes and failures. Specifically, innovation generates transient market power; this, in turn, is eroded by rival innovation and imitation. Thus, a truly Schumpeterian framework requires that both market structure and R & D be taken as endogenous variables. Seen in this perspective, the relation of R & D and market structure must be explained by an appeal to more fundamental factors that jointly determine concentration and R & D: the

Richard C. Levin is a professor in the Department of Economics and the School of Management, Yale University. Peter C. Reiss is an assistant professor in the Stanford University Graduate School of Business.

The authors gratefully acknowledge the support of the Division of Policy Research and Analysis of the National Science Foundation under grant PRA-8019779.

structure of demand; the richness of technological opportunities; and the technological and institutional conditions governing appropriability.

A number of recent theoretical papers have attempted to capture the essence of this Schumpeterian simultaneity. Among these contributions are Dasgupta and Stiglitz (1980a, 1980b), Futia (1980), Lee and Wilde (1980), Levin (1978), Loury (1979), and Nelson and Winter (1977, 1978, 1980).[1] The approach taken by Nelson and Winter is the most comprehensive in its representation of the relevant forces influencing market structure and R & D intensity. The price of this generality is high, since their models are analytically intractable and are open only to simulation. By contrast, many of the remaining models are stark and highly stylized; each omits aspects of technology and competition that are important for a broader understanding of the relationship between market structure and R & D. For example, none of the above models is truly dynamic; market structure is represented by the number of identical firms; only Futia explicitly recognizes R & D spillovers where firms benefit from the efforts of rivals; and finally, no attention is paid to other activities, such as advertising, that affect market structure.

The problems hampering theoretical treatments of the simultaneity issue have made rigorous tests of the Schumpeterian process extremely difficult. To date only Levin (1981) and Farber (1981) have explored the simultaneity issue in any detail. Only Levin has attempted to test for the presence of simultaneity among the relevant variables, but Levin's model (in which concentration, private and government, R & D, advertising, and price-cost margins are determined simultaneously) is specified in the loose, eclectic manner that is characteristic of most empirical work in industrial organization. Although the findings are encouraging, they can only be regarded as a preliminary test of the neo-Schumpeterian theories.

In this paper we provide a more exacting empirical test. Unfortunately, in our efforts to formulate precise hypotheses, we are forced to ignore several aspects of Schumpeterian dynamics. Nonetheless, we are able to incorporate several important features of reality which have heretofore been missing in formal models of the R & D process. We begin with what is perhaps the simplest of the theoretical models of R & D and market structure, the Dasgupta-Stiglitz (1980a) model of noncooperative oligopoly with free entry. We generalize this model by adding two significant features. First, we allow for spillovers in the knowledge generated by R & D. Hence, some fraction of the returns to each firm's R & D efforts are appropriated by its rivals. Although we are not able at this stage to

1. Although they do not discuss the simultaneous determination of R & D activity and market structure, the work of Pakes and Schankerman (this volume) on the determinants of R & D spending is akin to the papers cited here in clear identification of demand, technological opportunity, and appropriability as central exogenous variables.

incorporate a specific allocation of investment to imitative effort, as Nelson and Winter do, we believe that our characterization of R & D spillovers illuminates important aspects of the appropriability problem, as we explain below. Second, we include advertising expenditures among the decision variables of the firm. Like R & D, advertising is an instrument of competition, and in many industries it is of greater empirical consequence than R & D. By thus enriching the model, we hope to gain a more precise understanding of the role of both R & D and advertising in Schumpeterian competition.

An attractive feature of our approach is that the general model contains a family of nested models that can be tested using classical procedures. For example, by constraining the value of certain parameters (namely, the degree of R & D spillovers and the elasticity of demand with respect to advertising) the Dasgupta-Stiglitz model falls out as a special case of our own model. In addition, questions concerning the effects of imperfect appropriability of R & D returns can be reduced to considerations of the magnitudes or signs of certain parameters in the model.

We believe our approach to be an attractive one, but we must caution the reader at the outset that limitations of both theory and data still preclude a fully satisfactory treatment of the issues we explore here. The model remains quite stark, abstracting from at least four issues of undoubted importance. First, the model is one of a static industry equilibrium, which is hardly in the spirit of the Schumpeterian dynamic disequilibrium arguments. Second, we take no account of uncertainty, which even in the absence of risk aversion has important implications for the allocation of R & D effort. Third, to make the analysis tractable, we consider only symmetric industry equilibria, where all firms within an industry are identical. Fourth, we consider only cost-reducing R & D, neglecting the well-known fact that most industries devote a major share of their innovative efforts to developing new products and improving the quality of existing products. In future work, we plan to generalize our model to take account of dynamics, uncertainty, asymmetry, and product R & D.

Limitations of the data constrain our efforts as well. Our model is designed for testing on a cross section of industries, but R & D data are available for industries only at a highly aggregated level, requiring us to aggregate data on other variables up to this level. Even more serious are deficiencies in the operational measures of technological opportunity and appropriability. In this paper, we rely on relatively crude proxies. However, together with several colleagues, we have recently initiated work on the direct measurement of opportunity and appropriability conditions through a survey of R & D managers. When collected, these data should substantially improve the reliability of the empirical results obtainable with this or any related econometric model.

The remainder of the paper is organized as follows. Section 8.2 presents and analyzes the model. Section 8.3 develops the empirical specification. The data and the econometric techniques are discussed in section 8.4. In section 8.5 the econometric results are presented and interpreted, and section 8.6 indicates directions for future research.

8.2   A Model of R & D with Spillovers

We begin by considering a firm that seeks to maximize profit with the use of three instruments: output, R & D expenditures, and advertising expenditures. Following Dasgupta and Stiglitz, we assume that a firm's R & D expenditure serves to shift downward its unit cost function, which is independent of output, but we generalize their formulation to include the effect of industrywide R & D on cost. Thus, unit cost of firm i is

$$(1) \qquad c_i = c(x_i, Z),$$

where x_i is the R & D expenditure of the ith firm, and Z is the sum of the R & D expenditures of all firms in the industry, including firm i.[2] It is further assumed that $c_1, c_2 < 0$, and $c_{11}, c_{22} > 0$; that is, there are positive but decreasing returns to both own R & D and to industrywide R & D.

This formulation emphasizes an important aspect of R & D technology neglected in most analytic models—external economies. An increase in the R & D expenditures of firm i not only reduces its own cost, but it also, through its "spillover" effect on Z, reduces the cost of all other firms in the industry. This characterization captures a central aspect of the appropriability issue; specifically, the cost function indicates the *technological* dimension of appropriability. To the extent that unit cost reduction is very elastic with respect to increments in the industrywide pool of R & D (holding own R & D constant), we infer that costless imitation is relatively easy, or that, technologically, the returns to R & D are relatively inappropriable. We shall indicate shortly how appropriability also has *structural* and *behavioral* dimensions.

Since our primary interest attaches to the analysis of R & D, we take a somewhat simpler approach with respect to advertising and its effect on demand. In particular, we assume that advertising shifts the industry demand curve. In principle, we could model advertising as altering the firm's own demand curve as well as spilling over to alter the demands facing all other firms. Such a characterization would reveal clearly the complete analogy between the roles of appropriability and R & D, since the appropriability of advertising also has technological, structural, and behavioral dimensions. Instead, we simply represent the industry inverse demand curve as:

---

2. Equation (1) can be generalized to include spillovers of R & D undertaken in other industries.

(2) $$p = p(Q, A),$$

where $Q = \Sigma q_i$ represents industry output, and $A = \Sigma a_i$ represents total advertising expenditures of all firms in the industry. We further assume that $p_1 < 0$, $p_2 > 0$, and $p_{22} < 0$.

The problem facing the ith firm is, therefore,

(3) $$\underset{q_i, x_i, a_i}{\text{Max}} \ \Pi_i \ = \ [p(Q, A) - c(x_i, Z)]q_i - x_i - a_i.$$

We assume that firms entertain Cournot conjectures regarding the output and advertising levels of all other firms. That is, we assume that the ith firm conjectures that $dQ/dq_i = 1$ and $dA/da_i = 1$. However, to bring clearly to the fore the full range of issues regarding the appropriability of R & D, we parameterize the conjectural variation with respect to R & D. That is, we let $dZ/dx_i = \theta_i$. Later, we shall explore several special cases by fixing θ_i.

We now write down the first-order conditions for the maximization of the ith firm's profit:

(4) $$p\left(1 - \frac{s_i}{\epsilon}\right) = c,$$

(5) $$-(c_1 + c_2\theta_i)q_i = 1, \text{ and}$$

(6) $$p_2 q_i = 1,$$

where $\epsilon = p/Qp_1$, the price elasticity of demand,[3] and $s_i = q_i/Q$, the market share of the ith firm.

We assume free entry such that the maximized profits of all firms in the industry are greater than or equal to zero, while profits of all firms outside (potential entrants) are less than or equal to zero for all nonnegative levels of output. We further restrict our attention to *symmetric* equilibria, where all firms in the industry behave identically. This latter restriction is, of course, less than desirable in a model purporting to represent Schumpeterian competition, but we have found that consideration of asymmetric equilibria raises some extremely difficult analytical problems.

Dasgupta and Stiglitz have shown that, given certain restrictions on the parameters, symmetric free entry equilibria exist. Somewhat more complicated restrictions on the parameters are necessary in the present model, but a wide range of parameter values remain consistent with equilibrium. Dasgupta and Stiglitz have also derived conditions under which the profit of each firm in the industry is approximately equal to zero in a symmetric equilibrium. Of course, it is perfectly plausible that the

---

3. Dasgupta and Stiglitz (1980a) use the notation ϵ to represent the quantity elasticity of *inverse* demand. We find that intuition is facilitated by reverting to the more customary notation.

technology of R & D (or of advertising) would be such that equilibrium contains a very small number of firms earning nonnegligible profit, but the entry of one additional firm would result in negative profits for all.[4] In what follows, we assume that the zero-profit condition can be invoked. Thus, we close the model by noting that each firm earns zero profit in equilibrium:

$$(7) \qquad [p(Q, A) - c(x_i, Z)]q_i = x_i + a_i.$$

Before proceeding, it is worth pointing out (in the manner of Dasgupta and Stiglitz) that conditions (4)–(7) may be used to illustrate that a market equilibrium involving cost-reducing and demand-shifting activities, such as R & D and advertising, will not result in a socially optimal allocation of resources. Conditions (4) and (7) indicate that prices must deviate from marginal cost to sustain R & D and advertising expenditures. Moreover, the left-hand sides of (5) and (6) show that the marginal private benefit of R & D and advertising depends on the firm's scale of output, whereas social optimality would require that the market output, Q, be substituted for q_i in these expressions.

We now proceed to analyze the market equilibrium in greater detail. First, note that in a symmetric equilibrium with n identical firms, $q_i = q$, $x_i = x$, $a_i = a$, $\theta_i = \theta$, and $s_i = 1/n$, for all firms. The zero-profit condition can then be summed over all firms by multiplying both sides of (7) by n to yield:

$$(8) \qquad [p(Q, A) - c(x, Z)]Q = nx + na.$$

Dividing both sides by pQ, we have

$$(9) \qquad \frac{p - c}{p} = R + S,$$

where the left-hand side is the Lerner index of monopoly power, R, is the ratio of industry R & D to sales, and S is the ratio of advertising to sales. Combining (9) and (4), we have

$$(10) \qquad \frac{1}{n} = \epsilon(R + S),$$

where $1/n$ is the Herfindahl index of concentration. This striking result says that industry concentration is proportionate to the sum of R & D and advertising intensity, where the elasticity of demand is the factor of proportionality. Our result parallels that of Dasgupta and Stiglitz, who, by neglecting advertising ($S = 0$), arrive at $1/n = \epsilon R$.

We choose to treat (10) as a structural equation for concentration in the econometric work which follows. Since we derived (10) from the first-

---

4. The problem with the zero-profit condition arises from the requirement that n be integer valued.

order condition (4) and the zero-profit condition (7), we now use (5) and (6) to derive structural equations for R & D intensity and advertising intensity, respectively.

Multiplying both sides of (5) by $R = x/pq$ gives:

(11) $$-(c_1 + c_2\theta)\frac{x}{p} = R.$$

Simple manipulation of (11) yields

(12) $$-\left(\frac{c_1 x}{c} + \frac{c_2 Z\theta}{cn}\right)\frac{c}{p} = R.$$

But inspection of equation (9) reminds us that $c/p = 1 - (R + S)$. We also observe that $-(c_1 x/c)$ can be interpreted as the elasticity of unit cost with respect to x, holding Z constant, which we will denote as $\alpha = \alpha(x|Z)$, and that $-(c_2 Z/c)$ is the elasticity of unit cost with respect to Z, holding x constant, which we denote as $\gamma = \gamma(Z|x)$. It follows that (12) can be rewritten as:

(13) $$\frac{R}{1 - (R + S)} = \alpha + \frac{\gamma\theta}{n}.$$

The numerator of the left-hand side of (13) is the ratio of R & D to sales, while the denominator is the share of sales revenue that is devoted to neither R & D nor advertising. In zero-profit equilibrium this is just equivalent to the ratio of production cost to revenue. Thus, the left-hand side of (13) is simply the ratio of R & D to total production cost, a variant of the more familiar representation of R & D intensity, typically used in econometric work.

We can interpret (13) as a structural equation in which R & D intensity depends on two terms. The first, α, is the elasticity of unit cost with respect to own R & D, holding industrywide R & D constant. It seems reasonable to interpret α as a measure of technological opportunity, indicating the responsiveness of cost to own research effort. The second term, $\gamma\theta/n$, has three components, each of which represents one of the dimensions of appropriability mentioned previously. We noted that γ, the elasticity of unit cost with respect to industrywide R & D (holding own R & D constant), is a reasonable measure of the *technological* dimension of appropriability, since it represents the extent to which a firm benefits from an increase in the common pool of R & D effort. On the other hand, since a firm's own R & D effort augments the common pool, higher levels of γ are associated with greater R & D intensity as long as the conjectural variation, θ, is positive.

Appropriability has a *structural* dimension as well, which was emphasized by Schumpeter (1950) and later by Galbraith (1956). For any given technology of R & D and market size, a firm's appropriable benefits from augmenting the common pool of knowledge depend on its market share, which is $1/n$ in a symmetric equilibrium. On the conjecture that rivals do not respond ($\theta = 1$), a 1 percent increase in own R & D produces a $1/n$ percent increase in industrywide R & D. Thus, whereas a monopolist's costs fall by γ percent for a 1 percent increase in its contribution to the common pool, an oligopolist's costs fall only by γ/n percent. Through this mechanism, the intensity of R & D increases with the greater appropriability associated with a more concentrated market structure.

Finally, there is a *behavioral* dimension to appropriability, represented here by the conjectural variation parameter, θ. It is easy to show that the Cournot conjecture, $\theta = 1$, results in a Nash equilibrium when combined with our previous assumptions that firms have Cournot conjectures in output and advertising decisions. But some informal arguments concerning the disincentive effects of spillover in R & D implicitly contain presumptions that $\theta < 1$. To the extent that firm j can costlessly borrow knowledge from firm i's R & D effort, it may choose to be a free rider and cut back on its own R & D. If the free-rider effect is sufficiently strong, θ may even be negative; that is, a one dollar increase in R & D expenditure by firm i may produce cutbacks in the R & D expenditures of the remaining firms that exceed one dollar in the aggregate. Such negative conjectural variations produce market equilibria that are not Nash equilibria, but the idea that free-rider effects are important in R & D is sufficiently well entrenched in the literature (e.g., Nelson 1959) to warrant testing for evidence of its presence.

In the empirical work that follows, we will examine several special cases of the model in which θ takes on prespecified values, as well as the case in which θ is free to vary. In particular, we will examine cases in which the value of θ is assumed to be 0, 1, and n, respectively. In the first case, the free-rider effect is such that rivals exactly offset a change in firm i's R & D, leaving industry R & D constant. The second is the Cournot-Nash case. In the third case, each firm behaves as if it expects its action to be matched by all others. This is the R & D analog of the Chamberlinian "constant market shares" conjecture. We shall test whether the data permit rejection of any of these three hypotheses about the value of θ.

Equations (10) and (13) represent structural equations for concentration and R & D intensity, respectively. We now derive a third equation for advertising intensity. Multiplying both sides of the first-order condition (6) by A/p and dividing both sides by q_i gives:

$$(14) \qquad p_2 \frac{A}{p} = \frac{A}{pq_i}.$$

The left-hand side of (14) is the advertising elasticity of inverse demand, which we denote as η. Dividing both sides of (14) by n we have

(15)
$$\frac{\eta}{n} = S,$$

where S is the ratio of industry advertising to sales, A/pQ.

It is easily shown that this expression for advertising intensity is simply a generalization of the Dorfman-Steiner (1954) rule to the oligopoly case. Note that at any given price

(16)
$$\eta = \frac{\partial p}{\partial A}\frac{A}{p} = \frac{-\partial p}{\partial Q}\frac{Q}{p}\frac{\partial Q}{\partial A}\frac{A}{Q} = \frac{\phi}{\epsilon},$$

where ϕ is the advertising elasticity of demand, and ϵ is the price elasticity of demand or, equivalently, the reciprocal of the inverse elasticity of demand. Thus, substituting (16) into (15), we have

(17)
$$S = \frac{\phi}{\epsilon n},$$

which is, of course, identical to the familiar Dorfman-Steiner result when, in the monopoly case, $n = 1$.

To summarize, we have now derived three structural equations representing the simultaneous determination of industry structure, R & D, and advertising intensity:

(10)
$$\frac{1}{n} = \epsilon(R + S),$$

(13)
$$\frac{R}{1 - (R + S)} = \alpha + \frac{\gamma\phi}{n}, \text{ and}$$

(17)
$$S = \frac{\phi}{\epsilon n}.$$

These structural relations are reasonably general, although they do depend on the special assumptions of symmetry and of Cournot conjectures in advertising and output.[5] In addition, ϵ, ϕ, α, and γ need not be constant parameters. However, by specializing the cost and demand functions they become so, and it becomes possible both to solve for reduced form equations for n, R, and S and to operationalize the model for empirical work. Thus, we assume that both cost and inverse demand functions have constant elasticities, so that

(18)
$$C(x, Z) = \beta x^{-\alpha} Z^{-\gamma}, \text{ and}$$

---

5. It is easy to see that the Dasgupta-Stiglitz model is a special case of our own— implicitly they assume $\gamma = \theta = 0$. Under these conditions, the model reduces to two structural equations for market structure and R & D: $1/n = \epsilon R$ and $R/(1 - R) = \alpha$. It is readily verified that these equations are equivalent to those contained in Dasgupta-Stiglitz (1980a).

(19) $$p(Q, A) = \sigma^{1/\epsilon} A^{\eta} Q^{-1/\epsilon}.$$

The inverse demand function (19), of course, corresponds to the demand function:

(20) $$Q(p, A) = \sigma A^{\phi} p^{-\epsilon}.$$

8.3 Empirical Specification

The model we have developed consists of three endogenous variables: the Herfindahl index of concentration ($H = 1/n$), research intensity, and advertising intensity. These three variables are jointly determined by the parameters of the cost and demand functions, which presumably differ across industries, and by the behavioral parameter, θ. We propose to test the model using cross-section data at the industry level to determine whether interindustry differences in opportunity, appropriability, and demand satisfactorily explain differences in concentration, R & D, and advertising. For the present we assume that θ is constant across all industries, except in the special case where $\theta = n$, and we test various hypotheses concerning its value.

Although the endogenous variables H, R, and S are directly observable (at least in principle), the exogenous parameters ϵ, ϕ, α, and γ are not. Since estimates of the price elasticity of demand exist for a wide range of industries, we treat ϵ_k as an observable exogenous variable for the kth industry. Next, to make the estimation problem tractable, we assume that α_k, the elasticity of unit cost with respect to own R & D in the kth industry, is a function of a vector of observables representing technological opportunity. Similarly, γ_k, the elasticity of unit cost with respect to industrywide R & D, is assumed to be a function of observables representing the degree of R & D spillovers (the technological dimension of appropriability). Finally, ϕ_k, the advertising elasticity of demand, is assumed to be a function of observable attributes of the industry's product. In the absence of any strong theoretical presumption, $\alpha_k(\cdot)$, $\gamma_k(\cdot)$, and $\phi_k(\cdot)$ are each assumed to be linear in parameters with an additive error term of mean zero. These errors are assumed to be uncorrelated with each other and uncorrelated with the error terms in the model's structural equations.

We can now write down a system of equations that can be estimated. Assuming that (10) is observed with multiplicative error and taking logarithms, we have

(21) $$\log H = a_o + a_1 (\log \epsilon) + a_2 [\log (R + S)] + e_1.$$

The model makes very precise predictions about the signs and magnitudes of the coefficients of (21). Specifically, we will test the hypotheses, separately and jointly, that $a_o = 0$ and $a_1 = a_2 = 1$. It should be recalled that the Dasgupta-Stiglitz model is a special case of our own when $\phi = \gamma = 0$. We can compare estimates of (21) with estimates of

$$(22) \qquad \log H = a'_o + a'_1 (\log \epsilon) + a'_2 (\log R) + e'_1,$$

to determine whether anything is gained by adding advertising to the model.

The second equation takes the general form:

$$(23) \qquad \frac{R}{1 - (R + S)} = \alpha(\cdot) + \theta H \gamma(\cdot) + e_2,$$

where errors are contained in the $\alpha(\cdot)$ and $\gamma(\cdot)$ expressions as noted above. Substituting for $\alpha(\cdot)$ and $\gamma(\cdot)$, we have

$$(24) \qquad \frac{R}{1 - (R + S)} = \left[b_o + \left(\sum_{m=1}^{M} b_m \text{OPP}_m \right) + u \right]$$

$$+ \theta H \left[c_o + \left(\sum_{n=1}^{N} c_n \text{APP}_n \right) + v \right] + e_2,$$

where OPP_m is an element in an M-dimensional vector of variables measuring technological opportunity, and APP_n is an element in an N-dimensional vector of variables measuring the technological conditions of appropriability. Before specifying these opportunity and appropriability measures, two general comments about (24) are in order.

First, the combined error term in equation (24), $e_2 + u + v\theta H$, has undesirable properties, except in cases where θ is assumed to be either zero or n. In the general case, (24) implies that the concentration term, H, has a random coefficient equal to $\theta(c_o + v)$. We will discuss this problem in section 8.4.

Second, it will not be possible to identify θ from the parameters of (24), since θ multiplies each coefficient in $\gamma(\cdot)$. Nevertheless, it will be possible to test specific hypotheses about the value of θ. For example, the hypothesis that $\theta = 0$ implies that the coefficients of the last $N + 1$ terms of the estimated equation are jointly equal to zero. In fact, we cannot distinguish between the hypotheses that $\theta = 0$ and $\gamma = 0$, since the latter carries the same implication for the coefficients of (24). Rejection of this hypothesis, however, is equivalent to rejection of the Dasgupta-Stiglitz specification, in which $\gamma = 0$.

We will also be able to test whether $\theta = n$, the "constant R & D shares" conjecture, since $\theta H = 1$ under this assumption. The specification resulting from this hypothesis is not nested in the empirical specification of the general model, since the elements in the APP vector now enter directly rather than interactively with H. We can, however, test the hypothesis $\theta = n$ against $\theta = 0$. Unfortunately, the hypothesis that conjectures are Cournot ($\theta = 1$) cannot be tested.

Given the available data, we have at best crude proxies for technological opportunity and appropriability. As noted in section 8.1, we are engaged in an effort to develop better measures through a survey of R & D managers. For the present, however, we follow and extend somewhat the approach of Levin (1981).

Opportunities for technical advance depend in part on the particular "science base" of an industry's technology. This suggests adopting the approach of Scherer (1965, 1967), who used dummy variables to classify industries as mechanical, chemical, electrical, or biological. We add metallurgical to this list as the excluded category in the regressions reported below. To represent the "closeness" of an industry's link to science, we use the share of basic research expenditures in total industry R & D. Life cycle models of industry evolution suggest that opportunities may increase in the early years of technological development, because technology is "cumulative" (e.g., Nelson 1981; Nelson and Winter 1982). Later, technological opportunities may be exhausted as industries reach maturity. This suggests that a variable representing industry age should be included among the opportunity measures in both linear and quadratic form. Finally, government policy may affect technological opportunity. It seems reasonable to hypothesize that government-funded R & D is complementary to private effort, thus increasing the elasticity of unit cost with respect to private R & D. Thus, to summarize, we postulate that $\alpha(\cdot)$ is:

(25) $\alpha = b_o + b_1(\text{ELEC}) + b_2(\text{CHEM}) + b_3(\text{BIO}) + b_4(\text{MECH})$
$+ b_5(\text{BASIC}) + b_6(\text{AGE}) + b_7(\text{AGE}^2)$
$+ b_8(\text{GOVRDS}) + u,$

where the first four variables are the technology base dummies described above, BASIC is the ratio of basic R & D to total industry R & D, AGE is industry age, and GOVRDS is the ratio of government-funded R & D to sales. Our expectation is that $b_5, b_6, b_8 > 0$, and $b_7 < 0$.[6] We do not have strong prior beliefs about the relative opportunity of each of the technology types, although we expect $b_1 > 0$ over the period covered by our data (1963–72).

6. There is an alternative expectation for the parameters b_6 and b_7 consistent with our discussion of the life cycle of technological opportunity. Our operational measure of industry age is based on the number of years since each four-digit industry first appeared in the Standard Industrial Classification (SIC), and it is entirely possible that the "early" years of the life cycle in which technological opportunity is growing typically occur before an industry is classified by the Bureau of Census. Under these circumstances, we would expect α to decline with age. Thus, we would not be surprised to find $b_6 < 0$. It is even possible that $b_7 > 0$ would be consistent with our expectations, provided the age at which opportunity is minimized is beyond the range of our observations.

The extent of interfirm spillovers of R & D depends both on legal-institutional features of the environment and on characteristics of the product or technology. For example, there are probably important interindustry differences in the effectiveness of patent protection. The inherent complexity of products and processes is also relevant to the ease of imitation. But these aspects of appropriability remain unmeasured. It is widely believed, however, that products, which can be examined directly by rivals, are more easily imitated than process innovations. Thus, the share of R & D expenditures devoted to new or improved products should be positively related to γ, the degree of spillovers. Fortunately, McGraw-Hill collects data of this sort, and recent work of Scherer (1981) presents alternative estimates derived from patent data.

Scherer's work also suggests another possible measure of appropriability. Industries differ widely in the extent to which they develop their own process technology or "borrow" it in the form of R & D embodied in capital goods and intermediate products. It seems reasonable to presume that interfirm R & D spillovers within the industry are positively associated with the share of process R & D that is "borrowed" rather than generated internally. This presumption rests on the idea that R & D within the industry spills over more readily if the participants are using a common process technology. If most process R & D is developed "in-house," it is more likely that firms have idiosyncratic technologies, and spillovers are less important. Thus, we take the ratio of borrowed R & D to total R & D "used" by an industry as another determinant of γ.

Finally, government funding of R & D frequently carries restrictions on appropriability, such as mandatory licensing. Thus, we expect government R & D to increase the extent of spillovers in an industry.

To summarize, we specify $\gamma(\cdot)$ as:

(26) $\gamma = c_o + c_1(\text{PROD}) + c_2(\text{BORROW}) + c_3(\text{GOVRDS}) + v$,

where PROD is the share of R & D devoted to new or improved products, BORROW is the ratio of R & D embodied in inputs to total R & D used (where R & D used by an industry is the sum of R & D embodied in inputs and its own process R & D), and GOVRDS is the ratio of government R & D to sales. We expect c_1, c_2, and $c_3 > 0$.

We now move to the third equation of the model, which we express in log-linear form:

(27) $\log S = d_0 + d_1[\log \phi(\cdot)] - d_2 \log \epsilon + d_3 \log H + e_3$.

This equation resembles the concentration equation (21), in that the model dictates the precise hypotheses that $d_0 = 0$; $d_1, d_2, d_3 = 1$. Here our problem is complicated because ϕ, the advertising elasticity of demand, cannot be observed directly. We assume, however, that ϕ is a function of product attributes that can be observed. Specifically, we assume that ϕ depends on whether the product in question is a consumer or producer's good, which we measure by the ratio of personal consumption expenditures to total industry sales. Moreover, in line with arguments advanced in the literature on advertising, we expect the responsiveness of demand to advertising to depend in part on whether the product is a durable one. Advertising is likely to have greater impact on the demand for nondurables, since other sources, such as retailers and other customers, tend to be relied on for information about long-lived products. We also add an interaction term, since the latter distinction is likely to be more pronounced for consumer goods.[7] Thus, we specify $\phi(\cdot)$ as:

$$(28) \quad \phi = \exp[g_0 + g_1(\text{PCE}) + g_2(\text{DUR}) + g_3(\text{PCE*DUR}) + w],$$

where PCE is the share of personal consumption expenditures in industry sales, and DUR is a dummy variable representing a durable good. We expect $g_1 > 0$, and $g_2, g_3 < 0$.

We choose an exponential form of $\phi(\cdot)$ for reasons of tractability, since substitution of (28) into (27) will now yield an equation that is linear in the parameters. Nonetheless, estimates of d_0 and d_1 cannot be recovered, since they appear only in combination with the coefficients of $\phi(\cdot)$. On the maintained hypothesis that $d_0 = 0$ and $d_1 = 1$, however, it is possible to identify these latter coefficients and to generate estimates of ϕ for each industry.

One additional issue requires discussion before the empirical specification is complete. It would be possible to treat government R & D as an exogenous policy instrument influencing opportunity and appropriability and, hence, private R & D. Yet there is considerable plausibility to the view that government R & D decisions are influenced by opportunity and appropriability conditions, even if not in quite the same way as private decisions are affected. In particular, the government may attempt to compensate for the inappropriability of R & D returns by allocating R & D funds to industries where spillovers are high and where concen-

---

7. Our argument about durability is based on the insights of Porter (1974), which are employed in the subsequent work of Caves, Porter, and Spence (1980). Porter distinguishes between "convenience" goods, for which advertising is the principal form of selling effort, and "nonconvenience" goods, for which the buyer typically seeks information from other sources. In Porter's scheme, convenience goods are nondurable, purchased frequently, and sold at a low unit price. Operationally, at the highly aggregated level of our data, the classification of industries as producers of nondurables and durables is identical to the classification based on Porter's scheme.

Caves, Porter, and Spence claim that advertising intensity should be higher in convenience goods industries, but this conclusion does not strictly follow from their argument. Instead, they offer good reasons why the ratio of advertising to other forms of selling effort should be higher for convenience goods, but no reasons why the ratio of advertising to sales should be higher.

tration is low. This view receives support from the previous exploratory model of Levin (1981), where the hypothesis that GOVRDS is uncorrelated with the error term in the private R & D equation is decisively rejected in a test based on the work of Wu (1973), Hausman (1978), and Reiss (1981). Although a truly satisfactory model of government R & D allocation would give due weight to political forces, we observe that most of the variance of government differences in R & D expenditures across industries is explained by government procurement policy. Put simply, the government supports R & D in industries where it is a major customer; this holds especially for defense procurements. Therefore, we specify that the ratio of government R & D to sales is determined by:

$$(29) \quad \text{GOVRDS} = h_0 + h_1(\text{DEFSHR}) + h_2(\text{GOVSHR})$$

$$+ h_3\text{H} + \left(\sum_{M=1}^{M} h_{3+m}\text{OPP}_m \right)$$

$$+ \left(\sum_{n=1}^{N} h_{3+M+n}\text{APP}_n \right) + e_4,$$

where DEFSHR is the share of industry sales going to the federal government for defense purposes, and GOVSHR is the share of industry sales purchased by the federal government for other purposes.

8.4 The Data and Estimation Issues

Table 8.1 provides definitions, scalings, and the sources of the data used in this study. Table 8.2 furnishes sample statistics for these data. We have already discussed in some detail the difficulties involved in measuring opportunity and appropriability and the rationale behind our measures. Most of the remaining variables are conventional and require no further comment (for further reference, see Levin 1981). We will, however, comment briefly on the measurement of R & D expenditures, price elasticities of demand, and concentration before discussing estimation procedures.

The only definitionally consistent industry R & D expenditure data are those tabulated by the Bureau of the Census for the National Science Foundation (NSF). Unfortunately, they are only available for highly agregated NSF industry classifications. Our sample consists of the twenty basic manufacturing industries for which data are available since 1963. These industries are a composite of four-, three-, and two-digit SIC classifications (see table 8.2). Since industry data on such variables as concentration and the number of firms are only available for Census of Manufactures survey years (1963, 1967, and 1972), our sample consists of sixty observations.

On the demand side, we use the price elasticities calculated by Levin

Table 8.1   **Variable Definitions and Data Sources**

Variable	Definition (data sources in parentheses)
H	Herfindahl index of concentration (computed from COM)
R	Company-financed R & D expenditures divided by value of shipments (NSF, COM)[a]
S	Advertising expenditures divided by industry output (IO)
RDINT	Research intensity; $R/(1 - R - S)$
GOVRDS	Government-financed R & D expenditures divided by value of shipments (NSF, COM)[a]
ε	Price elasticity of demand (Almon, IO)
ELEC	Industry technology base predominantly electrical (scaled 0–1)
CHEM	Industry technology base predominantly chemical (scaled 0–1)
BIO	Industry technology base predominantly biological (scaled 0–1)
MECH	Industry technology base predominantly mechanical (scaled 0–1)
MET	Industry technology base predominantly metallurgical (scaled 0–1)
BASIC	Basic R & D expenditures divided by total R & D (NSF)
AGE	Years since industry first appeared in *Census of Manufactures* with substantially same definition as today (COM)
AGESQ	Square of AGE
PROD	Share of industry R & D expenditures devoted to new or improved products (McGraw-Hill)
BORROW	R & D embodied in inputs divided by total R & D "used," where latter is the sum of own expenditures on process R & D and R & D embodied in inputs (Scherer)
PCE	Personal consumption expenditures divided by industry sales (IO)
DUR	Dummy variable set equal to one for durable goods, zero otherwise
DEFSHR	Federal government purchases for national defense purposes divided by industry sales (IO)
GOVSHR	Federal government purchases for purposes other than national defense divided by industry sales (IO)

Key to Data Sources:

Code	Source
Almon	Almon, C., et al. 1974. *1985: Interindustry forecasts of the American economy*. Lexington, Mass.: Lexington Books.
COM	U.S. Bureau of the Census. 1963, 1967, 1972. *Census of manufactures*. Washington, D.C.: GPO.
IO	U.S. Department of Commerce, Bureau of Economic Analysis. 1963, 1967, 1972. *Input-output tables for the United States*. Washington, D.C.: GPO.
McGraw-Hill	McGraw-Hill. Economics Department. *Annual surveys of business plans for research and development*. Mimeo, annually.
NSF	National Science Foundation, *Research and development in industry*. Washington, D.C.: GPO, annually.
Scherer	Scherer, F. M. 1981. The structure of industrial technology flows. Northwestern University and FTC. Mimeo.

[a]R & D expenditure data were deflated by a salary index for chemists and engineers constructed from data in U.S. Bureau of Labor Statistics, *National Survey of Professional, Administrative, Technical, and Clerical Pay*. Washington, D.C.: GPO, annually. Industry-specific employment weights for chemists and engineers were taken from BLS Bulletin no. 1609, *Scientific and Technical Personnel in Industry, 1961–66*. Washington, D.C.: GPO, 1968. Value of shipment data were deflated by use of sectoral output price deflators made available on computer tape by the Bureau of Labor Statistics.

Table 8.2   Sample Industries, Selected 1972 Data, and Sample Statistics

		1972 Values of Selected Variables				
SIC (1967)	Industry	C4	H	R	S	GOVRDS
20	Food and kindred products	36	.062	.0023	.0193	.0000+
22, 23	Textile products and apparel	28	.044	.0011	.0060	.0000
24, 25	Lumber, wood products, and furniture	20	.030	.0016	.0057	.0000
26	Paper and allied products	31	.050	.0067	.0080	.0000+
281, 282	Industrial chemicals	45	.102	.0298	.0105	.0060
283	Drugs	29	.043	.0680	.1015	.0002
284–289	Other chemicals	37	.063	.0146	.0877	.0006
29	Petroleum refining and related industries	31	.047	.0158	.0115	.0005
30	Rubber and miscellaneous plastics products	31	.060	.0111	.0155	.0015
32	Stone, clay, glass, and concrete products	36	.071	.0075	.0075	.0001
331–332, 3391, 3399	Ferrous metals and products	41	.068	.0038	.0029	.0001
333–336, 3392	Nonferrous metals and products	49	.086	.0048	.0043	.0004
34	Fabricated metal products	29	.055	.0049	.0057	.0002
35	Machinery, except electrical	38	.066	.0245	.0044	.0053
366, 367	Communication equipment and electronic components	45	.092	.0963	.0022	.1104
361–365, 369	Other electrical equipment	53	.110	.0294	.0179	.0249
371	Motor vehicles and motor vehicle equipment	81	.251	.0263	.0103	.0043
372	Aircraft and parts	62	.129	.0559	.0021	.2385
381–382	Scientific and mechanical measuring instruments	36	.059	.0282	.0094	.0068
383–387	Optical, surgical, photographic, and other instruments	60	.123	.0519	.0268	.0142
Sample statistics:						
	Mean over all industries and years	39.6	.078	.0247	.0172	.0259
	Standard deviation	14.1	.045	.0237	.0253	.0656
	Minimum value	15	.025	.0006	.0010	.0000
	Maximum value	81	.251	.0963	.1015	.3654

(1981). These were computed using time-series estimates of constant elasticity demand functions made by Almon and his colleagues (1974). These estimates are available for fifty-six input-output sectors in which the predominant fraction of output goes to personal consumption. On the rather strong pair of assumptions that all output in the fifty-six consumer goods sectors goes to personal consumption and that all output of the remaining maufacturing industries finds its way into consumer goods industries, derived elasticities of demand were calculated for each of the input-output sectors in manufacturing. To the extent that some output of the fifty-six consumer goods industries is used as intermediate input or as investment goods and to the extent that some output of the remaining industries is consumed directly, the calculated elasticities will be biased in uncertain direction and magnitude. Nevertheless, the procedure produces no serious anomalies; the relative magnitudes of the elasticities across industries accord reasonably well with intuition.

The most difficult data problem we confront is to develop an operational analog of n, the number of firms. Obviously, the rather special assumption of symmetric firm size is not consistent with the observed heterogeneity of firm sizes. Our approach here is to instead view n as a numbers equivalent (see, for example, Hart 1971) and to regard (22) as an approximation to a world with heterogeneous firm sizes. Our practical problem is how to summarize the empirical size distribution of firms by a numbers equivalent. For this purpose we choose to treat n as the Herfindahl numbers equivalent.

To obtain an operational measure, we must therefore construct a Herfindahl index for each sample industry. Since the empirical size distributions of firms are available in an incomplete form (e.g., C4, C8, C20, C50, etc.), we chose to fit two distributions (the Pareto and the exponential) to the available data for each four-digit industry. Of the two distributions the exponential provided the more satisfactory approximation. Using the estimated parameters of these size distributions, we simulated the Herfindahl index for each four-digit industry and then used a weighted average of four-digit Herfindahl indexes to represent concentration at the more aggregated level of our sample. The resulting index values, which appear in table 8.2, are quite plausible. At the four-digit level, the correlation between the four-firm concentration ratio and our estimated Herfindahl index is .91.

We now turn to estimation problems. While estimation of the advertising and concentration equations involves straightforward application of nonlinear two-stage least-squares procedures, estimation of the R & D equation is not straightforward. As noted above, the problem arises because of the error in observing γ, which leads to a random coefficient on the concentration term, unless γ or $\theta = 0$, $\theta = n$, or $\sigma_v^2 = 0$. That is, the system (22), (24), and (27) will not require attention to the random coefficient problem in three cases. (1) when γ or $\theta = 0$, firms behave as Dasgupta-Stiglitz firms, and appropriability does not affect R & D intensity; (2) when $\theta = n$, concentration drops out of the R & D equation; and (3) if γ_k is measured without error, no randomness is in the concentration coefficient.

The third of these special cases is the least probable. Thus, if we wish to estimate the model without prespecifying θ, we must deal with the problem of random coefficients in a nonlinear simultaneous equations context. To date, only Kelejian (1974) has suggested a procedure for the linear simultaneous equations model with random coefficients, and for our model his specialized results are inapplicable.

Although we have not been able to find a fully efficient random coefficient estimator for our model, we are able to show that consistent estimates of the parameters in (24) can be obtained by nonlinear instrumental variables techniques. Details concerning the assumptions employed and a rigorous statement of this result are available from the authors on request.

The more serious problem arising from the random coefficient version of (24) is the possible inconsistency of conventional approximations to the asymptotic standard errors of the coefficients. This inconsistency results from the heteroscedastic disturbance term. Our approach to this problem is to apply a generalized least-squares (GLS) correction to our instrumental variables estimator. To do this we must estimate σ_ϵ^2 and σ_ν^2. We do so by an auxiliary regression technique that uses the squared residuals from the initial instrumental variables estimates (see Hildreth and Houck 1968).

8.5   The Results

Table 8.3 summarizes the full specification of our four-equation model, along with the expected signs and parameter restrictions derived from the analysis of section 8.3. All equations satisfy the conditions for identification of models involving nonlinearities in the endogenous variables. We estimated each equation over our twenty industry sample for the years 1963, 1967, and 1972. For each specification reported in this section we could not reject the hypothesis of homogeneity across time periods. We therefore limit discussion to the results obtained using the pooled sample.

Each equation was estimated using single equation instrumental variables techniques involving linear approximations to the reduced forms. For the private R & D equation we attempted the Hildreth-Houck GLS correction to take account of possible heteroscedasticity in the disturbance term. A decomposition of the residuals, however, revealed no evidence of heteroscedasticity. Furthermore, the GLS estimates and their standard errors were not much different from those obtained by the

Table 8.3   **Expected Signs and Magnitudes of Structural Coefficients**

Variable	LOG(H)	RDINT	LOG(S)	GOVRDS
LOG(ϵ)	+1.0		−1.0	
LOG($R + S$)	+1.0			
LOG(H)			+1.0	
ELEC		+		+
CHEM		?		?
BIO		?		?
MECH		?		?
BASIC		+		+
AGE		+?		+?
AGESQ		−?		−?
GOVRDS		+		
H		?		−
H*PROD		+		
H*BORROW		+		
H*GOVRDS		+		
PCE			+	
DUR			−	
PCE*DUR			−	
DEFSHR				+
GOVSHR				+
PROD				+
BORROW				+
CONSTANT	0.0	?	?	?

uncorrected instrumental variables procedure. For these reasons, we proceed as if $\sigma_v^2 = 0$, and we report the uncorrected parameter estimates and standard errors in table 8.5 below.

Overall the results are quite encouraging given the small sample size, the degree of aggregation, and the potential measurement errors in our data. The signs of virtually all coefficients in the private R & D equation are in agreement with our predictions in table 8.3. Further, the point estimates are remarkably robust to minor modifications in the specification, and many coefficients are significant at a size of .01. The results for the concentration, advertising, and government R & D equations are less encouraging than those of the company R & D equation. Here the specifications are quite sensitive to our implied restrictions. We now proceed to a more detailed discussion of the results.

8.5.1   The Concentration Equation

Table 8.4 reports the results of estimating two variants of the concentration equation (21). Clearly, the estimated coefficients fail to conform to our precise predictions about their magnitudes. In specifications (4-1), we decisively reject the hypothesis that the constant is equal to zero, and the strict hypotheses that $a_1 = a_2 = 1$ are also clearly rejected, separately

Table 8.4   **Parameter Estimates: Concentration Equation (asymptotic standard errors in parentheses)**

Variable	(4-1)	(4-2)
LOG(ϵ)	−0.064	0.040
	(0.039)	(0.056)
LOG($R + S$)	0.164***	0.704***
	(0.068)	(0.028)
CONSTANT	−2.146†††	—
	(0.262)	
Mean of dep.		
variable	−2.666	−2.666
Std. error	0.434	0.660

***Asymptotic t-ratio indicates significance at .01 level (one-tailed test).

†††Asymptotic t-ratio indicates significance at .01 level (two-tailed test).

and jointly. On the other hand, the term representing cost-reducing and demand-shifting activities [LOG $(R + S)$] is significantly greater than zero at the .01 level.

Matters improve substantially when we impose the restriction that the constant term equal zero. The price elasticity coefficient remains insignificantly different from zero and significantly below one, but it has the correct sign in (4-2). The coefficient on LOG($R + S$) jumps dramatically to a value not far from our prediction. We must nevertheless reject the hypothesis that $a_2 = 1$.

We also explored variants of equation (22) in which advertising is excluded from the model. These results were similar to those reported in table 8.4. The price elasticity coefficient was much too small, and it had the wrong sign when the constant was included. The coefficient of LOG(R) was below 0.2 when the constant was included, and it jumped to 0.57 when the zero restriction was imposed. These specifications are not nested in (4-1) and (4-2), so we did not test the restriction implicit in the Dasgupta-Stiglitz model that advertising has no effect on concentration.

The failure of (4-1) and (4-2) to conform precisely to theoretical expectations is hardly surprising, given our highly aggregated data and the potential for measurement error in the price elasticities. Moreover, it is likely that the concentration equation is particularly sensitive to our neglect of dynamics. A less stylized theory of Schumpeterian competition would surely model concentration as the outcome of a sequence of past and current investments in R & D and advertising, rather than as a contemporaneously determined variable.

8.5.2   The Company R & D Equation

The results for several variants of the company R & D equation (24) are presented in table 8.5. Note at the outset that the γ function reflecting

Table 8.5   Parameter Estimates: R & D Equation (asymptotic standard errors in parentheses)

Variable	θ, γ Unrestricted		γ = Cons.	γ, $\theta = 0$	$\theta = n$
	(5-1)	(5-2)	(5-3)	(5-4)	(5-5)
ELEC	−0.038	0.017*	0.033***	0.036***	0.029***
	(0.027)	(0.012)	(0.011)	(0.011)	(0.012)
CHEM	−0.006	−0.005	0.000	−0.001	−0.004
	(0.010)	(0.007)	(0.007)	(0.007)	(0.008)
BIO	0.064††	0.060†††	0.053††	0.043†	0.038†
	(0.031)	(0.022)	(0.023)	(0.022)	(0.022)
MECH	0.004	0.009	0.017†	0.021††	0.017†
	(0.014)	(0.010)	(0.010)	(0.009)	(0.010)
BASIC	0.101	0.259***	0.299***	0.349***	0.358***
	(0.153)	(0.100)	(0.104)	(0.098)	(0.098)
AGE/100	0.152	−0.106†	−0.155††	−0.147††	−0.121†
	(0.135)	(0.064)	(0.064)	(0.066)	(0.069)
AGESQ/100	−0.0016	0.0012†	0.0018††	0.0017††	0.0015†
	(0.0015)	(0.0007)	(0.0007)	(0.0007)	(0.0008)
GOVRDS	2.694***	0.078**	0.120***	0.121***	0.115***
	(1.005)	(0.043)	(0.042)	(0.043)	(0.043)
H	−0.750††	−0.362†	0.080	—	—
	(0.326)	(0.205)	(0.066)		
H*PROD	1.715***	0.742**	—	—	—
	(0.595)	(0.328)			
H*GOVRDS	−24.082†††	—	—	—	—
	(9.239)				
PROD	—	—	—	—	0.034
					(0.031)
CONSTANT	−0.054	0.009	0.020	0.021	0.008
	(0.033)	(0.016)	(0.016)	(0.017)	(0.030)
Mean of dep. variable	0.027	0.027	0.027	0.027	0.027
Std. error	0.022	0.016	0.017	0.017	0.017

*Asymptotic t-ratio indicates significance at .10 level (one-tailed test).

**Asymptotic t-ratio indicates significance at .05 level (one-tailed test).

***Asymptotic t-ratio indicates significance at .01 level (one-tailed test).

†Asymptotic t-ratio indicates significance at .10 level (two-tailed test).

††Asymptotic t-ratio indicates significance at .05 level (two-tailed test).

†††Asymptotic t-ratio indicates significance at .01 level (two-tailed test).

the technological dimension of spillovers is modeled by including only a constant term, PROD and GOVRDS, as arguments. Inclusion of both PROD and BORROW in the regressions produced less satisfactory results, probably because of their near collinearity. Since PROD had the more robust parameter estimates over a range of specifications, we report results for variants of (24) omitting BORROW.

The most general form of (24) is where the value of θ is arbitrary but constant across industries, as in (5-1) and (5-2). If R & D spillovers are assumed to be greater for products than for processes (i.e., PROD has a positive coefficient in the γ function), then the estimated coefficients imply that θ is greater than zero. Thus, firms appear to maintain positive conjectural variations with respect to R & D, although the free-rider effect could still be present if $0 < \theta < 1$.

A somewhat surprising result of (5-1) is that the coefficient on H*GOVRDS is negative, suggesting that government funding diminishes spillovers. Although this contradicts our earlier expectation, the result is plausible for several reasons. Most prominently, much government funding supports R & D for large-scale, capital-intensive defense systems which are not cheaply replicable despite mandatory licensing and technology transfer provisions.

Specification (5-2) omits H*GOVRDS on the hypothesis that another reason for its unexpected sign may be its near collinearity with the opportunity vector. Once again, GOVRDS is significantly positive, although its effect on the cost elasticity is relatively small. The magnitude of the coefficient indicates that on average a one dollar increase in government R & D spending leads to a seven cent increase in company R & D spending. Estimates for specifications (5-3) through (5-5) are about eleven cents; however, at the means (5-1) yields a predicted effect of seventy-four cents. The other opportunity variables in (5-2) come in strongly with the correct signs.

Specifications (5-3) and (5-4) represent two restricted versions of (5-1) and (5-2). In (5-3) we test the hypothesis that γ is a constant across all industries. The positive coefficient on the Herfindahl index indicates that θ is greater than zero under the implicit assumption that γ is greater than zero. In any case, Wald tests indicate that either (5-1) or (5-2) is to be preferred to (5-3).

Specification (5-4) corresponds to a Dasgupta-Stiglitz world with no R & D spillovers. Interestingly enough, this equation does quite well in that all the opportunity variables are of the correct sign and highly significant. Furthermore, the point estimates of the opportunity coefficients differ only slightly from those in (5-2). However, Wald tests on the hypothesis that either γ or θ equals zero lead to rejection of the Dasgupta-Stiglitz model in favor of (5-1) or (5-2). The χ^2 statistics are respectively $\chi^2(3) = 10.2$ and $\chi^2(2) = 6.8$.

The final specification we report is the "constant shares" case where θ equals n and thus varies across industries. Once again the results accord reasonably well with the previous versions of the company R & D equation. Interestingly, we cannot reject the Dasgupta-Stiglitz model in favor of the constant shares case, given the insignificance of the coefficient of PROD. Unfortunately, we are unable to test (5-5) against the other versions, since (5-5) is not nested in the θ, γ unrestricted cases.

Since it is widespread practice to treat concentration and government

R & D as exogenous variables in empirical models of the determination of company R & D, it is interesting to ask whether anything is gained by treating them as endogenous. We checked for the possibility of simultaneity bias using the test proposed by Wu (1973). For each specification in table 8.5, we decisively rejected the hypothesis that the regressors we take to be stochastic were in fact uncorrelated with the disturbances.

As further checks on the models (5-1)–(5-5), we have tested the plausibility of our parameterization of α and the reasonableness of the opportunity measures. Operationally this was done by excluding the opportunity measures from all five equations. The resulting Wald statistics, which are asymptotically distributed as χ^2 random variables with eight degrees of freedom, all exceeded the critical value at the .001 level. Thus, we decisively reject the α constant version of the model.

We also computed the implied cost function parameters in those instances where it was feasible to do so. Table 8.6 reports the estimated values of $\hat{\alpha}$ for each industry in 1972, derived from equation (5-4), and the estimated values of $\hat{\alpha} + \hat{\gamma}$, derived from equation (5-5). The results accord well with expectations. The implied elasticity of unit cost with respect to company R & D ranges from near zero in textiles, paper, and fabricated metal products to the 0.05–0.07 range in aircraft, drugs, and

Table 8.6   Estimated Cost Elasticities and Actual R & D Intensity for Sample Industries, 1972

Industry	$\hat{\alpha}$ (from eq. [5-4])	$\hat{\alpha} + \hat{\gamma}$ (from eq. [5-5])	R
Food and kindred products	.041	.038	.0023
Textiles products and apparel	.001	.001	.0011
Lumber, wood products, and furniture	.012	.016	.0016
Paper and allied products	−.001	−.002	.0067
Industrial chemicals	.035	.036	.0298
Drugs	.059	.061	.0680
Other chemicals	.012	.014	.0146
Petroleum refining and related industries	.020	.019	.0158
Rubber and miscellaneous plastic products	.003	.003	.0111
Stone, clay, glass, and concrete products	.013	.014	.0075
Ferrous metals and products	.004	.007	.0038
Nonferrous metals and products	.002	−.003	.0048
Fabricated metal products	−.001	.001	.0049
Machinery, except electrical	.021	.022	.0245
Communication equipment and electronic components	.066	.066	.0963
Other electrical equipment	.037	.039	.0294
Motor vehicles and motor vehicle equipment	.014	.009	.0263
Aricraft and parts	.049	.051	.0559
Scientific and mechanical measuring instruments	.020	.021	.0282
Optical, surgical, photographic, and other instruments	.030	.032	.0519

electronics. Table 8.6 also includes the 1972 values of R & D intensity for convenient reference.

Finally, before turning to the advertising equation, we shall interpret the coefficients on the industry age variables. The signs on the age and age-squared coefficients suggest that once an industry is defined by the Census, it already faces declining opportunities for R & D. The magnitudes of these coefficients imply that opportunities decline for forty to forty-five years after definition, at which point they increase again. Depending on the specification, the standard error for the estimate of this turning point is between five and six years.

8.5.3   The Advertising Equation

Estimation of the advertising equation yields results that are not entirely satisfactory. As table 8.7 reveals, in the theoretically preferred specification (7-1), the coefficient on the concentration term has the wrong sign; a result which corresponds to that reported in Levin (1981). Moreover, the price elasticity of demand has a coefficient smaller in absolute value than its expected magnitude. Indeed, the hypothesis that $d_2 = -1$ is decisively rejected at better than a .001 level. The determinants of ϕ, the advertising elasticity of demand, fare somewhat better. Both PCE and DUR have the expected signs, although the interaction term does not.

We have no prior expectation about the constant term, since it repre-

Table 8.7   **Parameter Estimates: Advertising Equation**

Variable	(7-1)	(7-2)
LOG(ϵ)	−0.012	0.191
	(0.091)	(0.096)
LOG(H)	−0.465	1.656***
	(0.455)	(0.125)
PCE	2.284***	2.949***
	(0.841)	(0.987)
DUR	−0.539*	1.173***
	(0.358)	(0.396)
PCE*DUR	2.373	−1.258
	(1.619)	(1.702)
CONSTANT	−6.308†††	—
	(1.317)	
Mean of dep. variable	−4.743	−4.743
Std. error	0.859	1.022

*Asymptotic t-ratio indicates significance at .10 level (one-tailed test).

***Asymptotic t-ratio indicates significance at .01 level (one-tailed test).

†††Asymptotic t-ratio indicates significance at .01 level (two-tailed test).

sents the sum of d_0, which is expected to be zero, and $d_1 g_0$, which involves the constant in the ϕ function. If we constrain both d_0 and g_0 to be zero, however, the results improve markedly. The price elasticity coefficient reverses sign, but the remaining coefficients in (7-2) have the correct signs. The concentration term is now significantly different from zero, although the hypothesis that $d_3 = 1$ must be rejected at the .001 level. Each of the arguments of the ϕ function now has the predicted sign, and the hypothesis that $\phi(\cdot)$ is simply a constant across all industries can be rejected for both specifications (7-1) and (7-2). In the former case, the hypothesis that ϕ is a constant yields a test statistic that is $\chi^2(3) = 37.1$. In the latter case, the test that $\phi = 0$ is $\chi^2(3) = 104.4$.

The poor performance of the price elasticity variable in both this and the concentration equation calls attention to the very strong assumptions under which the elasticities were computed. Some further work is needed here; in future work we intend to employ alternative elasticity estimates.

8.5.4 The Government R & D Equation

As expected, the results in table 8.8 indicate that the allocation of government R & D expenditures is influenced most strongly by government defense procurement; the government supports R & D in those industries in which it is a major customer. Technological opportunity appears to offer little incentive to the government; of our opportunity measures only AGE and AGESQ are statistically significant at conventional levels. Interestingly, the signs of these coefficients are the reverse of those in most specifications of the private R & D equation. Given our expectation that the age profile proxies opportunity by first rising and then falling, we might tentatively note that the pattern of signs in tables 8.5 and 8.8 is consistent with the view that the government reacts to technological opportunity with a substantial lag relative to private industry.

The technological dimension of spillovers, here proxied by BORROW, appears to have some effect on government R & D; as expected, a higher degree of spillover increases the likelihood of government support. Again collinearity among the appropriability measures leads to better results when either PROD or BORROW is excluded. In this case, BORROW has the more plausible parameter estimate and a lower relative standard error. The structural dimension of appropriability, here represented by concentration, has the expected sign but falls well short of statistical significance. The remaining coefficients in the equation are almost completely insensitive to the exclusion of this variable, as shown in (8-2). This suggests that GOVRDS is not strictly endogenous, although its dependence on opportunity and appropriability conditions indicates that it is correlated with the error term in the private

Table 8.8   **Parameter Estimates: Government R & D Equation**

Variable	(8-1)	(8-2)
ELEC	0.010	0.009
	(0.016)	(0.014)
CHEM	−0.003	−0.003
	(0.011)	(0.010)
BIO	0.028	0.033
	(0.034)	(0.031)
MECH	0.010	0.008
	(0.014)	(0.012)
BASIC	0.012	0.015
	(0.159)	(0.141)
AGE/100	0.224††	0.220††
	(0.099)	(0.097)
AGESQ/100	−0.003††	−0.003††
	(0.001)	(0.001)
H	−0.038	—
	(0.099)	
BORROW	0.056**	0.053*
	(0.034)	(0.032)
DEFSHR	0.066***	0.066***
	(0.006)	(0.006)
GOVSHR	0.001	0.002
	(0.048)	(0.047)
CONSTANT	−0.072††	−0.071††
	(0.030)	(0.030)
Mean of dep. variable	0.026	0.026
Std. error	0.024	0.024

*Asymptotic t-ratio indicates significance at .10 level (one-tailed test).

**Asymptotic t-ratio indicates significance at .05 level (one-tailed test).

***Asymptotic t-ratio indicates significance at .01 level (one-tailed test).

††Asymptotic t-ratio indicates significance at .05 level (two-tailed test).

R & D equation. Thus, the use of instrumental variables for GOVRDS seems appropriate.

8.6 Conclusions

Given the deficiencies of the variables used to measure technological opportunity and appropriability conditions, as well as the highly aggregated nature of the data, the results reported in section 8.5 are quite encouraging. Although the statistical tests are not entirely consistent with our theoretical model, on the whole the findings support the Schumpeterian view that R & D investment and market structure are appropriately regarded as jointly determined outcomes of the competitive process. The private R & D equation performs especially well, yielding results that are quite robust and yet sufficiently precise to reject decisively the hypotheses that opportunity and appropriability conditions do not matter. Indeed, the private R & D results are substantially better than those obtained in the looser, more inclusive, specification in the earlier work of Levin (1981). The parameters of the concentration equation fail to conform to the precise predictions of our model, but the results nevertheless suggest a strong and significant connection between cost-reducing and demand-shifting activities and market concentration.

We do not wish to make exaggerated claims for our highly stylized theoretical model, which abstracts from obviously important features of Schumpeterian competition, such as dynamics and the heterogeneity of firms. But we believe that the model does place proper emphasis on demand, technological opportunity, and appropriability as the central forces determining the allocation of R & D and the evolution of market structure. In particular, we believe our treatment of R & D spillovers, which distinguishes clearly the technological, structural, and behavioral dimensions of appropriability, exemplifies how useful insights may be gained from relatively stark and stylized models. Moreover, our model brings to the foreground a thread linking much of the theoretical literature of the "new industrial organization": the endogeneity of market structure. Indeed, the recognition that market structure is endogenous is an element common to the literatures on Schumpeterian competition, monopolistic competition, strategic entry deterrence, and contestability. As these theoretical literatures continue to revise our understanding of structure-conduct-performance relationships, we will undoubtedly see more empirical work of the type represented here.

It is well to keep in mind, however, both the strengths and weaknesses of empirical work based on highly stylized analytic models. On the one hand, such models have the virtue of simplicity, of clear and precise hypotheses. On the other hand, important features of reality are brushed aside. To this extent our insights are only partial truths.

The present paper exemplifies this dilemma. We have tested a simple model which captures much that is important. But a model of Schumpeterian competition without dynamics, without transient monopoly, without innovators and imitators, is, at best, only part of the story. Much remains to be done.

References

Almon, C., M. B. Buckler, L. M. Horwitz, T. C. Reinhold 1974. *1985: Interindustry forecasts of the American economy.* Lexington, Mass.: Lexington Books.

Arrow, K. J. 1962. Economic welfare and the allocation of resources for invention. In *The rate and direction of inventive activity: Economic and social factors,* ed. R. R. Nelson. Universities-NBER Conference Series no. 13. Princeton: Princeton University Press for the National Bureau of Economic Research.

Caves, R. E., M. E. Porter, and A. M. Spence. 1980. *Competition in the open economy.* Cambridge: Harvard University Press.

Dasgupta, P., and J. E. Stiglitz. 1980a. Industrial structure and the nature of innovative activity. *Economic Journal* 90:266–93.

———. 1980b. Uncertainty, industrial structure, and the speed of R & D. *Bell Journal of Economics* 11:1–28.

Dorfman, R., and P. O. Steiner. 1954. Optimal advertising and optimal quality. *American Economic Review* 44:826–36.

Farber, S. 1981. Buyer market structure and R & D effort: A simultaneous equations model. *Review of Economics and Statistics* 63:336–45.

Fellner, W. J. 1951. The influence of market structure on technological progress. *Quarterly Journal of Economics* 65:556–77.

Futia, C. 1980. Schumpeterian competition. *Quarterly Journal of Economics* 94:675–95.

Galbraith, J. K. 1956. *American capitalism: The concept of countervailing power.* 2d ed. Boston: Houghton Mifflin.

Hart, P. E. 1971. Entropy and other measures of concentration. *Journal of the Royal Statistical Society,* series A, 134:73–85.

Hausman, J. A. 1978. Specification tests in econometrics. *Econometrica* 46:1251–71.

Hildreth, C., and J. P. Houck. 1968. Some estimators for a linear model with random coefficients. *Journal of the American Statistical Association* 63:584–95.

Kamien, M., and N. Schwartz. 1975. Market structure and innovation: A survey. *Journal of Economic Literature* 13:1–37.

Kelejian, H. H. 1974. Random parameters in a simultaneous equation framework: Identification and estimation. *Econometrica* 42:517–27.

Lee, T., and L. L. Wilde. 1980. Market structure and innovation: A reformulation. *Quarterly Journal of Economics* 94:429–36.

Levin, R. C. 1978. Technical change, barriers to entry, and market structure. *Economica* 45:347–61.

———. 1981. Toward an empirical model of Schumpeterian competition. Working paper series A, no. 43. Yale School of Organization and Management.

Loury, G. C. 1979. Market structure and innovation. *Quarterly Journal of Economics* 93:395–410.

Nelson, R. R. 1959. The simple economics of basic scientific research. *Journal of Political Economy* 67:297–306.

———. 1981. Balancing market failure and governmental inadequacy. Working paper. Yale University.

Nelson, R. R., and S. G. Winter. 1977. Dynamic competition and technical progress. In, *Economic progress, private values and public policy: Essays in honor of William Fellner*, ed. B. Balassa and R. R. Nelson. Amsterdam: North-Holland.

———. 1978. Forces generating and limiting concentration under Schumpeterian competition. *Bell Journal of Economics* 9:524–48.

———. 1982. The Schumpeterian trade-off revisited. *American Economic Review* 72:114–32.

Phillips, A. 1966. Patents, potential competition, and technical progress. *American Economic Review* 56:301–10.

———. 1971. *Technology and market structure: A study of the aircraft industry*. Lexington, Mass.: Lexington Books.

Porter, M. E. 1974. Consumer behavior, retailer power, and market performance in consumer goods industries. *Review of Economics and Statistics* 56:419–35.

Reiss, P. 1981. Small and large sample equivalences of Hausman's *m* tests to the classical specification tests. Working paper. Yale University.

Scherer, F. M. 1965. Firm size, market structure, opportunity, and the output of patented inventions. *American Economic Review* 55:1097–1125.

———. 1967. Market structure and the employment of scientists and engineers. *American Economic Review* 57:524–31.

———. 1980. *Industrial market structure and economic performance*. 2d ed. Chicago: Rand-McNally.

———. 1981. The structure of industrial technology flows. Northwestern University and FTC. Mimeo.

Schumpeter, J. A. 1950. *Capitalism, socialism, and democracy*. 3d ed. New York: Harper and Row.

Wu, D. 1973. Alternative tests of independence between stochastic regressors and disturbances. *Econometrica* 41:733–50.

Comment   Pankaj Tandon

This is an interesting paper, since it seeks to initiate the difficult task of building bridges between recent theoretical work on R & D and the empirical work on this subject, of which we have seen many fine examples at this conference. I welcome this bridge-building activity, but believe, as the authors do, that much remains to be done. I will therefore concentrate my remarks on the general modeling approach and on some of the problems that need special attention.

The authors are obviously aware of many of the modeling problems involved here, and they have commented on several. I will reemphasize only one—the question of timing. When one considers timing, it is probable—as the authors point out—that the effect of technological change on market structure is a long-term effect, one that cannot be easily captured in tests that span only ten years. Seen in this light, it is not surprising that the concentration equation fails to be convincing. On the other hand, the influence of market structure on R & D investment decisions is more likely to be captured in a short panel. As a short-run approximation, it is probably reasonable to assume a fixed market structure. Thus, the relative success of the R & D intensity equation was also to be expected.

Let me comment on what the authors mention as the two major contributions of their approach. The first relates to the inclusion of advertising as another variable influencing and being influenced by market structure. Though the authors comment on the similarity between advertising and R & D as instruments of competition, they treat them in quite an asymmetric manner in their model. Specifically, the way advertising is included is to model the aggregate industry demand curve as depending on the aggregate level of advertising in the industry. Thus, if there are n firms in the industry, any firm doing advertising gets in effect only $1/n$th of the increase in demand. Clearly this does not really capture the *competitive* characteristics of advertising; rather, advertising acquires the characteristics of a public good—with a serious attendant free-rider problem. The corresponding treatment for R & D would be to say that a firm's unit cost is a function only of Z (aggregate industry R & D). This doesn't seem reasonable. Likewise, I don't find the treatment of advertising convincing. The results from the advertising equation are not satisfactory in any case and might be better explained by a story that concentrated on the role of advertising in product differentiation. I don't know if I'm just revealing my bias as someone primarily interested in R & D, but I think I'd be happier with advertising left out of the model. Another

Pankaj Tandon is an assistant professor in the Department of Economics, Boston University.

reason for this is that it is far from clear to me that R & D and advertising are jointly determined. In a model with uncertain R & D outcomes, advertising may be *sequentially* determined, conditioned on the results of R & D.

Since the primary interest of this group is R & D, let me move on to the second, more interesting point, namely, the model of R & D spillovers. This is surely welcome. The public goods aspect of R & D has received insufficient attention in the theoretical and empirical literature, yet it is one of the things that makes R & D particularly interesting and distinct from traditional investment. I am, however, a little uncomfortable with the way spillovers have been modeled here. The authors have taken the unit cost of production to be a function of the firm's own R & D and of aggregate industry R & D. I have two problems with this. First, as is clearly evidenced by Professor Scherer's work on interindustry technology flows, spillovers are also important *across* industries. This will be of special importance when, as the authors promise us, more disaggregated data at the four-digit level are used. Further, such cross-industry externalities may be quite important in determining market structure in user industries. Two obvious examples spring to mind. The development of small, powerful computers will surely be important, although it is not clear whether the major influence will be to increase or decrease concentration. It seems that minimum efficient scale may be significantly reduced in many industries. On the other hand, small computers may combine with the second example, improved telecommunications capabilities, to considerably ease the control problem in large corporations. This would clearly tend to increase concentration. Some attention must be paid to these interindustry externalities.

The second problem with the spillover analysis concerns the logical consistency of the model. All firms in an industry are assumed to face the same technological opportunities. We then concentrate on symmetric equilibria. In the absence of uncertainty, the only logical interpretation of such an equilibrium is that every firm has discovered the same things. But then, why should the R & D of other firms be of any value to me, since I already know what they know? (In fact, this logic would say that the only permissible spillovers would be interindustry!) I grant that a model with asymmetric information and uncertainty—even with symmetry—is going to be analytically much less tractable, and it is not obvious that the estimable equations would be structurally different, but I do worry about the logical anomaly.

The spillovers were nevertheless an interesting addition to the model, particularly because of a side product that was generated. Including Z in the cost function enabled the authors to begin to model different behavioral hypotheses about firms, specifically: What are firms' conjectures about the response of their rivals to their own R & D? Unfortunately, the authors are not able to get clear answers on this, mainly because the conjectural parameter θ appears multiplicatively with the appropriability parameter γ. I have two suggestions. First, Z could be measured excluding the firm's own R & D. This seems conceptually preferable anyway, although I grant that at this level of aggregation it may not appear important. However, the interpretation of θ will be quite different and, in particular, its sign will matter. Now $\theta - 0$ will correspond to the Cournot conjecture, θ will be negative for free-rider situations, and θ will be positive for what might be called Schumpeterian conjectures. Under the assumption that γ is positive, it will be possible to run tests on the sign of θ. In the present version, θ is always positive, and it is not possible to find its magnitude distinctly from γ. Of course, equation (13) would have to be modified—the n in the denominator would have to be replaced by $(n - 1)$—and perhaps other things would change in the model. The second suggestion is that perhaps a model that pays more explicit attention to product differentiation may be more successful and desirable. In other words, we might get more directly at the conjectural hypotheses by thinking about the firm's demand or price being influenced by other firms' costs.

A specific comment here: In regression (5-4), the R & D equation for the Dasgupta-Stiglitz case, the authors do not tell us if the left-hand side was modified to be $R/(1 - R)$. Advertising surely ought to be excluded if the equation is to represent the Dasgupta-Stiglitz model.

I have a couple of other brief comments. We have not seen the conditions that ensure the existence of symmetric equilibria, but we can easily guess the parameters that must appear there. Since it is well known that R & D causes nonconvexities, it is of some interest to make sure that the existence conditions are being satisfied. It would be useful if the authors provided that information. Further, since α, the elasticity of unit cost with respect to R & D, is estimable in their model, it would be useful to report it and to compare it with the many other estimates of this parameter or its variants that exist in the literature.

One noticeable feature of this paper is how distinct it is methodologically from other papers presented at this conference. I think it may be useful for the authors in their bridge-building activity to look again at the work of Pakes and Schankerman (this volume) on the determinants of research intensity. This is perhaps the clearest exposition of the traditional approach, and it might be productive to attempt to modify this approach with an endogenous market structure.

Finally, let me say what I think is most attractive about this line of work. The significant way in which the endogenous market structure approach differs from the traditional approach is that the new approach introduces a zero-profit condition. This corresponds to what Scherer, in the context of the optimal patent literature, has called the Lebensraum effect. I think that it is entirely possible that this Lebensraum effect actually is the dominant consideration in R & D allocations, especially because of the serious possibility of nonconvexities. Consideration of the Lebensraum effect might also have implications for patent policy or antitrust. For example, some very crude Harberger-type calculations that I have done indicate that in industries where technological opportunity is high, the optimal market structure may be quite concentrated, precisely because of the Lebensraum effect. The trade-off between static and dynamic efficiency is very explicit in Schumpeter's work, and the approach of Levin and Reiss is a first step toward the integration of this trade-off into empirical work on R & D.

9   An Exploration into the Determinants of Research Intensity

Ariel Pakes and Mark Schankerman

This paper explores the economic factors that determine the distribution of research effort across firms. Our main objectives are to provide a general framework for analyzing the demand for research by private firms and to document empirically certain stylized facts about R & D intensity and its determinants at different levels of aggregation.

Three competing explanations of the distribution of research are in the literature, each emphasizing a different aspect of the problem. Schmookler (1966) and Griliches and Schmookler (1963) emphasize the importance of expected market size as an inducement to research effort. They recognize that the cost of reproducing the knowledge generated by research is low relative to the original cost of producing it, and therefore, that the private return to research varies directly with the number of units of output embodying the knowledge or with the size of the market. Differences across industries in the cost of producing knowledge are downplayed, based on the argument that scientific knowledge is sufficiently well developed to make the supply of new industrial knowledge highly elastic at the same level of costs for all industries. Rosenberg (1963, 1969, 1974) and Scherer (1965), while granting the importance of market size, argue that the body of scientific and engineering knowledge

Ariel Pakes is a lecturer in the Department of Economics in the Hebrew University of Jerusalem, and a faculty research fellow of the National Bureau of Economic Research. Mark Schankerman is an assistant professor in the Department of Economics at New York University, and a faculty research fellow of the National Bureau of Economic Research.

This paper is a consolidated and extended version of work originally contained in Pakes and Schankerman (1977) and Pakes (1979). The authors would like to thank Zvi Griliches for many constructive discussions of the issues in this paper, and Gary Chamberlain and Bronwyn Hall for econometric and computing suggestions. The authors gratefully acknowledge the financial assistance of the National Science Foundation, grants GS-2726X and PRA-8108635, and the C. V. Starr Center for Applied Economics at New York University.

grows at different rates in different areas and suggest that these differences in the cost of producing industrial knowledge, or technological opportunity, are a major determinant of the observed distribution of research effort. Schumpeter (1950), on the other hand, argues that research effort generates temporary monopoly power for the innovating firm and that the private benefits from the production of knowledge must be a result of quasi-rents appropriated by the producer of the innovation. Schumpeter therefore emphasizes the determinants of the degree of appropriability, such as entrepreneurial ability, industrial market structure, and the general institutional framework (including patent rights) in which the firm operates.

We develop a simple model consistent with the theoretical argument that the output of research activities (industrial knowledge) possesses unique economic characteristics. Our model implies that research intensity depends on three factors: appropriability, technological opportunities, and expected market size or demand inducement. This is a richer set of determinants than those underlying the demand for traditional inputs and is therefore consistent with the empirical observation that the coefficient of variation of research intensity is an order of magnitude larger than those of traditional inputs. We specify an explicit set of stochastic disturbances in a set of factor demand equations and estimate the model both at the intraindustry and interindustry levels of aggregation. The empirical results of the intraindustry analysis imply that, though part of the variance in R & D intensity is attributable to measurement and decision errors, the bulk of the variance in observed research intensity is structural in the sense that it is consistent across factor demand equations. Growth rates of output account for very little of this structural variance. We then explicitly aggregate the micro relations to the interindustry level of observation, and the empirical results at this level of aggregation are strikingly different. In particular, differences in industry growth rates account for well over 50 percent of the interindustry variance in research intensity. We explain how these differences can arise and demonstrate their empirical importance.

Section 9.1 specifies the production relationships in the model. In section 9.2 a model of the private returns to R & D is presented. Section 9.3 specifies the stochastic structure and discusses identification. Section 9.4 applies the model to the intraindustry variance in research intensity. In section 9.5 we explicitly aggregate and estimate the model at the industry level of observation. Brief concluding remarks follow in section 9.6.

9.1 Production Relationships

This paper is based on an extended Cobb-Douglas production function in which research resources enter the production process by raising the productivity of traditional factors of production in a disembodied manner. Problems involved in the construction of R & D variables have made this specification the most widely used framework in the empirical analysis of the role of research resources in production (see Griliches 1973, 1980). Our specification differs from the conventional one in two respects: (1) we decompose R & D resources into research capital and research labor, and (2) we explicitly incorporate an R & D gestation lag and a rate of obsolescence of produced knowledge, both of which influence the optimal R & D intensity of the firm.

We begin with the traditional production function

$$(1) \qquad Q = \gamma_0 K N^{\gamma_1} H^{\gamma_2},$$

where Q is output (value added), γ_0 is a constant (which may be both firm and time specific), K is the stock of accumulated and still productive knowledge produced by the firm, N and H are traditional labor and capital services, and all firm and time subscripts have been omitted for convenience. Since K is not observable, its units are arbitrary, and we normalize it so that a 1 percent increase in K raises output by 1 percent.

The generation of knowledge is summarized by its production function

$$(2) \qquad \dot{K}_t^G = A_1 L_{t-\theta}^a C_{t-\theta}^b,$$

where $\dot{K}_t^G$ is the gross increment in produced knowledge in period t, A_1 is a constant (which may be both firm and time specific), $L_{t-\theta}$ and $C_{t-\theta}$ are research labor and research capital services in $t - \theta$, and θ is the mean lag between the time research is undertaken and its embodiment in the traditional production processes of the firm.[1] The parameters a and b are the elasticities of research labor and research capital in the production of increments to K, and they will be assumed not to differ among firms in a given industry (though differences among industries are permitted). These parameters are indices of the technological opportunities of the industry, that is, they reflect the ease with which the underlying scientific and engineering knowledge permits firms in a given industry to transfer their research inputs into cost-reducing innovations (see Scherer 1965 and Rosenberg 1969).

Assuming geometric decay of knowledge at the rate of δ_1 and taking the growth rates of research capital and research labor to be constant both during and prior to the period of analysis (as required by our data), the net increment to knowledge $\dot{K}_t^N$ is

---

1. Since our data cannot sustain an investigation of the distributed lag between the expenditure of research resources and the resultant increases in a firm's productivity, we use the simplification of a mean lag which applies to all units of research resources equally. Note also that research capital refers to an aggregate of all research resources other than research labor, and that the constant term A_1 captures both the effects of "learning by doing" and of other firms' research as inputs in the production process of the firm in question.

(3) $$\dot{K}_t^N = Ae^{(a\hat{L}+b\hat{C})t} - \delta_1 K_t,$$

where $A = A_1 L_0 C_0 e^{-(a\hat{L}+b\hat{C})\theta}$, and a caret denotes a rate of growth. Solving this differential equation and assuming that $\lim\limits_{t \to -\infty} K_t = 0$, the stock of productive knowledge becomes:

(4) $$K_t = \frac{A_1 L_{t-\theta}^a C_{t-\theta}^b}{\delta_1 + a\hat{L} + b\hat{C}}.$$

This concludes the specification of the production relationships. However, we will require expressions for the reduction in unit costs attributable to an increase in research labor and research capital. Assuming that the firm is a cost minimizer facing fixed input prices, the unit cost function associated with (1) can be expressed as

(5) $$Z = \frac{h(w, p_H)}{K},$$

where Z represents unit costs, and w and p_H denote the (fixed) wage and rental rates for traditional labor and capital services, respectively. Substituting (4) into (5) and differentiating the cost function at time $t + \theta$ with respect to research labor and research capital services at time t, we obtain

(6) $$-\frac{\partial Z_{t+\theta}}{\partial L_t} = \frac{aZ_{t+\theta}}{L_t}, \text{ and } -\frac{\partial Z_{t+\theta}}{\partial C_t} = \frac{bZ_{t+\theta}}{C_t}.$$

9.2   Optimal Factor Intensities for Research Inputs

If private firms are motivated by potential profits, the level of their research effort will be determined by the expected net income generated by investment in research resources. The large observed variance across firms in research intensity should be attributable to the variance in the expected private returns to research. The objective factors that could cause differences in the expected net income generated by the use of research resources are: (1) variation in the costs of research inputs; (2) differences in the productivity of research resources in generating usable industrial knowledge; and (3) differences in the ability to derive monetary benefits from a given unit of produced knowledge. Variation in costs of research inputs will be incorporated into the model and discussed in section 9.3. In connection with the productivity of research resources, the basic model assumes that all firms in a given industry produce a single homogeneous output subject to the same production conditions (as specified in section 9.1), and the model is tested separately for each industry in our data set. Consequently, differences in the expected returns from research beyond those caused by differences in the cost of research inputs will be associated with differences in the ability to derive monetary benefits from a given unit of produced knowledge. However, the industries in the data set are defined quite broadly, so there could be some intraindustry differences in the output elasticity of research resources. At this stage, we do not separate these supply-side differences from those differences in ability to capture the monetary benefits from knowledge. We return to this problem later in this paper.

The difficulty in specifying a mechanism that determines the stream of private benefits accruing to new industrial knowledge is a result of the fact (stressed by Arrow 1962) that knowledge has no, or a very small, cost of reproduction. Since any economic agent aware of the information embodied in the innovation can exploit it, the private benefits from the production of industrial knowledge must be a result of quasi-rents or temporary monopolies accruing to the producer of the innovation (Arrow 1962; Machlup 1962; Nordhaus 1969a, 1969b). The strength of these monopolies, that is, the abilities of firms to appropriate the benefits from the knowledge which they have developed, will determine the private return to research resources and therefore the research intensity of firms. The private return to the development of a new cost-reducing technique will depend on the number of units of output embodying this new knowledge and the fraction of the cost reduction attributable to the innovation apropriated by the innovating firm.

We begin by reviewing the "maximum appropriability environment," first described by Arrow (1962) and later adapted to determine the rate of return to research resources by Nordhaus (1969b). Consider a constant cost industry in competitive equilibrium and an innovation which reduces the cost of production for the firms in the industry and only for such firms. The maximum appropriability environment is based on the assumption that the innovator patents the innovation costlessly and leases the cost-reducing technique to all firms in the industry (including itself), subject to the condition that the final product must sell at a uniform price to consuming units. The lease can be defined in terms of a royalty per unit of output produced with the innovation, ρ_0. The lessor acts as a monopolist and sets the initial royalty to maximize profits subject to the constraint that the royalty plus the new cost of production ($\rho_0 + Z_1$) is less than or equal to the preinnovation cost of production ($Z_0 = P_0$). In virtually all cases the profit-maximizing royalty at the date of introduction will be $\rho_0 = Z_0 - Z_1 = \Delta Z.$[2] The revenue collected in the first year of the

2. If P_1 is the profit-maximizing price for a monopolist with constant unit cost Z_1, the Arrow royalty described in the text will yield maximum profits if and only if $P_1 > P_0$. If the industry demand is price inelastic over the relevant range, the Arrow royalty will be optimal regardless of the magnitude of the cost reduction from the innovation. If the industry demand is price elastic, the condition $P_1 > P_0$ can be written $\Delta Z/Z_0 < |\eta(P_1)^{-1}|$, where η denotes the price elasticity of industry demand. It is apparent from this inequality that the Arrow royalty will be optimal for all but the most major innovations and will certainly be optimal for the cost reduction resulting from the employment of the marginal research resource.

innovation in that appropriability environment is $\rho_0 Q_0^I = \Delta Z Q_0^I$, where Q_0^I denotes the industry output in the year the innovation is introduced. We now extend the analysis to firm-specific, nonmaximal appropriability environments and to the calculation of the entire revenue stream accruing to the innovation. Let $k_{i\tau}$ be the fraction of industry output from which firm i receives royalties on its innovation of age τ, $\rho_{i\tau}$ the royalty per unit of output, and $B_{i\tau}$ the total revenues accruing to the innovation in year τ. Then the discounted value of the stream of revenues generated by the innovation is

$$(7) \qquad \Pi = \int_0^\infty B_{i\tau} e^{-r\tau} d\tau = \int_0^\infty \rho_{i\tau} k_{i\tau} Q^I e^{-r\tau} d\tau,$$

where r is the discount rate.

The specification of the appropriability environment is based on the following two assumptions:

1. It is easier, or less costly, for a firm to capture the benefits of the knowledge it produces through embodiment in its own output (internal appropriation) than through embodiment in the output of other firms. Internal appropriation is less costly because of the difficulties involved in establishing an effective market for information (for more discussion see Arrow 1962).

2. The revenues accruing to an innovation decline with the age of the innovation. This occurs because new techniques are developed by the firm and its competitors which substitute for the original innovation and because the use of the information in any productive way reveals and spreads it. This tends to erode both the unit royalty that can be charged and the part of industry output from which royalties accrue.

To maintain a specification that is both as general as possible and consistent with the preceding two assumptions, we let

$$(8) \qquad \rho_{i\tau} k_{i\tau} = (\rho_{i0} e^{-\delta\tau}) e^{\bar{k} + k_i} \frac{Q_{i\tau}}{Q_\tau^I} = \Delta Z e^{\bar{k} + k_i - \delta\tau} \frac{Q_{i\tau}}{Q_\tau^I},$$

where $\Sigma_{i=1}^n k_i = 0$ by construction (n is the number of firms in the industry), and $Q_{i\tau}$ denotes the expected output of firm i at time τ. Revenues in period τ become

$$(9) \qquad B_{i\tau} = \rho_{i\tau} k_{i\tau} Q_\tau^I = (\Delta Z e^{-\delta\tau}) e^{\bar{k} + k_i} Q_{i\tau}.$$

We interpret the parameters in the following manner: δ is the rate of decay in the unit royalty, $\exp(\bar{k} + k_i)$ is the proportion of firm i's share of industry output from which the firm receives this royalty, and $\exp(\bar{k})$ is the (geometric) mean of this proportion over all firms in the industry. However, it is impossible to distinguish empirically between a rate of decay in the proportion $\exp(\bar{k} + k_i)$ and δ, or between a firm-specific component in the rate of decay and $\exp(k_i)$. Since appropriable revenues alone suffice to determine the private benefits from an innovation, it is immaterial whether the firm-specific component applies to the royalty (the price side of revenues) or to the number of units from which the firm receives these royalties (the quantity side of revenues). Hence, these relationships may be interpreted as saying that the revenues generated by a given innovation of age τ depend on: (1) the importance of the innovation, ΔZ; (2) the age of the innovation, τ (through the rate of obsolescence of the *private* returns from knowledge, δ); (3) a firm-specific structural parameter, $\exp(k_i)$, which determines the extent to which the firm can monopolize the information produced by its research resources; and (4) the expected output of the innovating firm, $Q_{i\tau}$, because of the relative ease of internal appropriation.

To obtain the present value of revenues generated by the employment of the marginal unit of research labor (Π_ℓ), substitute (6) and (8) into (7). Setting the price of output equal to one (as it implicitly is in our data) and recalling that the cost reduction does not occur until θ years after the employment of the unit of research labor, we have

$$\Pi_\ell = \int_\theta^\infty \frac{a_0}{L_0} e^{\bar{k}+k_i-r\tau-\delta(\tau-\theta)} Q_{0i} e^{g_{i\tau}^*} d\tau,$$

where g_i^* is the expected rate of growth of output of firm i. Equating Π_ℓ to the wage rate for research labor (w_r), taking a first-order expansion of $\log(\delta + r - g^*)$ around $\log(\delta + r)$, and rearranging terms, we can express the optimal research labor intensity (and following an analogous procedure, the optimal research capital intensity) as:[3]

(10) $$\log(w_r L/Q) = \log \beta_0 + \alpha g^* + k_i,$$

$$\log(p_c C/Q) = \log \beta_1 + \alpha g^* + k_i,$$

where $\log \beta_0 = \log[a/(r + \delta)] - r\theta + \bar{k}$, $\log \beta_1 = \log[b/(r + \delta)] - r\theta + \bar{k}$, p_c is the price of research capital services, and $\alpha = (r + \delta)^{-1} + \theta$.

Several features of equation (10) are worth noting. First, since the returns from both research labor and research capital are derived from the returns to industrial knowledge, any factor that affects the returns to knowledge will influence the optimal intensities of both research variables. This fact permits econometric identification of the relative importance of the unobserved structural parameter (k_i) in determining the research intensities of firms. Also noted that the indices of technological opportunity at the industry level and the average degree of appropriability (a, b, and $\bar{k}$) affect the research intensities of firms since they appear in

---

3. Since only the moment matrix of the variables was available, we were limited to linear combinations of the original variables and forced to use Taylor approximations. The approximation error evaluated at the means is about 2 percent, and if g is distributed symmetrically this will not affect the estimate of α.

the constant terms in (10). These parameters are assumed not to vary within a given industry, but they may vary across industries and, in fact, could be endogenously determined in a more complete model. Second, equation (10) indicates that the firm's employment of research resources will vary directly with its expected market size (its appropriability base) and inversely with the rate of obsolescence and the rate of discount. The importance of expected market size in determining the optimal level of research resources follows directly from the fact that knowledge has a low cost of reproduction.[4] For a given value of initial revenues accruing to an innovation, the higher rate of obsolescence, the smaller the total value of private benefits from the innovation, and therefore, the less intense the research effort will be. Moreover, since research produces a stock (knowledge) whose benefits accrue over the future, the optimal research intensity will vary inversely with the rate of discount (see Lucas 1967 for an empirical test on aggregate data). Finally, the longer the gestation lag, the larger the influence of the future is in determining the returns to R & D, and hence, the more important the effect of expected growth on the optimal R & D intensity.[5]

The model presented here posits a set of firms that produce knowledge from research resources and produce output by combining this knowledge with traditional factors of production. The price of output is determined by the cost of traditional factors plus quasi-rents generated by temporary monopoly power over the information produced by the research resources. It is important to realize that there will be no private benefits from the employment of research resources without some degree of monopoly power. The unique characteristics of knowledge as a commodity imply that the private rate of return to research resources must be determined jointly by the parameters of the production function for knowledge and the ability of the firm to internalize the benefits from the knowledge it produces.

We would like to clarify the relationship between our model and Schmookler's (1966) celebrated work on demand inducement. Schmookler argued that the *level* of inventive activity is directly related to the absolute size of the market for the output of such activity. By focusing on the determination of *research intensity*, our model normalizes by the current level of output, and further differences in expected market size are associated with the expected rate of growth in demand. Of course, one would not expect that equations relating research intensity to ex-

---

4. This should be distinguished from the role of market size in models of the demand for traditional capital. The level of investment in traditional capital is related to the expected growth of output (accelerator models), whereas in our model the level of investment in the stock of R & D depends on the expected level of output.

5. This does not mean, however, that an increase in θ raises the optimal R & D intensity, since θ affects both β_0 and β_1.

pected growth would fit as well as those relating the level of research to the current level of output. Nevertheless, we will show that the explanatory power of growth rates remains substantial, at least at higher levels of aggregation.

9.3 Stochastic Specification and Identification of the Model

The equations for the optimal intensities of research capital and research labor (10), together with that for traditional labor, form the basis of the model to be estimated. In this section we add appropriate disturbance terms and consider the identification of the model's parameters. Letting asterisks denote the optimal levels of each variable, we have

(11) $\log(w_r L^*/Q^*) = \log \beta_0 + \alpha g^* + k_i,$

$\log(p_c C^*/Q^*) = \log \beta_1 + \alpha g^* + k_i,$

$\log(w N^*/Q^*) = \log \gamma_1.$

The variable denoted by Q^* is expected output, that is, the value of output on which input decisions are made. We follow Mundlak and Hoch (1965) in assuming "partial transmission" of the error in output to the input decision-making process. Letting the superscript o denote the observed value of a variable, η_q represent disturbances and firm-specific characteristics known before input decisions are made, and ν_q reflect transitory disturbances realized after inputs are chosen, we have[6]

(12) $Q^o = \gamma_0 K N^{\gamma_1} H^{\gamma_2} e^{\eta_q + \nu_q}$, and $Q^* = E(Q^o | \eta_q),$

so that $Q^o = Q^* e^{\nu_q}$, and we define $\sigma_q^2 = E(\nu_q^2)$.

The observed level of each factor of production differs from its optimal level by an error which has two components: a decision component resulting from an inoptimal choice of factor levels and a pure measurement component. Letting ϵ_j be the sum of the two errors for factor j, we have

(13) $C^o = C^* e^{\epsilon_c}, L^o = L^* e^{\epsilon_\ell}$, and $N^o = N^* e^{\epsilon_n},$

where $E(\epsilon_j) = 0$ and $V(\epsilon_j) = \sigma_j^2$ for $j = c, \ell, n$.

To complete the model two further assumptions need to be made, one on the structure of the covariance matrix of the error components (ν_q and ϵ_j for $j = \ell, c, n$) and one providing an empirical measure of the expected growth rate (g^*). For expositional clarity we first describe the identification scheme under the assumption that all the error components are

---

6. Familiar special cases of partial transmission are full transmission ($\nu_q = 0$; Marschak and Andrews 1944) and zero transmission ($\eta_q = 0$; Zellner, Kmenta, and Drèze 1966).

mutually uncorrelated. The model actually estimated allows for free covariances between the $\epsilon_j (j = \ell, c, n)$ and a test of whether it is reasonable to assume that ν_q is uncorrelated with them. The extensions required to estimate the more general model are briefly summarized at the end of this section. Finally, the empirical results in sections 9.4 and 9.5 are based on the assumption that a firm's expected growth rate equals its average past growth rate plus a component reflecting common expectational changes in the trend of industry demand, that is,

$$(14) \qquad g_i^* = \Delta g + g_i \text{ for } i = 1, \dots, n,$$

where g_i is the average past growth rate of firm i, and Δg is the commonly held, expected difference between the average past and the expected future growth rates. In section 9.5, where we use more flexible data seta than the ones used in section 9.4, we try alternative empirical specifications of g^*, but these alternatives do not change our basic conclusions.

For the remainder of this section it will prove convenient to redefine all variables as deviations from their sample means. With this understanding, substitution of (12), (13), and (14) into (11) yields the following system of factor share equations:

$$(15a) \qquad \log(w_r L^o / Q^o) = \alpha g + k_i + \epsilon_c - \nu_q,$$

$$(15b) \qquad \log(p_c C^o / Q^o) = \alpha g + k_i + \epsilon_\ell - \nu_q,$$

$$(15c) \qquad \log(w N^o / Q^o) = \epsilon_n - \nu_q.$$

Assuming that the structural parameter, k_i, is uncorrelated with the various error components and with g (see below), maximum likelihood estimation provides consistent and asymptotically efficient estimates of α and of the variance-covariance matrix of disturbance Ω, where

$$(16) \qquad \Omega = \begin{bmatrix} \sigma_k^2 + \sigma_c^2 + \sigma_q^2 & & \\ \sigma_k^2 + \sigma_q^2 & \sigma_k^2 + \sigma_\ell^2 + \sigma_q^2 & \\ \sigma_q^2 & \sigma_q^2 & \sigma_q^2 + \sigma_n^2 \end{bmatrix}.$$

The identification of the various components from (16) is straightforward. Any factor which affects the returns to the production of knowledge will affect the optimal intensities of both research resources. Consequently, the covariance between the disturbances in the two research intensity equations will capture σ_k^2. However, this covariance also picks up any measurement or expectational error in output, σ_q^2. Since the traditional labor intensity equation will also contain the error in output, σ_q^2 can be identified by the covariance between the research intensity and the traditional labor demand equations. Finally, the variances of the errors in the research resource variables are calculated as the residual portion of the research intensity equations.

The parameters from (15) and (16) permit a decomposition of the variance in research intensity into three components: (1) variance caused by differences in the expected growth rate of the internal appropriability base of the firm ($\alpha^2\sigma_g^2$), or demand inducement; (2) variance caused by differences in the structural parameter (σ_k^2), which determines the private benefits accruing to a cost-reducing innovation, given the internal appropriability base of the firm; and (3) variance caused by measurement and decision errors in research resources and in expected output.

We would like to explore briefly the economic interpretation of this decomposition and indicate caveats concerning the distinction between technological opportunity, appropriability, and demand inducement as determinants of R & D intensity. Put simply, the optimal R & D intensity depends on the supply of new knowledge and the effective demand for that knowledge. Given factor prices, differences across firms or industries in the supply curve for new knowledge reflect differences in the parameters of the underlying knowledge production function. This is the precise meaning we have given to technological opportunity in the earlier discussion. The effective demand for new knowledge depends on the current level (by which we normalize) and the expected rate of growth in the demand for products that embody the new knowledge (demand inducement) and on the ability of the firm to capture the benefits from the market (appropriability). If a measure of the expected shift in the product demand curve were available, it would serve to identify empirically the role of demand inducement separately from the joint contribution of technological opportunity and appropriability. However, the available measures are based on realized growth rates of output reflecting shifts in both the product supply and demand curves for the firm. Therefore, these growth rates will be positively correlated with the other structural determinants of R & D intensity in our model, namely, technological opportunity and appropriability. We have developed a model which endogenizes the firm's expected growth in output, and it generates structural equations similar to those in this paper. The main difference is the positive correlations referred to above, in particular, k may be correlated with g. As a result, our empirical estimates in section 9.4 represent the reduced form association between growth rates of output and R & D intensity and overstate the importance of pure demand inducement.[7] In section 9.5 we explore some aspects of the association among the structural determinants and demonstrate their empirical importance.

Our main focus is on total research intensity and we now derive an

7. If $E(kg) \neq 0$, the estimates of α derived from the models relying on zero correlation should differ from estimates based on other techniques. An assortment of exogenous information on the components of α (described in section 9.4) yields estimates of α similar to those obtained from our models, at least in the intraindustry regressions, and this may be interpreted as an indirect test of the assumption $E(kg) \simeq 0$. At the interindustry level this problem has more empirical content, and we explore it in greater depth in section 9.5.

equation for that variable. The observed research expenditures of a firm are calculated as the sum of the firm's expenditures on research labor and research capital. That is,

$$(17) \qquad\qquad R^o = w_r L^o + p_c C^o.$$

Since we define research capital to include all R & D expenditures other than payments to scientists and engineers, (17) is an identity. Analogously, we define the optimal level of research expenditures, R^*, as

$$(18) \qquad\qquad R^* = w_r L^* + p_c C^*.$$

It follows from (15a), (15b), (17), and (18) that

$$(19) \qquad\qquad \log(R^o/Q^o) = \alpha g + k_i + \epsilon_r - v_q,$$

where $\epsilon_r = \psi\epsilon_\ell + (1 - \psi)\epsilon_c$, and $\psi = a/(a + b)$. Since equations (15a), (15b), and (19) are definitionally related, only two of these equations contain independent information. The form of the data made it simpler to estimate (15a) and (19) together with (15c).

We have proceeded on the assumption that the various error components are mutually uncorrelated. As indicated earlier, we actually estimate a more complicated six-equation model which allows for free correlation among the ϵ_j $(j = c, \ell, n)$ and a test of whether v_q is correlated with them. Details of this model and its identification scheme are contained in Pakes (1978) and Schankerman (1979). Briefly, the six-equation model is constructed by adding the factor demand equations for research expenditures, research labor, and traditional labor in year $(t - 1)$ to those same equations for year t. Each error component is assumed to be generated by an arbitrary, stationary stochastic process. The model allows for a χ^2_{20} test of the stationarity assumptions (T_1), a χ^2_8 test of the assumption of no correlation between v_q and the factor errors (T_2), and a χ^2_4 test of the intertemporal stability of the coefficient of g (T_3). T_1, T_2, and T_3 test for consistency between the data and the assumptions used to identify the model. There are also nonnegativity restrictions on all the estimated error variances. Since these restrictions are equivalent to a ranking of the elements of the covariance matrix and as such are not guaranteed by our estimating procedure, the nonnegativity conditions constitute an informal test of the model.

The six-equation model also allows us to investigate two aspects of the intertemporal stability of the unobservable structural parameter, k. First, we provide a test of whether the interfirm variance in k is constant over time. Second, we can estimate the correlation coefficient between the values of the structural parameter for a given firm between two adjacent years, which we denote by λ.

9.4   Empirical Results at the Intraindustry Level

The data were gathered jointly by the National Science Foundation and the Bureau of the Census (for a more complete description see Griliches 1980). They contain company information on R & D expenditures, the number of scientists and engineers, total employment, value added, and a variety of other company economic indicators. The data include observations on one level year value and a corresponding growth rate for most variables. The sample used here consists of 433 large firms which account for 48 percent of all R & D performed in American industry in 1963, and 78 percent of all R & D excluding aircraft and missiles.[8] The firms are broken down into four broad industry groups—chemicals and petroleum, electrical and communications equipment, fabricated metals products and machinery, and motor vehicles and other transport equipment—and the analysis is performed on each of these industries separately.

The six-equation version of equations (15a), (15c), and (19) is the model which we estimate.[9] Before presenting the empirical results, however, exogenous information is used to derive a plausible range for α. Recall that $\alpha = [1/(r + \delta)] + \theta$, where r, δ, and θ are the discount rate, the decay rate in appropriable revenues accruing to the innovation, and the mean lag between the outlay of research resources and the beginning of the associated revenue stream, respectively. A comparison of exogenous information on the value of α with the direct estimates here will provide an informal test of the assumptions of the model. The estimates of δ and θ (taken from Pakes and Schankerman, chap. 4 in this volume) range between 0.18–0.36 and 1.2–2.5 (years). Based on a discount rate of 0.15, these estimates provide an approximate range of $3 < \alpha < 5$.

The model was estimated using a full information maximum likelihood technique developed by Jöreskog (1973). A summary of the empirical results is presented in table 9.1. The computed values (pooled across industries) for the test statistics T_1 (χ^2_{20}), T_2 (χ^2_8), and T_3 (χ^2_4) are 29.80,

8. The original sample consists of 883 firms. We discarded the data for the "aircraft and missiles" and the "all others" industries. The first was dropped because of inconsistencies in the data and because it is dominated by government-financed R & D (74 percent versus 20 percent in the other industries). Our market-inducement model has limited applicability for government-financed R & D unless it were known that privately financed and government-financed R & D are close substitutes and that the supply of the latter is very elastic. The "all others" category was discarded on the grounds that it contains both intraindustry and interindustry variance in R & D intensity, a critical distinction as we show in section 9.5.

9. Two points should be noted. First, the data include both the average past growth rate in sales and in value added for each firm. To allow each variable to contain measurement error, we use both variables and identify σ_g^2 from the covariance between the two (see Pakes and Schankerman 1977 for details). Second, the parameter $\psi = a/(a + b)$ is measured as the share of scientists and engineers in total R & D expenditures, constructed for each industry from information in National Science Foundation (1963).

Table 9.1   **Summary of Results of the Six-Equation Model**

	Industry			
Parameter	Chemicals and Petroleum	Metal Products and Machinery	Electrical and Communications Equipment	Motor Vehicles and Transport Equipment
1. α	4.10	2.62	5.13	3.49
2. Standard error of α	2.26	1.30	2.40	3.40
3. σ_r^2	0.33	0.20	0.37	0.58
4. $\sigma_r^2/\sigma_{\log R^O}^2$	0.12	0.07	0.08	0.09
5. $(\sigma_r^2 + \sigma_q^2)/\sigma_{\log R^O/Q^O}^2$	0.23	0.28	0.21	0.40
6. $\alpha^2\sigma_g^2/\sigma_{\log R^*/Q^*}^2$ [a]	0.04	0.04	0.06	0.03
7. $\sigma_k^2/\sigma_{\log R^*/Q^*}^2$ [b]	0.96	0.96	0.94	0.97
8. λ	1.00	0.99	0.99	0.99
9. n	110	187	102	34

[a]$\sigma_g^2 = \text{cov}(g_1 g_2)$, where g_1 and g_2 are the measured average past growth rates of sales and value added, respectively.

[b]$\sigma_{\log R^*/Q^*}^2 = \sigma_{\log R^O/Q^O}^2 - \sigma_r^2 - \sigma_q^2$.

0.32, and 1.32, respectively. None of these values is surprising under the null hypothesis that the constraints are indeed satisfied. It is noteworthy that the value of T_2 indicates strong acceptance of the assumption of a zero covariance between the transitory error in output and the factor errors in this sample. The difference between the sum of squared residuals in the model using all three test constraints and in the totally unconstrained model can be used to produce a χ_{32}^2 test of the validity of the model as a whole. The computed value of the χ_{32}^2 statistics is 32.76, which is about equal to the expected value of a χ_{32}^2 deviate under the null hypothesis. As noted above, there is an additional test of whether the interfirm variance in k is stable over time. The observed value of the χ_4^2 deviate (combined over the four industries) for this test is 6.64. While this indicates acceptance of the hypothesis at the 5 percent level, a sample with more than two time periods would be required to determine more conclusively whether the variance in the structural parameter is in fact constant over time. We also note that of the twenty-four error variances estimated in the model (σ_j^2 for $j = \ell, c, r, n, q, k$ in each industry), only two violated the nonnegativity restriction (see Pakes and Schankerman 1977 for details).

All of the estimated α coefficients are of the right sign, and three are statistically significant. Moreover, all of the four point estimates of α are within or very near the interval predicted by the prior information summarized earlier. To derive a summary measure of α, we tested the null hypothesis that the differences between the various estimates of α are simply a result of random differences in the estimators. The hypothesis is accepted. The value of α for the combined sample is 3.37 with a standard error of 0.95. On the whole, the data and the exogenous information provide mutually consistent information on the magnitude of the parameters determining α.

We now turn to the basic decomposition of the intraindustry variance in research intensity. Line 5 in table 9.1 indicates that an average of 28 percent of the variance in observed research intensity is attributable to errors (of measurement and decision), and hence, 72 percent of the variance is accounted for by the structural determinants in the model. Of special interest is the effect of the firm's past growth rate. Though this variable is neither statistically nor economically insignificant in determining the firm's R & D intensity,[10] line 6 indicates that differences in growth rates account for only a minor portion (3–6 percent) of the structural intraindustry variance in R & D intensity. Morevoer, this finding is robust to different specifications of the expected growth rate variable (see discussion in section 9.5). It is evident that a pure demand inducement mechanism does not do well in explaining the *intraindustry* variance in R & D intensity. As noted earlier, this finding does not contradict Schmookler's argument for demand inducement, which is cast in terms of the level of R & D and the absolute size of the market.

As indicated in line 7, over 95 percent of the structural variance in R & D intensity is picked up by differences in the firm-specific structural parameter, k, which we have interpreted as reflecting appropriability conditions facing the firm (but which may also include intraindustry differences in technological opportunity). It is also noteworthy that the value of the structural parameter associated with a given firm seems to be stable, at least over short periods of time, since the correlation coefficient (λ) between the values of k_t and k_{t-1} is essentially unity in all industries.

Finally, line 4 provides the fraction of the variance in measured R & D expenditures attributable to errors in research resources. The average value is quite large, 9 percent. Unfortunately, it is not possible at this stage to determine what fraction of this error variance is caused by pure measurement error, as opposed to other factors that cause inoptimal choices of R & D intensity in the context of our model.

9.5   Aggregation Effects and the Interindustry Variance in R & D Intensity

In the previous sections we presented a model of R & D intensity at the micro level and explored the empirical determinants of the intrain-

---

10. The elasticity of R & D intensity with respect to past growth rates, evaluated at the sample mean of the growth rate, is about 0.25.

dustry variance in research intensity. We now explicitly aggregate the micro equation and empirically examine the determinants of the interindustry variance in research intensity (i.e., the variance in the average R & D intensities of different industries). Based on the NSF industrial classification (which is roughly at the two-digit SIC level), the total variance in R & D intensity is about equally divided between intraindustry and interindustry variance.

The Griliches-Census data (GD) used in section 9.4 do not contain sufficient industrial detail to investigate the interindustry variance in research intensity. The main data set used here is constructed from information contained in *Business Week* (see Pakes 1979 for details). These data (BWD) contain the ratio of company-financed R & D to sales, five-year average past growth rates of sales, and a fairly detailed industrial classification for 536 firms, which account for the vast majority of company-financed R & D in the United States in 1976. The BWD do not contain the full set of variables required to estimate the entire model in section 9.3. However, the main points we wish to emphasize in this section can be demonstrated by focusing on the R & D intensity equation.[11]

To analyze these data we let the index (i, j) refer to firm i in industry j ($i = 1, \ldots , N_j$ and $j = 1, \ldots , J$), and for simplicity we define $y_{ij} = \log(R^o_{ij}/Q^o_{ij})$. Then the R & D intensity equation of the model (see eq. [19]) can be written

$$(20) \qquad y_{ij} = \alpha_{0j} + \alpha_j g_{ij} + \mu_{ij},$$

where $\mu_{ij} = k_{ij} + \epsilon_{r,ij} - \nu_{q,ij}$, and by assumption, $E(\mu_{ij}) = E(\mu_{ij}g_{ij}) = 0$, and $E(\mu_{ij}\mu_{i'j'}) = \sigma^2_\mu$, if $i = i'$ and $j = j'$, and zero otherwise.[12]

Recall that the micro (intraindustry) coefficient on the expected growth rate is $\alpha_j = 1/(r_j + \delta_j) + \theta_j$, and that the GD used in section 9.4 indicated acceptance of $H^0_1 : \alpha_j = \alpha$ for available j. Seventeen NSF industries are available in the BWD, and the computer χ^2_{16} test statistic for H^0_1 on the BWD is 14.71, also indicating acceptance of H^0_1. Hence, in the remainder of the discussion we maintain H^0_1. The parameter α_{0j} depends on the determinants of α_j and, in addition, on the indices of the technological opportunities and the average degree of appropriation (a_j, b_j, and $\bar{k}_j$) in the industry. Testing $H^0_2 : \alpha_{0j} = \alpha_0$ for $j = 1, \ldots , J$ on the BWD

---

11. In previous work we estimated the entire three-equation model on aggregate data (industry means) on a data set we constructed by combining information from the annual reports of the NSF and the Census of Manufactures (see Pakes and Schankerman 1977). There are no essential differences between those estimates and the ones reported in this section.

12. We are assuming that σ^2_μ does not vary across industries even though α_{0j} and α_j may. This simplifies the presentation without affecting our major results. In general we keep the discussion of technical details in this section very brief since they can be found in Pakes (1978, 1983) and the literature cited there.

yields an $F(16,512)$ test statistic of 24.20. The 5 percent critical value is 1.67, so H_2^0 is clearly rejected. We conclude that, though there are no perceptible interindustry differences in the micro growth rate coefficient, there are clear interindustry differences in the constant terms.

Since we accept H_1^0 and reject H_2^0, equation (20) is formally identical to econometric models that allow for group effects. Under the assumption that the average past growth rate provides a reasonable approximation to the expected growth rate relevant to R & D decisions (which we discuss below), the presence of a group effect indicates that there is a determinant of R & D intensity which is common to all firms in an industry but differs across industries. Since we show later that this group effect is an important determinant of the interindustry variance in R & D intensity, we now explore its characteristics in more detail. To do so we note that one can always define α_0 and ϕ such that

$$(21) \qquad \alpha_{0j} = \alpha_0 + \phi g_{\cdot j} + \zeta_j,$$

where $g_{\cdot j} = N_j^{-1} \Sigma_{i=1}^{N_j} g_{ij}$, and the mean and the sample covariances of ζ_j with g_{ij} and $g_{\cdot j}$ are all zero by construction. We assume that ζ_j are random draws from a common population that satisfy the mean and the covariance restrictions stated above and define $\sigma_\zeta^2 = E[\zeta_j^2]$. Equation (21) partitions the group effect into a part correlated with the past industry growth rate ($\phi^2 \sigma_{g_{\cdot j}}^2$) and into a part not correlated (σ_ζ^2). The parameter ϕ may be interpreted as the reduced form response of R & D intensity of a firm to a unit increase in its industry growth rate, holding constant its own growth rate.

Using H_1^0 and substituting (21) into (20), the micro R & D intensity equation becomes

$$(22) \qquad y_{ij} = \alpha_0 + \alpha g_{ij} + \phi g_{\cdot j} + v_{ij},$$

where $v_{ij} = \zeta_j + \mu_{ij}$, $E(v_{ij}) = E(v_{ij} g_{ij}) = E(v_{ij} g_{\cdot j}) = E(\zeta_j \mu_{ij}) = 0$, and the covariance matrix of v_{ij} has a standard error components structure. Summing (22) over i and dividing by N_j, we obtain the corresponding interindustry R & D intensity equation:

$$(23) \qquad y_{\cdot j} = \alpha_0 + (\alpha + \phi) g_{\cdot j} + \zeta_j + \mu_{\cdot j},$$

$$i = 1, \ldots, J.$$

Clearly, the determinants of the interindustry variance in R & D intensity are a mixture of the determinants of the intraindustry variance and the determinants of the variance in α_{0j}. The growth rate coefficient from the intraindustry regression (20) (with $\alpha_j = \alpha$ for all j) provides an unbiased estimate of the firm's response in R & D intensity to an increase in its own growth rate, holding constant the group effect. The interindustry growth rate coefficient from (23) provides an unbiased estimate of the

Table 9.2   **Aggregation Effects in the BWD**[a]

Parameter	Intraindustry (industry specific constant terms) (1)	Weighted Aggregate between Industry[b] (2)	Mixed[c] Effects (3)
Micro growth rate, α	1.04 (0.31)	—	1.04 (0.31)
		10.96 (1.88)	
Industry growth rate, ϕ	—	—	9.93 (1.91)
σ^2	0.39[d]	n.r.	n.r.[e]
R^2	0.02	0.66	n.r.
Degrees of freedom	512	15	527

[a]Numbers in parentheses are standard errors. The letters n.r. mean "not relevant."
[b]The weight for industry j is $(\hat{\sigma}_\zeta^2 + \hat{\sigma}_\mu^2/N_j)^{-\frac{1}{2}}$.
[c]Estimated by generalized least squares using $\hat{\sigma}_\mu^2$ and $\hat{\sigma}_\zeta^2$ from notes d and e.
[d]$\hat{\sigma}_\mu^2 = 0.39$.
[e]First stage $\hat{\sigma}^2 = 0.51$. It follows that $\hat{\sigma}_\zeta^2 = 0.51 - \hat{\sigma}_\mu^2 = 0.12$.

sum of this response plus the response of the group effect to a unit increase in the industry's past growth rate. The intraindustry and interindustry coefficients would be similar only in the special case where the determinants of the α_{0j} are uncorrelated with the past industry growth rate.

Table 9.2 summarizes the empirical results for the intraindustry regression under H_1^0 (column [1]), the interindustry regression (column [2]) and the "mixed effects" model in equation (23) (column [3]). Column (1) indicates that the intraindustry growth rate coefficient is similar to (but somewhat smaller than) the estimates obtained with the GD in section 9.4.[13] It indicates that growth rates account for very little (about 2 percent) of the intraindustry variance in research intensity, which confirms our earlier results with the GD.

The interindustry regression yields very different results. The estimate of the aggregate mean response in R & D intensity to a unit increase in the industry growth rate is about ten times as large as the firm's response to its own growth rate in the intraindustry regression. As a consequence, growth rates account for over 65 percent of the interindustry variance in

13. There are several differences between the two data sets which could account for this difference: the BWD use the ratio of company-financed R & D to sales while the GD use the ratio of total R & D to sales; the GD are at a slightly higher level of aggregation than the BWD; the estimating technique on the GD allows for an error in the measurement of g (see note 9 and the discussion below); the g used for the GD is based on a slightly longer average of past years than the BWD; and the two data sets are for different years.

research intensity. As the results in the mixed effects model indicate, however, this is not a result of firms' responses to their own growth rates. Rather, it reflects the fact that the industry growth rate is highly correlated with factors at the industry level which stimulate research activity of all firms in the same industry (compare α and ϕ in column [3]).

The BWD also contain the three-digit industrial classification of the firm. We use this information to examine whether the three-digit assignment of the firm exerts any independent influence (beyond the two-digit classification) on its choice of R & D intensity and whether the factors underlying this influence are correlated with the past industry growth rate at the three-digit level. The procedure we use is a generalization of the one outlined in equations (20)–(23). We allow the constant term in the R & D intensity equation to vary across both two-digit and three-digit industries and partition the constant term into a part correlated with the past two- and three-digit industry growth rates (with coefficients ϕ and ϕ_*, respectively) and a part uncorrelated with those growth rates. The test of whether there is any variation in the constant terms across three-digit (within two-digit) industries yields a computed $F(25,484)$ test statistic of 1.98, marginally significant at the 1 percent level of significance. There is weak evidence of variation in the constant terms across three-digit industries. Estimation of the mixed effects model with both two-and three-digit industry growth rates yields estimates of ϕ and α which are almost identical to those reported in table 9.2, while the estimate of ϕ_* is 0.02 (standard error 1.31). Hence, there is not much evidence of a group effect varying among three-digit (within two-digit) industries, and whatever effect there is does not seem to be related to the past growth rate of the three-digit industry.

We next ask whether the results in table 9.2 could reflect nothing more than a misspecification of the expected growth rate relevant to a firm's R & D decisions, g^*. We consider two alternative specifications that may predict the type of discrepancy between the micro and aggregate growth rate coefficients which we observe in table 9.2.

In the first we allow the average past growth rate to measure expected growth with an independently distributed and uncorrelated error, $g_{ij} = g_{ij}^* + v_{ij}$, where v_{ij} has a zero mean and is uncorrelated with g_{ij}^*. This is the classical "errors in variable" model (EVM), suggested in the literature as an alternative explanation of differences in estimated coefficients at different levels of aggregation (Aigner and Goldfeld 1974; Eisner 1978). The motivation for the EVM is that under the stated assumptions $\underset{N_j \to \infty}{\text{plim}}$ $\gamma_{\cdot j} = 0$, so that $g_{\cdot j}$ converges in distribution to g_j^*. In the interindustry regression the error in $g_{\cdot j}^*$ averages out so the estimated coefficient does not contain errors in variable bias, but the intraindustry coefficient is biased downward.

We use two approaches to this problem. The first is to use an instrumental variables estimator with past rates of growth of employment (g_n) as instruments (see Pakes 1979 for details). This approach relies on the assumption that $E(g_n v) = 0$. The second technique uses a three-digit industrial classification as a grouping device to average out the measurement error, and then obtains an estimate of ϕ from a comparison of the interindustry regression at the NSF (two-digit) level with the regression between three-digit (but within two-digit) industries (see Pakes 1983). This provides an asymptotically unbiased estimate of ϕ regardless of errors in variables and a test of the presence of such measurement error. The results indicate acceptance of the hypothesis that there are no errors in variables. Both estimation procedures yield estimates of ϕ which are nearly identical to the estimate in table 9.2. The instrumental variables estimate of α is 1.55 (standard error 0.40), which is slightly larger than the estimate in table 9.2 but does not change our basic conclusions.

The alternative specification we consider assumes that the firm forms a rational forecast by taking the expectation of its future growth rate conditional on the information available to it in period t, $g^* = F_{t}g_{t+1}$. This implies $g_{t+1} = g^* + \omega$, where $E_{t}\omega = 0$, so that the actual future growth rate measures expected growth subject to an error uncorrelated with all variables known to the firm in period t (including the firm's and the industry's past growth rates). To obtain asymptotically unbiased estimators, we substitute the future growth rate for g_{ij} in (23), and use g_{ij} and $g_{.j}$ as instruments on the future growth rate. Since the BWD did not contain future growth rates, additional sources of information were used and the analysis was conducted at a somewhat different level of aggregation (see Pakes 1979 for details). Nonetheless, this rational expectations formulation of g^* yields the same basic results as those reported in table 9.2.

We conclude that the evidence does not support the hypothesis that the differences between the intraindustry and interindustry results reported in table 9.2 are the result of a misspecification in the measure of the expected growth rate. These experiments may not dispose of the issue entirely, but they do indicate that misspecifications which average out over firms (within an industry) and the use of the past industry growth rate by firms to predict their own growth do not explain the results in table 9.2.

We now summarize the empirical characteristics of the group effect, which appears to be the dominant determinant of the interindustry variance in research intensity. First, it is associated with the NSF (roughly two-digit) industry to which the firm belongs. The three-digit industrial classification of the firm has little independent influence on the firm's R & D intensity. Of course, there may be other more detailed classification schemes that group firms with common industry factors affecting their R & D intensities. Second, the factors in the industry environment affecting the firms' R & D intensities are highly and positively correlated with the industry's past growth rates. Various experiments on other data sets (not reported here) indicate that the industry growth rate coefficient is larger, the longer the term of the past growth rate used (we experimented with values from four to eight years), and that the values of the group effect for different industries are fairly stable over time. The factors affecting the R & D intensities of firms in an industry appear to be associated with sustained, long-term past growth. Third, the evidence does not support the hypothesis that the observed group effect is simply a result of the industry growth rate acting as an indicator of the firm's expected growth rate. Finally, the group effect provides an explanation for the basic empirical anomaly that growth rates account for only a minor portion of the intraindustry variance in R & D intensity but for about 65 percent of the interindustry variance.

We have established both the importance of and certain empirical characteristics of the group effect. At this stage, we cannot determine the underlying mechanisms generating it. Within the context of our model, the industry constant terms contain indices of technological opportunity and the average degree of appropriability. This suggests a reduced form, empirical association between technological opportunity, appropriability, and a broader concept of demand inducement. As Schmookler (1966, pp. 176–77) put it:

> . . . science and engineering appear as given, to be used to explain but not themselves to be explained. In the larger context, however, these too would require explanation. I believe that their explanation, at least for modern times, would probably not differ greatly from that advanced here for invention. The rate and direction of scientific and engineering process are probably greatly affected by demand, subject to the constraints imposed by man's innate abilities and by nature. . . . If this view is approximately correct, then even if we choose to regard the demand for new knowledge for its own sake as a non-economic phenomenon, the growth of modern science and engineering is still primarily a part of the economic process.

9.6 Concluding Remarks

In the literature on the determinants of research demand by private firms one can identify three leading hypotheses: expected market size for the output of R & D activities, the degree to which firms can appropriate the benefits from the industrial knowledge they produce, and the technological opportunities facing firms reflecting the set of production possibilities for transforming research resources into innovations. The model we presented provides a first step toward integrating these hypotheses into a formal framework capable of investigating the empirical determinants of R & D demand. Our framework is only a first step because of its partial equilibrium character and because it does not model explicitly the underlying mechanisms determining the degree of appropriability, technological opportunities, and the expected market size of firms. The model results in equations determining the intensity of use of research inputs as a function of these three factors and disturbance terms. The market size relevant to current R & D decisions depends on current output and the expected growth rate. Since R & D intensity measures research effort relative to current output, further differences in expected market size are associated in our model with differences in expected growth rates. The limitations of the model and of the available data require us to specify technological opportunity and appropriability as unobservable factors in a way that permits us to assess their (joint) empirical contribution to the observed variance in research intensity within and across industries.

The empirical results indicate that at the intraindustry level of aggregation about three-quarters of the large observed variance in research intensity is structural, in the sense that it is consistent across diffrent factor demand equations. However, while the growth rate coefficient is broadly consistent with the predictions of the micro model, the variance in growth rates accounts for very little (less than 5 percent) of the intraindustry structural variance in research intensity. Most of the variance is attributed to differences in appropriability and technological opportunities within industries. The results at the interindustry level of aggregation are strikingly different. Growth rates account for the majority of the interindustry variance in research intensity. The evidence suggests that this finding is not due to differences in firms' responses to their own growth rates. Rather, it appears to be due to factors in the industry environment that affect the research intensities of all firms within the industry and that are highly and positively correlated with past industry growth.

The theoretical framework and these stylized empirical facts suggest certain fruitful lines for future research. The first is to model the determinants and disentangle the empirical contributions of appropriability and technological opportunity at the intraindustry level of aggregation. Second, work is needed to understand the causal nexus underlying the empirical association between the industry effect on the choice of R & D intensity and past industry growth.

References

Aigner, Dennis, and Stephen Goldfeld. 1974. Estimation and prediction from aggregate data when aggregates are measured more accurately than their components. *Econometrica* 42:113–34.

Arrow, Kenneth. 1962. Economic welfare and the allocation of resources for invention. In *The rate and direction of inventive activity: Economic and social factors*, ed. R. R. Nelson. Universities-NBER Conference Series no. 13. Princeton: Princeton University Press for the National Bureau of Economic Research.

Eisner, Robert. 1978. *Factors in business investment*. National Bureau of Economic Research General Series no. 102. Cambridge, Mass.: Ballinger.

Griliches, Zvi. 1973. Research expenditures and growth accounting. In *Science and technology in economic growth*, ed. B. R. Williams, 59–83. London: Macmillan.

————. 1980. Returns to research and development expenditures in the private sector. In *New developments in productivity measurement and analysis*, ed. J. W. Kendrick and B. N. Vaccara, 419–54. Conference on Research in Income and Wealth: Studies in Income and Wealth, vol. 44. Chicago: University of Chicago Press for the National Bureau of Economic Research.

Griliches, Zvi, and Jacob Schmookler. 1963. Inventing and maximizing. *American Economic Review* 53:725–29.

Jöreskog, Karl. 1973. A general computer program for estimating a linear structural equation system involving multiple indicators of unmeasurable variables (LISREL). Princeton University.

Lucas, Robert. 1967. Tests of a capital-theoretic model of technological change. *Review of Economic Studies* 34:175–89.

Machlup, Fritz. 1962. *The production and distribution of knowledge in the United States*. Princeton: Princeton University Press.

Marschak, Jacob, and W. H. Andrews. Random simultaneous equations and the theory of production. *Econometrica* 12:143–205.

Mundlak, Yair, and I. Hoch. 1965. Consequences of alternative specifications in estimation of Cobb-Douglas production functions. *Econometrica* 33:814–28.

National Science Foundation. 1963. *Basic research, applied research, and development in industry*. Washington, D.C.: GPO.

Nordhaus, William D. 1969a. *Invention, growth, and welfare: A theoretical treatment of technological change*. Cambridge: MIT Press.

————. 1969b. An economic theory of technological change. *American Economic Review* 59:18–28.

Pakes, Ariel. 1978. Economic incentives in the production and transmission of knowledge: An empirical analysis. Ph.D. diss., Harvard University.

————. 1979. Aggregation effects and panel data estimation problems: An investigation of the R & D intensity decision. Discussion Paper 699. Harvard Institute of Economic Research.

————. 1983. On group effects and errors in variables in aggregation. *Review of Economics and Statistics* 64:168–73.

Pakes, Ariel, and Mark Schankerman. 1977. A decomposition of the intraindustry and interindustry variance in research intensity in American manufacturing: Demand inducement, technological opportunity, and appropriability. Harvard University. Manuscript.

Rosenberg, Nathan. 1963. Technological change in the machine tool industry, 1840–1910. *Journal of Economic History* 23:414–43.

————. 1969. The direction of technological change: Inducement mechanisms and focusing devices. *Economic Development and Cultural Change* 18, no. 1: 1–24.

————. 1974. Science, invention, and economic gowth. *Economic Journal* 84:90–108.

Schankerman, Mark. 1979. Essays on the economics of technical change: The determinants, rate of return, and productivity impact of research and development. Ph.D. diss., Harvard University.

Scherer, F. M. 1965. Firm size, market structure, opportunity, and the output of patented inventions. *American Economic Review* 55: 1097–125.

Schmookler, Jacob. 1966. *Invention and economic growth*. Cambridge, Mass.: Harvard University Press.

Schumpeter, Joseph. 1950. *Capitalism, socialism, and democracy*. 3d ed. New York: Harper and Row.

Zellner, A., J. Kmenta, and J. Drèze. 1966. Specification and estimation of Cobb-Douglas production function models. *Econometrica* 34:784–95.

10 Firm versus Industry Variability in R & D Intensity

John T. Scott

In this paper I show that (1) company effects as well as industry effects explain a substantial proportion of the variance in R & D intensity, (2) the apparent impact of seller concentration is collinear with, and apparently a result of differences in, types of products, and (3) government subsidization of R & D does not displace private R & D spending.

10.1 The Inverted-U in Theory

Empirical studies of nonprice competition have hypothesized and found an "inverted-U" relation between media advertising or company-financed R & D and seller concentration. These studies include Greer (1971), Strickland and Weiss (1976), Scott (1978), Martin (1979), and Scherer (1967; 1980a, p. 437), as well as others cited by these authors. There is nonetheless good reason to question the *cause* of the relation in conventional cross-sectional studies where firms operating in many different industries contribute to the variance in nonprice competition. The relation could be explained by variance across industries in (1) the value or cost of nonprice competition, (2) the opportunity (the odds for success) for it, (3) the condition of entry, or (4) the ability to "collude" (tacitly or otherwise) on nonprice competition while holding constant the ability to "collude" on price competition (Scott 1981b).

John T. Scott is an associate professor in the Department of Economics, Dartmouth College.

The author is especially indebted to Zvi Griliches and also wishes to thank Alan L. Gustman, Meir Kohn, Albert Link, William F. Long, James M. Lowerre, Stephen Martin, George A. Pascoe, Jr., David Ravenscraft, and Leonard W. Weiss for discussions or comments. The Federal Trade Commission has ensured that no individual company line of business data are revealed in this paper. The conclusions here are the author's and not those of the commission.

This paper presents evidence about the *cause* of the inverted-U relation between seller concentration and nonprice competition. If observed inverted-U's result for reasons (1) or (2) only, then they do not imply collusion and, therefore, do not imply concern with "wasteful competition." The variance in nonprice competition across firms and industries may have nothing to do with the conjectural interdependence and mutual dependence recognized among sellers, but instead may reflect differing prospective rewards to R & D or advertising *in the absence of* conjectural interdependence. The observations studied here are for the 3388 manufacturing lines of business of the 437 firms reporting for 1974 to the Federal Trade Commission's Line of Business (LB) program.[1] The results are that a statistically significant inverted-U relation exists for these observations if one does not attempt to control for differences across firms and industries in the value, costs, and "opportunity for" nonprice competition apart from that correlated with concentration, but once controls are added in *the form of a fixed-effects model* the relation disappears. The fixed-effects model may not, of course, be the appropriate way to control for the varying economic potential for nonprice competition. As discussed later, one cannot unambiguously conclude that the "collusion" hypothesis is rejected. But despite considerable variation in concentration within two-digit industries, it apparently has no impact on behavior within them.

10.2   The Inverted-U in Fact

Table 10.1 shows that the inverted-U relation between company-financed R & D intensity and seller concentration is statistically significant in the LB sample. Table 10.2 controls for company and two-digit industry effects as well as Weiss's (1980) adjusted four-firm seller concentration ratio at the four-digit FTC industry level and its square. The inverted-U relation does not remain once differences in value, costs, and opportunity for nonprice competition are controlled for *with a fixed-effects model*.

Apparently, the inverted-U results because firms face different opportunities apart from those inherent in concentration and because, for example, breakfast cereals and cold-rolled steel and chemicals are very different products for which the value of innovative investment differs even without consideration of the extent of sellers' interdependence. I believe my interpretation is valid even though the two-digit industry dummies will capture the variance in concentration to the extent that

1. See U.S. Federal Trade Commission (1979, 1981) for a description of the program. I also present the R & D model for the 3550 manufacturing lines of business of the 474 firms reporting for 1975 in 260 FTC four-digit manufacturing categories. Scott (1981a) presents evidence for advertising intensity analogous to that presented here for R & D intensity.

Table 10.1   The Inverted-U for 1974: 3388 Manufacturing LB's[a]

Company-financed R & D Intensity: (R/S) for an LB (the operations of a firm in a FTC four-digit manufacturing industry) as a function of four-firm seller concentration $(C4)$ in the four-digit FTC industry.

$$R/S = .00094 + .00049(C4) - .0000038(C4)^2.$$
$$ (.43) \qquad (4.5)^b \qquad (-3.1)^b$$

Since one extraordinary outlier was excluded from the sample, degrees of freedom = 3384; F-value for significance of the equation as a whole = 25, significant at the .0001 level; R^2 = .015. R/S reaches its predicted maximum when $C4 = 64$.

[a]The intensity variables are ratios with LB sales as the denominator. I scale by LB sales to control for types of company-specific effects that are correlated with the firm's LB sales in the manufacturing category. $C4$ is Weiss's (1980) adjusted ratio in percentage form. The t-ratios are in parentheses below the coefficients.

[b]Significance level (two-tailed test): $b = .01$.

Table 10.2   Controlling for Company and Industry Effects: 3388 Manufacturing LB's, 437 Companies, 20 FTC Two-Digit Industries[a]

Company-financed R & D Intensity: (R/S) for an LB (the operations of a firm in a FTC four-digit manufacturing industry) as a function of four-firm seller concentration $(C4)$ in the FTC four-digit industry.

$$R/S = b_0 + \sum_{c=1}^{436} b_c + \sum_{i=1}^{19} b_i + f(C4) + g(C4)^2.$$
$$ 3.7^{b,c} \quad 7.5^b \qquad .02 \quad .06$$

F-value for null hypothesis of no effect in complete model given below the coefficients. Here and throughout the paper, whenever coefficients are shown as letters, it is because the statistical package used provided only the F-values for the effects. Since one extraordinary outlier was excluded from the sample, degrees of freedom = 2929; F-value for significance of the equation as a whole = 3.8, significant at the .0001 level; $R^2 = .37$.

[a]See note a of table 10.1. Also note that to reduce the size of the $X'X$ matrix for computational purposes, the company effects were absorbed. Hence, the F-test for their significance in the complete model is not computed. It is computed later for the "best" model.

[b]Significance level: $b = .0001$.

[c]F-test for the significance of the company effects alone—given only the intercept—not controlling for the other variables. That is, the reduction in the sum of squares because of the company effects is what results when they are fitted first, not last. See note a.

concentration is homogeneous within two-digit industries. In the extreme case, one could not control at the two-digit level for different types of goods, say food in general versus chemicals in general, *and* seller concentration at the four-digit level. In fact, such control is possible. In general, for the 259 four-digit FTC industries, 74 percent of the variance in concentration is within two-digit industries. In the specific 3388 observation sample, 68 percent of the variance in concentration is within two-digit industries.

The inference that collusion does not cause observed inverted-U's may well be inappropriate for two reasons. First, industry seller concentration alone may not be the appropriate control for mutual dependence recognized. Market power may have more to do with *firm share*, given that seller concentration is sufficiently great for recognized mutual dependence. Long (1981) is currently exploring the implications of such Cowling-Waterson (1976) conjectural interdependence equilibria for nonprice competition. In any case, the results here at least imply that traditional interpretations of the collusion hypothesis do not hold up once the fixed effects are used as controls.

The second reason for cautious interpretation of the results is that fixed effects may simply not be the appropriate way of controlling for the differentiability of products, the potential for R & D, the need for informative advertising, and so forth. True, the fixed-effects models explain more variance than the simple structural models. Table 10.3 shows the pure fixed-effects model unalloyed with the structural variables.[2] But these fixed-effects models have many times the number of regressors as the simpler models. In terms of a probability-of-F-test comparison of the models, structural models look quite good. Further, and related to the first concern, the company effects may well be picking up firm-specific aspects of market power.

As noted above, even the significance of firm-specific effects and the insignificance of seller concentration need not imply that industrywide recognition of mutual interdependence is absent. Cowling-Waterson (1976) conjectural interdependence equilibria can imply company-specific differences in R & D intensity caused by "collusion" within industries. Table 10.4 explores that possibility and, perhaps more importantly, provides descriptive information about "part-of-company" effects and narrow-industry effects within broad industry categories by estimating the fixed-effects model for each two-digit FTC industry. The twenty models for R & D intensity provide strong support for the importance of company effects within industries. Of the twenty models for R & D, the company effects are significant at the .05 level for nine cases rather than the one "expected."

In conclusion, the evidence from the fixed-effects models suggests caution when interpreting cross-sectional, multi-industry, inverted-U relations between seller concentration and nonprice competition. This is not to say that with intricate interactive simultaneous-equations modeling of various factors other than firm effects, opportunity classes at the

---

2. I tested the sensitivity of the results to the inclusion or exclusion of the observations for which R & D is zero. The models in table 10.3 were rerun, dropping all observations for which R & D was zero from the R & D model. The results were virtually the same as those in table 10.3.

Table 10.3   **Pure Company and Industry Effects Model: 3388 Manufacturing LB's, 437 Companies, 259 FTC Four-Digit Manufacturing Industries, 20 FTC Two-Digit Industries**[a]

	Dependent Variable	
	R/S^d	R/S
F-value for null hypothesis of no effect in the complete model		
436 company dummies	$3.7^{b,c,d}$	$3.7^{b,c}$
258 FTC four-digit industry dummies	$3.5^{b,d}$	—
19 FTC two-digit industry dummies	—	7.8^b
Degrees of freedom	2817^d	2931^e
F-value for equation as a whole	$3.6^{b,d}$	3.9^b
R^2	$.49^{d,f}$	$.37$

[a]See note a of table 10.2.

[b]Significance level: $b = .0001$.

[c]F-test for significance of company effects alone—given only the intercept—not controlling for industry effects. That is, the reduction in the sum of squares due to the company effects is what results when they are fitted first, not last. See note a of table 10.2.

[d]This result is for the 3550 manufacturing LB's of the 474 firms reporting in 260 FTC four-digit manufacturing industries in 1975. Thus, there were 473 company dummies and 259 FTC four-digit industry dummies. The 1975 observations were used because there were no extraordinary outliers in 1975, and thus, the validity of the 1974 results could be checked.

[e]Since one extraordinary outlier was dropped, only 3387 LB's were used.

[f]The explanatory power (but see note c) was divided as follows: For R/S 32.1 percent from company effects, 16.4 percent from industry effects. Inspection of table 10.5 shows that one can get roughly the 32 percent with either the company effects or the industry effects and get the remaining 16 percent with whichever is left. One could say then that about 16 percent is clearly from company effects, about 16 percent is clearly from industry effects, and about 16 percent is confounded in the two types of effects. Note that, as shown below, the company effects *are* significant in the complete model.

broad industry level, and concentration, or simply more control variables, that the positive correlation or the inverted-U would not be found. Rather, since we find a strong inverted-U in the data without control for variance in opportunity across observations, but eliminate that relation once the opportunity controls are added, there is the presumption that all such previously adduced correlations may be artifacts of insufficient control for opportunity. On the positive score, the results suggest that company-specific and FTC industry-specific effects can explain a large amount of the variance in nonprice competition. Clearly, the evidence suggests that company policy may influence the technological progress of the economy. One cannot explain R & D activity simply by observing the industries within which a company operates. There is more to be understood.

Table 10.4A Company and Industry Fixed-Effects Model of R & D Intensity[a] for Two-Digit Manufacturing Industries[b]

	20	21	22	23	24	25	26	27	28	29
FTC two-digit industry	20	21	22	23	24	25	26	27	28	29
No. of companies sampled = c	60	10	65	38	44	44	68	58	52	41
No. of FTC four-digit industries sampled = i	10	3	12	7	5	6	11	9	5	3
No. of LB's sampled = n	119	14	125	83	91	57	161	98	63	50
Degrees of freedom[c] $= n - c - i + 1$	50	2	49	39	43	8	83	32	7	7
R^2	.99	.81	.85	.71	.91	.90	.82	.60	.95	.999
F-value for the model[b]	57	.76	3.8	2.2	8.9	1.6	4.8	.75	2.5	144
	.0001	.69	.0001	.007	.0001	.26	.0001	.84	.11	.0001

F-value for the null hypothesis: all company effects are zero in complete model[b]	62	.72	3.8	2.3	9.5	1.6	3.7	.76	2.4	145
	.0001	.70	.0001	.005	.0001	.24	.0001	.82	.11	.0001
F-value for null hypothesis: all industry effects are zero in complete model[b]	3.0	1.3	1.2	.27	1.3	1.6	2.6	1.6	1.5	1.9
	.007	.44	.34	.95	.28	.28	.008	.17	.30	.22

[a]R & D intensity is defined as company-financed R & D/sales for the LB, i.e., the company's operations in the FTC four-digit industry category. To reduce the cost of estimating these models, subsets of the two-digit industry's four-digit industries were sampled randomly when the number of companies and industries were otherwise large.

[b]Below the F-values are the related probability statements. Each statement gives the probability of $F \geqq$ the observed value if in fact the null hypothesis of zero coefficients in the linear model specified were true.

[c]In those special cases where an industry and a company dummy coincide (for example, where a single-manufacturing-LB company is the sole producer in its manufacturing industry category), the degrees of freedom will be greater since the number of dummies is reduced until no linearly dependent columns are in the X matrix.

Table 10.4B   Company and Industry Fixed-Effects Model of R & D Intensity[a] for Manufacturing Industries[b]

	30	31	32	33	34	35	36	37	38	39
FTC two-digit industry	30	31	32	33	34	35	36	37	38	39
No. of companies sampled $= c$	10	17	76	81	63	61	73	70	73	56
No. of FTC four-digit industries sampled $= i$	3	4	18	5	5	10	10	5	9	8
No. of LB's sampled $= n$	14	21	134	103	70	107	119	86	104	69
Degrees of freedom[c] $= n - c - i + 1$	2	1	41	18	4[c]	37	37	12	23	7[c]
R^2	.83	.98	.76	.79	.99	.83	.84	.71	.81	.99
F-value for the model[b]	.91	2.8	1.5	.83	4.5	2.5	2.4	.40	1.2	16
	.63	.44	.093	.72	.075	.0014	.002	.99	.31	.0008
F-value for the null hypothesis: all company effects are zero in complete model[b]	.39	2.1	1.2	.71	4.5	2.3	1.8	.34	1.2	17
	.87	.50	.22	.85	.076	.004	.033	.997	.35	.0006
F-value for null hypothesis: all industry effects are zero in complete model[b]	.38	.96	.72	2.9	8.6	2.5	2.6	.88	.53	82
	.73	.62	.76	.05	.033	.026	.019	.50	.83	.0001

[a]See note a of table 10.4A.

[b]See note b of table 10.4A.

[c]See note c of table 10.4A.

10.3   Government R & D Financing on Company R & D Expenditures

Several studies have suggested low private returns to government-financed R & D. Scherer (1980b, pp. 19–20, 26–27, 29; 1981, pp. 16–17) finds that patent output per dollar of private R & D spending is significantly lower when government financing of R & D is high. Griliches (1980, pp. 439, 445–46) and Terleckyj (1974; 1980, p. 362, 367) find that the rate of return to government-financed R & D appears far lower than that for company R & D.

Griliches (1980, pp. 445–46, and note 14) and Scherer (1980b, p. 20) explain that such results may be because of externalities and restrictions on the appropriability of innovations. Levy and Terleckyj (1981) suggest further that government-financed R & D may have an indirect influence on productivity by increasing the amount of private R & D above what it would be in the absence of government funding.

Here I present a simple test of the extent to which government R & D spending is a substitute for, and therefore displaces, *or* is a complement to, and therefore increases, company-financed R & D spending. Although the methodology is different, the question is the same as one of Mansfield's (this volume). The test uses the 3388 observations on lines of business for the 437 companies reporting in 259 FTC four-digit manufacturing categories in 1974 to the FTC's Line of Business (LB) program. The company and industry fixed-effects model for these data discussed in section 10.2 shows that, with 1975 data, 49 percent of the variance in the ratio of company-financed R & D to LB sales is explained by the company and industry effects. Here, the 1974 data are used and one additional explanatory variable is added—the ratio of government-financed R & D to LB sales. The question is, other things equal: Is company-financed R & D intensity greater or lower in LB's where government-financed R & D is greater?

Table 10.5 provides the answer, although it is certainly possible to question causal stories since random disturbances in company- and government-financed R & D might reasonably be correlated. Equation (1) shows that government financing goes to firms that do a lot of R & D in a LB. Equation (2) shows that the relation is not simply the result of funds going to firms that characteristically do a lot of R & D. Equation (3) shows that we are not simply observing that funds go to firms in R & D-intensive industries. Equation (4) shows that the relation is not simply the result of government funds going to R & D intensive firms in R & D intensive industries. The substitution hypothesis is rejected. There appears to be stimulation rather than substitution.

There is, however, the possibility of spurious results in table 10.5's specifications because sales appear in the denominator on both sides of the equation (see Kuh and Meyer 1955). Table 10.6 presents results that show that the positive relation, other things equal, between government-

Table 10.5   **Company R & D Intensity (R/S) and Government-Financed R & D Intensity (G/S) for an LB[a]**

(1)
$$R/S = .013 + .10\,(G/S).$$
$$(29)^{b} \qquad (9.9)^{b}$$

The t-ratios are in parentheses below the coefficients. Degrees of freedom = 3385; F-value for significance of the equation = 98^{b}; R^2 = .028.

(2)
$$R/S = b_0 + \sum_{c=1}^{436} b_c + .076\,(G/S).$$
$$(3.6)^{b,c} \quad (58)^{b}$$

The F-values are in parentheses below the coefficients. Degrees of freedom = 2949; F-value for the significance of the equation = 3.7^{b}; R^2 = .36.

(3)
$$R/S = b_0 + \sum_{i=1}^{258} b_i + .091\,(G/S).$$
$$(6.0)^{b,c} \quad (62)^{b}$$

The F-values are in parentheses below the coefficients. Degrees of freedom = 3127; F-value for significance of the equation = 6.2^{b}; R^2 = .34.

(4)
$$R/S = b_0 + \sum_{c=1}^{436} b_c + \sum_{i=1}^{258} b_i + b_g\,(G/S).$$
$$(4.4)^{b,c} \quad (3.6)^{b} \quad (40)^{b}$$

The F-values are in parentheses below the coefficients. Degrees of freedom = 2691; F-value for significance of the equation = 4.2^{b}; R^2 = .52.

[a]A line of business (LB) is the operations of a company in a FTC four-digit manufacturing industry. The intensity variables are ratios with LB sales as the denominator. I scale by sales to control for types of company-specific effects that are correlated with the firm's LB sales in the manufacturing category. One extraordinary outlier was excluded from the sample; thus 3387 LB observations were used.

[b]Significance level is .0001.

[c]To reduce the size of the $X'X$ matrix for computational purposes, these effects were absorbed. Hence the F-test for their significance in the *complete* model is not computed. That is, the reduction in sum of squares due to these effects is what results when they are fitted first, not last. The F-test for the significance of the company effects in the complete model is computed later for the "best" model.

financed R & D and company-financed R & D is not spurious. The relation is, however, far less significant in the specifications used in table 10.6, suggesting that variance in the denominators of the intensity variables did affect the high level of significance for the intensity of government-financed R & D.

In table 10.6, all LB observations for which company-financed R & D was zero were dropped. We are, after all, interested in whether companies that do private R & D do more when government financing is present. The following variables are used:

LR = the natural logarithm (ln) of company-financed R & D (R) in the line of business (LB);

LS = ln of LB sales (S);

Table 10.6   **Regressions with LR as the Dependent Variable**

Independent Variables	Equation (1)	Equation (2)	Equation (3)	Equation (4)
Intercept	-3.5 $(t = -9.6)^a$	b_0	b_0	b_0
LS	.91 $(t = 40)^a$	.88 $(t = 41)^a$	.99 $(t = 43)^a$	$F = 1709^a$
LG	.081 $(t = 2.3)^b$	.075 $(t = 2.3)^b$	.060 $(t = 1.6)^c$	$F = 2.9^c$
X	$-.49$ $(t = -1.8)^c$	$-.20$ $(t = -.78)$	$-.10$ $(t = -.41)$	$F = .15$
Company effects		$F = 7.7^{a,d}$ 394 effects		$F = 10^{a,d}$ 394 effects
Industry effects			$F = 8.5^{a,d}$ 254 effects	$F = 3.5^a$ 253 effects^e
Degrees of freedom	2476	2082	2222	1829
R^2	.44	.70	.66	.80
F-value for equation	655^a	12^a	17^a	11^a

[a,b,c]Significance levels: [a] $= .0001$; [b] $= .05$; [c] $= .10$.

[d]See note c, table 10.5.

[e]An additional industry dummy had to be dropped. See note c, table 10.4A.

$LG = \ln$ of government-financed R & D (G) if $G > 0$, $\ln(1)$ if $G = 0$; $X = 0$, if $G > 0$, 1 if $G = 0$; and various fixed effects.

The difference in significance in tables 10.5 and 10.6 for the government-financed R & D variable is a result of the new functional form and *not* a result of dropping the zero R & D observations, since the specifications in table 10.5 yield similar results with and without the zero observations.

It must be noted that I have said nothing more than what is apparent. In particular, I have said nothing about productivity. It is entirely possible that government subsidies stimulate "wasteful" private spending. Nonetheless, the results suggest some complementarity between private and government-financed R & D and are remarkably similar to the results in Mansfield (this volume).[3] An incidental finding in table 10.6 is that there is no evidence that company-financed R & D increases more than proportionately with LB size.

One more fact provides a useful conclusion. Although my primary interest is in whether an impact for seller concentration or government-financed R & D remains after controlling for company effects (either idiosyncratic reporting or real R & D activity) and industry effects, many

3. When $R = G$ so that slope and elasticity are the same, my results in table 10.6 are very close to Mansfield's result. Holding constant S and therefore using the estimate of $d(R|S)/d(G|S)$ to estimate dR/dG, my results in table 10.5 are also quite close to Mansfield's results, although somewhat larger.

readers may want to see the F-test *in the complete model* for company effects. I have therefore computed one such test (for what seems to me to be the "best" model). All such tests would be similar, so there is no need to do them all.

Using sums of squares from equations (3) and (4) from table 10.6, the test of the hypothesis that *in the complete model* $b_c = 0$ for all 394 effects is given by:

$$F = \frac{\left[\begin{array}{c}\text{reduction in residual sum of squares from fitting the} \\ \text{394 effects after fitting the other variables}/394\end{array}\right]}{[\text{the residual sum of squares for the complete model}/1829]}$$

= 3.3 with 394 and 1829 degrees of freedom, highly significant by classical standards.

References

Cowling, Keith, and Michael Waterson. 1976. Price-cost margins and market structure. *Economica* 43:267–74.

Greer, Douglas. 1971. Advertising and market concentration. *Southern Economic Journal* 38:19–32.

Griliches, Zvi. 1980. Returns to research and development expenditures in the private sector. In *New developments in productivity measurement and analysis*, ed. John W. Kendrick and Beatrice N. Vaccara, 419–54. Conference on Research in Income and Wealth: Studies in Income and Wealth, vol. 44. Chicago: University of Chicago Press for the National Bureau of Economic Research.

Kuh, Edwin, and John R. Meyer. 1955. Correlation and regression estimates when the data are ratios. *Econometrica* 23:400–416.

Levy, David M., and Nestor E. Terleckyj. 1981. Government-financed R & D and productivity growth: Macroeconomic evidence. National Planning Association. Mimeo.

Long, William F. 1981. R & D in a static Cournot oligopoly model. Federal Trade Commission. Mimeo.

Martin, Stephen. 1979. Advertising, concentration, and profitability: The simultaneity problem. *Bell Journal of Economics* 10:639–47.

Scherer, F. M. 1967. Market structure and the employment of scientists and engineers. *American Economic Review* 57:524–31.

———. 1980a. *Industrial market structure and economic performance*. 2d ed. Chicago: Rand-McNally.

———. 1980b. The propensity to patent. Northwestern University and FTC. Mimeo.

———. 1981. The structure of industrial technology flows. Northwestern University and FTC. Mimeo.

Scott, John T. 1978. Nonprice competition in banking markets. *Southern Economic Journal* 44:594–605.

———. 1981a. Nonprice competition: Theory and evidence. Dartmouth College and FTC. Mimeo.

———. 1981b. The inverted-U for nonprice competition: A theory with policy implications. Dartmouth College. Mimeo.

Strickland, Allyn D., and Leonard W. Weiss. 1976. Advertising, concentration, and price-cost margins. *Journal of Political Economy* 84:1109–21.

Terleckyj, Nestor E. 1974. *Effects of R & D on the productivity growth of industries: An exploratory study*. Washington, D.C.: National Planning Association.

———. 1980. Direct and indirect effects of industrial research and development on the productivity growth of industries. *New developments in productivity measurement and analysis*. ed. John W. Kendrick and Beatrice N. Vaccara, 419–54. Conference on Research in Income and Wealth: Studies in Income and Wealth, vol. 44. Chicago: University of Chicago Press for the National Bureau of Economic Research.

U.S. Federal Trade Commission. 1979. *Statistical report: Annual line of business report, 1973*. Report of the Bureau of Economics. Washington, D.C.: GPO.

———. 1981. *Annual line of business report, 1974*. Mimeo.

Weiss, Leonard W. 1980. Corrected concentration ratios in manufacturing—1972. University of Wisconsin and FTC. Mimeo.

Comment   Albert N. Link

Professor Scott presents the results from two interesting, and competently done, empirical experiments. Using the FTC's line of business data for the domestic manufacturing sector, he examined the separate relationships between company-financed applied research and development (AR & D) expenditures per unit of sales and both seller concentration and government-financed AR & D expenditures per unit of sales. This study is useful because it illustrates some fundamental R & D–related relationships that have previously been clouded by the inability of researchers to disaggregate by line of business.

Company AR & D Expenditures and Concentration

The first of Professor Scott's two analyses is motivated by the economic literature on nonprice competition: nonprice competition, it is hypothesized, will increase with seller concentration to a point, and then will

Albert N. Link is professor of economics and department head in the School of Business and Economics, University of North Carolina at Greensboro.

decline as a result of collusive-like activities that hamper all forms of rivalry. He illustrates in tables 10.1 and 10.2 that the line of business AR & D per unit of sales to concentration relationship exhibits an inverted-U; however, once differences in value, costs, and opportunities for nonprice competition are controlled for this relationship is no longer significant.

I have two comments on this section of the paper. The first is a general comment on the use of line of business AR & D data for measuring nonprice competition, and the second is a specific comment on the particular specification used in the estimation.

Professor Scott's argument for examining nonprice competition at the line of business level is quite valid, I think. There is no a priori reason to believe that firms' nonprice competition is homogeneous across industry bounds. But when AR & D is the form of nonprice competition, several conceptual questions arise that should be considered. The first relates to a characteristic of AR & D. AR & D often leads to a secondary product, knowledge. If, for example, AR & D expenditures are process related and are allocated in one line of business, might the knowledge obtained from implementing the associated technology be adaptable to a second line of business? If the answer is yes, and I think it is, then the ratio of AR & D to sales in the second line of business may understate the relevant degree of technology-related competition. The second question relates to the source of technology. Companies can invest in technology-related nonprice competition in at least two ways: they can induce the technology through their own AR & D activities, or they can purchase the technology in the form of new or improved capital equipment. As I have shown (Link, Tassey and Zmud 1983), this decision to induce or purchase (make versus buy) is related, in part, to the stage of the industry's product life cycle. Consequently, if at a given time industries (lines of business) with the same degree of seller concentration are at different stages in their life cycles, then a comparison of just AR & D expenditures may be misleading.

Finally, I have a general comment about the line of business file. I realize that at the time of Professor Scott's study data were available for only one year. Consequently, his cross-sectional analysis assumes that AR & D expenditures can accurately reflect interline business difference across firms in AR & D effort. Economists using cross-sectional data often make such an implicit assumption, but some statistical controls seem warranted if this study were to be extended. Not all firms have a centralized R & D unit, but those that have such a unit conduct long-term applied research that often influences the level and successfulness of divisional AR & D as reported in the FTC's file. (A case in point is the research done at General Electric on synthetic industrial diamonds.) Consequently, two firms could have the same level of AR & D expenditures for comparable lines of business, but one could be significantly more capable of nonprice competition because of previous research conducted in its central R & D unit. Would firm dummy variables indicating the presence of a central R & D unit control for this problem?

More germane to the actual empirical analysis is the issue of how accurately 436 company and 19 two-digit dummy variables (see table 10.2) can control for differences in the value, costs, and opportunities for nonprice competition. Technological opportunities are clearly a difficult, yet important, concept to quantify. I suspect, however, that the company dummies are controlling for many more things than intended. One suggestion for future work might be to replace the company effects with a single vector of the percentages of company-financed R & D allocated to basic research which can be calculated from the FTC file. Perhaps firms, through their own profit calculus, engage in relatively more basic research the fewer the direct opportunities for nonprice competition?

And, of course, there is the issue of simultaneity between R & D (and presumably AR & D) and concentration, which has already been discussed by Professors Levin and Reiss at this conference.

Company AR & D Expenditures and Government AR & D Financing

In section 10.3 of his paper, Professor Scott again imaginatively employs a fixed-effects model to test whether government AR & D spending is a substitute for, or a complement to, company-financed AR & D expenditures. This is an important issue, and there is a paucity of evidence related to it. Some speculations can be traced to the "pump priming" versus "substitution" hypothesis of Blank and Stigler (1957). On the one hand, new knowledge resulting from federal R & D enlarges the firm's scientific base and thus expands the opportunities for additional company-financed R & D. On the other hand, increases in federal allocations may displace private investments if (1) the resulting R & D output can be internalized by the firm, or if (2) federal obligations cause the firm to reach its capacity for technical operations. Professor Scott's finding (table 10.5) of a positive correlation between the two variables leads him to conclude that government-financed AR & D does indeed stimulate the level of company AR & D.

My comments here are similar to those directed to this issue in the first section of Professor Scott's paper. To what extent do the results of government-financed AR & D spill over to other lines of business within the firm and thus augment the efficiency of that AR & D? In other words, does government-financed AR & D contracted to the company's first line of business impact on company-financed AR & D in its second line of business? Concomitantly, how mobile are R & D scientists between line of business activities? To what extent might they embody AR & D related human capital?

My own work in this area, using firm level data, suggests that the relationship between government-financed and company-financed R & D (or even company productivity growth) is more subtle than Professor Scott's analysis reveals (Higgins and Link 1981; Link 1981). This subtlety can be detected by disaggregating R & D by category of use. I have found that government-financed R & D not only augments the level of company-financed R & D, but also alters the composition of that R & D. For example, a marginal dollar of government-financed R & D increases the probability of a private dollar of R & D being allocated to process-related, as opposed to product-related, R & D activities. Also, some interesting results have been obtained by analyzing the composition of government-financed R & D. For example, I have estimated a negative relationship between company-financed basic research and government-financed basic research expenditures while holding the financing capabilities of the firm constant (Link 1982). Perhaps, then, federal funds not only increase the level of corporate R & D but also alter its relative composition as well.

I realize my comments raise questions that are beyond the scope of the FTC's line of business file. Still, these questions should be discussed if we are to better comprehend the robustness of Professor Scott's findings and if we are to model corporate R & D decisions more accurately in future research.

References

Blank, D. M., and G. J. Stigler. 1957. *The demand and supply of scientific personnel*. New York: National Bureau of Economic Research.

Higgins, R. S., and A. N. Link. 1981. Federal support of technological growth in industry: Some evidence of crowding out. *IEEE Transactions on Engineering Management* EM-28:86–88.

Link, A. N. 1981. *Allocating R & D resources: A study of the determinants of R & D by character of use*. National Science Foundation. Final report.

Link, A. N. 1982. An analysis of the composition of R & D spending. *Southern Economic Journal* 49:342–49.

Link, A. N., G. Tassey, and R. W. Zmud. 1983. The induce versus purchase decision: An empirical analysis of industrial R & D. *Decision Sciences* 14:46–61.

11   Market Value, R & D, and Patents

Zvi Griliches

To the extent that R & D investments create "intangible" capital for a firm, it should show up in the valuation of the firm by the market. Such a valuation need not occur only after the long lag of converting an invention into actual product sales. It will, instead, reflect the current present value of expected returns from the invention (and from the R & D program as a whole). Thus, it is both possible and interesting to use the market value of the firm as a partial indicator of the expected success of its inventive efforts.

In a first effort to explore this topic, I start from the simplest "*definitional*" model:

$$V = q(A + K),$$

where V is the current market value of the firm (equity plus debt) as of the end of the year, A is current value of the firm's conventional assets (plant, equipment, inventories, and financial assets), K is the current value of the firm's intangible "stock of knowledge," to be approximated by different distributed lag measures of past R & D and the number of patents, and q is the current market valuation coefficient of the firm's assets, reflecting its differential risk and monopoly position.

Zvi Griliches is professor of economics at Harvard University, and program director, Productivity and Technical Change, at the National Bureau of Economic Research.

This is a preliminary report from a more extensive study of the returns to R & D at the firm level, supported by NSF grants PRA79–13740 and SOC78–04279. The author is indebted to John Bound, Bronwyn Hall, and Ariel Pakes for comments and research assistance with this work.

This note originally appeared in Zvi Griliches, "Market Value, R & D and Patents," *Economics Letters* 7 (1981): 183–87. Reprinted here with permission of North-Holland Publishing Company.

Writing

$$q_{it} = \exp(m_i + d_t + u_{it}),$$

where m_i is the permanent firm effect, d_t is the overall market effect at time t, and u_{it} is an individual annual disturbance or error term assumed to be distributed independently across firms and time periods. Defining $Q = V/A$, substituting, and taking logarithms, we get

$$\ell n Q = \ell n V/A = m + d + \ell n(1 + K/A) + u.$$

Substituting $\Sigma b_h R_{t-h}$, a distributed lag term of past R & D expenditures (R) for the unobserved K (or a similar additional term involving patents) and approximating $\ell n(1 + x)$ by x, we get

$$\ell n Q \simeq m + d + (\Sigma b_h R_{-h})/A + u,$$

which is the general form of the equation estimated in the first round of this study.

For a sample of 157 firms, we constructed the Q measure from the information given in Standard and Poor's Compustat tape, adapting for this purpose both the procedures and the program used by Brainard, Shoven, and Weiss (1980).[1] Using annual observations for the years 1968–74 but excluding observations with missing R & D data, large mergers, and large outliers ($|\ell n Q - \ell n Q_{-1}| > 2$), the final sample consisted of approximately 1000 observations with up to six lags for R & D and two lags on patents (we do not have valid patent data before 1967).

The results of estimating such an equation are given in table 11.1. To allow for interfirm differences in other unmeasured capital components, such as advertising or monopoly power, all the estimates are based on "within" regressions, on deviations around the individual firm means. Because of this preprocessing, what remains reflects largely shorter run fluctuations in R & D intensity and patent behavior and may be affected by errors of measurement and other transitory influences. In spite of these problems, we do find significant and positive effects of R & D and the number of patents applied for on the value of the firm.

To allow for both serial correlation and a more complicated lag structure, I added the lagged value of Q to the regressions. (This raises statistical problems because of the potential endogeneity of Q_{-1}, the Nerlove-Balestra problem, but that is a relatively minor issue when T is not too small. Moreover, all this is at an exploratory stage anyway.) The lagged value of Q is highly significant, and the models reported in lines (2)–(4) of table 11.1 imply that the long-run effect of $1 of R & D is to add about $2 to the market value of the firm (above and beyond its indirect effect via patents), while a successful patent is worth about $200,000.

1. See Pakes and Griliches (this volume) for a more detailed description of the provenance of this sample.

Table 11.1   Market Value as a Function of R & D and Patent Variables: 1968–74

| | Coefficients of | | | | | | | | |
| Equation and Dependent Variable | RD | | | Patents | | | $\log Q_{-1}$ | Other Variables in Equation | SEE |
	$\Sigma_0^5 b$	SRD	PRD	P_t	SP_t	PP_t			
$\log Q$ (1)	1.33	—	—	24.9 (4.8)	—	—	—	TD's & FD's	.319
(2)	2.63	—	—	10.8 (3.9)	—	—	.54 (.02)	TD's I FD's	.261
(3)	—	1.17 (.39)	—	—	11.6 (3.5)	—	.56 (.02)	TD's & FD's	.263
(4)	—	1.08 (.38)	—	—	8.4 (3.4)	—	.55 (.02)	TD's & FD's β's	.251
(5)	—	1.58 (.44)	1.23 (.54)	—	10.1 (3.9)	5.5 (5.2)	.54 (.02)	TD's & FD's β's	.251
$d\log Q$ (6)	—	1.93 (.45)	—	—	9.4 (4.1)	—	—	TD's & FD's	.304

Note: Estimated standard errors in parentheses. $N = 1091$; approximate degrees of freedom, 920.
Σb = sum of the coefficients of six R & D variables, e.g., in equation (2):

$$2.0 r_0 - .8 r_{-1} + .5 r_{-2} + .2 r_{-3} - 1.7 r_{-4} + 2.4 r_{-5}.$$
$$\quad (.5) \quad (.6) \quad (.6) \quad (.6) \quad (.8) \quad (.7)$$

SRD = "Surprise" in R & D. $RD - PRD = SRD$.
PRD = Predicted R & D. Predicted from a regression using RD_{-1}, P_{-1}, and $\log Q_{-1}$.
$SP_t = P_t - PP_t$; "Surprise" in patents.
PP_t = Predicted patents using P_{-1}, RD_{-1}, and $\log Q_{-1}$.
TD's = Year dummies.
FD's = Firm dummies.
β's = Individual estimated market β's (from *Value Line*, as of 1973) times the TD's.
SEE = Standard Deviation of the Estimated Residuals.

The estimated lag structure makes little sense, however, at least on first sight (see note at the bottom of the table 11.1). Except for the first coefficient, which is positive and highly significant, the subsequent coefficients change sign and are often insignificant. One possible interpretation of these results is that, in the presence of the lagged value of the firm in the equation, past R & D should not have any direct effect on V or Q. All of its anticipated effect should already be reflected in Q_{-1} as well as the effects of current and future R & D to the extent that they can be anticipated. What should *change* the market value is the inflow of news about new actual or potential discoveries. In short, only *unanticipated* R & D expenditures and patents should have positive effects in such an equation.

Equations (3)–(6) follow up this idea by using as their main variables the "surprise" components of R & D and patents, the changes that could not be predicted given historical information alone (the actual variables are constructed using a relatively simple equation containing lagged values of R & D, patents, and Q). Lines (3)–(4) show that such constructs do about as well in terms of fit as six separate lagged terms. Line (6), which is perhaps the easiest to interpret, says that a surprise $1 move in R & D results is equivalent to a $2 change in other assets. For patents there is little difference between using the actual number versus the nonpredictable component. Apparently most of the relevant variance in patents is unpredictable.

We do have the problem, though, that our patent variable is by the date applied for rather than by the date granted. The first is not fully public information and there may be quite a bit of uncertainty ex ante if a particular application will be, in fact, successful. We intend, therefore, to experiment also with the patents by date granted variable.

Current work is proceeding along the lines of incorporating rational expectation assumptions explicitly into our model and using modern time-series methods (a la Sims) to estimate it.[2] Such work needs to be based on larger samples than we have used to date. The second thrust of our work, therefore, is to expand our sample size significantly.

Reference

Brainard, W., J. Shoven, and L. Weiss. 1980. The financial valuation of the return to capital. *Brookings Papers on Economic Activity* 2:453–502.

2. See Pakes (this volume).

12 Patents, R & D, and the Stock Market Rate of Return: A Summary of Some Empirical Results

Ariel Pakes

This is an abstract of the results obtained in an earlier paper (Pakes 1981). That paper was motivated by the recent computerization of the U.S. Patent Office's data base. This has provided us with perhaps the most direct, and certainly the most detailed, indicator of inventive activity currently available. These data, then, ought to enable us to perform more detailed investigations of the causes and effects of invention and innovation than have been possible to date. To use the patent data effectively, however, requires an empirical understanding of the relationship between successful patent applications and measures of the inputs to, and the outputs from, the inventive process.

The study summarized here is designed to investigate the dynamic relationships between the number of successful patent applications of firms, a measure of the firm's investment in inventive activity (its R & D expenditures), and a measure of firm performance (its stock market values). There is a particular reason for using stock market values as the performance indicator in this context. As noted by Arrow (1962), the public-good characteristics of inventive output make it extremely difficult to market. Returns to innovation are mostly earned by embodying it in a tangible good or service which is then sold or traded for other information that can be so embodied (Wilson 1975; von Hippel 1982). There are, therefore, no direct measures of the value of invention, while indirect measures of current benefits (such as profits or productivity) are likely to react to the output of the firm's research laboratories only slowly and erratically. On the other hand, under simplifying assumptions, changes in the stock market value of the firm should reflect (possibly with error) changes in the expected discounted present value of the firm's entire,

Ariel Pakes is a lecturer in the Department of Economics at the Hebrew University of Jerusalem, and a faculty research fellow of the National Bureau of Economic Research.

uncertain, net cash flow. Thus, if an event does occur that causes the market to reevaluate the accumulated output of the firm's research laboratories, its full effect on stock market values ought to be recorded immediately. This is, of course, the expected effect of the event on future net cash flows and need not be equal to the effect which actually materializes. Measuring expectations rather than realizations has its advantages. In particular, expectations ought to determine research demand, so the use of stock market values will allow us to check whether the interpretation we give to our parameter estimates is consistent with the observed behavior of the research expenditure series.

This study is organized around a model serving both to interpret the parameters estimated and to provide a set of testable restrictions which indicate whether this interpretation is consistent with the observed behavior of the data. Two behavioral assumptions underlie the model (see Lucas and Prescott 1971 for a more detailed discussion of their theoretical implications). First, the firm is assumed to choose a research program to maximize the expected discounted value of its future net cash flows. A research program is defined as a sequence of random variables determining the firms current and future research expenditures conditional on (or as a function of) the information that will be available to the firm when those research expenditures must be made. The program is modified yearly on the basis of information accumulated on: the success and failure of the firm's R & D program; conditions in the markets relevant to the output of the firm's R & D activities; and input prices. The second behavioral assumption of the model determines the formation of stock market values. The stock market value of the firm is assumed to be an error-ridden measure of the expected discounted value of the firms future net cash flows. This assumption provides the interrelationship between the forces that drive the R & D expenditure series and those that drive stock market values, and it implies certain restrictions on the behavior of the data.

To be more precise, one can show that if the stock market provided an exact evaluation of the expected discounted value of the firm's future net cash flows based on the same information set used by management, then the one-period excess rate of return on the firm's equity (capital gains plus dividends on $1 invested in the firm minus the interest rate) would equal the percentage change in the expected discounted value of future net cash flows caused by the information accumulated over the given period. We are assuming that the observed rate of return on the firm's equity equals this change plus a disturbance uncorrelated with information publicly available at the beginning of the period. The latter assumption ensures that the process generating this disturbance does not allow agents operating on the stock market to use publicly available information and a simple linear trading rule to make excess returns on that market.

To the equations that determine the one-period rate of return and research demand, an equation is added that determines patent applications. Patents are assumed to react to the same events that cause revisions in the firm's R & D program and changes in the market's evaluation of that program, and, in addition, to be subject to a separate disturbance process. The disturbance process in the patent equation reflects differences in what Scherer (1965a; 1965b) has termed the propensity to patent; that is, it reflects differences in the number of patents applied for given the history of the inputs (the firm's current and past R & D expenditures) and the outputs (the firm's current and past stock market values) of the firm's inventive activity. This disturbance is allowed to be freely correlated over time but has the distinguishing characteristic that changes in it never affect the firm's R & D program or its stock market value. This provides an additional set of testable restrictions and underlies the interpretation of the variance in the disturbance process as differences in the propensity to patent.

The econometric structure of the model is that of a dynamic factor analysis (or index) model (see Geweke 1977 or Sargent and Sims 1977). There are three equations: one for each of the three observed variables (the stock market rate of return, R & D expenditures, and patent applications). The dynamic factor is a stochastic process affecting all three variables, though the time pattern of this factor's effect on the different variables differs. This factor is built up from current and past events that have caused changes in the expected discounted value of the firm's formal inventive endeavors. In addition to the dynamic factor, the patent and stock market rate of return equations are affected by the disturbance processes outlined above. It can be shown that, if these assumptions provide an adequate description of the data, the trivariate stochastic, process generating patents, R & D, and the stock market rate of return have a particularly simple recursive form in which all the restrictions of the model appear in the form of exclusion restrictions. This makes the recursive form particularly simple to estimate and interpret.

The parameters estimated from the recursive form can be used to calculate: the change in the stock market value of the firm associated, on average, with given changes in patent applications and in R & D expenditures; the change in R & D expenditures associated with given changes in patent applications; the time pattern of the effect of events changing the stock market value of the firm's inventive endeavors on patent applications and on R & D expenditures; the percentage of the variance in the stock market rate of return attributable to the factors causing changes in the firm's inventive activity; the percentage of the variance in the patent variable that is caused by differences in the propensity to patent (i.e., that never affects either stock market values or the firm's R & D program); and the serial correlation properties of this propensity in a given firm over time.

Before going on to a brief description of the empirical results, it is worth elaborating on the implications of two of the restrictions that were imposed. Neither of these restrictions were necessary. Rather, both were maintained because the data could not distinguish between the simpler models they implied and the more complicated models that would result without imposing them. First, no distinction was made at this stage between the different kinds of events likely to cause changes in inventive activity (say demand shocks versus supply shocks, where demand shocks are only transformed into more patents as a result of the R & D expenditures they induce, while supply shocks have a direct effect on patenting as well as an indirect effect via induced R & D activity). In principle, the techniques and data used here should be sufficient to isolate the effect of these different events. The data indicated, however, that to accomplish this task empirically one is likely to require a measure that distinguishes more effectively between demand and supply shocks than R & D does (perhaps investment expenditures). This is a topic I intend to pursue further. Second, in the model estimated, no allowance was made for a disturbance process in the R & D demand equation; that is, a process that does affect R & D but does not affect either the stock market value of the firm or its patent applications. Here the data indicated that such a process was simply not necessary. This was comforting since it indicates that once we move away from indirect measures of current benefits (such as productivity) there is less need to worry about measurement error in the R & D series.

The data set used to estimate the model contained patent applications, R & D expenditures, and stock market rates of return for 120 firms over an eight-year period (this data set is described more fully in Pakes and Giliches, chap. 3 in this volume). The restrictions of the model were accepted, and, on the whole, parameters were estimated with a great deal of precision. The qualitative nature of the empirical results can be summarized quite succinctly. First, it is clear that there is a highly significant correlation between the stock market rate of return and unexpected changes in both patent applications and R & D expenditures (unexpected changes here refer to changes that could not be predicted from the history of the variables in our data set). Moreover, the estimates imply that the unexpected changes in the patent and R & D series are associated with quite large movements in stock market values. On the other hand, the estimates imply that the vast majority of the variance in the stock market rate of return is determined by factors that have little to do with inventive activity. Thus, if one were to use movements in the stock market rate of return as an indicator of changes in the private value of inventive output (and there are strong theoretical and empirical reasons for doing so), one ought to be careful to allow for a disturbance process to intercede between them (as noted above, we do have information on the properties of that disturbance process).

The events that do cause the market to reevaluate the firm's inventive endeavors have long-lasting effects on both the patents and the R & D expenditure series of the firm. In fact, most of the cross-sectional variance in patents is caused by them; that is, differences in patent applications between firms seem to be mostly determined by the same factors which cause differences in the market's evaluation of the firm's inventive endeavors. On the other hand, most of the variance in patent applications within a given firm over time is determined by intertemporal differences in the propensity to patent; that is, by factors that never cause changes in its R & D program or its market value. As a result, the patent variable is likely to be less useful in studies of changes occurring in the inventive output of a given firm over time. This last statement must be modified somewhat when one considers longer term differences in the patents applied for by a given firm (say differences over a five- or a ten-year interval), since a larger proportion of their variance is caused by events that lead the market to reevaluate the firm's inventive output during these periods.

The timing of the impact of the events that cause unexpected changes in the market value of a firm's inventive activity on patents is very close to the timing of their impact on R & D. In fact, one gets the impression from the estimates that an event which causes a 1 percent change in the market value of a firm's inventive activity starts a chain reaction leading to more R & D expenditures far into the future, with the firm patenting around the links of the chain almost as quickly as they are completed. Thus, if one were to use the estimates to compute a distributed lag from R & D to patents, most of the weight in the lag coefficients would be concentrated in the first three R & D variables. This lag distribution also has a long slim tail which probably represents the effect of the basic research done in the past on current patented innovations. Finally, these timing patterns imply that current patent applications are highly correlated with the factors setting current R & D demand.

To date our understanding of the role of invention and innovation in economic processes has been severely hampered by a lack of empirical evidence on its causes and its effects. In large part this reflects the difficulty in finding (or constructing) meaningful measures of invention. This paper investigated whether (and in which dimensions) the patent data are likely to alleviate this problem. The answers are somewhat mixed. There is a large variance in the patent applications of different firms, and this variance is mostly determined by events that have changed both the market value of the firm's research program and its research input. Though, in the cross-sectional dimension, differences in current patent applications are closely related to differences in current research investments, both of these variables have long memories; that is their levels reflect events that have occurred over a long time period. In several situations R & D (and, for that matter, market value) data are simply not available (of particular interest is when one wants to study the research investments of different firms in particular product fields). Use of the patent data as a proxy for R & D in these cases, together with some of the qualitative results derived here, is likely to be quite fruitful. On the other hand, much of the variance in the patent applications of a given firm over time is simply a result of noise (differences in the propensity to patent). Of course, some information is still in the time-series dimension. If one were to observe, for example, a sudden burst in the patent applications of a given firm, one would be quite sure that events have occurred causing significant change in the market value of its R & D program, but smaller changes in the patents of a given firm are not likely to be very informative. To establish that one can use the patent and R & D data together to distinguish between the different kinds of events that can cause changes in research activity, one requires the addition of more variables, and perhaps more structure, to the model used here.

References

Arrow, Kenneth J. 1962. Economic welfare and the allocation of resources for invention. In *The rate and direction of inventive activity: Economic and social factors*, ed. R. R. Nelson, 609–25. Universities-NBER Conference Series no. 13. Princeton: Princeton University Press for the National Bureau of Economic Research.

Geweke, John. 1977. The dynamic factor analysis of economic time-series models. In *Latent variables in socio-economic models*, ed. D. J. Aigner and A. S. Goldberger, 365–83. Amsterdam: North-Holland.

Lucas, Robert E., Jr., and Edward C. Prescott. 1971. Investment under uncertainty. *Econometrica* 39 (September): 659–81.

Pakes, Ariel. 1981. Patents, R & D, and the stock market rate of return. National Bureau of Economic Research Working Paper no. 786. Cambridge, Mass.: NBER.

Sargent, Thomas J., and Christopher A. Sims. 1977. Business cycle modeling without pretending to have too much a priori economic theory. In *New methods in business cycle research: Proceedings from a conference*, ed. Christopher A. Sims, 45–109. Minneapolis: Federal Reserve Bank of Minneapolis.

Scherer, Frederic M. 1965a. Corporate inventive output, profits, and growth. *Journal of Political Economy* 73 (June): 209–97.

———. 1965b. Firm size, market structure, opportunity, and the output of patented inventions. *American Economic Review* 55 (December): 1097–1125.

von Hippel, Eric. 1982. Appropriability of innovation benefit as a predictor of the functional locus of innovation. *Research Policy* 11:95–115.

Wilson, Robert Woodrow. 1975. The sale of technology through licensing. Ph.D. diss., Yale University.

13 R & D and the Market Value of the Firm: A Note

Andrew B. Abel

In his paper, Ariel Pakes (1981) uses stock market valuation data to study the relation between patents and the economic value of inventive activity. The econometric technique is based on dynamic factor analysis (Geweke 1977) and index models (Sargent and Sims 1977). Although these techniques often involve the frequency domain, Pakes's model can be estimated in the time domain and thus leads to more intuitively interpretable results.

Pakes's model is based on an intertemporally optimizing firm which chooses a research and development program to maximize the expected present value of its net cash flow. One approach to the problem would be to specify the net cash flow function and solve for the optimal decision rules. Pakes shuns this approach because it does not relate changes in research activity to the expected present value of cash flow. Instead, he uses the stock market to evaluate the expected present value of cash flow and associates changes in a firm's market value with the value of new research. This choice between explicitly solving a dynamic optimization problem, on the one hand, and using stock market valuation to determine the level of investment in new capital, or R & D, on the other hand, reflects a dichotomy in the investment literature between structural models and market valuation models based on Tobin's q (1969). Recent research has established a link between these approaches, and, after

Andrew B. Abel is an associate professor in the Department of Economics, Harvard University, and a faculty research fellow of the National Bureau of Economic Research.

The author thanks Zvi Griliches and Ariel Pakes for helpful comments on an earlier draft.

This chapter originated as a comment on the paper presented by Ariel Pakes at the conference. References to footnotes and equation numbers are to those in Pakes (1981) which was presented at the conference.

261

commenting briefly on Pakes's paper, I will present a simple model of R & D illustrating this link.

Pakes defines q_t to be the excess one-period rate of return on a firm's equity and relates q_t to R & D activity. One must proceed with caution in associating changes in the value of a firm with the value of R & D activity. The value of the firm can change for a variety of reasons other than successful research activity. The stock market values the firm as an ongoing concern and reflects the value of all the firm's capital assets, both tangible and intangible. Even in the absence of any changes in the level or composition of the capital stock, the value of the firm may change because of changes in demand or in the supply of variable factors of production. Even ignoring these sources of variation in the valuation of the firm, the value of the firm may increase by less than the present value of quasi-rents associated with new R & D if the new R & D makes obsolete some previously valued process (see Bailey 1981). It should be pointed out that Pakes is well aware of the implications of these points. Indeed, he calculates that 95 percent of the variance in q_t is unrelated to either R & D or patents.

The formal tests in the paper are based on the assumed stochastic structure of the innovations a_t, b_t, and g_t. Since Pakes assumes a form for the value function in his equation (2) rather than deriving it from an optimization problem, the interpretation of the shock B_t in this equation is unclear. One interpretation of B_t might be that for some reason (either attitude toward risk or systematic bias) the stock market places a value B_t on an uncertain cash flow with expected present value equal to one. In this case B_t should multiply the expected present value of future cash flows but should not multiply the known current cash flow, R_t. If equation (2) is modified as suggested here, then the optimal value of R_t would not be independent of B_t, and this would have implications for the set of testable restrictions.

Since all of the empirical work is based on the time-series behavior of the trivariate process generating (q_t, r_t, p_t), the timing of these variables is extremely important. As explained in Pakes's footnote 13, q_t refers to the stock market rate of return from date $t - 1$ to date t, whereas r_t and p_t refer to R & D and patents from date t to date $t + 1$. Letting Ω_{t-1} denote the set of information available at date $t - 1$, market efficiency implies $E(q_t | \Omega_{t-1}) = 0$. Pakes assumes that r_{t-1} and p_{t-1} are contained in Ω_{t-1} and tests the implication that $E(q_t | r_{t-1}, p_{t-1}) = 0$. However, the assumption that p_{t-1} is in Ω_{t-1} means that investors know at date $t - 1$ the number of successful patent applications between date $t - 1$ and date t. Since successful patent applications are not a decision variable of the firm (as is r_{t-1}), but are rather the outcome of a process with a variable lag and an uncertain outcome, the assumption that p_{t-1} is in Ω_{t-1} is quite strong.

That is, the restriction $E(q_t | r_{t-1}, p_{t-1}) = 0$ is stronger than implied by market efficiency. Nonetheless, the data fail to reject this restriction.

As a final direct comment on Pakes's paper, I would point out that patents are essentially a sideshow in Pakes's theoretical model. As Pakes observes, patents must have some economic value since resources are expended to acquire them. However, patents do not enter into the value function for the firm. In fairness to Pakes, it must be noted that his research strategy at this stage is merely to determine whether the newly available patent data are related to anything that might be called inventive output. This appears to be a reasonable strategy at this early stage to search the data for correlations and then at a later stage to impose the discipline of a structural model.

A Model of R & D and the Value of the Firm

I will devote the remainder of this chapter to the development of a simple model of the R & D activity and valuation of a firm. Although the model presented below is probably too simple to use directly for empirical work, it does provide an explicit optimizing framework with which to analyze optimal R & D activity and firm value. Furthermore, the stochastic elements are incorporated directly into the firm's optimization problem.

We consider the production and R & D decisions of an intertemporally optimizing, risk-neutral firm. Suppose that the firm uses labor, L_t, and accumulated technology, T_t, to produce output, Q_t, according to the Cobb-Douglas production function $Q_t = L_t^\alpha T_t^{1-a}$. Labor is hired in a spot market at a fixed wage rate, w. Technology accumulates over time as a result of the firm's R & D activity, R_t, and can decumulate over time from obsolescence. We specify the technology accumulation equation as

$$(1) \qquad T_{t+1} = \eta_{1t+1} R_t + \eta_{2t+1} T_t,$$

where

$$E(\eta_{1t}) = 1 \text{ and } 0 < E(\eta_{2t}) = \delta < 1,$$

where η_{1t} and η_{2t} are each serially uncorrelated random variables. In equation (1) we explicitly recognize that the outcome of R & D activity is uncertain by including the random variable η_{1t}. The assumption that $E(\eta_{1t}) = 1$ is simply a convenient normalization. The random variable η_{2t} reflects the fact that obsolescence also occurs randomly. The assumption that $E(\eta_{2t}) = \delta$ implies that, in the absence of R & D activity, the expected proportional rate of "depreciation" of technology is $1 - \delta$. We assume that η_{2t} takes on only positive values.

We have defined R_t as R & D activity which, according to (1), is the expected gross addition to the stock of technology. We assume that the marginal cost of R_t is an increasing function of R_t. Specifically, it will be convenient to model R & D expenditures as a quadratic function of R & D activity, aR_t^2. Finally, we allow the price of output, p_t, to be random. Specifically, we suppose

(2) $$p_{t+1}/p_t = \epsilon_{t+1}^*,$$

where ϵ_{t+1}^* is serially independent. The random term ϵ_{t+1}^*, which is equal to one plus the rate of inflation, is assumed to take on only positive values.

The firm's decision problem at time t is to maximize the expected present value of its net cash flow. Assuming that the discount rate r is constant, and defining the discount factor $\beta = 1/(1 + r)$, we can write the decision problem as

(3) $$\max_{L_t, R_t} E_t \sum_{j=0}^{\infty} \beta^j (p_{t+j} L_{t+j}^a T_{t+j}^{1-a} - wL_{t+j} - aR_{t+j}^2),$$

subject to the technology accumulation equation (1), the price process (2), and the initial condition that T_t is given. Let $V(T_t, p_t)$ denote the maximized value of the present value of expected cash flow in (3), and observe that under risk neutrality $V(T_t, p_t)$ is the value of the firm. The value of the firm is a function of only p_t and T_t, since p_t is a sufficient statistic for the history of price shocks, ϵ_{t-i}^*, and T_t is a sufficient statistic for the history of shocks η_{1t-i} and η_{2t-i}.

We will solve the maximization problem in (3) using stochastic dynamic programming. The equation of optimality, known as the Bellman equation, is

(4) $$V(T_t, p_t) = \max_{L_t, R_t} E_t [p_t L_t^a T_t^{1-a} - wL_t - aR_t^2 + \beta V(T_{t+1}, p_{t+1})].$$

The Bellman equation merely states that the value of the firm is the maximized sum of the current cash flow plus the expected present value of the firm next period.

Since the labor input, L_t, affects only current cash flow, we can easily "maximize it out" of the equation (4). Recognizing that the optimal value of L_t equates the marginal revenue product of labor, $ap_t(L_t/T_t)^{a-1}$, with the wage rate, w, we obtain

(5) $$\max_{L_t} (p_t L_t^a T_t^{1-a} - wL_t) = hp_t^{\frac{1}{1-a}} T_t,$$

where

$$h = (1-a) \left(\frac{a}{w} \right)^{\frac{a}{1-a}}$$

Substituting (2), (3), and (5) into (4), we obtain

(6) $V(T_t, p_t) = \max_{R_t} E_t \left[h p_t^{\frac{1}{1-a}} T_t - a R_t^2 + \beta V(\eta_{1t+1} R_t + \eta_{2t+1} T_t, p_t \epsilon_{t+1}^*) \right].$

Equation (6) is a functional equation which must be solved for the value function $V(T, p)$. The solution to this functional equation is

(7) $$V(T_t, p_t) = \gamma_1 p_t^{\frac{1}{1-a}} T_t + \gamma_2 p_t^{\frac{2}{1-a}},$$

where

(7a) $$\gamma_1 = \frac{h}{1 - \beta(\delta \bar{\epsilon} + \sigma_{\epsilon \eta_2})},$$

and

(7b) $$\gamma_2 = \frac{1}{4a(1 - \beta \overline{\epsilon^2})} [\beta \gamma_1 (\bar{\epsilon} + \sigma_{\epsilon \eta_1})]^2,$$

and where we have defined $\epsilon_t \equiv \epsilon_t^{*\frac{1}{1-a}}$, $E(\epsilon_t) = \bar{\epsilon}$, $E(\epsilon_t^2) = \overline{\epsilon^2}$, $\text{cov}(\epsilon_t, \eta_{1t}) = \sigma_{\epsilon \eta_1}$, and $\text{cov}(\epsilon_t, \eta_{2t}) = \sigma_{\epsilon \eta_2}$. We assume that the random vector $(\eta_{1t}, \eta_{2t}, \epsilon_t)$ is serially uncorrelated. This solution can be obtained using the method of undetermined coefficients by assuming that the value function has the form in (7) and then solving for the coefficients γ_1 and γ_2. Rather than derive the solution here, we will verify that the function in (7)–(7b) is the solution to the Bellman equation.

To show that (7) is indeed the solution to the Bellman equation, we must calculate $E_t V(T_{t+1}, p_{t+1})$. Substituting (1) and (2) into (7) and calculating expected values, we obtain

(8) $E_t V(T_{t+1}, p_{t+1}) = \gamma_1 p_t^{\frac{1}{1-a}} R_t (\bar{\epsilon} + \sigma_{\epsilon \eta_1}) + \gamma_1 p_t^{\frac{1}{1-a}} T_t (\bar{\epsilon} \delta + \sigma_{\epsilon \eta_2})$

$$+ \gamma_2 p_t^{\frac{2}{1-a}} \overline{\epsilon^2}.$$

From equation (4) it is clear that the optimal value of R_t is such that the marginal cost, $2aR_t$, is equal to $\beta E_t \{[\partial V(T_{t+1}, p_{t+1})]/\partial R_t\}$. Using (8) to calculate $\beta E_t \{[\partial V(T_{t+1}, p_{t+1})]/\partial R_t\}$, the optimal value of R_t is

(9) $$R_t = \frac{1}{2a} \beta \gamma_1 (\bar{\epsilon} + \sigma_{\epsilon \eta_1}) p_t^{\frac{1}{1-a}}.$$

Note that the optimal rate of R & D activity is independent of the level of accumulated technology, T_t. We will discuss further properties of the optimal R & D activity later.

Given the optimal rate of R & D activity in (9), we can now express $E_t V(T_{t+1}, p_{t+1})$ as a function of the state variables T_t and p_t. Substituting (9) into (8) and using (7b), we obtain

$$(10) \quad E_t V(T_{t+1}, p_{t+1}) = \gamma_1 p_t^{\frac{1}{1-a}} T_t(\bar{\epsilon}\delta + \sigma_{\epsilon\eta_2}) + \gamma_2 \left(\frac{2}{\beta} - \overline{\epsilon^2}\right) p_t^{\frac{2}{1-a}}.$$

Finally, we can express the current cash flow under optimal behavior, $hp_t^{1/1-a} T_t - aR_t^2$, as function of the state variables T_t and p_t by using (9) to substitute for R_t,

$$(11) \qquad hp_t^{\frac{1}{1-a}} T_t - aR_t^2 = hp_t^{\frac{1}{1-a}} T_t - \gamma_2(1 - \beta\overline{\epsilon^2}) p_t^{\frac{2}{1-a}}.$$

The value of the firm is the sum of the current cash flow in (11) and the expected present value of the firm's value next period in (10). Thus, the value of the firm is obtained by adding β times the right-hand side of (10) plus the right-hand side of (11). Performing this operation verifies that (7)–(7b) is indeed the solution to the Bellman equation (6).

Observe from (7) that the value function is a linear function of the stock of accumulated technology. The economic intuition underlying the value function is quite straightforward. Let λ_t denote the expected present value of marginal revenue products accruing to technology from time t onward. That is,

$$(12) \qquad \lambda_t = hp_t^{\frac{1}{1-a}} + E_t\left[\sum_{j=1}^{\infty} \beta^j \left(\prod_{i=1}^{j} \eta_{2t+i}\right) hp_{t+j}^{\frac{1}{1-a}}\right].$$

Recognizing that

$$E_t\left(p_{t+j}^{\frac{1}{1-a}}\prod_{i=1}^{j} \eta_{2t+1}\right) = p_t^{\frac{1}{1-a}} E_t\left(\prod_{i=1}^{j} \epsilon_{t+i}\eta_{2t+i}\right)$$

$$= p_t^{\frac{1}{1-a}} (\bar{\epsilon}\delta + \sigma_{\epsilon\eta_2})^j,$$

we can (if $\beta(\bar{\epsilon}\delta + \sigma_{\epsilon\eta_2}) < 1$) rewrite (12) as

$$(13) \qquad \lambda_t = hp_t^{\frac{1}{1-a}}\left[1 + \sum_{j=1}^{\infty} \beta^j\left(\bar{\epsilon}\delta + \sigma_{\epsilon\eta_2}\right)^j\right] = \frac{hp_t^{\frac{1}{1-a}}}{1 - \beta(\bar{\epsilon}\delta + \sigma_{\epsilon\eta_2})}.$$

Observe that λ_t is equal to the slope coefficient (with respect to T_t) of the value function (7). Thus, the first term in the value function (7), $\gamma_1 p_t^{1/1-a} T_t$, represents the present value of net cash flow accruing to technology existing at time t.

To interpret the second term in the linear function (7), we observe that the quadratic form of the R & D expenditure function implies the existence of inframarginal rents to R & D activity. In figure 13.1 we illustrate the first-order condition for the optimal rate of R & D activity, namely, that marginal R & D expenditure, $2aR_t$, is equal to the expected present value of the marginal contribution of R & D activity. The shaded area in figure 13.1 represents the rents accruing to the inframarginal

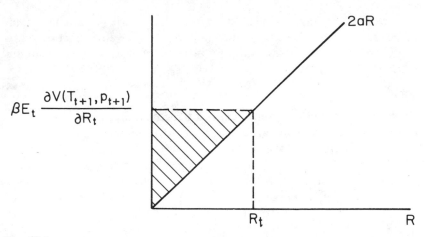

Fig. 13.1

R & D activity. The area of the shaded region is equal to aR_t^2, which, using (9), is equal to

$$\frac{1}{4a}\beta^2\gamma_1^2\left(\bar{\epsilon}+\sigma_{\epsilon\eta_1}\right)^2 p_t^{\frac{2}{1-a}}.$$

The expected present value of these rents over the entire future is

$$(14)\qquad E_t\sum_{j=0}^{\infty}\beta^j\frac{\beta^2\gamma_1^2(\bar{\epsilon}+\sigma_{\epsilon\eta_1})^2}{4a}p_{t+j}^{\frac{2}{1-a}}$$

$$=\frac{\beta^2\gamma_1^2(\bar{\epsilon}+\sigma_{\epsilon\eta_1})^2 p_t^{\frac{2}{1-a}}}{4a}\left[1+\sum_{j=1}^{\infty}\beta^j E_t\left(\prod_{i=1}^{j}\epsilon_{t+i}^2\right)\right].$$

Observe that $E_t(\prod_{i=1}^{j}\epsilon_{t+i}^2)=(\overline{\epsilon^2})^j$. Therefore, we obtain (if $\beta\overline{\epsilon^2}<1$)

$$(15)\qquad 1+\sum_{j=1}^{\infty}\beta^j E_t\left(\prod_{i=1}^{j}\epsilon_{t+i}^2\right)=\frac{1}{1-\beta\overline{\epsilon^2}}$$

Substituting (15) into (14), the expected present value of rents to inframarginal R & D activity is

$$(16)\qquad \frac{\beta^2\gamma_1^2(\bar{\epsilon}+\sigma_{\epsilon\eta_1})^2}{4a(1-\beta\overline{\epsilon^2})}p_t^{\frac{2}{1-a}}.$$

Note that the expression in (16) is equal to the second term in the value function (7).

To summarize our description of the value function, we have shown that it is a linear function of the stock of accumulated technology, T_t. The term which is proportional to T_t represents the expected present value of net revenues accruing to the existing stock T_t over the remaining lifetime of this technology. The constant term in the linear value function represents the expected present value of inframarginal rents to present and future R & D activity.

The Effects of Uncertainty

The model developed above explicitly incorporates three channels for uncertainty: the price of output (ϵ_t), the contribution of R & D activity to the gross increase in the stock of technology (η_{1t}), and the rate of obsolescence (η_{2t}). Note from equations (7)–(7b) and (9) that increased uncertainty in either of the two shocks to technology accumulation (η_{1t} and η_{2t}) will affect optimal R & D activity and the value of the firm only if these shocks are correlated with the price shock $\epsilon_t \equiv \epsilon_t^{*1/(1-a)}$. If $\sigma_{\epsilon\eta_1}$ and $\sigma_{\epsilon\eta_2}$ are zero, then only the expected values, but not the variances, of η_{1t} and η_{2t} are relevant.

The optimal rate of R & D activity is an increasing function of both $\sigma_{\epsilon\eta_1}$ and $\sigma_{\epsilon\eta_2}$. A higher value of $\sigma_{\epsilon\eta_1}$ increases $E_t(hp_{t+1}^{1/(1-a)}\eta_{1t+1}R_t)$, the expected value of the next period's net revenue accruing to current R & D activity. Therefore, the optimal level of R & D activity is an increasing function of $\sigma_{\epsilon\eta_1}$. A higher value of $\sigma_{\epsilon\eta_2}$ increases $E_t[hp_{t+j}^{1/(1-a)}$ $(\Pi_{i=2}^{j}\eta_{2t+i})\eta_{1t+1}R_t]$, $j = 2, 3, \ldots$, which is the expected net revenue in period $t + j$ accruing to R & D activity undertaken in period t.

The model developed above can be used to study the stochastic behavior of the value of the firm and of R & D activity. For instance, recall that Pakes analyzes the stochastic properties of the one-period excess rate of return, q_t. Pakes calculated q_t empirically as

$$(17) \qquad q_t = \frac{V(T_t,p_t) - V(T_{t-1},p_{t-1}) + (1+r)D_{t-1}}{V(T_{t-1},p_{t-1})} - r,$$

where $D_{t-1} = p_{t-1}L_{t-1}^a T_{t-1}^{1-a} - wL_{t-1} - aR_{t-1}^2$ is the dividend earned at time $t - 1$. In words, q_t is simply the excess of the sum of the dividend and capital gain over the rate of interest. (As noted by Pakes, since dividends are paid at the beginning of the period, we must include within-period interest earnings on the dividend. Empirically, Pakes found that inclusion of rD_{t-1} made no substantial difference.) We can rewrite (17) using the Bellman equation (4) to obtain

$$(18) \qquad q_t = \frac{V(T_t,p_t) - E_{t-1}V(T_t,p_t)}{V(T_{t-1},p_{t-1})}.$$

From (18) it is clear that q_t is completely unforecastable as of time $t - 1$. This finding is simply the well-known implication of efficient markets theory which states excess returns are uncorrelated with any past information.

Recalling that R & D expenditures are equal to aR_t^2, we can easily calculate optimal R & D expenditures using (9),

$$(19a) \qquad aR_t^2 = \frac{1}{4a}[\beta\gamma_1(\bar{\epsilon} + \sigma_{\epsilon\eta_1})]^2 p_t^{\frac{2}{1-a}}$$

$$(19b) \qquad\qquad = (aR_{t-1}^2)\epsilon_t^2.$$

Thus, current R & D expenditures, aR_t^2, depend only on the previous period's R & D expenditure, aR_{t-1}^2, and the contemporaneous shock to the price of output, ϵ_t. Of course, this simple relation in (19b) is a consequence of the simple structure of the model. However, even this simple model illustrates that much of the variance in q_t can be unrelated to R & D expenditures. Recall that the variation in q_t is from ϵ_t, η_{1t}, and η_{2t}, whereas the variation in R & D expenditures is from ϵ_t^2. To the extent that η_{1t} and η_{2t} account for much of the variation in q_t and to the extent that they are uncorrelated with ϵ_t^2, we would expect only a small part of the variation in q_t to be related to R & D activity. This situation is consistent with Pakes's finding that only 5 percent of the variation in q_t is related to either R & D or patents.

The model presented above provides a useful stochastic framework for analyzing the value of the firm and R & D activity. However, this model, like the model in Pakes's paper, fails to account explicitly for patents. Although it would be straightforward to model the cost of obtaining a patent, modeling the benefits accruing to a patent requires further work.

References

Bailey, Martin N. 1981. Productivity and the services of capital and labor. *Brookings Papers on Economic Activity*, 1:1–50.

Geweke, John. 1977. The dynamic factor analysis of economic time-series models. In *Latent variables in socio-economic models*, ed. D. J. Aigner and A. S. Goldberger, 365–83. Amsterdam: North-Holland.

Pakes, Ariel. 1981. Patents, R & D, and the stock market rate of return. National Bureau of Economic Research Working Paper no. 786. Cambridge, Mass.: NBER.

Sargent, Thomas J., and Christopher A. Sims. 1977. Business cycle modeling without pretending to have too much a priori economic theory. In *New methods in business cycle research: Proceedings from a conference*, ed. Christopher A. Sims, 45–109. Minneapolis: Federal Reserve Bank of Minneapolis.

Tobin, James. 1969. A general equilibrium approach to monetary theory. *Journal of Money, Credit and Banking* (February): 15–29.

14 An Extended Accelerator Model of R & D and Physical Investment

Jacques Mairesse and Alan K. Siu

14.1 Introduction

The purpose of the present study is to investigate the determinants of both R & D and physical investment using a panel of firm data. In a standard neoclassical model of investment, the firm is assumed to choose an investment plan to maximize the present discounted value of net cash flow subject to the production technology, cost of adjustment function, initial capital stocks, and other appropriate constraints (or else to minimize the present discounted total cost of production subject to the same constraints and an expected production plan). In full generality, this involves considering nonlinear stochastic control problems, and explicit solutions of the first-order conditions are intractable without very restrictive assumptions. Assumptions such as static expectations about prices, a simple form of the production function, the absence of an explicit cost of adjustment function, and the imposition of a given lag structure are usually made to derive the specification of the investment function.

In view of the complexities of a formal model of investment decisions, and also because of a lack of data on factor prices at the firm level, we

Jacques Mairesse is a professor at Ecole des Hautes Etudes en Sciences Sociales and Ecole Nationale de la Statistique et de l'Administration Economique, and a research affiliate of the National Bureau of Economic Research. Alan K. Siu is an assistant lecturer in the Department of Economics at the Chinese University of Hong Kong.

The authors are particularly indebted to Zvi Griliches for his interest and suggestions, and they have benefited from reading Uri Ben-Zion's and Ariel Pakes's related studies (in this volume) and from helpful discussions with them and with John Bound. The authors also thank Edmond Malinvaud, Alain Monfort, and Pascal Mazodier for their comments on a preliminary draft. This work is an outgrowth of a larger National Bureau of Economic Research project on R & D and Productivity supported by the National Science Foundation (PRA 79–13740). Additional financial support by the Centre National de la Recherche Scientifique (ATP 070199) and by Maison des Sciences de l'Homme is gratefully acknowledged.

have to settle for a looser approach in the spirit of data analysis as advocated by Sims (1972a, 1972b, 1977, 1980; Sargent and Sims 1977). A priori, expected demand and expected profitability are important determinants for investment decisions. Both are unobservable. Following Pakes (this volume), we propose to use the stock market one-period holding rate of return, q, as an indicator of changes in expectation about the firm's future profitability. For expected demand, we have used a more traditional distributed lag formulation of the rate of growth of sales, s. These two variables plus the rates of growth of R & D and physical investment, r and i, are embedded in a multivariate autoregressive model. We perform a series of exogeneity tests to investigate the appropriateness of restricted versions of this general model which are of interest. In particular, we vindicate an extended form of the traditional accelerator model: extended both because it applies to R & D as well as to physical investment and because it takes expected profitability, not only demand, as a major explanatory factor. The specification of our model is discussed in section 14.2, while our results are presented in section 14.3. We end with a few remarks in section 14.4.

14.2 Model Specification: Statistical and Economic Considerations

We start from what we call our general model and derive our extended accelerator model, discussing the meaning and specification of each equation in turn.

14.2.1 A General Model

First, let us denote the four variables our study concentrates on by q_{nt}, s_{nt}, r_{nt}, and i_{nt}, where n and t represent firm and year subscripts ($n = 1$ to N; $t = 1$ to T), respectively. To simplify matters we shall suppress the firm subscript n in general and, when convenient, we shall also represent by y_{nt} or y_t the column vector of our four variables, that is, $y_t = (q_t, s_t, r_t, i_t)'$.

The variable q_t is the stock market one-period holding rate of return, defined as $q_t = (p_t - p_{t-1} + d_t)/p_t$, where p_t is the price of a share at the end of year t, and d_t is the dividend per share paid during this year. Thus, q_t is equal to the rate of change of the value of a one dollar share over the year plus the corresponding dividend. Variables s_t, r_t, and i_t denote the first difference between year t and year $(t - 1)$ of the logarithms of sales, R & D expenditures, and gross investments, respectively, and are thus approximately equal to their rate of change from year to year: $s_t = \log(S_t/S_{t-1})$; $r_t = \log(R_t/R_{t-1})$; $i_t = \log(I_t/I_{t-1})$.[1]

---

1. In the empirical implementation, q_t is adjusted for stock splits when they occur. Sales are deflated using industry price indexes; R & D and investment expenditures are also deflated by an overall price index. There is the possibility of some mismatch in timing between s_t, r_t, and i_t, which are based on the companies' fiscal year, and q_t, which is based on

Given our focus on these four variables, we are interested in investigating thoroughly their mutual dynamic interrelationships. Without pretending too much a priori knowledge about these interrelations, we start by assuming that they can be represented by an autoregressive model:

(1) $$y_t = A(L)y_{t-1} + \lambda_t + \eta_t,$$

where $A(L)$ is a matrix of polynomials in the lag operator (L), λ_t is a vector of time-specific effects or year dummies, and η_t is a vector of disturbances assumed to be normally distributed, uncorrelated over time but correlated across equations: η_t serially uncorrelated $N(0, \Sigma)$. The vector η_t is called the vector of "innovations" in the variables. We can write (1) more simply as:

(1') $$y_t = A(L)y_{t-1} + \eta_t,$$

if we take care of the year effects λ_t by measuring our variables relative to their year means, as we shall assume from now on.[2]

With an adequate number of lags, the autoregressive model is flexible enough to account well for the correlation structure of our variables and simulate their dynamic behavior. From a purely statistical standpoint, equivalent formulations can be obtained by multiplying both sides of (1') by any nonsingular (four by four) matrix B_0. Among them, recursive formulations may be of practical interest, especially one that corresponds to the causal ordering we are going to hypothesize between our variables; that is, causality running from q to s, and from both q and s to r and i. This particular recursive formulation can be written as:

(1'') $$B_0 y_t = B(L)y_{t-1} + \zeta_t,$$

where $B(L) = B_0 A(L)$, and $\zeta_t = B_0 \eta_t$, B_0 being a triangular matrix with 0 above the diagonal and 1 in the diagonal, such that the transformed disturbances ζ_{jt} are othogonal (i.e., uncorrelated across equations). In fact, B_0 is uniquely determined; its inverse, B^{-1}, has the exact same

---

the calendar year. From previous work, we know that fiscal and calendar years do not coincide for a large enough proportion of firms; an attempt to correct for this problem had, however, very little impact on our results. We preferred not to make any such correction in the present study.

2. Our adoption of a formulation in terms of the rates of growth of the variables or log differences results from a number of considerations. Using first differences is usually advised in the time-series literature to get more stationary processes (Granger and Newbold 1977). Actually, when we tried to estimate the autoregressive model in the levels of variables, the results suggested a first difference formulation (some of the roots of the characteristic equation associated with the model being close to one in absolute values). Going to first differences is also a simple way to avoid dealing with firm-specific effects, while the formulation in terms of levels raises the well-known difficulties of estimating a dynamic model with such effects (Balestra and Nerlove 1966). First differences have, however, the drawback of magnifying the problems of errors in the variables (augmenting the ratio of error to true variance.)

lower triangular form with 1 on the diagonal and can be obtained from the appropriate Cholewski decomposition of the original variance-covariance matrix σ. This can be written as $\eta_t = B^{-1}\zeta_t$, and amounts, in practice, to successive projections of the original disturbances η_{jt}, which transform them into ζ_{jt}'s:

$$\eta_{1t} = \zeta_{1t}; \eta_{2t} = a\zeta_{1t} + \zeta_{2t}; \ldots \ldots$$

Among the many statistically equivalent formulations, we endeavor to give a specific structural economic meaning to the pure autoregressive form (1), and we therefore refer to it as our general model. All four equations of the general model (q, s, r, and i) can be interpreted and motivated by more or less precise economic considerations, and we can test whether the restrictions suggested by such considerations are compatible with our data.

14.2.2 Interpretation and Motivation

We can justify our i equation as an investment demand equation, referring directly to Malinvaud's recent book, *Profitability and Unemployment* (1980; see also Malinvaud 1981). In his book, Malinvaud studies the implications of an investment model in which net investment depends on expected capacity need and expected profitability. While the influence of capacity needs corresponds to the well-known accelerator phenomenon and is supported by the bulk of the vast number of econometric studies of investment, he stresses the importance of profitability as another major determinant. If we assume the investment equation to be log-linear and take first differences, we get:

$$i_t^* = \phi q_{t-1}^e + \gamma s_{t-1}^e,$$

where $i_t^* = \log(NI_t/NI_{t-1})$ is the log change in desired net investment between periods $(t-1)$ and t, $s_{t-1}^e = \log(S_{t-1}^t/S_{t-2}^{t-1})$ and $q_{t-1}^e = \log(Q_{t-1}^t/Q_{t-1}^{t-1})$ are the log changes or revisions of capacity need and profitability between these same periods and as expected one period before.

The revision in the expected profitability q_{t-1}^e is presumably because of new information about the future which becomes available between $(t-2)$ and $(t-1)$. Such revisions should have direct bearing on the movements of stock prices during the same period and, hence, will be reflected in the lagged values, q_{t-2} and q_{t-1}, of our stock market holding rate of return variable. Therefore, we will interpret q_{t-1} and q_{t-2} as reasonable indicators of the unobservable q_{t-1}^e in the investment equation.[3]

3. The usefulness of stock market valuation as an indicator of expectations about future profitability in an investment function can be traced back to Grunfeld (1960), and more recently to the literature on "Tobin's Q" (Tobin 1971). Our q variable will be equal to the

In the absence of any direct information on expectations about capacity, the usual and simple procedure in most econometric studies is to treat them as a function of past levels of output or sales. We can likewise take the revision in the expected capacity need s^e_{t-1} as a distributed lag function of past changes in sales $s_{t-\tau}$, thereby justifying why lagged values $s_{t-\tau}$ should appear in the investment equation. More generally, we can consider s^e_{t-1} as a forecast function depending not only on the past $s_{t-\tau}$, but also on the past values of other relevant variables. Assuming rational expectations, the actual change in sales s_t itself should be an unbiased "forecast" of the expected s^e_{t-1}, conditional on all the information available in period $(t-1)$, and s^e_{t-1} should only differ from s_t by an uncorrelated forecast error. In particular, one would think that q_{t-1}, being a forward-looking variable, has a predictive value for both s^e_{t-1} and s_t, and therefore, will enter significantly in the forecast function even in the presence of lagged $s_{t-\tau}$ terms. Thus, one should find that q_{t-1} influences investment both directly and indirectly via its effect on expected sales.

Finally, the change in the desired net investment variable i^*_t itself is also unobservable, and its relationship with the actual change in gross investment must be specified. The various kinds of delays occurring between the decision and the execution of investment plans, as well as an approximate proportionality of retirements to past investments, suggest reasons why lagged investment terms should also appear in the investment equation.

In sum, starting from Malinvaud's (1980) theoretical equation and taking into account all the necessary transformations for its empirical implementation, we get to an equation that is very close to the investment equation of our general model. Clearly, such a tentative and informal derivation involves many problematic assumptions and issues. Be that as it may, our investment equation consists of two main factors: scale and intensity, as indicated by sales and stock market profitability, respectively, and allows for a quite flexible lag structure. The standard objection one could raise is that more explanatory variables should have been included, mainly the relative cost of labor and capital and the financial liquidity of the firm. It is difficult, though, to get relevant information about factor prices at the firm level; it is also plausible that they tend to move roughly parallel for all firms, and that will be taken care of by the year dummies in the equation. As for financial liquidity of the firm, it

---

percentage change in Tobin's Q variable, if debts are proportional to equity and there is no change in the replacement value of the firm. Actually, the correlation between our q variable and the change in Tobin's Q variable, as computed otherwise, is quite high in our sample. Our study is thus related to the studies investigating Tobin's Q as a determinant of investment. See, for example, Engle and Foley (1975), Von Fustenberg (1977), and Summers (1980), among others.

could be gauged by the importance of past profits, and it may be worth-while to consider this possibility in further research.

The r equation can be justified along the same lines as the i equation and interpreted in terms of an R & D demand equation. One of our basic topics of interest is to assess whether R & D and physical investment behave more or less similarly.

From what we have already said, the s or sales equation can be understood as a forecast function purporting to account for the expectations of firms about their future sales. It seems plausible, however, that these expectations might also depend on other variables besides the ones already included in the equation.

The q or stock market holding rate of return equation has little economic justification. For the sake of symmetry with the s equation, it could also be viewed as a forecast function of expectations on q_t. However, it is usually admitted that q_t cannot be predicted by its own past values or that of any other variable. This property is known as Fama's semistrong test of stock market efficiency (Fama 1970, 1976). Conditional on the information available at the beginning of period t, the expected value of q_t should, by standard arbitrage argument, equal the prevailing market rate of interest. In other words, a trading rule based on public information alone would not allow traders to achieve any excess return on average.

14.2.3 An Extended Accelerator Model

The considerations we have just developed suggest a causal ordering of the variables and specific restrictions on the equations.

We have touched on the issue of stock market efficiency. The hypothesis of stock market efficiency simplifies our general model importantly, the q equation reducing itself to $q_t = \eta_{1t}(= \zeta_{1t})$. In other words, q is exogenous relative to the other variables, or s, r, and i do not cause q in the sense of Granger (Pierce and Haugh 1977, 1979; Granger, 1980). Such a hypothesis has been generally accepted in empirical work, but rather than taking it for granted, it seems better to test it on our data.[4]

Our central interest, however, is in the appropriateness of the traditional formulation of the accelerator model. This formulation postulates that sales or expected sales are exogenous relative to investment, thus ruling out feedback effects from investment to sales. This is a major assumption, since without it not only the usual estimates of the so-called accelerator effect might be biased, but the whole notion itself might not be very meaningful. Within our general model, the accelerator assump-

---

4. Doubts have recently been expressed about the efficiency of stock markets. Schiller (1981) pointed out that the actual stock prices fluctuate too much to reconcile with the stable and smooth series of the present value of subsequent real dividends. See also Malinvaud (1981) and Summers (1982).

tion is directly testable, requiring i, as well as r by analogy, to not appear in the s equation.

Besides the questions of stock market efficiency and the appropriateness of the accelerator assumption, we have also considered two other issues of lesser significance. The first concerns the interrelations of physical and R & D investment. There seems to be no reason why physical investment should influence R & D investment per se. One might expect, however, that the converse would not be true. A successful R & D program would lead to product or process innovations, which could result in new programs of investment. There is, however, little evidence in our data of such a causal ordering from R & D to investment. While we do not find any significant influence of past i on r, the influence of past r on i is not significant either, and at best appears to be rather weak.

The second issue relates to the existence of contemporaneous reciprocal influences between our variables, or "instantaneous causality". In our general model (1), this amounts to testing the diagonality of the variance-covariance matrix Σ (i.e., no correlation across equations among the disturbances η_{jt}), while in the transformed recursive formulation (1''), it becomes the test of the restriction that the contemporaneous value of a variable does not enter as a regressor (i.e., B_0 is an identity matrix, otherwise $\eta_{jt} = \zeta_{jt}$). A year being a long enough period for interactions between variables to develop, one would expect instantaneous causality to occur and, hence, the diagonality restriction to be strongly rejected. This is indeed what happens. Another explanation of why the disturbances in our model may be correlated across equations is of course the omission of relevant (common or correlated) variables. One would thus expect the disturbances in the investment and R & D equations (η_{3t} and η_{4t}) to be correlated with each other and also with the disturbance in the sales equation (η_{2t}). Indeed, this last disturbance can proxy for variables influencing sales expectations but actually omitted from our forecast equation; as such it should enter in both the investment and the R & D equations, accounting partly for the correlation of their disturbances. The structure of the disturbances and their correlations is clearly revealed by the appropriate Cholewski decompostion, $\eta_t = B^{-1}\zeta_t$, as previously indicated.

We can focus our interest primarily on two restricted versions of the general model: the first one assuming only stock market efficiency; the second one also assuming the appropriateness of the accelerator formulation. We call the latter restricted model the accelerator model or the *extended accelerator model* since it extends the traditional investment accelerator to research and development expenditures, and because it tries, through the use of the q variable, to incorporate expected profitability as an important determinant of investment and R & D. Since interactions between investment and R & D do not appear to be significant, we generally consider the extended accelerator model without them, but this need not be so in principle.

14.2.4 Moving Average Representation and Multipliers

Changing slightly our notation but still measuring variables relatively to their year means, the extended accelerator model can be written:

(2)
$$q_t = \eta_{1t},$$
$$s_t = \beta(L)q_{t-1} + \alpha(L)s_{t-1}\eta_{2t},$$
$$r_t = \phi(L)q_{t-1} + \gamma(L)s_{t-1} + \theta(L)r_{t-1} + \eta_{3t}$$
$$i_t = \psi(L)q_{t-1} + \delta(L)s_{t-1} + \mu(L)i_{t-1} + \eta_{4t},$$

where the η_{jt} are mutually correlated across equations (but uncorrelated over time). The causal structure of the model is simple and can be illustrated by the path diagram in figure 14.1. Changes in q induce variations in s, r, and i, and changes in s move r and i, but there is no feedback from r and i to s or from s to q; there is also no interaction between r and i. As we already stated, in view of this specific structure, there is one appropriate and economically meaningful decomposition of the correlated η_t in terms of uncorrelated ζ_t. Renaming these ϵ_t, u_t, v_t, and w_t (instead of ζ_{jt}), we can write

(2')
$$\eta_{1t} = \epsilon_t,$$
$$\eta_{2t} = a\epsilon_t + u_t,$$
$$\eta_{3t} = b\epsilon_t + cu_t + v_t,$$
$$\eta_{4t} = d\epsilon_t + eu_t + fv_t + w_t.$$

In this form the independent errors, ϵ_t, u_t, v_t, and w_t, are intrinsically related to the different equations of the accelerator model. They can be regarded as the exogenous and unobservable (or unobserved) basic factors of our model accounting for the evolution of our observed variables. A change or "shock" in ϵ_t, or an innovation in q_t, can thus be interpreted as a shift in the firm's future profitability as expected by the traders on the stock market. We shall call such a shock an expected profitability shock, or q shock, and the dynamic responses of our variables to it the q effects of q multipliers. Similarly, a change or a "shock" in u_t, or an independent innovation in s_t, can be viewed as a shift in the expectation of the rate of growth of sales, and we shall speak of a demand shock, or s shock, and of the s effects or s multipliers. It is of some interest to separate in the (total) q or s effects the own effects and the additional or cross effects. The own effects are computed in the absence of instantaneous causality (i.e., $a = b = c = d = e = f = g = 0$ or $\eta_t = \zeta_t$); they result directly from the

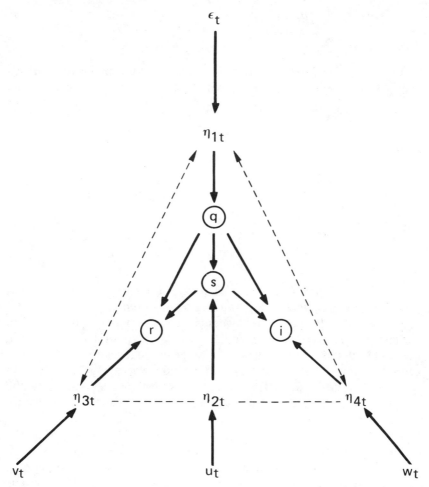

Fig. 14.1   Path diagram of the extended accelerator model.

initial change in q_t or s_t corresponding to a shock in ϵ_t or u_t, as if there was no other immediate impact of such shocks.[5]

To illustrate the q and s multipliers and how a shock in ϵ or u actually affects the movements of our variables, we can consider a simplified version of the accelerator model in which we keep only one lagged variable (i.e., a first-order autoregressive model), ignore the correlations of the disturbances across equations (i.e., Σ is diagonal), and drop the i equation (since i and r behave in the same way). It is enough to consider:

5. The formulations (1′) and (1″) of the general model can also be written: $y_t = P(L)\eta_t$ and $y_t = T(L)\zeta_t$, $P(L)$ and $T(L)$ being respectively the matrix of the own and total effects with: $P(L) = [I - A(L)L]^{-1} = [I - B^{-1}B(L)L]^{-1}$, and $T(L) = P(L)B_0^{-1}$.

$$q_t = \epsilon_t,$$
$$s_t = \beta q_{t-1} + \alpha s_{t-1} + u_t,$$
$$r_t = \phi q_{t-1} + \gamma s_{t-1} + \theta r_{t-1} + v_t,$$

with $|\alpha| < 1, |\theta| < 1$, and ϵ_t, u_t, and v_t mutually uncorrelated. For this simple system, we can write the moving average representation explicitly as:

$$q_t = \epsilon_t,$$

$$s_t = \beta \sum_{\tau=1}^{\infty} \alpha^{\tau-1} \epsilon_{t-\tau} + \sum_{\tau=0}^{\infty} \alpha^{\tau} u_{t-\tau},$$

$$r_t = \sum_{t=1}^{\infty} \omega_{\tau} \epsilon_{t-\tau} + \sum_{\tau=0}^{\infty} \rho_{\tau} u_{t-\tau} + \sum_{\tau=0}^{\infty} \theta^{\tau} v_{t-\tau},$$

where

$$\rho_{\tau} = \theta \rho_{\tau-1} + \gamma \alpha^{\tau-1} \quad \text{and} \quad \omega_{\tau} = \phi \theta^{\tau-1} + \beta \rho_{\tau-1},$$

with $\rho_o = 0$, and for $\tau = 1, 2, \ldots$. The response pattern of our variables is described completely by this moving average representation. For example, ω_{τ} is the effect on r after τ years of a one-period, one-unit shock in ϵ. Thus, $\Sigma_{\tau=1}^{k} \omega_{\tau}$ is the cumulative effect on r over a period of k years from this shock, that is, the proportional change in the level of R & D after k years from this shock. A shock appears to induce decaying fluctuations in growth rates and to put the levels on higher growth paths. Essentially, the effects on growth rates are transitory, while the changes in levels are permanent.

The long-run effects of a one-period, one-unit shock in ϵ or u on the levels of sales and R & D can be easily computed and are given in table 14.1. A 1 percent increase in u will induce sales and R & D to increase respectively by $\Sigma_{\tau=1}^{\infty} \alpha^{\tau} = 1/(1 - \alpha)$ and $\Sigma_{\tau=1}^{\infty} \rho_{\tau} = \gamma/(1 - \theta)(1 - \alpha)$. The ratio of these two effects, $\gamma/(1 - \theta)$, is the elasticity of R & D with respect to sales, and thus can be called the long-run accelerator effect or multi-

Table 14.1   **Long-Run Multipliers in the First-Order Autoregressive Accelerator Model**

Shock or Innovation	Percentage Change in Level	
	$\Delta S/S$	$\Delta R/R$
ϵ or q	$\dfrac{\beta}{1-\alpha}$	$\dfrac{\phi}{1-\theta} + \dfrac{\beta\gamma}{(1-\theta)(1-\alpha)}$
u or s	$\dfrac{1}{1-\alpha}$	$\dfrac{\gamma}{(1-\theta)(1-\alpha)}$

plier. The long-run elasticity of R & D with respect to q is $\Sigma^{\infty}_{\tau=1}\omega_\tau = \phi/(1 - \theta) + [\beta\gamma/(1 - \theta)(1 - \alpha)]$. This expression indicates clearly that q can affect R & D both directly and indirectly through its impact on sales: the direct effect being $\phi/(1 - \theta)$; the indirect effect being the product of the impact of q on sales $\beta/(1 - \alpha)$ and the long-run accelerator $\gamma/(1 - \theta)$.

14.3 Empirical Results

14.3.1 Tests and Estimates

The empirical implementation of our study is based on a sample of ninety-three firms with data from 1962 to 1977. This sample derives from the Griliches and Mairesses (this volume) restricted sample of 103 firms with no major merger problems. We had to discard ten firms because of the lack of all the necessary information to construct the q variable. Although our sample may seem small in terms of number of firms and cannot be taken as representative of the corporate sector in any definite sense, it is, in fact, about the largest size possible for firms doing R & D over a sufficiently long period (at least ten good years for our type of time-series cross-section analysis).

The sample means and standard deviations of our variables over the twelve-year period, 1966–77, as well as the standard deviations of our variables measured relative to their year means, are the following:

$$q = .104, \quad s = .062, \quad r = .025, \quad i = .036$$
$$(.433) \qquad (.120) \qquad (.217) \qquad (.465)$$
$$[.362] \qquad [.107] \qquad [.211] \qquad [.444]$$

As could be expected, the stock market rate of return is extremely variable. So is physical investment; it is not rare for a firm's physical investment to double (or go down by half) from one year to the next. Note that R & D expenditures are also quite variable, though much less so than physical investment.

We have estimated all our models by Zellner's seemingly unrelated regression least-squares method (based on the variance-covariance matrix Σ estimated once and for all for the general model case). The parameter estimates of the general model, the extended accelerator model, and its simplified first-order autoregressive version are given in tables 14.2 and 14.3, while all the different test results are brought together in table 14.4

The general model uses four lagged values of each of the four variables and is therefore estimated over the twelve-year period, 1966–77, including also twelve-year dummies. We have experimented some with shorter lags, but four lags seemed to be necessary to capture the dynamic behavior of our variables adequately. We have also checked for the possibility of serial correlation of the disturbances. It is apparently negligible, the first- and second-order autocorrelation coefficients of the residuals $\hat{\eta}_{jt}$ in each equation being rather small uniformly ($-.01$ and $-.06$ for the q equation residuals, respectively; $-.02$ and $-.03$ for the s equation residuals; $-.03$ and $-.01$ for the r equation residuals; and $-.01$ and $-.07$ for the i equation residuals).

Table 14.2   **Parameter Estimates, General Model**[a]

	q	s	r	i
q_{-1}	−.005	.044	.067	.172
	(.027)	(.008)	(.015)	(.030)
q_{-2}	.002	.008	.034	.171
	(.025)	(.007)	(.014)	(.028)
q_{-3}	−.009	.004	.021	.055
	(.022)	(.006)	(.012)	(.025)
q_{-4}	.037	.011	−.006	.051
	(.027)	(.006)	(.012)	(.025)
s_{-1}	−.161	.116	.335	.288
	(.109)	(.031)	(.060)	(.121)
s_{-2}	−.043	−.028	.097	−.006
	(.108)	(.031)	(.060)	(.120)
s_{-3}	−.076	.089	.102	.097
	(.109)	(.031)	(.060)	(.121)
s_{-4}	.069	.050	.072	.112
	(.107)	(.031)	(.059)	(.119)
r_{-1}	−.012	.023	−.243	.140
	(.055)	(.016)	(.031)	(.061)
r_{-2}	−.047	−.015	−.132	−.013
	(.062)	(.018)	(.034)	(.068)
r_{-3}	−.016	.026	.142	−.103
	(.065)	(.019)	(.036)	(.072)
r_{-4}	−.106	−.016	−.009	−.054
	(.057)	(.016)	(.031)	(.063)
i_{-1}	−.064	−.001	.003	−.344
	(.028)	(.008)	(.015)	(.031)
i_{-2}	−.062	.012	.003	−.332
	(.029)	(.008)	(.016)	(.032)
i_{-3}	−.004	−.007	−.023	−.209
	(.029)	(.008)	(.016)	(.032)
i_{-4}	−.019	−.004	.002	−.143
	(.028)	(.008)	(.015)	(.031)

Weighted residuals sum of squares = 4464
Degrees of freedom = 4346

[a]The parameter estimates of the s, r, and i equations do not differ in the general model without market efficiency nor with market efficiency, while the q equation vanishes in the latter case.

Conversely, the contemporaneous correlations of the residuals $\hat{\eta}_{jt}$ across equations are rather high (.19, .07, and .07 between the q equation and the s, r, and i equation residuals, respectively; .18 and .26 between the s equation and the r and i equation residuals; .18 between the r and i equation residuals). The test of diagonality is indeed strongly rejected. Using the Cholewski decomposition, we can write:

Table 14.3   **Parameter Estimates, Extended Accelerator Model**

	Extended Accelerator Model			First-Order Autoregressive Accelerator Model		
	s	r	i	s	r	i
q_{-1}	.043 (.008)	.068 (.015)	.174 (.030)	.041 (.007)	.063 (.015)	.194 (.029)
q_{-2}	—	.034 (.013)	.170 (.026)	—	—	—
q_{-3}	—	.020 (.012)	.052 (.024)	—	—	—
q_{-4}	—	−.012 (.012)	.038 (.024)	—	—	—
s_{-1}	.143 (.029)	.345 (.058)	.354 (.119)	.154 (.028)	.384 (.057)	.256 (.114)
s_{-2}	−.000 (.028)	.108 (.058)	.047 (.116)	—	—	—
s_{-3}	.106 (.028)	.095 (.059)	.074 (.117)	—	—	—
s_{-4}	.052 (.028)	.072 (.058)	.077 (.115)	—	—	—
r_{-1}	—	−.258 (.029)	—	—	−.227 (.028)	—
r_{-2}	—	−.125 (.033)	—	—	—	—
r_{-3}	—	.138 (.034)	—	—	—	—
r_{-4}	—	−.002 (.030)	—	—	—	—
i_{-1}	—	—	−.337 (.029)	—	—	−.190 (.027)
i_{-2}	—	—	−.346 (.030)	—	—	—
i_{-3}	—	—	−.203 (.030)	—	—	—
i_{-4}	—	—	−.146 (.029)	—	—	—

Weighted residuals sum of squares = 4519   Weighted residuals sum of squares = 4777
Degrees of freedom = 4381   Degrees of freedom = 4402

Table 14.4   Test Statistics[a]

Hypothesis	Weighted Residuals Sum of Squares	Degrees of Freedom	Test against the General Model, H_0			Test the General Model with Market Efficiency, H_1		
			Number of Restrictions	Value f of F-statistic	Prob $(F>f)$	Number of Restrictions	Value f of F-statistic	Prob $(F>f)$
H_0: general model	4464	4346	—	—	—	—	—	—
H_0': general model with the diagonality restriction	4650	4352	6	30.2	.000	—	—	—
H_1: general model with stock market efficiency	4490	4362	16	1.6	.060	—	—	—
H_2: extended accelerator with r and i interactions	4508	4373	27	1.6	.030	11	1.6	.092
H_3: extended accelerator without r and i interactions	4519	4381	35	1.5	.030	19	1.5	.075
H_4: first-order autoregressive accelerator	4777	4402	56	5.4	.000	40	7.0	.000

[a]The test statistics are the standard F statistics computed from the weighted residuals sum of squares. These are based on the Σ matrix estimated under the alternative hypothesis of the general model.

$$\hat{\eta}_{1t} = \hat{\epsilon}_t,$$
$$\hat{\eta}_{2t} = .055\,\hat{\epsilon}_t + \hat{u}_t,$$
$$\eta_{3t} = .040\,\hat{\epsilon}_t + .329\,\hat{u}_t + \hat{v}_t,$$
$$\hat{\eta}_{4t} = .079\,\hat{\epsilon}_t + .974\,\hat{u}_t + .284\,\hat{v}_t + \hat{w}_t,$$

the standard deviations of the uncorrelated ϵ_t, u_t, v_t, and w_t being .358, .101, .194, and .380, respectively. It appears from these estimates that u, the independent innovation in s, has an immediate and strong impact on i and a more moderate one on r, while the immediate effect of ϵ, the innovation in q, is quite weak. Note also that the independent innovation in r has a sizeable effect on i as well.

Considering the estimated equations of the general model in turn, it is clear that all the implications suggested by the economic interpretation are by and large supported. All the coefficients of the q equation (i.e., the sixteen coefficients of the lagged values of q, s, r, and i except for the time dummies) are insignificant and, even taken together, the hypothesis of their joint nullity cannot be rejected at the 5 percent significance level. This is another confirmation of the unpredictability of q from past information and thus also of the hypothesis of stock market efficiency.

All eight coefficients of the lagged r and i terms are insignificant in the s equation. Assuming stock market efficiency, their joint nullity (together with that of the coefficient of q_{-2}, q_{-3}, and q_{-4} which are also individually insignificant) cannot be rejected at a 5 percent level of significance. We can thus accept the hypothesis that s and q are exogenous relative to r and i, and that the accelerator model is a reasonable specification, even though at first it appeared to be a rather strong simplification.[6]

In the r equation, the four lagged i terms and likewise the four lagged r terms in the i equation are all insignificant, except for the coefficient of r_{-1} on i, which is on the verge of individual significance at the 5 percent level. As a group, they are insignificant at the 5 percent level. We can accept the absence of interactions, other than instantaneous, between r and i and hence we can accept the accelerator model without such interactions. On the other hand, the hypothesis (considered by way of illustration) that the accelerator is first-order autoregressive is strongly rejected.

14.3.2 Dynamic and Long-Run Multipliers

The implications of our results are best described by the dynamic responses of our variables to the different shocks and the q and s effects or multipliers. All long-run multipliers are given in table 14.5, while the q and s dynamic multipliers are represented in figures 14.2 to 14.5. We shall comment on them in turn.

6. The fact that we cannot reject exogeneity tests of both q (stock market efficiency) and s (accelerator model) is all the more meaningful since our sample has a large number of observations (see, for example, Leamer 1978, chap. 4).

Table 14.5 Long-Run Multipliers

Model	Shocks on	Total Effects				Own Effects			
		ε	u	v	w	ε	u	v	w
General model without stock market efficiency	q	.915	−.528	−.162	−.069	.949	−.414	−.143	−.069
	S	.155	1.267	.004	−.006	.085	1.271	.006	−.006
	R	.190	.826	.791	−.015	.127	.579	.795	−.015
	I	.276	.658	.093	.477	.229	.208	−.043	.477
General model with stock market efficiency	q	1	0	0	0	1	0	0	0
	S	.169	1.357	.031	.005	.093	1.342	.029	.005
	R	.208	.936	.824	−.001	.138	.665	.825	−.001
	I	.301	.818	.142	.497	.244	.333	.001	.497
Extended accelerator model	q	1	0	0	0	1	0	0	0
	S	.141	1.431	0	0	.062	1.431	0	0
	R	.191	.978	.805	0	.119	.714	.805	0
	I	.288	.849	.140	.492	.229	.369	0	.492
First-order autoregressive accelerator model	q	1	0	0	0	1	0	0	0
	S	.114	1.181	0	0	.049	1.181	0	0
	R	.120	.637	.814	0	.067	.369	.814	0
	I	.253	1.072	.239	.840	.173	.254	0	.840

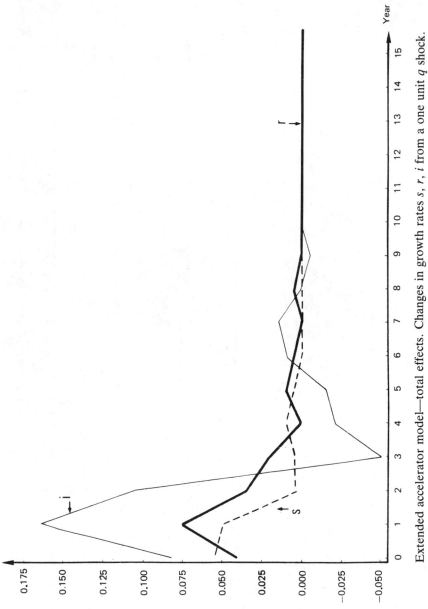

Fig. 14.2   Extended accelerator model—total effects. Changes in growth rates s, r, i from a one unit q shock.

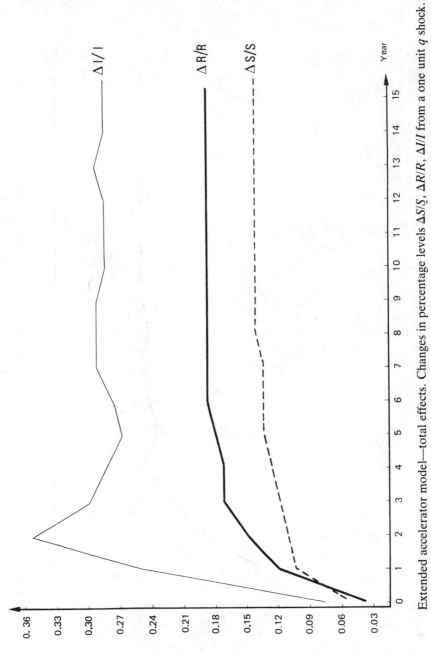

Fig. 14.3 Extended accelerator model—total effects. Changes in percentage levels $\Delta S/S$, $\Delta R/R$, $\Delta I/I$ from a one unit q shock.

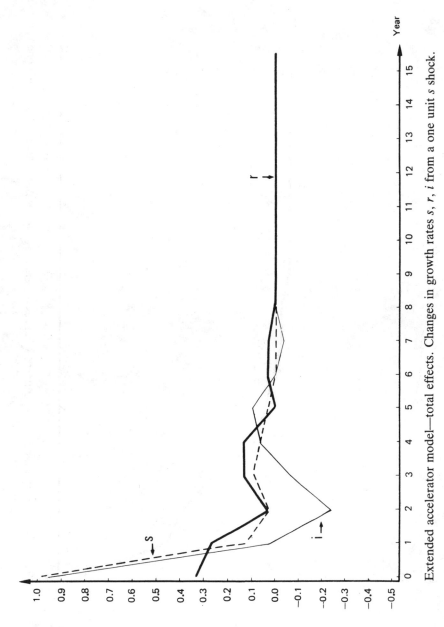

Fig. 14.4 Extended accelerator model—total effects. Changes in growth rates *s*, *r*, *i* from a one unit *s* shock.

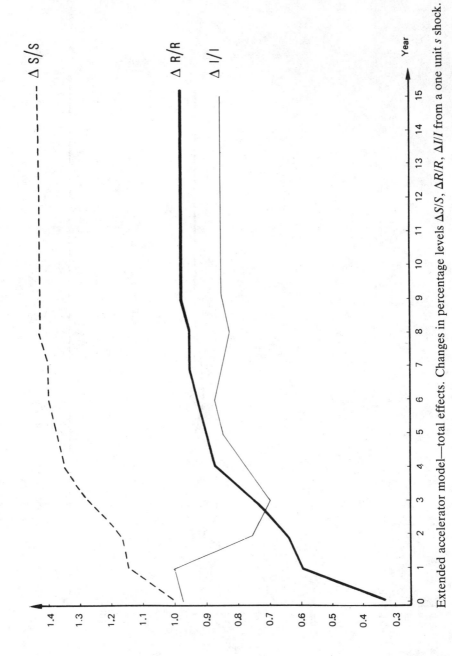

Fig. 14.5 Extended accelerator model—total effects. Changes in percentage levels $\Delta S/S$, $\Delta R/R$, $\Delta I/I$ from a one unit s shock.

The eight matrices in table 14.5 consist of the own and total effects estimated for the general model with and without stock market efficiency, the extended accelerator model (without r and i interactions), and the first order autoregressive accelerator model. We have not endeavored to compute the standard deviations of these coefficients.[7] However, the comparison of their values for the four different specifications gives us a feeling for their precision. As we have seen, the general model with market efficiency and the extended accelerator are not statistically different at the 5 percent significance level; indeed all the estimated effects for these two models are very close. The general model without market efficiency differs mainly from that with market efficiency by the estimated effect of s (or u) on q; however, this effect should not be statistically significant, corresponding mainly to the large insignificant coefficient of s_{-1} in the q equation (see table 14.2). The largest discrepancies between the extended accelerator model and its first-order autoregressive version occur in the estimated effects of s (or u) on r and i (and also of i [or w] on itself); these discrepancies are probably significant since they correspond to the significant coefficients of s_{-2} and s_{-4} in the r and i equations (and also of i_{-2}, i_{-3}, and i_{-4} in the i equation).

The comparison of the own and total effects shows the importance of the contemporaneous influences of q and s on r and i (i.e., the importance of instantaneous causality). This was already clear from the Cholewski decomposition given above, showing the correlation structure of the innovations in our variables. Consider the one very striking case: the long-run impact of a 1 percent s or u shock on the level of physical investment would amount only to .35 percent, instead of about .85, if the contemporaneous dependence between s and i were eliminated.

Figures 14.2 to 14.5 each consist of three graphs, depicting the yearly q or s (total) effects of the three rates of growth: s, r, and i, or on the three percentage changes in levels: $\Delta S/S$, $\Delta R/R$, $\Delta I/I$ (these effects being estimated for the extended accelerator model). The responses of s and r to the q or s shocks are similar enough, damping down rapidly with most of the effects dissipating in three years. The investment growth rate i reacts more strongly and irregularly. In response to a 1 percent q shock, it goes up to about .15 in the first year and down to .10 and $-.05$ in the second and third years, then cycles down quickly to zero. In response to a 1 percent s shock, after an immediate impact of about 1, it plunges to 0 and $-.25$ in the first and second years, then cycles back quickly to zero. In coherence with these patterns of response, the levels of sales and R & D expenditures increase steadily toward their new long-run values while

---

7. The total and own effects are highly nonlinear and complicated expressions of the estimated parameters, making the derivation of their standard deviations a problematic task (see note 5).

investment starts by overshooting its own, all cumulated effects being practically completed in five years.

The long-run (total) effects of a 1 percent q shock on sales, R & D, and investment levels are respectively about .15, .20, and .30. These elasticities appear to be rather small; however, gauged in terms of the standard deviations of the corresponding rates of growth, they are quite sizeable. A one standard deviation q shock induces changes in the levels of sales, R & D, and investment of about .55, .40, and .25 of their respective standard deviations.

The absolute long-run effects of a 1 percent s shock are much larger than those of a 1 percent q shock, moving the levels of sales, R & D, and investment by about 1.4, .95, and .85 repectively. Yet, measured in units of standard deviations, s shocks are not more effective than q shocks in driving R & D and physical investments: the changes induced by the former being about .50 and .20, compared to .40 and .25 by the latter. In this regard it should be noted that only 30 percent of the q effect on R & D and 55 percent of the q effect on investment relies on the direct influence of q, the remaining effect resulting from the impact of q on s. This remark shows that in considering an R & D or investment equation in isolation, one might be led to a serious underestimate of the significance of the q variable.

For comparison with the results of other investment studies, it is interesting to translate the long-run s effects into the usual accelerator elasticities $(\Delta I/I)/(\Delta S/S)$ or $(\Delta R/R)/(\Delta S/S)$: they are about .6 ($\sim .85/1.4$) and .7 ($\sim .98/1.4$) for physical investment and R & D, respectively. The latter estimate of .7 accords well with the elasticity of R & D capital stock reported to be around .5 to .8 by Nadiri and Bitros (1980) in the only other study investigating investment and R & D demand jointly. The former estimate of .6 is, however, lower than their estimated elasticity of around 1 for physical capital stock. A unitary elasticity is implied by the standard Jorgensonian factor demand framework (i.e., the inverse of the returns to scale in the production function, which presumably are not very far from being constant) and is in fact found in many econometric studies (for example, Jorgenson and Stephenson 1967; Jorgenson 1971). Because of the various differences in specification, it is difficult to pinpoint the actual reasons for our relatively low accelerator estimate. It probably arises from our rate of growth formulation. Using a similar formulation, Eisner found an even lower estimate of about .4 (Eisner 1978a, 1978b; see also Oudiz 1978).[8] Eisner's explanation, which is similar to Friedman's permanent income hypothesis, may also be applicable

---

8. To be precise, Eisner's dependent variable is the deviation from the firm mean of the investment-capital ratio, or the rate of growth of the capital stock plus its rate of depreciation.

to our results. In our specification of the accelerator model, the q and s shocks are assumed to be free from errors or contamination by any noise. In reality, the fluctuations in q and s have large transitory components, which will have presumably little impact on i and r. Our estimates of the accelerator elasticity and, more generally, of the q and s effects might be larger if we could disentangle the transitory variations from the permanent changes in q and s.

14.4 Final Remarks

Using a multivariate autoregressive framework, we have found a simple causal structure for the variables of interest, q, s, r, and i, which is consistent with our data. As expected from the stock market efficiency hypothesis, q, the stock market one-period holding rate of return, is exogenous relative to the other three variables (or Granger causes them). As postulated in the traditional accelerator model of investment, the rate of growth of sales, s can also be treated as exogenous to the rates of growth of R & D and physical investment, r and i. Moreover, no strong feedback interaction is detected between r and i.

Within the simple structure of the extended accelerator model, the substantive conclusion is that R & D and physical investment react very similarly to the growth of sales and to movements in q; however, the response of R & D is more stable or less irregular than that of physical investment. Both expected demand and expected profitability thus appear to be important determinants for R & D expenditures and physical investment.

It will be important to check our findings against other data. Also, our study could be improved by incorporating other variables of interest (see Ben-Zion, this volume). In future work, it would be particularly interesting to go further in two directions:

1. The multivariate autoregressive setup proved to be useful and convenient for studying the dynamic relationships between variables. However, a more elaborate specification might help to filter out the permanent from the transitory components of the variables. This issue is related to our choice of growth rates formulation, which has many advantages but also tends to magnify the relative importance of transitory components or errors in the variables.

2. The fact that past q's, though probably error ridden, are significantly correlated with s, r, and i confirms that movements in stock prices carry valuable expectational information about future profitability. This interpretation of the q variables should be more rigorously substantiated and its relation to "Tobin's Q" clarified. More generally, the extended accelerator model should be grounded more firmly in theory and provided with a more definite behavioral interpretation.

References

Balestra, P., and M. Nerlove. 1966. Pooling cross section and time series data in the estimation of a dynamic model. *Econometrica* 34:585–612.

Eisner, R. 1978a. *Factors in business investment.* National Bureau of Economic Research General Series no. 102. Cambridge, Mass. Ballinger.

———. 1978b. Cross section and time series estimates of investment functions. *Annales de l'INSEE*, no. 30–32:99–129.

Engle, R., and D. Foley. 1975. An asset price model of aggregate investment. *International Economic Review* 16:625–47.

Fama, E. F. 1970. Efficient capital markets: a review of theory and empirical work. *Journal of Finance* 25:383–417.

———. 1976. *Foundation of finance.* New York: Basic Books.

Granger, C. W. 1980. Testing for causality: A personal viewpoint. *Journal of Economic Dynamics and Control* 2:329–52.

Granger, C. W., and P. Newbold. 1977. *Forecasting economic time series.* New York: Academic Press.

Grunfeld, Y. 1960. The determinants of corporate investment. In *The demand for durable goods*, ed. A. C. Harberger, 211–66. Chicago: University of Chicago Press.

Jorgenson, D. W. 1971. Econometric studies of investment behavior: A survey. *Journal of Economic Literature* 9:1111–47.

Jorgenson, D. W., and J. A. Stephenson. 1967. Investment behavior in U.S. manufacturing, 1947–1960. *Econometrica* 35:169–220.

Leamer, E. L. 1978. *Specification searches: Ad hoc inferences with nonexperimental data.* New York: Wiley.

Malinvaud, E. 1980. *Profitability and unemployment.* Cambridge: Cambridge University Press.

———. 1981. Profitabilité et investissement. Document de travail ENSAE et Unité de Recherche. Paris: INSEE.

Nadiri, M. I., and G. C. Bitros. 1980. Research and development expenditures and labor productivity at the firm level: A dynamic model. In *New developments in productivity measurement and analysis*, ed. J. W. Kendrick and B. N. Vaccara, 387–412. Conference on Research in Income and Wealth: Studies in Income and Wealth, vol. 44. Chicago: University of Chicago Press for the National Bureau of Economic Research.

Oudiz, G. 1978. Investment behavior of French industrial firms: A study on longitudinal data. *Annales de l'INSEE*, no. 30–32: 511–41.

Pierce, D. A., and L. A. Haugh. 1977. Causality in temporal systems: Characterizations and a survey. *Journal of Econometrics* 5:265–93.

———. 1979. The characterization of instantaneous causality: A comment. *Journal of Econometrics* 7:257–59.

Sargent, T. J., and C. A. Sims. 1977. Business cycle modeling without pretending to have too much a priori economic theory. In *New methods in business cycle research: Proceedings from a conference*, ed. C. A. Sims, 45–109. Minneapolis: Federal Reserve Bank of Minneapolis.

Schiller, R. J. 1981. Do stock prices move too much to be justified by subsequent changes in dividends? *American Economic Review* 71:421–36.

Sims, C. A. 1972a. Money, income and causality. *American Economic Review* 62:540–52.

———. 1972b. Are there exogenous variables in short-run production relations? *Annals of Economic and Social Measurement* 1:17–36.

———. 1977. Exogeneity and causal ordering in macroeconomic models. In *New methods of business cycle research*, ed. C. A. Sims, 23–43. Minneapolis: Federal Reserve Bank of Minneapolis.

———. 1980. Macroeconomics and reality. *Econometrica* 48:1–48.

Summers, L. H. 1981. Taxation and corporate investment: A *Q* theory approach. Brookings Papers on Economic Activity 1: 67–140.

———. 1982. Do we really know that financial markets are efficient? National Bureau of Economic Research Working Paper no. 994.

Tobin, J. 1971. Cambridge, Mass.: NBER. A general equilibrium approach to monetary theory. In *Essays in economics: Macroeconomics*, vol. 1, 332–38. Chicago: Markham.

Von Fustenberg, G. 1977. Corporate investment: Does market valuation matter in the aggregate? *Brookings Papers on Economic Activity* 2:347–408.

Comment   John J. Beggs

This paper is an extensive attempt at data analysis of the relationships between market value, sales, research and development, and investment at the firm level. The sample is large, being a cross section of 103 firms for a fifteen-year period. The now familiar vector autoregressive formulation of the dynamic process is employed, and unrestricted and restricted formulations of the lagged variable interactions are estimated. The resulting discussion in section 14.3 of the paper provides a thoughtful interpretation of empirical results.

Within the Mairesse-Siu framework at least three major methodological issues must be addressed, though I believe these comments extend to a good number of the papers presented in this volume. The first, and most fundamental, is the complete lack of recognition of the competitive environment in which a firm exist. For instance, it seems quite incon-

John J. Beggs is a professor in the Department of Statistics, Faculty of Economics, Australian National University.

gruous that a firm's R & D should depend on its own R & D four years lagged, yet not be made to depend on that of its major competitor's R & D in the recent year. Both the dictates of fashion and the availability of modeling apparatus have led to this neglect. It should be emphasized that the "fix-up" of adding dummy variables does accommodate this critique, through recentering the data, but the essential interaction among firms remains unaddressed. To illustrate the consequences of the interpretation of so-called multipliers extensively used in the vector autoregressive context (table 14.5; figs. 14.2–14.5), consider the following two simple, extreme cases:

(1) $$Y_i = \alpha + \beta X_i,$$

(2a) $$Y_i = \alpha + \beta \left(X_i - \frac{\Sigma X_i}{N} \right)$$

(2b) $$Y_i = \left[\alpha - \beta \left(\frac{\Sigma X_i}{N} \right) \right] + \beta X_i, \, i = 1, \ldots N.$$

Think of N as the number of firms in the industry, X_i as the R & D of each firm, and y_i as profits. Models (1) and (2b) are observationally equivalent. In the case of model (1) a Δ increase in R & D results in a $\beta\Delta$ increase in profits. In model (2) a Δ increase in R & D only affects profits to the extent that other firms respond by altering their R & D. In the case where all firms respond equally, the multiplier will be exactly *zero*.

The second issue is that the goal of much recent research effort has been to explain the manner in which R & D effort affects the fortunes of a company, but it remains true that R & D represents only a small proportion of the operating budget of most firms.[1] It is reasonable to question to what extent the R & D tale can wag the VAR. Further, concern about R & D often focuses on the essential uncertainty of the research venture, the implication having been drawn that many unsuccessful attempts must be made before a successful invention is identified. This notion does not fit well in the linear model employed here (equation [2] in Mairesse and Siu) and in other papers in this volume. The model presumes a marginalist-type relationship between the variables, that is, a little more R & D results in a little more sales or investment or profit. Since R & D is a small part of the operating budget and bears such uncertain fruits, it seems that those year-to-year relationships of R & D to other firm-level variables must be swamped by the consequences of wage settlements, cost of raw materials, strikes, advertising, and the response of competitors. Those identified links may more strongly reflect the "continuity" of operations of the firm than a causal-link chain of events.

---

1. These data are not reported by Mairesse and Siu but seem essential for understanding relative magnitudes in the analysis.

The third comment draws heavily from what has been said above. The vector autoregressive framework for examining links between variables fails to recognize the explicit capacity of a firm to *think*. Why should a firm's investment depend on sales four periods lagged and on R & D four periods lagged? Perhaps there are adjustment costs; perhaps there is some information in this old data. However, how rich is this information in relation to other knowledge available to the firm? Corporate expenditure decisions must reflect, for example, how close the firm is running to full capacity, or what the relative prices of labor and capital and fuel might become in the near future. These factors determine, in a calculated fashion, the levels of R & D and investment and the relative mix in the current year and future years. In the vector autoregressive formulation, this "thinking" is reduced to a sad series of stochastic disturbances in the equation system.[2]

2. Adjustment costs are a "thin" explanation of lagged R & D's ability to explain current R & D.

15   The R & D and Investment Decision and Its Relationship to the Firm's Market Value: Some Preliminary Results

Uri Ben-Zion

15.1 Introduction

Earlier work on R & D expenditures in firm cross sections has suggested that the market value of a firm reflects the R & D stock of R & D intensive firms (see Ben-Zion 1978 and Griliches 1979).

Recently Griliches (1981) has extended this analysis to a time-series context, using a time-series, cross-sectional model of a firm's market valuation. He tested the within-firm effects of R & D investment (and changes in R & D) on changes in market value of a firm and found that only "unexpected" changes in R & D affect the market value of a firm. These results are consistent with the financial literature on "market efficiency."

The purpose of this paper is to focus on the relationship between R & D, patents, net investment, and the market value of the firm. It is part of a broader plan to analyze the interrelationships between production, investments, market demand, and financial variables (see Ben-Zion 1980).

Grunfeld (1960) emphasized the role of expectational variables, such as the market value, in the determination of investments by individual

Uri Ben-Zion is associate professor of economics, Faculty of Industrial Engineering and Management, Technion.

In revising the paper, the author benefited from the comments of John Beggs, Robert Evenson, Edwin Mansfield, Ariel Pakes, Maurice Teubal, and participants in the NBER Summer Workshops. Zvi Griliches has followed the paper closely and made many valuable comments and suggestions that significantly improve the paper. The author is solely responsible for the remaining errors. The author would also like to thank John Bound and Avner Aleh for their efficient programming support, and to acknowledge the financial support of the University of Pennsylvania, the National Bureau of Economic Research, and the National Science Foundation (grant no. PRA79–13740).

firms. His work was extended to a "macro" model of investment by Griliches and Wallace (1965) and Ben-Zion and Mehra (1980). Schmookler (1966) looked at the interrelationship between series of patent output and investment in different industries, emphasizing the time pattern of different series, on the basis of which he conjectured the direction of causality.

Another direction of research, more microdata based, has been pursued by Mansfield and his associates. Their research uses more detailed data as well as more informal "inside information."[1]

In this paper, I present a simple framework for the analysis and testing of the interaction between corporate decision variables in response to changing market conditions. The emphasis here is on the determinants of market value and the rate of return. It is closely related to the work of Ariel Pakes (this volume) and Jacques Mairesse and Alan Siu (this volume), who also deal with the interaction between related sets of a firm's variables.

15.2 Interaction between the Firm's Decision Variables

A firm is assumed to maximize a "target" function subject to market and production constraints.[2] As a solution to the above optimization problem, a firm simultaneously determines its plans with repect to investment, R & D, and other variables. At the same time, on the basis of the available relevant information for the firm and the general market (e.g., interest rate, inflation, tax policy), the stock market determines the price of the firm's securities and its market value.

With changes in the market condition (e.g., changes in the demand for its products and in market prices), the firm revises its plans with respect to the above variables, and simultaneously, the firm's market value is revised. Initially, changes in the market value are mainly a response to changes in exogenous information. There is, however, a possible interaction between stock market variables and the firm's decisions. For example, the market may respond to "news" (formal announcement or leakage) about investment plans, while investment plans may respond to changes in the market value.[3]

The response of different variables to new information is inherently different. On the one hand, studies of investment suggest a built-in time lag between investment and new information. Parts of this time lag depend on the decision-making process in the corporation and on the lag between the "ordering of machines" and capital expenditures.

1. See Mansfield (this volume) for additional details and references.
2. For simplicity, we assume that the firm maximizes its market value. However, the same framework could be applied to other target functions.
3. An increase in the market value may reduce the cost of new equity capital.

On the other hand, studies of capital markets have suggested that market efficiency will lead to an instantaneous adjustment of the stock market to new information. If we could find an empirical variable that measures the flow of information to a firm, we could estimate the time patterns of response to such information. A comparison of the estimated time patterns may provide a clue to the structural relationship between the different variables. In this paper I look at the market value of the firm and its relationship to current and past measures of the firm's activity.

15.3 The Firm's Policy with Respect to Investment, R & D, and Patents

Economic theory suggests that R & D and investment are based on similar considerations, and that one could use the discounted present value of future income streams to evaluate the desirability of R & D investment. There are many differences between the two types of investment:

(a) The future net income stream resulting from an R & D project is subject to more uncertainty with respect to the cost of the R & D project and to the potential cash flows compared with an investment in plant and equipment. R & D projects are subject to major uncertainty about both the probability of their scientific success and the cost required for economically successful commercialization. Even a successful scientific completion (e.g., a patent) does not ensure business profit; not all patents result in the production of a new profitable product (see fig. 15.1). The risk of a given project is somewhat reduced by use of portfolios of R & D projects by big manufacturing firms.

(b) The uncertainty about an R & D project, as well as the business secrets involved in details of the project, may require the firm to rely more on internal financing (or financing by a single investor, e.g., the owner) rather than use the financial markets for borrowing. Thus, a firm's current earnings (net income inflow) may be a crucial source for financing an R & D project. This view is consistent with the casual observation that R & D projects normally represent a much smaller percentage of sales than does investment in fixed assets. (The accounting definition of R & D as current expense also encourages positive association between profits and R & D expenditures to stabilize accounting earnings.)

(c) Positive association between a firm's "success" and a change in R & D expenditures, viewed as an "extraordinary" expense, is also consistent with psychological experiments as well as "administrative" theories of a firm's behavior.[4] In other words, if we do not regard R & D as a "necessary" expenditure, but rather as a luxury item, then expendi-

---

4. I have benefited from discussion of this point with Professor Abraham Meshulach and Amos Tversky.

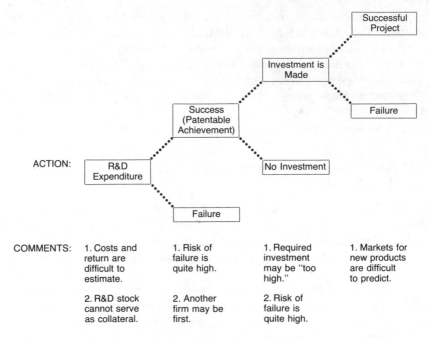

Fig. 15.1   R & D process and investment.

ture on R & D can be more easily justified in a period of comparable prosperity.

15.4 Patents

The patent variable is a random variable that measures the output of R & D activity (for a given "patenting policy"). The randomness of patents as a measure of inventive output results from several factors:

First, as mentioned by Pakes and Griliches (1980 and this volume), "There is a great deal of randomness in the timing of both the successful output of the R & D process and the decision as to when and whether to patent it."

Second, the value of a patent varies significantly between patents, and it is possible that a 100 percent increase in the number of patents by a firm is consistent with a decline in the real production of knowledge.

Since patents can be a proxy for an increase in technical knowledge, it is possible that a firm's patents are also relevant for other firms in the industry. First, patents by another firm may indicate a potential for new lines of research in the industries in which the firm operates.[5] Second,

5. In the same way that finding oil in a given area is positive information for other oil-seeking firms in the neighborhood.

patents may sometimes be imitated and improved at a fraction of the cost to the original inventor. Furthermore, patents by one firm may also create a new market which may affect the demand of current and new products by other firms. On the other hand, the effects of patents by other firms may also reduce the market value of a given firm, if the achievement of such patents reduces the firm's ability or increases its cost to obtain another patent.[6] It is possible that the net effects may be of different magnitudes and signs in different industries.

15.5 Framework for Empirical Work

The market value valuation approach, which was developed informally by Lester Telser, was applied in Ben-Zion (1972). This approach compares the market value of a firm to the value of its tangible and intangible assets. The tangible assets, which are a result of investment and net increase in assets, are represented by the value of common equity. The intangible assets are: a firm's "monopoly" power allowing it to obtain above normal return, the stock of R & D resulting from past R & D investments,[7] and a level of additional knowledge that may be partially reflected by patents.[8]

We can formalize this approach in the following equation, which relates the market value of a firm to its assets:[9]

$$(1) \qquad MV = \alpha_1 BV + \alpha_2 KM + \alpha_3 KRD + \alpha_4 KN,$$

where BV is the book value of the firm; KM is a measure of the firm's monopoly power; KRD is a measure of the stock of R & D; and KN is a measure of the firm's patent-based stock of knowledge.

Since the stocks of the different kinds of intangible capital (KM, KRD, and KN) are not observed directly, we use the following approximations:

$$(2) \qquad KRD_t = \sum_{k=0}^{\infty} RD_{t-k}(1 - \delta)k = \beta_2 RD_t,$$

where t denotes the current period, $t - k$ the past periods, and δ the rate of depreciation of R & D. The stock of R & D is thus viewed as a net value of past investment in R & D using a constant depreciation rate δ, but is in fact approximated by the current level of R & D, assuming a constant growth rate, with $\beta_2 = 1/[1 - (g - \delta)]$, where g is the rate of growth of R & D in the past.

6. This will also depend on the effectiveness of the patent system in the industry and on the possibility of imitation.
7. Since R & D expenditure is regarded by accountants as current expenditure, the stock of R & D is not included in the book value of common equity.
8. We do not know in advance whether the stock of R & D summarizes all the accumulated knowledge in the firm, or whether patents—given R & D—supply additional information. This is an empirical question.
9. This framework is somewhat related to Tobin's q model.

We assume that the monopoly power KM is proportional to the "above normal" income of the firm:

(3) $KM = \beta_1 (E + R \& D - \rho BV - \delta KRD) = \beta_1 E^*,$

where ρ is the "normal" return to equity, and E is a measure of the firm's earnings. The term $E^* (= E + R \& D - \rho BV - \delta KRD)$ is the above normal income of the firm. The R & D expenditures are added to earnings since accounting earnings (E) do not include R & D.

Finally, the capital value of patents for a given industry is assumed to be proportional to the number of patents of the firm (P) and the number of patents by the industry (PT):

(4) $KN = \beta_3 P + \beta_4 PT.$

Substituting the above approximation in equation (1), we get:

(5) $MV = \alpha_1 BV + \alpha_2 [\beta_1 (E + RD) - \rho BV]$
            $+ \alpha_3 \beta_2 (RD) + \alpha_4 \beta_3 P + \alpha_4 \beta_4 PT.$

Assuming for simplicity that $\alpha_1 = \alpha_2 = \alpha_3 = \alpha_4$, equation (6) can be written as:

(6) $MV = \alpha_1 BV \left[(1 - \rho) + \beta_1 \dfrac{E^*}{BV} + \beta_2 \dfrac{RD}{BV} + \beta_3 \dfrac{P}{BV} + \beta_4 \dfrac{PT}{BV} \right],$

which can be approximated as:

(7) $\ln MV = (\ln \alpha_1 - \rho) + \beta_0 \ln BV + \beta_1 \dfrac{E^*}{BV} + \beta_2 \dfrac{RD}{BV}$

            $+ \beta_3 \dfrac{P}{BV} + \beta_4 \dfrac{PT}{BV},$

where β_0 is expected to be unity.

The empirical framework outlined above is also consistent with viewing current flows of R & D and patents as indicators of market opportunity. If a firm's manager takes on an R & D project, it indicates that he foresees an expansion of the market for the firm's products. In this interpretation the coefficient of current R & D will be positive, but it will not be necessarily related to the stock value of R & D capital but rather to future opportunities. While in principle one could distinguish between these two hypotheses, data limitations will not allow such a distinction.

A similar argument could be made for the inclusion of investment variables in the valuation framework. The inclusion of the investment rate in the market equation is also consistent with the Modigliani-Miller [1961] classical work, which suggests that if a firm has an opportunity to invest in a project with above normal returns, such investment opportunities increase the value of the firm. On the basis of the above, we can rewrite equation (7) as:

$$(8) \qquad \ln MV = (1n\alpha_1 - \rho) + \beta_0 \ln BV + \beta_1 \frac{E^*}{BV} + \beta_2 \frac{RD}{BV}$$

$$+ \beta_3 \frac{P}{BV} + \beta_4 \frac{PR}{BV} + \beta_5 \frac{INV}{BV}.$$

Following the previous discussion, the current market value of a firm should incorporate expectation about R & D, patents, and investment variables. Assuming rational expectation, one could use future values of the above variable to approximate the predicted values.

We have not yet considered the effect of risk and leverage on a firm's market value. We use the "beta coefficient" as a measure of risk, even though some previous studies (e.g., Ben-Zion 1972) have shown that beta may not be a good measure of risk in a valuation framework. As to leverage, Modigliani and Miller (1963) have shown that it increases the value of a firm because of the tax advantage of debt in a world with a corporate income tax. Their results, however, are not unambiguous in the presence of a personal income tax, which treats interest income in a less favorable way compared to capital gains.

15.6 Data and Results

The data used in this empirical work consist of a sample of 157 industrial firms reporting R & D expenditures for the period 1955–77. This sample was constructed by Griliches and has been used extensively in other related work.[10] The empirical results reported here use a subsample of ninety-three firms for which we have continuous data series for all variables in the period 1969–77.[11]

The specific variables used in this study are based on Standard and Poor's Compustat tape and are the market value and book value of common equity and the earnings of each firm (after interest and taxes).[12] The beta risk measure used in this analysis is taken from the *Value Line Investment Survey* for 1977.

To treat the issue of leverage, we consider the basic market value equation for the firm as a whole (i.e., "unlevered" firm) as well as for the common equity part. The main difference between the two versions is that for the common equity version, the market value, book value, and

10. See Pakes and Griliches (this volume) for a more detailed description of this sample.
11. This sample is closely related to the sample used in Mairesse and Siu (this volume).
12. The market value of each firm is divided by Standard and Poor's Stock Price Index (S & P 425) to take into account the effect of the market trend.

earnings are calculated from the point of view of the firm's stockholders (which is the standard accounting and financial treatment), in the whole firm approach we have considered the market value of bonds, preferred stocks, and common stocks. The book value is the current value of the firm's assets, while earnings are measured as net operating income before interest and taxes. (See Griliches 1981 for details of this approach.)

In the empirical framework we also consider two approaches to test the effect of the industry patents. First, as summarized by equation (8), we consider the patents by other firms in the industry as a possible addition to the given firm's stock of knowledge in the same way as the firm's own patents. In this approach, patents by other firms in the industry have stronger relative effects on the market value of the smaller firms in the industry. As an alternative empirical approach, we can measure the effect of other own patents as a ratio of patents to book value of the other firms in the industry, or the "patent intensity" of the industry. The use of the patent intensity variable can be justified on two counts:

(a) A firm's ability to benefit from patents and knowledge created by other firms in the same industry may depend on its relative size. The size of a firm may restrict its ability to imitate and utilize in production the patents developed by other firms in the same industry. Furthermore, if the industry definition is quite wide (two- or three-digit SIC), a firm's relative size may be a proxy for the percentage of industry products in which the given firm is actually involved.

(b) The overall industry patent intensity may affect a firm indirectly as a proxy for the potential increase in the markets for the industry's product, from which firms may benefit whether or not they currently have patents.

The alternative approach to industry patents is summarized by:

$$(9) \quad \ln MV = \alpha_0 + \beta_0 \ln BV + \beta_1 \left(\frac{E}{BV} \right) + \beta_2 \left(\frac{RD}{BV} \right) + \beta_3 \left(\frac{P}{BV} \right)$$

$$+ \beta_4 \left(\frac{PT}{TBV} \right) + \beta_5 \left(\frac{INV}{BV} \right) + \beta_6 \cdot \text{beta},$$

where the ratio (PT/TBV) represents the patent intensity of other firms in the industry, and beta represents a measure of risk.

15.7 Empirical Results

The specific variables used in this study are as follows: In the common equity version, market value and book value of common equity as well as net earnings (available to common) and net investment are taken from the Compustat data tape. R & D expenditures and "patents applied for" were collected (and constructed) by Griliches. For the beta risk measure we have used the *Value Line Investment Survey* for 1977. (A value of one was assigned to missing beta values.) The market value of each firm was divided by Standard and Poor's Stock Price Index (S & P 425; 1959 = 100) to take into account the trend of the overall market. Industry classification for ten industry groups is based on the work of Griliches and Mairesse (this volume).

For the whole firm version we use the market value of all the firm's securities (bonds, preferred stocks, and common stocks). The book value variable reflects the value of all assets evaluated at current prices. These two variables were constructed by Griliches (1981). As the earnings variable, we use the net operating income from the Compustat tape, which is a measure of gross income of the firm and includes net earnings, preferred dividends, interest payment, operating expenses, and taxes. Other similar firm variables in the common equity version are R & D expenditures, investment, and patents.

Because of some randomness in the timing of patents, we use a two-year moving average of their annual numbers. The limited time span of patent data availability and the reporting of patent applications only on their approval restrict our sample period to the years 1969–76. We also use industry dummy variables to control for differences between industries.

The results for the common equity version (for the period 1969–76) are given in table 15.1, while the results for the whole firm version are given in table 15.2. The results indicate that—given the book value—earnings, investment, and R & D are important determinants of the market value for both the common equity and the firms as a whole. These results seem to support the simple theoretical framework developed in this paper.

Patents seem to have significant positive effects (as expected) only in the industry totals version. The effect of a firm's patents on its market value is positive but not statistically significant. The weak effect of a firm's own patent numbers may result because the market does not normally know at the time of application whether a patent applied for will in fact be approved. In this respect the variable we use to measure the "perceived" number of patents is subject to error. This error may be less important in the aggregate industry data.[13]

The effect of industry patents on the market value of individual firms seems to be quite high, which may indicate that patents are proxies for other omitted variables (e.g., expected market growth). To arrive at a meaningful estimate of the magnitude of such effects, we will have to estimate separate equations for each industry.

---

13. The results reported here are for the "patent intensity" variable, that is, industry patents/industry book value. Results of the alternative version, industry patents/firm book value, are less significant.

Table 15.1   Determinants of the Market Value of the Firm's Common Equity, 1969–76 (regression includes also industry dummies)

Variable	Coefficient[a]	Variable	Coefficient[a]
ln BV	1.012	P/BV	0.166
	(61.7)		(1.60)
RD/BV	0.654	TP/TPB	5.244
	(1.8)		(10.70)
INV/BV	0.409	Beta	0.226
	(1.6)		(1.67)
EAR/BV	2.119	R^2	0.9272
	(8.4)	N	727

[a]t-values are given in parentheses.

Table 15.2   Determinants of the Market Value of the Firm, 1969–76 (regression includes also industry dummies)

Variable	Coefficient[a]	Variable	Coefficient[a]
ln BV	1.027	P/BV	0.065
	(88.9)		(0.8)
RD/BV	3.376	TP/TPB	4.673
	(8.7)		(11.4)
INV/BV	2.623	R^2	0.955
	(7.3)	N	728
EAR/BV	2.143		
	(14.30)		

[a]t-values are given in parentheses.

Looking at the size of the coefficient of earnings, an increase in current earnings of one dollar makes the same impact on the market value as a two-dollar increase in the book value.[14] This indicates that an increase in current earnings also affects the market's expectation about future earnings.

The results in table 15.2 indicate that a one-dollar expenditure in R & D or investment increases the firm's market value by a magnitude of 2.6 or 3.3, respectively, relative to the market's valuation of one dollar of its current book value. The results for the common equity version in table 15.1 indicate that the effects of R & D expenditure and investment are much smaller. A one-dollar expenditure on R & D or investment is equivalent to a 0.6 or 0.4 increase in the book value, respectively.[15] When we do not include earnings in the equation, the R & D coefficient is much

14. Because of the use of different dependent and independent variables, the two versions are not strictly comparable.

15. As noted before, these results cannot be compared in a simple way because of different definitions of variables in the two versions.

higher, indicating that some of the effects of R & D are captured by the firm's earnings.

In evaluating these results, one should note, however, that they are based on a relatively small number of firms in very different industries. We would like to delay any concrete conclusions until we have estimated these models using a new larger sample, where we shall be able to have better control for differences between industries. The current findings are not very stable, the magnitude of the estimated coefficients varying significantly with changes in estimation techniques or the exact specification of the equation. The main qualitative conclusions, however, are quite robust. In particular, it is interesting that the results are similar when we use the future values of these variables (in a two-stage least-squares framework, not reported here).

On the basis of the finding that patents by other firms in the industry affect the market value of the firm, we also tried to test the effect of R & D and investment expenditures by other firms. The results indicate that R & D investment expenditures by other firms do not have a significant effect on a firm's own market value.

Finally, to test a dynamic version of our model, we have considered a first difference version of the basic equation for common equity. In this approach we relate the change in the market value to changes in R & D, investment, earnings, and patents. The theory of efficient financial markets assumes that the current price incorporates expectations of the future values of the relevant variable. Thus, one would expect that only the unexpected part should be related to the change in prices.[16]

The theory of the capital asset pricing model (CAPM), developed by the works of Sharpe (1964) and Lintner (1965), emphasized that the return on an individual security is related to the overall market return.

$$(10) \qquad R_i = \alpha_i + \beta_i R_M + u_i,$$

where R_i is the return on the individual security, β_i is the beta risk measure, and R_M is the return on the market portfolio, for which we use Standard and Poor's Stock Price Index (S & P 425). In the dynamic version, we have combined this approach by regressing the return on common equity on the general market return and the percentage change in the firm's earnings, R & D, and investment during that year.

To combine cross-section and time-series data, we have multiplied the observed market return by the firm-specific beta risk measure. We have estimated the following equation:

$$(11) \qquad R_i = \gamma_0 + \gamma_1 (\beta_i R_M) + \gamma_2 \left(\frac{\Delta E}{E} \right) + \gamma_3 \left(\frac{\Delta RD}{RD} \right) + \gamma_4 \left(\frac{\Delta INV}{INV} \right),$$

---

16. For a similar approach, see Griliches (1981) and Ben-Zion and Rozenfeld (1979).

Table 15.3   The Effect of Changes in the Earnings, R & D, and Investment on the Company Rate of Return, 1967–76

Variables	Coefficients in Actual Version[a]	Coefficients in Unexpected Version[a]
DE	0.938	1.357
	(7.8)	(10.2)
DRD	−0.402	0.097
	(2.0)	(1.9)
DINV	0.219	0.001
	(1.5)	(0.0)
Beta	1.17	1.17
	(4.3)	(25.0)
R^2	0.407	0.434
DF	914	914

[a]t-values are given in parentheses.

where we have considered (alternatively) actual changes and unexpected changes. The unexpected change ("surprise" in Griliches 1981 terminology) is obtained by subtracting a predicted change based on past data from the overall change.[17]

Results for 1967–76 are presented in table 15.3. They suggest that changes in earnings are significant in explaining market returns (either actual changes or expected changes). The unexpected components of R & D are also significant in explaining market returns. Unexpected investment seems to be less important. It is important to note that our predicted values for earnings, R & D, and investment are based solely on published data from previous annual reports (balance sheet and income statement) as reported in the Compustat tape. Stock market investors, particularly financial analysts and managers of large portfolios (mutual funds, pensions funds), are better informed. At any point in time an analyst's prediction for each of the above variables is based on additional information and should outperform our technical predictions.[18] It thus seems that if we could use the true (unobserved) predictions by investors, results would be much sharper.

15.8 Concluding Comments

The results in this paper suggest the market value of a firm is affected by its R & D and investment policy. The patent intensity of the industry as

17. To calculate the predicted value in each equation, we have regressed the current rate of change in each variable on lagged value of changes in the variable as well as on changes in sales. This procedure should be viewed as an illustration rather than a complete model of prediction.

18. For a study supporting this claim, see Brown and Rozeff (1978).

a whole has a positive effect on a firm's market value, while the effect of a firm's own patents is weaker. The results also support the common notion that earnings are probably the most important factor in the market value equation for both levels and rates of change. A more detailed analysis (by industry groups) could yield sharper results and would increase our understanding of the determinants of market value.

References

Ben-Zion, U. 1972. Measures of risk in the stock market and valuation of corporate stock. Ph.D. diss. University of Chicago.

———. 1978. The investment aspect of nonproduction expenditures: An empirical test. *Journal of Economics and Business* 30, no. 3:224–29.

———. 1980. The R & D and investment decision and its relationship to the firm's market value and sales: A research proposal. Mimeo, December.

Ben-Zion, U., and Y. P. Mehra. 1980. Risk and the determinants of aggregate investment. *Applied Economics* 12 (June):209–22.

Ben-Zion, U., and Ahron Rozenfeld. 1979. The relationship of stock prices to earnings, interest rate and risk: A time series test of market efficiency. Manuscript, September.

Brown, Lawrence D., and Michael Rozeff. 1978. The superiority of analysts' forecasts as a measure of expectation: Evidence from earnings. *Journal of Finance* 33 (March):1–16.

Griliches, Z. 1979. Issues in assessing the contribution of research and development to productivity growth. *Bell Journal of Economics* 10, no. 1:92–116.

———. 1981. Market value, R & D, and patents. *Economics Letters* 7:183–87.

Griliches, Z., and J. Mairesse. 1981. Productivity and R & D at the firm level. This volume.

Griliches, Z., and N. Wallace. 1965. The determinants of investment revisited. *International Economic Review* 6, no. 3:311–29.

Grunfeld, Y. 1960. The determinants of corporate investment. In *The demand for durable goods*, ed. A. C. Harberger, 211–66. Chicago: University of Chicago Press.

Lintner, J. 1965. The valuation of risk assets and the selection of risky investments in stock portfolios and capital budgets. *Review of Economics and Statistics* 47:13–37.

Modigliani, F., and M. H. Miller. 1961. Dividend policy, growth, and the valuation of shares, *Journal of Business* 34:411–33.

————. 1963. Corporate income taxes and the cost of capital: A correction. *American Economic Review* 53:433–43.

Pakes, A., and Z. Griliches. 1980. Patents and R & D at the firm level: A first look. This volume.

Schmookler, J. 1966. *Invention and economic growth*. Cambridge, Mass.: Harvard University Press.

Sharpe, W. F. 1964. Capital asset prices: A theory of market equilibrium under conditions of risk. *Journal of Finance* 19:425–42.

Tobin, J. 1969. A general equilibrium approach to monetary theory. *Journal of Money, Credit and Banking* 1:15–29.

Comment   Robert E. Evenson

Professor Ben-Zion reports estimates of the effects of R & D investment, other capital investment, and patenting on the market value of the firm. He finds support for the hypothesis that R & D investment produces an increase in the market value of the firm. Its effect appears to be slightly larger than the effect of capital investment. Perhaps the most interesting finding in the paper is that the patent intensity of other firms in the same industry affects the market value of the firm. While this finding is open to a range of interpretations, it represents at least a potential recognition of some of the factors influencing the productivity of R & D investment.

A number of the papers in this volume have treated invention in a very restricted way. The probability that invention by a particular firm conducting R & D is altered by the inventions of other firms, the scientific and technological discoveries of public research centers, or the acquisition of scientific human capital by the firm is often not considered in these studies. Nor is R & D investment by competing firms, which is likely to produce inventions and patents that will block certain lines of invention and cause diversionary or "inventing around" R & D strategies, taken into account. Samples of large firms (as in the case of Ben-Zion's sample of ninety-three large firms) can also provide a very biased picture of the industry equilibrium. Smaller firms in the industry are likely to have different levels of R & D spending and different R & D strategies. They may be purchasing technology through licensing arrangements and engaging in more adaptive or derivative invention than large firms.

Ben-Zion's paper at least considers some of these factors and justifies the inclusion of the industry patent variable on the grounds that patents

Robert E. Evenson is a professor of economics at the Economic Growth Center, Yale University.

by another firm "may sometimes be imitated and improved at a fraction of the cost to the original inventor." This perspective is consistent with the legal scholars' perception of the bargain inherent in the patent laws. In return for the grant of a limited monopoly, the inventor is required to remove from secrecy the essentials of his invention by adequately disclosing them in the patent document. This disclosure is deemed to be of value to society for the reasons given by Ben-Zion.

Ben-Zion also notes that patents by other firms have a blocking effect, lowering the value of the firm's own R & D by forcing more diversionary R & D to invent around the other firms' patents. The patent intensity of other firms may also index a more general industry effect. Industries which are technologically dynamic will tend to have high patent/book value ratios and high market value/book value and market value/earnings ratios. Of course, if the market is valuing all firms in the industry relatively highly regardless of their own patenting, it is presumably expecting strong disclosure effects. These disclosure effects may not induce patented inventions but could induce imitations and minor unpatented inventions.

Ben-Zion's obtaining own R & D effects but not own patenting effects on market value is consistent with the notion that firms can take advantage of disclosure effects by investing in R & D but that this R & D may not lead to patenting inventions. The industry equilibrium (and Ben-Zion's sample) may well include firms with a heavily adaptive, imitative R & D strategy along with firms with a more pioneering invention strategy. The definition of the industry variable as the patent intensity of the firms seems appropriate. It is a bit puzzling that the counterpart R & D intensity of other firms was not significant. It appears that patents per unit of R & D vary significantly across industries, calling into question their treatment as a common metric of real invention. If patents in a given industry tend to be relatively "small," that is, have few claims, a given R & D investment in that industry will produce more patents than investment in an alternative industry but not necessarily more real invention. R & D intensity should then be the better measure of real invention.

If one is to pursue this line of analysis further, one would wish to consider patenting in the industry by foreigners as well as by U.S. firms. Ed Mansfield's work on R & D by overseas affiliates of U.S. firms demonstrates that much of this R & D produces transferable technology, presumably with strong disclosure effects as well. This extension would require more attention to the blocking or competitive effects of patenting. Other measures of patent quality are now available as well. One can obtain data on subsequent citations of a given patent, which would be a meaningful quality weight for older patents. One can also obtain data on the granting of patent protection in other countries based on any U.S. patent.

Ben-Zion's paper has the merit of both being quite readable and not providing unnecessary algebra. His results generally fit well in the market value framework utilized, and this appears to be a useful methodology for investigating the questions posed. The model appears to be more restricted than necessary in some respects. I would think that with data on R & D from 1955–1977 it would not be necessary to approximate KRD_t, as in equation (2), by assuming a constant growth rate. It also isn't clear why a two-year average of patents is used. I would have thought that this variable should be in a "stock" form, as is R & D. Ben-Zion notes that the market test of the real value of invention requires time (even though the stock market forms expectations instantaneously). The use of the term "patent applications" is also misleading, since the patents have actually been granted.

Ben-Zion offers a dynamic version of the model which he regards as supportive of the model. He does not provide a rationale for excluding patenting from the dynamic version, even though the patented invention has a natural interpretation in the context of unexpected or surprise events. As he notes, further detailed analysis is required to clarify a number of unanswered questions. Given the findings of this paper, one can say that further analysis along these lines has promise.

16 Investment in R & D, Costs of Adjustment, and Expectations

Mark Schankerman and M. Ishaq Nadiri

This paper develops a simplified cost of adjustment model of R & D investment by private firms in which expectations play a central role. Our main objective is to provide a dynamic equilibrium framework in which alternative hypotheses of expectations formation can be tested empirically.

Most of the existing empirical work on R & D investment at the micro level is based on static equilibrium models, sometimes modified by arbitrary distributed lags, and on the assumption that firms hold static or myopic expectations on the exogenous variables in the model (e.g., Goldberg 1972; Nadiri and Bitros 1980; for a cost of adjustment model, see Rasmussen 1969). It seems clear that static expectations are inadequate as an untested maintained hypothesis, and they have the additional serious drawback of making it difficult to interpret the empirically determined lag distribution in a meaningful way. It is virtually impossible to disentangle the part of the observed lag structure caused by costs of adjustment from the lags reflecting expectational formation. Partly as an attempt to rectify this problem and to give estimated lag distributions an economically meaningful interpretation, recent work on aggregate investment in physical capital integrates rational expectations (in the sense of Muth 1961) into investment models and in some cases tests that

Mark Schankerman is an assistant professor in the Department of Economics, New York University, and a faculty research fellow of the National Bureau of Economic Research. M. Ishaq Nadiri is professor of economics at New York University, and a research associate of the National Bureau of Eonomic Research.

The authors thank Roman Frydman, Zvi Griliches, Ariel Pakes, and Ingmar Prucha for constructive suggestions on an earlier version of this paper. All remaining errors and interpretations are the authors' responsibility. Gloria Albasi provided steady research assistance. The financial assistance of the National Science Foundation, grants PRS-7727048 and PRA-8108635, is gratefully acknowledged.

expectations hypothesis (Abel 1979; Kennan 1979; Meese 1980). However, this approach has not been applied to R & D investment, and even more important, no attempt has been made to formulate and empirically test other less restrictive mechanisms of expectations formation. This paper represents a first attempt at these important tasks.

Our model is based on the assumption that a firm selects an R & D investment profile (i.e., a current investment decision plus a stream of future planned investment) which minimizes the present value of its costs, given its expectations of the future price of R & D and the level of output. If there are convex adjustment costs (i.e., a rising marginal cost of R & D investment, either because of capital market imperfections or internal adjustment costs), this yields a determinate rate of current R & D and of multiple-span, planned R & D. The optimal R & D profile is determined by the firm equating the marginal cost of adjustment to the shadow price of R & D expected to prevail at the time the investment is actually made. We show that the marginal cost of adjustment depends on the anticipated price of R & D, while the shadow price (which reflects the present value of savings in variable costs because of investment in R & D) depends on the anticipated demand for output. This links the optimal investment profile directly to the firm's expectations of these economic variables. The model of R & D investment also generates a realization function relating the difference between actual and planned R & D to revisions over time in the firm's expectations of the exogenous variables. This integration of the investment profile, the firm's expectations, and the realization function represents a formalization and extension of earlier work by Modigliani (1961) and Eisner (1978).

The general investment framework is designed to accommodate arbitrary expectations hypotheses, but to provide the model with empirical content, a specific forecasting mechanism for the price of R & D and the level of output must be postulated. We consider three alternative specifications and develop a set of empirical tests for each. The first, rational expectations, is based on the idea that the firm formulates its forecasts according to the stochastic processes (presumed to be) generating the exogenous variables. Using a third-order autoregressive specification for these variables, we derive a set of testable, nonlinear parameter restrictions in the actual and planned investment equations and some additional tests on the realization function. This represents an application to R & D of the methodology developed by Sargent (1978, 1979a), with some extensions to planned investment and the associated realization function. Next, the model is formulated under adaptive expectations according to which the firm adjusts its forecasts by some fraction of the previous period's forecast error. We show that this hypothesis also delivers a set of testable, nonlinear restrictions on the R & D investment equations. Finally, we consider the conventional hypothesis of static expectations and show that, since it is a limiting case of adaptive forecasting, it can be tested directly by exclusion restrictions on the model under adaptive expectations.

The model under each expectations mechanism is estimated using a set of pooled firm data containing both actual and planned (one year ahead) R & D. The empirical results indicate strong rejection of the parameter restrictions implied by rational expectations, and general support for the adaptive expectations hypothesis. The hypothesis most favored by the evidence appears to be a mixed one, with adaptive forecasting on the level of output and static expectations on the price of R & D. We provide some discussion of the possible reconciliation of rational expectations with this mixed forecasting hypothesis.

Section 16.1 develops the general model of R & D investment. The specifications of the model under rational, adaptive, and static expectations are provided in section 16.2. Section 16.3 provides a brief description of the data and presents the empirical results and their interpretation. Brief concluding remarks follow in section 16.4.

16.1 Investment Model for R & D

Consider a firm with a production function exhibiting constant returns to scale in traditional inputs (labor and capital) and facing fixed factor prices for those inputs. The firm's decision problem is to select an R & D investment profile that minimizes the discounted value of costs, given its expected factor prices and levels of output. This "certainty equivalence" separation of the optimization problem and the formation of expectations is justified by the separable adjustment costs specified below. Formally, the decision problem is:

(1)
$$\underset{<\bar{R}_{t,s}>}{\text{Min}} \sum_{s=0}^{\infty} \alpha^s [C(K_{t,s}, Q_{t,s}, w_{t,s}) + \bar{R}_{t,s} h(\bar{R}_{t,s})]$$

$$\text{s.t. } K_{t,s+\theta} - K_{t,s+\theta-1} = \bar{R}_{t,s} - \delta K_{t,s+\theta-1},$$

where $\bar{R}_{t,s}$ is R & D investment in real terms planned in period t for $t + s$ (we refer to t as the base period, $t + s$ as the target period, and s as the anticipations span), $C(\cdot)$ is the restricted cost function defined over the stock of knowledge K and the vector of prices for variable inputs w, $\alpha = 1/(1 + r)$ and δ are the (constant) discount factor and the rate of depreciation of the stock of knowledge, θ is the mean gestation lag between the outlay of R & D and the production of new knowledge, and $h(\cdot)$ describes the unit cost of R & D investment.

Specific functional forms are assumed for $h(\cdot)$ and $C(\cdot)$. First, we assume that the unit cost of R & D rises linearly with the level of R & D:

(2)
$$h(\bar{R}_{t,s}) = P_{t,s}(1 + A\bar{R}_{t,s}), \qquad A > 0,$$

where $P_{t,s}$ is the anticipated price of R & D. This formulation implies that total costs of R & D, $\bar{R}h(\bar{R})$, are a quadratic function of the level of R & D. Second, the assumption of constant returns to scale implies that $C(K, Q, w) = QF(K, w)$. We also assume that $F(\cdot)$ is separable and can be written $F(K, w) = f(w) - vK$, where $v > 0$, whence $C(K, Q, w) = Q[f(w) - vK]$.[1]

Two limitations of the basic model should be noted. First, the model treats the stock of knowledge as the only (quasi-fixed) capital asset and implicitly views traditional capital as variable in the short run. A more general model would treat both capital and R & D as quasi-fixed assets with associated costs of adjustment, but such a model would be considerably more complicated. The advantage of the present formulation is that it obviates the need to introduce the capital constraint in the determination of the level of the firm's output. The second limitation is the assumption that the parameter "v" is known and is the same for all firms. This parameter is one determinant of the savings in variable costs because of a marginal investment in the stock of knowledge ($\partial C/\partial K = vQ$). One might expect differences across firms or uncertainty about the "productivity" of R & D (for example, technological opportunity) to be reflected in the parameter "v". This important aspect of the problem is not treated in the present model.

With these qualifications in mind, we proceed with the derivation of the optimal R & D profile. Using the specific forms for $h(\cdot)$ and $C(\cdot)$ and the constraint in (1), the decision problem can be expressed

$$
(3) \qquad \underset{<K_{t,s}>}{\text{Min }} V_t = \sum_{s=0}^{\infty} \alpha^s \{ Q_{t,s} [f(w_{t,s}) - vK_{t,s}]
$$
$$
+ P_{t,s} [K_{t,s+\theta} - (1 - \delta)K_{t,s+\theta-1}]
$$
$$
+ AP_{t,s} [K_{t,s+\theta} - (1 - \delta)K_{t,s+\theta-1}]^2 \},
$$

where we note that the decision variable is the stock of knowledge. The first-order (Euler) conditions are:

$$
(4) \qquad \frac{\partial V_t}{\partial K_{t,j}} = -v\alpha^j Q_{t,j} + \alpha^{j-\theta} P_{t,j-\theta}
$$
$$
+ 2A\alpha^{j-\theta} P_{t,j-\theta} [K_{t,j} - (1 - \delta)K_{t,j-1}]
$$
$$
- (1 - \delta)\alpha^{j+1-\theta} P_{t,j+1-\theta}
$$
$$
- 2A\alpha^{j+1-\theta} P_{t,j-\theta} [K_{t,j+1} - (1 - \delta)K_{t,j}]
$$
$$
= 0,
$$

1. This assumption implies that the marginal savings in variable costs due to R & D is a constant, i.e., $\partial^2 C/\partial K^2 = 0$. This violates the standard second-order condition for restricted cost functions that $\partial^2 C/\partial K^2 < 0$ and, in a static context, generates an infinitely elastic demand for R & D (and hence an indeterminate level of R & D). In a cost of adjustment framework the analog is an infinitely elastic shadow price of R & D, but an optimal level of R & D is ensured by an upward sloping marginal cost of investment schedule (see fig. 16.1).

for $j \geq \theta$. Noting that $K_{t,j} - (1 - \delta)K_{t,j-1} = \bar{R}_{t,j-\theta}$ and defining $R_{t,j} = P_{t,j}\bar{R}_{t,j}$ for $j \geq 0$, the Euler conditions can be written:

$$(5) \qquad (1 - \frac{1}{\beta}L)R_{t,j+1-\theta} = -aP_{t,j+1-\theta} + \frac{a}{\beta}P_{t,j-\theta} - \frac{av\alpha^{\theta}}{\beta}Q_{t,j},$$

where $a = 1/2A$, $\beta = (1 - \delta)/(1 + r)$, and L denotes the lag operator.

Since $\beta < 1$ we can obtain the forward solution to the difference equation in (5) (see Sargent 1979b, chap. 9). Letting $s = j + 1 - \theta$ for simplicity, this yields, after some manipulation,

$$(6) \qquad R_{t,s} = -aP_{t,s} + a[\alpha^{\theta} \sum_{j=s+\theta}^{\infty} \beta^{j-s-\theta} vQ_{t,j}].$$

This equation, which we refer to as the structural investment equation, says that planned R & D depends on the expected price of R & D investment goods and the stream of future expected levels of output. To gain more insight into the solution, note that the term $vQ_{t,j}(j \geq s + \theta)$ represents the expected savings in variable costs in period $t + j$ due to a unit increase in the stock of knowledge in $t + j$. This, in turn, reflects the marginal dollar of R & D planned for $t + j - \theta$. Hence, the bracketed expression in (6) is the discounted value (in terms of period $t + j - \theta$) of cost savings from planned R & D and may be interpreted as the expected shadow price of R & D, $q_{t,s}$. Then (6) expresses the optimal planned expenditures on R & D as a linear function of the anticipated price of R & D investment goods and the implicit shadow price of R & D.

The model is illustrated in figure 16.1. The marginal cost of R & D schedule rises linearly with the level of R & D, and is shifted by anticipated changes in the price of R & D investment goods. The shadow price relevant to investment planned for year $t + s$ in year t, $q_{t,s}$, depends on the expected future stream of output (which determines the cost savings from R & D investment), but it is independent of the level of R & D. The optimal amount of planned R & D, $\bar{R}^{*}_{t,s}$, is fixed by the intersection of the shadow price and marginal cost schedules. Both the supply and demand schedules of R & D are driven by the firm's expectations. Any shift in expected output or the anticipated price of R & D will alter the optimal level of planned R & D.

An alternative form of the investment equation can be obtained in which the infinite series of expected output does not appear. Leading the target period in (6), multiplying by β, and subtracting the result from (6), we obtain

$$(7) \qquad R_{t,s} = -aP_{t,s} + a\beta P_{t,s+1} + bQ_{t,s+\theta} + \beta R_{t,s+1},$$

where $b = av\alpha^{\theta}$. We refer to equation (7) as the reduced form investment equation. (The terminology is somewhat unconventional since the equation contains a simultaneous anticipation as a regressor, but we retain it for simplicity.) One advantage of the reduced form in (7) is that it contains a testable implication of the cost of adjustment formulation, conditional on the particular specification of expectations. Specifically, the coefficient on the leading R & D anticipation, $R_{t,s+1}$, should be approximately equal to the gross discount factor $\beta = (1 - \delta)/(1 + r)$.

The realization function relates the difference between actual and planned investment in R & D for a given target period (the realization error) to its determinants. Using (6), the general form for the realization function is

$$(8) \qquad D_{t,s} \equiv R_{t,0} - R_{t-s,s} = -a(P_{t,0} - P_{t-s,s})$$

$$+ b \sum_{j=0}^{\infty} \beta^j (Q_{t,j+\theta} - Q_{t-s,j+\theta+s}).$$

Note that the realization error depends on the error in predicting the price of R & D and the discounted value of the revisions in expected output (i.e., the revision in the shadow price of R & D). Hence, the realization function reflects the use of new information regarding the exogenous variables in the investment model which becomes available between the formation and implementation of the investment plans. However, the precise form of the realization function (and of the underlying investment function) depends critically on how the new information is used, that is, on the manner in which expectations are formed.

One general point of interest is that the realization errors will have zero mean under a variety of expectational mechanisms. It follows from (8) that $E_t D_{t,s} = 0$ if two conditions hold: (i) $E_t P_{t,0} = E_t P_{t-s,s}$ and (ii) $E_t Q_{t,j+\theta} = E_t Q_{t-s,j+\theta+s}$, where E_t is the expectation operator over t. A sufficient condition for (i) and (ii) to hold is that the firm forms unbiased predictors of the price of R & D and the level of output.

16.2 Model under Specific Expectations Hypotheses

In this section we derive estimable forms of the investment and realization functions under three alternative expectations hypotheses. The available data set (described in section 16.3) contains actual and one-span, planned R & D expenditures; no multiple-span anticipations are provided. Though the model applies to multiple-span investment decisions, we are limited in the empirical work to the actual and one-span structural investment equation, the reduced form equation for actual R & D, and the one-span realization function (refer to [6]–[8] above).

16.2.1 Rational Expectations

The test of the rational expectations hypothesis is based on the assumption that the firm forms expectations of the price of R & D and the level of output according to the stochastic processes (presumed to be) generating these exogenous variables. We assume that each variable evolves according to an autoregressive process:

(9) $P_t = b_1 P_{t-1} + \ldots + b_m P_{t-m} + \epsilon_t,$

(10) $Q_t = c_1 Q_{t-1} + \ldots + c_n Q_{t-n} + u_t,$

where ϵ_t and u_t are mutually uncorrelated white noise disturbances.[2] Define

$$
x_t = \begin{bmatrix} P_t \\ P_{t-1} \\ \cdot \\ \cdot \\ \cdot \\ P_{t-m+1} \end{bmatrix}, \quad
z_t = \begin{bmatrix} Q_t \\ Q_{t-1} \\ \cdot \\ \cdot \\ \cdot \\ Q_{t-n+1} \end{bmatrix}, \quad
B = \begin{bmatrix} b_1 & \ldots & & & b_m \\ 1 & 0 & & & 0 \\ & & & & \cdot \\ 0 & 1\ 0 & \ldots & & \cdot \\ & & & & \cdot \\ & & & & \cdot \\ 0 & & \ldots & 1 & 0 \end{bmatrix},
$$

$$
C = \begin{bmatrix} c_1 & \ldots & & c_n \\ 1 & 0 & & 0 \\ & & & \cdot \\ 0 & 1\ 0 & \ldots & \cdot \\ & & & \cdot \\ & & & \cdot \\ 0 & & \ldots 1 & 0 \end{bmatrix}, \quad
\epsilon_t^* = \begin{bmatrix} \epsilon_t \\ 0 \\ \cdot \\ \cdot \\ \cdot \\ 0 \end{bmatrix}, \quad
u_t^* = \begin{bmatrix} u_t \\ 0 \\ \cdot \\ \cdot \\ \cdot \\ 0 \end{bmatrix},
$$

and the $1 \times m$ and $1 \times n$ vectors $d = (1, 0 \ldots 0)$ and $e = (1, 0 \ldots 0)$. If the eigenvectors of B and C are distinct, we can write $B = M\Lambda M^{-1}$ and $C = N\Omega N^{-1}$, where Λ and Ω are diagonal matrices of eigenvalues, and M and N are matrices of associated eigenvectors. Then one can show that under the rational expectations hypothesis the following set of equations results:[3]

2. The following setup is based on Sargent (1978), but we extend the argument to planned investment and realization functions. The assumption that u_t and ϵ_t are contemporaneously uncorrelated simplifies the prediction formulas for P_t and Q_t. This assumption is subjected to an empirical test (see note 8).

3. The procedure to derive (11a)–(11c) is as follows: From the assumption $E_t(\epsilon_{t+j}) = E_t(u_{t+j}) = 0$ for $j > 0$, we obtain $P_{t,s} = dM\Lambda^s M^{-1} x_t$, and $Q_{t,s} = eN\Omega^s N^{-1} z_t$. Substitutions of these expressions into (6) and (7), with some manipulation, yields (11a) and (11b). To derive (11c), note from (9)–(10) that $x_t = B^s x_{t-s} + B^{s-1}\epsilon_{t-s+1}^* + \ldots + \epsilon_t^*$ and $z_t = C^s z_{t-s} + C^{s-1} u_{t-s+1}^* + \ldots + u_t^*$. Using these and the expressions for $P_{t,s}$ and $Q_{t,s}$ in (8) yields (11c).

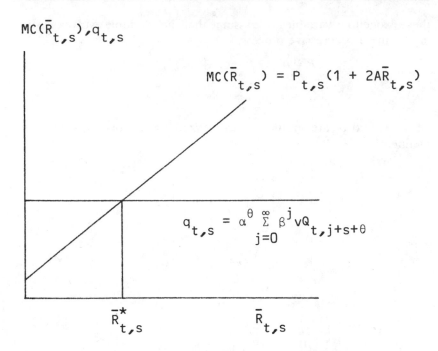

Fig. 16.1   Determination of optimally planned R & D.

(11a) $R_{t,s} = [-daB^s]x_t + [ebN\Omega^{s+\theta}JN^{-1}]z_t, \qquad s = 0, 1,$

(11b) $R_{t,0} = [da(\beta B - I)]x_t + [ebC^\theta]z_t + \beta R_{t,1},$

(11c) $D_{t,1} = -da\epsilon_t^* + [ebN\Omega^\theta JN^{-1}]u_t^*,$

where $R_{t,s}$ denotes the R & D planned in period t for period $t + s$, $D_{t,1}$ is the one-span realization error for R & D, J is a diagonal matrix with elements $(1 - \beta\omega_i)^{-1}$ and ω_i as the eigenvalues of Ω, and the bracketed terms represent the vector of coefficients under rational expectations.

The structural equation for planned R & D s periods ahead in (11a) is simply a distributed lag against m past prices of R & D and n past levels of output, where m and n are the orders of the autoregressions in (9) and (10). The reduced form equation in (11b) includes these determinants plus the leading R & D anticipation (i.e., planned R & D for one period ahead). Equation (11c) relates the one-span realization error to the unanticipated components (or "surprises") in the price of R & D and the level of output realized between the formulation and the implementation of the planned R & D investment. Since under the rational expectations hypothesis the firm exploits the available information on the exogenous variables fully (i.e., according to their true stochastic structures), the realization error should be determined solely by these surprises.[4] The rational expectations hypothesis delivers a set of nonlinear parameter restrictions both within and across equations (given by the bracketed terms in [11a]–[11c]) which serve to identify the parameters a, b, and β. These restrictions are related to the parameters in the underlying stochastic representations of the exogenous variables in the model. However, since the realization function in (11c) is definitionally related to the investment equation (11a), the parameter restrictions in (11c) contain no independent information. Therefore, the basic system of equations which we estimate consists of the autoregressions in (9) and (10), and (11a) and (11b). First the unconstrained system is estimated and then the parameter restrictions are imposed and tested. In addition to these parameter restrictions, the rational expectations hypothesis implies two testable propositions on the realization function. First, only the contemporaneous surprises in the price of R & D and the level of output should matter, since earlier surprises are known when the R & D plans are formed and should already be reflected in those plans. Hence, lagged surprises should be statistically insignificant when added to (11c). Second, since the unanticipated components ϵ_t and u_t have zero means by construction, the mean of the realization errors must be zero under rational expectations. This simply reflects the unbiasedness of rational forecasts and the linearity of the model in the stochastic exogenous variables.

16.2.2 Adaptive Expectations

Suppose that the firm forms its forecasts of exogenous variables according to an adaptive expectations mechanism, revising its single-span forecast by some fraction of the previous period's forecast error:

(12a) $$P_{t,1} - P_{t-1,1} = \gamma (P_t - P_{t-1,1}), \qquad 0 < \gamma < 1,$$

(12b) $$Q_{t,1} - Q_{t-1,1} = \lambda (Q_t - Q_{t-1,1}), \qquad 0 < \lambda < 1.$$

It is well known that this procedure implies forecasts that are geometrically weighted averages of all past realizations:

(13a) $$P_{t,1} = \gamma \sum_{i=0}^{\infty} (1 - \gamma)^i P_{t-i},$$

(13b) $$Q_{t,1} = \lambda \sum_{i=0}^{\infty} (1 - \lambda)^i Q_{t-i}.$$

We also note that if (and only if) P_t and Q_t are (mean) stationary

---

4. Similar implications appear in the literature on the efficient market hypothesis (Fama 1970) and recent work on the permanent income hypothesis under rational expectations (Bilson 1980; Hall 1978).

processes, the adaptive forecasts in (13a) and (13b) are unbiased predictors.[5]

For present purposes we also need multiple-span forecasts, since they appear in the expression for the shadow price of R & D. However, the adaptive expectations hypothesis is silent on how agents form multiple-span forecasts. Muth (1960) has shown that if the underlying stochastic process is of a particular form for which adaptive forecasts are also rational, then the (minimum mean squared error) multiple- and single-span forecasts are identical. This line of argument, however, erases the distinction between adaptive and rational forecasts. An alternative way of linking single- and multiple-span forecasts would be to construct an explicit model of learning in which agents do not know the true stochastic structure but form adaptive expectations which are "optimal" predictors on the basis of some subjectively assumed structure, and then somehow update their knowledge of that structure and the associated coefficient of adaptation. Models of this type, however, are not yet available in the literature, and to construct one here would take us far afield. In the absence of a learning model, we adopt the arbitrary assumption that a firm which forms its single-span expectation adaptively also holds that forecast for multiple spans, that is, $P_{t,s} = P_{t,1}$ and $Q_{t,s} = Q_{t,1}$ for $s \geq 1$.[6] Although this assumption is formally identical to Muth's result, it is *not* assumed here that the multiple-span forecasts are minimum mean squared error predictions.

Using this assumption and (13a) and (13b), we obtain the following system of structural (14a)–(14b), reduced form (14c), and realization functions (14d) under adaptive expectations:[7]

(14a) $R_{t,0} = -aP_t + a(1 - \lambda)P_{t-1} + \dfrac{b\lambda}{1 - \beta}Q_t + (1 - \lambda)R_{t-1,0}.$

(14b) $R_{t,1} = -a\gamma P_t + a\gamma(1 - \lambda)P_{t-1} + \dfrac{b\lambda}{1 - \beta}Q_t - \dfrac{b\lambda(1 - \lambda)}{1 - \beta}Q_{t-1}$

$+ (2 - \gamma - \lambda)R_{t-1,1} - (1 - \gamma)(1 - \lambda)R_{t-2,1}.$

5. If the forecasted variable, say P_t, is trended, then the adaptive forecast in (13a) will be biased. If the series is growing at the rate g, then an unbiased predictor is obtained from the modified adaptive forecast $P_{t,1} = (1 + g)\gamma\sum_{i=0}^{\infty}(1 - \gamma)^i P_{t-i}$. Given that the agent forecasts adaptively and that g is ascertainable, it is reasonable to assume that the agent uses the modified formula.

6. If P_t and Q_t are growing at rate g_t and g_q and the firm uses the unbiased modified version of adaptive forecasting (note 5), we have $P_{t,s} = (1 + g_P)^{s-1}P_{t,1}$ and $Q_{t,s} = (1 + g_q)^{s-1}Q_{t,1}$. Then the coefficients in the system of equations in (14a)–(14d) are slightly modified.

7. Equation (14a) is obtained by substituting (13b) into (6) for $s = 0$ and performing a Koyck transformation on Q_t (to remove the infinite past series on Q_t). To obtain (14b), substitute (13a)–(13b) into (6) for $s = 1$ and perform two sequential Koyck transformations on P_t and Q_t. Equation (14c) is derived by a similar procedure using (13a)–(13b) in (7). Finally, (14d) is obtained by lagging (14b) and subtracting it from (14a).

(14c) $R_{t,0} = -a(1 - \beta\gamma)P_t + a[2 - \gamma - \lambda - \beta\gamma(1 - \lambda)]P_{t-1}$

$\qquad - a(1 - \lambda)(1 - \gamma)P_{t-2} + b\lambda Q_t - b\lambda(1 - \lambda)Q_{t-1}$

$\qquad + \beta R_{t,1} + (2 - \gamma - \lambda)R_{t-1,0} - (1 - \lambda)(1 - \gamma)R_{t-2,0}$

$\qquad - \beta(2 - \gamma - \lambda)R_{t-1,1} + \beta(1 - \lambda)(1 - \gamma)R_{t-2,1}.$

(14d) $D_{t,1} = -aP_t + a(1 + \gamma - \lambda)P_{t-1} - a\gamma(1 - \lambda)P_{t-2} + \dfrac{b\lambda}{1 - \beta}Q_t$

$\qquad - \dfrac{b\lambda}{1 - \beta}Q_{t-1} + \dfrac{b\lambda(1 - \lambda)}{1 - \beta}Q_{t-2} + (1 - \lambda)R_{t-1,0}$

$\qquad - (2 - \gamma - \lambda)R_{t-2,1} + (1 - \gamma)(1 - \lambda)R_{t-3,1}.$

The model provides qualitative predictions on the coefficients of all variables in the unconstrained system. Note also that the adaptive expectations hypothesis implies a set of fifteen nonlinear parameter restrictions in (14a)–(14c) serving to identify the five underlying parameters a, b, β, γ, and λ. Estimation of the realization function (14d) is redundant since it is a linear combination of (14a) and (14b). Therefore, the basic set of estimating equations consists of (14a)– (14c). We first estimate these equations unconstrained, and then impose and test the identifying restrictions. Finally, it was noted earlier that adaptive forecasts are unbiased if the stochastic exogenous variables are (mean) stationary. This property implies the testable proposition that the realization errors have a zero mean.

16.2.3 Static Expectations

Under the static expectations hypothesis, the firm assumes that the future values of exogenous variables will remain at their current levels, that is $P_{t,s} = P_t$ and $Q_{t,s} = Q_t$ for $s \geq 1$. It is clear from (12a) and (12b) that this hypothesis is a limiting case of adaptive expectations where $\gamma = \lambda = 1$. By substituting this condition into (14a) and (14b) we observe that, under static expectations, the structural investment equation depends only on the contemporaneous price of R & D and level of output, while the realization error depends solely on the most recent, actual (not unanticipated) changes in these exogenous variables.

The most straightforward way of testing static expectations is to impose the constraints $\gamma = \lambda = 1$ in the system of equations under adaptive expectations. This procedure generates thirteen exclusion restrictions in (14a)–(14c) that can be tested directly. In addition, we estimate the realization function under static expectations (by regressing the realization error against the most recent actual change in the price of R & D and the level of output) and test the joint significance of lagged changes in these variables.

16.3 Data and Empirical Results

16.3.1 Description of Data

The data set used in this study is drawn from annual surveys (conducted by McGraw-Hill) of actual and planned investment expenditures on plant and equipment and R & D by firms (for a fuller description, see Eisner 1978 and Rasmussen 1969). There was a problem of sporadic missing observations in the data for different firms. Using some supplementary information, we were able to construct a set of data on actual and one-span planned R & D for the period 1959–69 and on sales for 1954–69 for forty-nine manufacturing firms, subject to the requirement that no firm have more than two missing observations. Because the missing data vary by firm and by variable, the usable sample depends on the model being estimated. It is not entirely clear whether the reported data on planned R & D should be interpreted as expressed in current or antici-pated prices. Since the McGraw-Hill surveys request information on planned R & D *expenditures* and do not indicate that these should be in present prices, we interpret them as in anticipated (one-year ahead) prices (which is consistent with the definition of $R_{t,s}$ in the model; see seciton 16.1). The sales data are deflated by the Wholesale Price Index for total manufacturing. We also require (as an independent variable) a price index for R & D investment goods. To construct a firm-specific index would require information on the firm's composition of R & D expenditures, which is not available. We therefore chose to use an aggregate index for manufacturing constructed on the basis of the mix of R & D inputs at the (roughly two-digit SIC) industry level (Schankerman 1979). This is essentially equivalent to using time dummies in the regres-sions.

Estimation of the model under rational expectations is conducted on detrended variables. Each variable is regressed on a constant, a linear trend, and trend squared (for each firm separately), and the residuals from these regressions are used as data in estimating the R & D invest-ment model. This is frequently done (Sargent 1978, 1979a; Meese 1980) to ensure stationarity of the stochastic variables in the model and on the argument that the theory under rational expectations predicts that the deterministic components (presumed to be known) of the process linking endogenous and exogenous variables will not be characterized by the same distributed lag model as their indeterministic components. De-trending prior to estimation is an attempt to isolate the indeterministic components. We also estimated the model without detrending, and the major conclusions reported later did not change. These arguments in favor of detrending do not apply to the model under adaptive and static expectations because these forecasting devices are not based on the underlying stochastic processes generating the exogenous variables, and hence they do not distinguish between the deterministic and indeterministic components. We therefore estimate the model under adaptive and static expectations without prior detrending. This means, of course, that the fits of the equations under rational expectations cannot be compared directly, since the dependent variables are measured differently.

All models were estimated by Zellner's seemingly unrelated equations technique (Zellner 1962), which is generalized least squares allowing for free correlation in the errors across equations. It should be noted that the estimated system of equations under each expectations hypothesis is structurally recursive. That is, the leading anticipation $R_{t,1}$ appears on the right-hand side of the investment equation for $R_{t,0}$, but not vice versa. Hence, if the disturbances in the equations are mutually uncorrelated, instrumental variables on $R_{t,1}$ are not required to obtain consistent estimates. This approximately holds in the system under rational expectations, but under adaptive and static expectations the disturbances exhibit considerable correlation across equations. We tried using instrumental variables for $R_{t,1}$ in these cases (consisting of both firm-specific and more aggregative variables), but the results were not robust, apparently because the instruments were not strongly correlated with $R_{t,1}$. However, the general compatibility of the parameter estimates with theoretical expectations (see section 16.3.3) suggests that the problem of inconsistency may not be serious.

16.3.2 Empirical Results under Rational Expectations

Table 16.1 presents the unconstrained estimates of the model under rational expectations using a third-order autoregressive specification for the price of R & D and the level of output.[8] Because the means were removed in the detrending procedure, the results in table 16.1 represent within-firm, over-time regressions. We first note that the estimated autoregressions imply both real and complex roots satisfying the stationarity condition that the largest modulus be less than unity. The low R^2 in the

---

8. Two points should be noted. First, we checked the assumption that the disturbances ϵ_t and u_t in (9) and (10) are contemporaneously uncorrelated by testing the univariate autoregressive representations against a general bivariate specification. This involves testing the joint significance of three lagged values of Q_t in the autoregression for P_t and three lagged values of P_t in the autoregression for Q_t. The computed F statistics are 1.42 and 1.60, respectively, compared to the critical level $F(3,548) = 2.60$. The simplifying assumption $E(\epsilon_t u_t) = 0$ is accepted. Second, there is evidence that a higher order autoregression is appropriate, but including more than three lagged values of output and price would reduce the sample size unacceptably. These higher order terms affect only the last coefficient in the $AR(3)$ representation and they do not improve the equations in terms of serial correlation. Still, they probably do indicate that a moving average or mixed process is more appropriate, but the structure of our data does not permit use of such specifications. In section 16.3.4 we discuss the implications of these considerations for the interpretation of the empirical findings under rational expectations.

Table 16.1 Empirical Results under Rational Expectations

Equation/ Dependent Variable	Independent Variable											R^2	Durbin-Watson
	P_t	P_{t-1}	P_{t-2}	P_{t-3}	Q_t	Q_{t-1}	Q_{t-2}	Q_{t-3}	$R_{t,1}$	S_t^{pa}	S_t^{Qb}		
Structural $R_{t,0}$	.11* (.13)	—	—	—	.071 (.027)	.16 (.03)	−.17 (.03)	—	—	—	—	.13	2.45
Reduced form $R_{t,0}$	−.27 (.15)	.45 (.24)	−.15* (.14)	—	.020 (.010)	.039 (.011)	−.011* (.011)	—	.85 (.002)	—	—	.88	1.44
Structural $R_{t,1}$	.70 (.22)	−.92 (.28)	.37 (.17)	—	.060 (.029)	.140 (.032)	−.19 (.034)	—	—	—	—	.11	2.51
Realization $D_{t,1}$	—	—	—	—	—	—	—	—	—	1.62* (1.13)	.23 (.05)	.05	2.10
Autoregression P_t	—	1.08 (.03)	.12 (.05)	−.55 (.03)	—	—	—	—	—	—	—	.95	2.38
Autoregression Q_t	—	—	—	—	—	.33 (.05)	−.27 (.05)	.053* (.055)	—	—	—	.11	1.82

Notes: Estimated standard errors are in parentheses. An asterisk denotes statistical insignificance at the 0.05 level.

[a] $S_t^P = P_t - b_1 P_{t-1} - b_2 P_{t-2} - b_3 P_{t-3}$, where the b's are the estimated coefficients in the autoregression for P_t.

[b] $S_t^Q = Q_t - c_1 Q_{t-1} - c_2 Q_{t-2} - c_3 Q_{t-3}$, where the c's are the estimated coefficients in the autoregression for Q_t.

autoregression for output indicates a large unanticipated component in the prediction of output. The much higher R^2 in the autoregression for the price of R & D is not a statistical artifact reflecting the use of the same aggregate price index for all firms in the sample. Estimation of this autoregression on a single time series yields an $R^2 = .98$. There is, in fact, only a very small unanticipated component in the measured price of R & D.

Most of the estimated coefficients in the investment equations are statistically significant. The sum of the output coefficients is positive in two of the three investment equations, which is expected since a sustained increase in the level of output should raise the shadow price and hence to optimal level of R & D. By analogous reasoning, we expect the sum of the price coefficients to be negative, but it is essentially zero in the empirical results. Not much can be deduced from the particular pattern of coefficients, since under rational expectations this pattern is related in a highly nonlinear way to the eigenvalues from the autoregressions for price and output. We formally test these restrictions later. Also note that the structural investment equations account for only about 10 percent of the within-firm variance in actual and planned R & D. The much better fit of the reduced form equation for actual R & D is from the presence of the leading anticipation, $R_{t,1}$, as a regressor.

One notable result in table 16.1 is the coefficient on $R_{t,1}$ in the reduced form equation for $R_{t,0}$. We showed in seciton 16.1 that this coefficient should equal the gross discount factor $\beta = (1 - \delta)/(1 + r)$. Assuming $r = .10$ and $\delta = .10$, we expect to obtain $\beta = 0.8$, which is close to the actual estimated value $\hat{\beta} = .85$.[9] As we will see later, however, the estimate of β is robust to different specifications of expectations formation, and hence the result in table 16.1 should not be interpreted as evidence in favor of rational expectations.

The realization function in table 16.1 relates the (one-span) realization error to the contemporaneous unanticipated components in the price of R & D and the level of output (S_t^P and S_t^Q). These components are defined within the estimation procedure to ensure that they are consistent with the estimated autoregressions for price and output (see notes to table 16.1). The "surprise" in output has a significantly positive effect on the difference between actual and planned R & D, which is the expected result since a positive surprise in output raises the shadow price of R & D and hence the optimal R & D investment. The expected effect of a surprise in the price of R & D is negative, since an unexpected rise in its price shifts the marginal cost of R & D schedule upward and hence lowers

---

9. The assumed $\delta = .10$ is much lower than the rate estimated by Pakes and Schankerman (this volume). However, in our model δ is the rate of decline in the ability of R & D to "produce" cost reductions, not the rate of decline in appropriable revenues considered by Pakes and Schankerman. For more on the distinction, see Schankerman and Nadiri (1983).

the optimal investment in R & D. The estimate in table 16.1 has the wrong sign but is statistically insignificant.

We turn next to various tests of the rational expectations hypothesis. The first, and least stringent, test concerns the realization errors. It was pointed out in section 16.1 that the mean of the realization errors will be zero if the firm forms unbiased predictors of the price of R & D and the level of output. Since rational forecasts are unbiased, this is an implication of the rational expectations hypothesis. The mean of the realization errors (based on data prior to detrending) for the entire sample is not significantly different from zero (-0.83 with a standard error of 2.18). When computed separately for each firm, only three of the forty-nine firms exhibit nonzero means and each of these cases is only marginally significant. We conclude that the rational expectations hypothesis passes this weak necessary condition, but it is important to reiterate that any unbiased forecasting device would also satisfy this requirement.

The formal parametric tests are considered next. First, rational expectations implies a set of nonlinear restrictions on the parameters of the system of investment equations. These restrictions are expressed in terms of the eigenvalues of the autoregressive structures generating the price of R & D and the level of output. We use the following two-stage testing procedure: First the unconstrained system ([9]–[10] and [11a]–[11b]) is estimated and the eigenvalues are computed. The nonlinear restrictions embodied in (11a)–(11b) are then computed numerically, and the constrained system is estimated. We do not iterate on this procedure (using the new estimates for the autoregressions), but the second-stage constrained estimates are consistent in any case. The test requires an assumed value for the gestation lag, θ. The reported results are based on $\theta = 2$ (from Pakes and Schankerman, chap. 4 in this volume), but they are not sensitive to different values (we experimented with $1 \leq \theta \leq 4$).

The results are summarized in the first row of table 16.2. The parameter restrictions are strongly rejected. The computed F of 21.4 greatly exceeds the critical value of 1.62. Imposition of the restrictions reduces the total mean squared error by 11.2 percent. However, one may object to a simple F test at a fixed level of significance in a sample as large as ours (1444 observations in the system as a whole). The reason is that any null hypothesis (viewed as an approximation) will be rejected with certainty as the sample size goes to infinity if the Type I error is held constant. Leamer (1978, chap. 4) argues forcefully that the critical value of the F statistic should rise with sample size to avoid this interpretive problem. He proposes an alternative measure of the critical value (which we call the Bayesian F) which has the property that, given a diffuse prior distribution, the critical value is exceeded only if the posterior odds favor the alternative hypothesis. The Bayesian F is reported in the last column of table 16.2. In the case of rational expectations, the Bayesian F is 7.54,

Table 16.2   **Tests of Expectations Hypotheses**

Equations	Com-puted F	Critical $F_{.05}$	% Δ MSE	Bayesian F[a]
Rational				
(1) Investment equations	21.4	$F(18,1426) = 1.62$	11.2	7.54
(2) Realization function	10.5	$F(4,376) = 2.39$	11.0	6.05
Adaptive				
(3) Investment equations	4.32	$F(15,1201) = 1.67$	4.4	7.31
Static				
(4) Investment equations $\gamma = 1$	3.84	$F(5,1201) = 2.22$	1.5	7.22
(5) Investment equations $\gamma = \lambda = 1$	189.0	$F(13,1201) = 1.73$	201.0	7.29
(6) Realization function	12.8	$F(4,439) = 2.39$	10.4	6.23

[a]Bayesian $F = [(T - k)/P](T^{p/T} - 1)$, where T is the sample size, $T - k$ denotes degrees of freedom, and p is the number of restrictions.

which is far below the computed F of 21.4. We conclude that the parameter restrictions under rational expectations are rejected even after this adjustment for sample size.

The second row in table 16.2 summarizes the test of the joint significance of two lagged surprises in the price of R & D and the level of output in the realization function. Under rational expectations only the contemporaneous surprises should affect the realization error, since earlier surprises were known when the R & D plan was formulated. Again, the computed F statistic of 10.5 exceeds both the conventional and the Bayesian critical values (2.39 and 6.05, respectively), and the null hypothesis is rejected.

We conclude from these results that the evidence does not support the rational expectations formulation of the model, at least one based on a third-order autoregressive representation of the price of R & D and the level of output. Various qualifications and explanations for this negative finding will be discussed later, but first we examine the empirical results under alternative expectations hypotheses.

16.3.3 Empirical Results under Adaptive and Static Expectations

The unconstrained estimates of the model under adaptive expectations are reported in table 16.3. The fits of the regression are very good, especially since the data contain both cross-sectional and time-series variation (the cross-sectional variance compromises about 75 percent of the total variance in the sample). On the whole, the pattern of estimated

Table 16.3   Empirical Results under Adaptive Expectations

Equation/ Dependent Variable	Independent Variable											
	P_t	P_{t-1}	P_{t-2}	Q_t	Q_{t-1}	$R_{t-1,0}$	$R_{t-2,0}$	$R_{t,1}$	$R_{t-1,1}$	$R_{t-2,1}$	R^2	Durbin-Watson
Structural $R_{t,0}$	-.29* (.27)	.29* (.28)		.024 (.004)		.74 (.02)					.80	1.98
Reduced form $R_{t,0}$	-.15* (.11)	.20* (.19)	-.05* (.10)	.035 (.009)	-.033 (.010)	.49 (.05)	-.085* (.05)	.84 (.01)	-.34 (.04)	.037* (.053)	.99	2.03
Structural $R_{t,1}$	-.25* (.31)	.25* (.32)	(.017)	-.018*† (.017)	.045†			(.04)	.66 (.038)	.095†	.75	2.24

Notes: Estimated standard errors are in parentheses. An asterisk denotes statistical insignificance at the 0.05 level. A cross denotes an estimated coefficient which has the wrong sign according to the model.

coefficients is consistent with the adaptive expectations hypothesis. The estimated coefficients on the price variables are uniformly insignificant, which may reflect the inadequacy of the aggregate price index used in the estimation.[10] However, all but two of the other coefficients are statistically significant and seventeen of the twenty estimated parameters have the sign predicted by the model. Also note that the point estimate of the coefficient of $R_{t,1}$ in the reduced form equation for $R_{t,0}$ is 0.84, which is very close to its predicted value. This is almost identical to the estimate under rational expectations, and as we indicated earlier, it should be interpreted more as support for the cost of adjustment formulation of the model than for either specific expectations mechanism. The magnitudes of the other parameter estimates in table 16.3, however, do tend to support the adaptive expectations hypothesis. A comparison of these results with the corresponding parameters in (14a)–(14c) indicates that many of the parameter restrictions implied by adaptive expectations are satisfied approximately by the unconstrained point estimates.

Before turning to the formal tests of adaptive expectations, we first note that this hypothesis is not consistent with the zero mean of the realization errors. The reason is that the observed price of R & D and the level of output are not mean stationary, and hence adaptive forecasts as formulated in (13a)–(13b) are not unbiased. This violation should be qualified by two considerations. First, we have only single-span realization errors to test the hypothesis. Second, and more important, the adaptive forecasting device in (13a)–(13b) can be modified easily to account for (known) trends in the variables, and the modified version will produce unbiased forecasts (see note 5 for discussion).

The formal tests of adaptive expectations are presented in the third row of table 16.2.[11] There are fifteen nonlinear restrictions implied by the hypothesis. The computed F statistic is 4.32, compared to a critical value of 1.67, and the hypothesis is rejected formally. However, imposition of the constraints raises the mean squared error by only 4.4 percent. This suggests that the restrictions may not be a bad approximation in view of the large sample size. A testing procedure using the Bayesian F supports this view. The critical value is 7.31 and the adaptive expectations restrictions are not rejected. It is worth reiterating that the proper interpretation of this result is that, given a diffuse prior distribution on the pa-

10. One problematic result is that the price coefficients in each equation sum to zero. This suggests that the true model should relate the stock of knowledge to the price of R & D, since the first-differenced version (involving R & D flow) would then yield the observed result. On the other hand, the result may just reflect the rather poor price index used.

11. We also reformulated the model in (14a)–(14c), using the modified version of adaptive expectations, and estimated the unconstrained and restricted systems. This required estimates of the trends in P_t and Q_t obtained from regressions of the logs of these variables against time. The formal tests of the parameter restrictions were qualitatively similar to those reported in the text.

rameters, the posterior odds favor the null hypothesis that the restrictions hold.

As indicated in section 16.2.3, static expectations are a special case of adaptive expectations where $\gamma = \lambda = 1$. Inspection of the unconstrained estimates in table 16.3 suggests that the constraint $\gamma = 1$ is more reasonable than $\lambda = 1$, and we therefore test the former separately. The results are summarized in rows (4) and (5) of table 16.2. The computed F for the five restrictions implied by $\gamma = 1$ is 3.84, while the critical value is 2.22. The restrictions are marginally rejected, but the change in the mean squared error is a negligible 1.5 percent. When judged against the Bayesian F of 7.22, the hypothesis $\gamma = 1$ is easily accepted. However, the restrictions implied by the joint hypothesis $\gamma = \lambda = 1$ (completely static expectations) are strongly rejected. The computed F of 189.0 greatly exceeds both the conventional and Bayesian critical values, and the mean squared error more than doubles when the constraints are imposed. As an additional check, we also estimated the realization function under full static expectations and tested the joint significance of two lagged changes in the price of R & D and the level of output. Under static expectations only the contemporaneous changes in these variables should influence the realization error. As row (6) in table 16.2 indicates, the hypothesis is rejected at both conventional and Bayesian critical values.

We conclude from these tests that the evidence generally supports the adaptive expectation hypothesis and decisively rejects the strong version of static expectations. Actually, the hypothesis most favored by the data is a mixed one with static expectations on the price of R & D and an adaptive mechanism on the level of output.

We can use the constrained estimates from the adaptive version to identify the underlying parameters in the model. The estimates (standard error) are: $\hat{a} = -.003\ (.0009)$, $\hat{\beta} = .85\ (.015)$, $\hat{\lambda} = .17\ (.032)$, $\hat{\gamma} = 1.28$ (.080), and $\hat{b} = .013\ (.017)$ The estimate $\hat{a}$ has the right sign but is insignificant, and $\hat{\gamma}$ lies outside the required range $0 < \gamma \leq 1$ but not substantially so. (This violation can occur because the restrictions are rejected under classical testing criteria, but accepted after a Bayesian adjustment for sample size.) The $\hat{\lambda}$ implies an average lag of about five years in the formation of output expectations $[(1 - \hat{\lambda})/\hat{\lambda} = 4.9]$. The estimate $\hat{b}$ can be used to compute the elasticity of R & D with respect to the shadow price of R & D, η_{rq}. Using equations (6) and (7), we can write $\eta_{rq} = b(\Sigma_{j=s+\theta}^{\infty} \beta^j Q_{t,j})/R$. Evaluating at the sample means (denoted by bars) and letting $Q_{t,j} = \bar{Q}$, $\eta_{rq} = [b/(1 - \beta)]\bar{Q}/\bar{R}$. This yields the point estimate (standard error) $\hat{\eta}_{rq} = 1.45\ (0.82)$. The point estimate is imprecise (which may not be surprising since $\hat{\eta}_{rq}$ is a nonlinear function of estimated parameters), but it indicates that a 10 percent increase in the shadow price of R & D raises the optimal level of R & D by about 15 percent. It is interesting to note that this estimate of η_{rq} is broadly similar to estimates of the elasticity of the investment-capital ratio with respect to Tobin's q for traditional capital (Abel 1979; Ciccolo 1975). Also note that our model of investment in R & D is based on cost minimization, and as a result, the shadow price of R & D is proportional to the expected levels of output in the future. Therefore, η_{rq} may also be interpreted as the elasticity of R & D with respect to a sustained increase in all future levels of output. The estimate $\hat{\eta}_{rq} = 1.45$ then implies that R & D rises somewhat more than proportionally to the ("permanent" or sustained) level of output. Given its statistical imprecision, this finding is not inconsistent with the empirical literature on the relationship between R & D and output (for a review, see Scherer 1980).

16.3.4 Adaptive versus Rational Expectations

The statistical tests conducted in sections 16.3.2 and 16.3.3 yield two main conclusions. First, the data do not support a rational expectations formulation based on third-order autoregressive representations of the exogenous variables (price of R & D and level of output). Second, the evidence is generally consistent with adaptive expectations and especially favors adaptive forecasting on output and static expectations on the price of R & D. Why would a firm employ two different forecasting devices for the two exogenous variables? The simple answer that the empirical confirmation of this mixed hypothesis is weak and should not be taken too seriously seems at odds with the statistical tests. A more interesting explanation might argue that this finding reflects rational forecasting for the true stochastic processes generating the exogenous variables and that the rejection of rational forecasting in section 16.3.2 is the result of a misspecification of these processes. Is the mixed static-adaptive expectations hypothesis consistent with rational expectations?

As indicated earlier (note 8, section 16.3.2), there is some evidence that a moving average specification of the stochastic processes might be more appropriate than a third-order autoregressive one. However, for this alternative explanation to work the true stochastic processes must be of a particular form: (1) Q_t must be an IMA (1, 1) (integrated moving average) process $Q_t = Q_{t-1} + \zeta_t - \psi\zeta_{t-1}$, where ζ_t is a white noise error, since Muth (1960) shows that for this process rational forecasts are also adaptive; (2) P_t must be a random walk process, $P_t = P_{t-1} + v_t$, where v_t is a white noise error, since for this model static expectations are rational.

We cannot test this explanation rigorously with the available data, but several pieces of indirect evidence are worth noting. First, Muth (1960) shows that for an IMA (1,1) process the adaptation coefficient in the rational forecast (λ in our notation) equals the ratio of the variance of the permanent component to the total variance. A consistent estimator of this ratio is given by the R^2 from the fitted IMA (1,1) regression. Under this hypothesis the estimated autoregression on Q_t in section 15.3.2 is of course misspecified, but it is interesting to note that the $R^2 = .11$ from that regression is quite close to (and within two standard errors of) the constrained estimate of the adaptation coefficient $\hat\lambda = .17$. Similarly, the $R^2 = .98$ from the autoregression on P_t is very close to the restricted value $\gamma = 1$, which was accepted by the data. These observations lend some credence to this alternative explanation.

On the other hand, if this alternative were true, one would expect the adaptive expectations formulation to be confirmed on detrended data (where the nonstationarity in the observed price and output series has been removed). However, reestimation of the model under adaptive expectations on detrended data indicates that the parameter restrictions are rejected both at conventional and Bayesian critical values of the F statistic.[12] As a further check, we estimated a first-order autoregressive process for detrended P_t. Under this explanation, the coefficient on lagged P_t should be unity and the errors should be serially uncorrelated. The estimated coefficient is essentially unity, but there is strong evidence of serial correlation (Durbin Watson = 0.57), and in this respect the first-order specification is distinctly worse than higher order autoregressive processes.

We conclude that the evidence is mixed on whether rational expectations can be reconciled wtih the empirically supported adaptive-static expectations scheme.

16.4 Concluding Remarks

This paper proposes a framework that integrates convex costs of adjustment and expectations formation in the determination of acutal and multiple-span planned investment decisions in R & D at the firm level. The framework is based on cost minimization subject to the firm's expectations of the future stream of output and the price of R & D. The model results in equations for actual and multiple-span, planned R & D investment and for the realization error as a function of these expectations. One of the unique features of the model is that it accommodates alternative mechanisms of expectations formation and provides a methodology for testing these hypotheses empirically. To give the model empirical content, a specific mechanism of expectations formation must be specified. We investigate the three leading forecasting hypotheses—rational, adaptive, and static expectations. Estimable equations and a set of testable parameter restrictions are derived under each of these three hypotheses.

---

12. The computed F is 8.59, compared to the conventional $F(15,1171) = 1.67$ and the Bayesian $F = 7.29$. Imposition of the restrictions raised the mean squared error by 10.0 percent.

The models are estimated on a set of pooled firm data covering the period 1959–69. The empirical results indicate that the parameter restrictions implied by both the rational and (fully) static expectations hypotheses are strongly rejected. The evidence generally supports adaptive expectations, both in terms of qualitative consistency of the unconstrained estimates with the predictions of the model and in terms of formal tests of the parameter restrictions. Actually, it appears that the hypothesis most favored by the data is a mixed one, with adaptive forecasting on the level of output and static expectations on the price of R & D. We also investigate whether this basic empirical finding could be reconciled with rational expectations and the formal rejection of this hypothesis explained by a misspecification of the stochastic processes generating the exogenous variables in the model. The available evidence for this interpretation is mixed. We emphasize that the basic empirical conclusion of this paper is that adaptive (or mixed adaptive-static) expectations are confirmed by the data. The appropriate interpretation of this result, however, remains an open question.

The theoretical framework and the empirical findings suggest directions for future research. The model could be improved by endogenizing the level of output and proceeding from profit maximization rather than cost minimization, and by treating both R & D and physical capital as quasi-fixed assets subject to costs of adjustment. On the empirical side, richer data sets are needed to explore the formation of expectations more fully, specifically to establish whether the adaptive expectations hypothesis constitutes a substantive alternative to or simply a guise for rational expectations.

References

Abel, A. 1979. Empirical investment equations: An integrative framework. Report no. 1932, Center for Mathematical Studies in Business and Economics, University of Chicago.

Bilson, J. 1980. The rational expectations approach to the consumption function: A multi-country study. *European Economic Review* 13:272–308.

Ciccolo, J. 1975. Four essays on monetary policy. Ph.D. diss, Yale University.

Eisner, R. 1978. *Factors in business investment*. National Bureau of Economic Research General Series no. 102. Cambridge, Mass.: Ballinger.

Fama, E. F. 1970. Efficient capital markets: A review of theory and empirical work. *Journal of Finance* 25:383–417.

Goldberg, L. 1972. The demand for industrial R & D. Ph.D. diss., Brown University.

Hall, R. 1978. Stochastic implications of the life cycle–permanent income hypothesis: Theory and evidence. *Journal of Political Economy* 86:971–88.

Kennan, J. 1979. The estimation of partial adjustment models with rational expectations. *Econometrica* 47:1441–55.

Leamer, E. E. 1978. *Specification searches: Ad hoc inferences with nonexperimental data.* New York: Wiley.

Meese, R. 1980. Dynamic factor demand schedules for labor and capital under rational expectations. *Journal of Econometrics* 14:141–58.

Modigliani, F. 1961. *The role of anticipations and plans in economic behavior and their use in economic analysis and forecasting.* Urbana: University of Illinois Press.

Muth, J. F. 1960. Optimal properties of exponentially weighted forecasts. *Journal of the American Statistical Association* 55:299–306.

———. 1961. Rational expectations and the theory of price movements. *Econometrica* 29:315–35.

Nadiri, M. I., and G. C. Bitros. 1980. Research and development expenditures and labor productivity at the firm level: A dynamic model. In *New developments in productivity measurement and analysis*, ed. J. W. Kendrick and B. N. Vaccara. Conference on Research in Income and Wealth: Studies in Income and Wealth, vol. 44. Chicago: University of Chicago Press for the National Bureau of Economic Research.

Rasmussen, J. A. 1969. Research and development, firm size, demand and costs: An empirical investigation of research and development spending by firms. Ph.D. diss., Northwestern University.

Sargent, T. J. 1978. Rational expectations, econometric exogeneity, and consumption. *Journal of Political Economy* 86:673–700.

———. 1979a. Estimation of dynamic labor demand schedules under rational expectations. *Journal of Political Economy* 86:1009–14.

———. 1979b. *Macroeconomic theory.* New York: Academic Press.

Schankerman, Mark. 1979. Essays on the economics of technical change: The determinants, rate of return, and productivity impact of research and development. Ph.D. diss., Harvard University.

Schankerman, Mark, and M. I. Nadiri. 1983. Restricted cost functions and the rate of return to quasi-fixed factors with an application to R & D and capital in the Bell System. New York University. Mimeo.

Scherer, F. M. 1980. *Industrial market structure and economic performance.* 2d. ed. Chicago: Rand-McNally.

Zellner, A. 1962. An efficient method of estimating seemingly unrelated regressions and tests for aggregation bias. *Journal of the American Statistical Association* 57:348–68.

17 Productivity and R & D at the Firm Level

Zvi Griliches and Jacques Mairesse

17.1 Motivation and Framework

17.1.1 Introduction

Because of worries about domestic inflation and declining international competitiveness, concern has been growing about the recent slowdowns in the growth of productivity and R & D, both on their own merit and because of their presumed relationship. This paper tries to assess the contribution of private R & D spending by firms to their own productivity performance, using observed differences in both levels and growth rates of such firms.

A number of studies have been done on this topic at the industry level using aggregated data, but ours is almost the first to use time-series data for a cross section of individual firms, that is panel data.[1] The only similar

Zvi Griliches is professor of economics at Harvard University, and program director, Productivity and Technical Change, at the National Bureau of Economic Research. Jacques Mairesse is a professor at Ecole des Hautes Etudes en Sciences Sociales and Ecole de la Statistique et de l'Administration Economique, and a research affiliate of the National Bureau of Economic Research.

A first draft of this paper was presented at the Fifth World Congress of the Econometric Society at Aix-en-Provence, August 1980. This work is part of the National Bureau of Economic Research Program of Productivity and Technical Change Studies. The authors are indebted to the National Science Foundation (PRA79-1370 and SOC78-04279) and to the Centre National de la Recherche Scientifique (ATP 070199) for financial support. The authors are also thankful to John Bound, Bronwyn Hall, and Alan Siu for very able research assistance.

1. M. Ishaq Nadiri and his associates have done important related investigations. In their work at the firm level they have estimated factor demand equations (including demand for R & D) but did not pursue the direct estimation of production functions (see, for example, Nadiri and Bitros 1980).

study at the firm level is Griliches's (1980a) use of pooled NSF and Census data for 883 R & D performing companies over the 1957–65 period. This study had to rely on various proxies (and on corresponding ad hoc assumptions) for the measurement of both physical (C) and R & D (K) capital. Furthermore, because of confidentiality requirements, the data were provided only in moment-matrices form, which made it both impossible to control for outliers and errors and difficult to deal with the special econometric problems of panel data. In spite of these limitations, the results were very (and somewhat surprisingly) encouraging, yielding an elasticity of output with respect to R & D capital of about .06 in both the time-series and cross-section dimensions of the data.

A major goal of our work described in this paper, was to confirm these findings using a longer and more recent sample of firms, while paying more attention to the definition and measurement of the particular variables and to the difficulties of estimation and specification in panel data. In spite of these efforts, under close scrutiny our results are somewhat disappointing. This paper includes, therefore, two very different parts: section 17.2 documents the various estimates in detail, while section 17.3 attempts to rationalize and circumvent the problems that are evident in these estimates. First, however, we shall set the stage in this first section by explaining our data and our model. A more detailed description of the variables used and a summary of results using alternative versions of some of these variables can be found in the appendix.

17.1.2 The Data and Major Variables

We started with the information provided in Standard and Poor's Compustat Industrial Tape for 157 large companies which have been reporting their R & D expenditures regularly since 1963 and were not missing more than three years of data. Because of missing observations on employment and of questionable data on other variables, we first had to limit the sample to 133 firms (complete sample), and then, in response to merger problems, to restrict it further to 103 firms (restricted sample). The treatment of mergers has an impact on our estimates. These two overlapping samples are fully balanced over the twelve-year period, 1966–77.[2]

Our sample is quite heterogeneous, covering most R & D performing manufacturing industries and also including a few nonmanufacturing

---

2. We also consider two corresponding subsamples (96 firms and 71 firms) with no data missing for the entire eighteen-year (1960–77) period. We focus in this paper on the larger, shorter samples because of potential errors in our R & D measures in the earlier years. Most of the interpolation and doctoring of R & D expenditures (for missing observations or changes in definition) occurred in the years before 1966. Also, we had to estimate an initial R & D capital stock level in 1958 by making various and somewhat arbitrary assumptions whose impact vanishes by 1966.

firms (mainly in petroleum and nonferrous mining). Since the number of firms is too small to work with separate industries, we have dealt with the heterogeneity problem by dividing our sample into two groups: *scientific firms* (firms in the chemical, drug, computer, electronics, and instrument industries) and *other firms*.

The measurement of the variables raises many conceptual issues as well as practical difficulties. These problems have been discussed at some length in Griliches (1979, 1980a), and we shall only allude to the most important ones in our context. We think of the unobservable research capital stock (K) as a measure of the distributed lag effect of past R & D investments on productivity: $K_{it} = \Sigma_\tau w_\tau R_{i(t-\tau)}$, where R is a deflated measure of R & D, and the subscripts t, $(t - \tau)$, and i stand for current year, lagged year, and firm, respectively. Ideally, one would like to estimate the lag structure (w_τ) from the data, or at least an average rate of R & D obsolescence and the average time lag between R & D and productivity. Unfortunately, the data did not prove to be informative enough. Various constructed lag measures and different initial conditions made little difference to the final results. We focused, therefore, on one of the better and most sensible looking measures based on a constant rate of obsolescence of 15 percent per year and geometrically declining weights $w_\tau = (1 - \delta)^\tau$.

We measure output by deflated sales (Q) and labor (L) by the total number of employees. There is no information on value added or the number of hours worked in our data base. This raises, among other things, questions about the role of materials (especially energy in the recent period) and about the impact of fluctuations in labor and capacity utilization and the possibility that ignoring these issues may bias our results—see section 17.3 where we address these questions and the related question of returns to scale. Sales are deflated by the relevant (at the two- or three-digit SIC level) National Accounts price indexes.[3] We assume that intrasectoral differences in price movements reflect mostly quality changes in old products or the development of new products. Accordingly (and to the extent that this assumption holds), we are in principle studying here the effects of both process- and product-oriented R & D investments.

---

3. At least two problems arise in applying these price indexes to our data. First, our firms are diversified and a significant fraction of their output does not fall within the industry to which they have been assigned. Second, observations are based on the companies' *fiscal* years which often do not coincide with price index calendar years. Experiments performed to investigate these problems indicated that our conclusions are not affected by them. We used 1978 Business Segment data to produce weighted price indexes for about three-quarters of our sample, with the results changing only in the second decimal place. Similarly, a separate smoothing of the price indexes, to put them into fiscal year equivalents, has very little impact on the final results.

Finally, we have used gross plant adjusted for inflation as our measure of the physical capital stock (C). This variable (as in some of our previous studies) performs reasonable well; however, it tends to be collinear over time with the R & D capital stock K, especially for some sectors and subperiods. We have tried various ways of adjusting gross plant for inflation and have also experimented with age of capital and net capital stock measures. Since random errors of measurement are another issue, we made various attempts to deal with the errors in variables problem by going to three-year averages. All these experiments resulted in only minor perturbations to our estimates.

Table 17.1 provides general information on our samples and variables, while more detail is given in the appendix. Note the much more rapid productivity growth and the higher R & D intensiveness in the "scientific firms" subsample.

17.1.3 The Model and Stochastic Assumptions

Our model, which is common to most analyses of R & D contributions to productivity growth (see Griliches 1979, 1980b), is the simple extended Cobb-Douglas production function:

$$Q_{it} = Ae^{\lambda t}C_{it}^{\alpha}L_{it}^{\beta}K_{it}^{\gamma}e^{e_{it}},$$

or in log form:

$$q_{it} = a + \lambda t + \alpha c_{it} + \beta \ell_{it} + \gamma k_{it} + e_{it},$$

where (in addition to already defined symbols) e_{it} is the perturbation or error term in the equation; λ is the rate of disembodied technical change; α, β, and especially γ are the parameters (elasticities) of interest—in addition to the weights w_{τ} or the rate of obsolescence δ implicit in the construction of the R & D capital stock variable.

One could, of course, also consider more complicated functional forms, such as the CES or Translog functions. We felt, based on past experience and also on some exploratory computations, that this will not matter as far as our main purpose of estimating the output elasticities of R & D and physical capital (α and γ), or at least their relative importance (α/γ), is concerned. However, two related points are worth making.

First, an important implication of our model in the context of panel data is that in the cross-sectional dimension differences in levels explain differences in levels, while in the time dimension differences in growth rates explain differences in growth rates. An alternative model would allow γ to vary across firms and impose the equality of marginal products or rates of return across firms, $\partial Q/\partial K = \rho$, implying that the rate of growth in productivity depends on the intensity of R & D investment (rewriting $\gamma k = (\partial Q/\partial K)(K/Q)(\dot{K}/K) = \rho \dot{K}/Q = \rho(R - \delta K)/Q \approx \rho R/Q$ for

Table 17.1   Sample Composition and Size, R & D/Sales Ratio, and Labor Productivity Growth Rate[a]

SIC Industry Classification	Complete Sample			Restricted Sample		
	Number of Firms	R & D Sales (%)	Productivity Growth Rate (%)	Number of Firms	R & D Sales (%)	Productivity Growth Rate (%)
Scientific firms:						
28(−283)—chemicals	19	3.4	6.7	16	3.6	5.1
283—drugs	19	6.5	3.3	10	7.5	4.6
357—computers	10	5.3	7.8	6	5.3	8.0
36—electronic equipment	14	4.6	3.3	10	4.7	3.8
38—instruments	15	5.5	3.6	15	5.5	3.6
Subtotal	77	5.0	4.3	57	5.2	4.7
Other firms:						
29—oil	6	0.7	5.1	6	0.7	5.1
35(−357)—machinery	13	2.8	0.7	10	2.8	0.7
37—transportation equipment	8	2.2	1.8	8	2.2	1.8
Other manufacturing—mostly 20–32–33	20	2.3	0.2	17	2.3	0.8
Nonmanufacturing—mostly 10	9	2.0	−0.6	5	2.2	0.5
Subtotal	56	2.2	0.9	46	2.2	1.5
Total	133	3.8	2.9	103	3.8	3.3

[a]The restricted sample excludes firms with large jumps in the data, generally caused by known merger problems.

small δ). We have not pursued such an alternative here, but we may consider it again in future work.[4]

Second, we also have the choice of assuming constant returns to scale (CRS) in the Cobb-Douglas production function: $\alpha + \beta + \gamma + \mu = 1$, or not—which amounts to estimating the regression

$$(q_{it} - \ell_{it}) = a + \lambda t + \alpha(c_{it} - \ell_{it}) + \gamma(k_{it} - \ell_{it}) + (\mu - 1)\ell_{it} + e_{it},$$

with $(\mu - 1)$ left free or set equal to zero. In our data the constant returns to scale assumption is accepted in the cross-sectional dimension, but is rejected in the time dimension in favor of significantly decreasing returns to scale. Because of the large effects of this restriction on our estimates of γ, we shall report both the estimates obtained with and without imposing constant returns to scale.

A distinct issue, which may explain why not assuming constant returns to scale and freeing the coefficient of labor in the regressions causes a problem, is that of simultaneity. Actually, it seems to provide a better explanation of our results than left-out variables or errors of measurement. We have, therefore, estimated a two, semireduced form, equations model in which output and employment are determined simultaneously as functions of R & D and physical stocks, based on the assumption of short-run profit maximization and predetermined capital inputs. These estimates yield plausible estimates of the relative influence of R & D and physical capital on productivity in both the cross-sectional and time dimensions. We elaborate on this line of research in section 17.3.

These different specification issues are, of course, related to the assumptions made about the error term, e_{it}, in the production function. When working with panel data, it is usual to decompose the error term into two independent terms: $e_{it} = u_i + w_{it}$, where u_i is a permanent effect specific to the firm and w_{it} is a transitory effect. In our context u_i may correspond to permanent differences in managerial ability and economic environment, while w_{it} reflects short-run changes in capacity utilization rates, in addition to other sources of perturbation. The habitual and convenient way to abstract from the u_i's is to compute the *within-firm regression* using the deviations of the observations from their specific firm means: $(y_{it} - y_{i.})$, which is equivalent to including firm dummy variables in the regression using the original observation (y_{it}). While the way to eliminate the w_{it}'s (in a long enough sample) is to compute the *between-firm regression* using the firm means $(y_{i.})$. The least-squares estimates of the *total regression* are in fact matrix-weighted averages of the least-squares estimates of the within and between regressions. If most of the

4. An important practical advantage of this alternative approach is that by assuming $\delta = 0$ a priori it does not require the construction of an R & D capital stock. See Griliches (1973), Terleckyj (1974), and Griliches and Lichtenberg (this volume) for estimates based on this approach.

variability of the data is between firms rather than within, as is the case here, the total and between estimates will be very close.[5]

Another manner of viewing the decomposition of the overall error into permanent and transitory components, and of interpreting the between and the within estimates, is to consider them as providing cross-sectional and time-series estimates, respectively. Both estimates will be consistent and similar if the u_i's and the w_{it}'s are uncorrelated with the explanatory variables. Very often, however, the two are rather different, implying some sort of specification error. This is, unfortunately, our case. Following the early work of Mundlak (1961) and Hoch (1962), the general tendency is to hold the u_i's responsible for the correlations with the explanatory variables and to assume that the within estimates are the better, less biased ones.[6] This leads to the discarding of the information contained in the variability between firms, which is predominant (at least in our samples), relying thereby only on the variability within firms over time, which is much smaller and more sensitive to errors of measurement. In fact, there are also good reasons for correlations of the w_{it}'s with the explanatory variables and, therefore, putting somewhat more faith in the between estimates. These reasons have been sketched in Mairesse (1978); they will be considered further in section 17.3 when we discuss the potential influence of misspecifications on our results.

17.2 Overall and Detailed Estimates

17.2.1 First Look at Results

Our first results were based on the complete sample of 133 firms for the 1966–77 period and various variants of our variables, especially R & D capital. Although the use of different measures had little effect, disappointing our hope of learning much about the lag structure from these data, the actual estimates looked reasonably good even if far apart in the cross-section and time dimensions. Table 17.2 gives the total, between, and within estimates (and also the within estimates with year dummies instead of a time trend), using our main variants for output, labor, and physical and research capital, both with and without the assumption of constant returns to scale. The total estimates of the elasticities of physical and R & D capital (α and γ) are about .30 and .06, respectively, similar to Griliches's (1980a) previous estimates. The more purely cross-sectional between estimates are nearly identical to the total estimates, .32 and .07, respectively. This follows from the fact that most of the relevant variability in our sample is between firms (about 90 percent, see table 17.A.1 in

---

5. An independent year effect $v_t (e_{it} = u_i + v_t + w_{it})$ can also be taken into account by adding year dummies instead of a time trend to the regression.

6. The model is then equivalent to the so-called fixed effects model.

Table 17.2   Production Function Estimates (complete sample, 133 firms, 1966–77)

Total Regressions

α	γ	$(\mu - 1)$	λ	R^2	MSE
0.319	—	—	0.012	0.499	0.099
(0.009)			(0.002)		
0.310	0.073	—	0.011	0.514	0.097
(0.008)	(0.011)		(0.002)		
0.332	0.054	−0.032	0.011	0.524	0.094
(0.009)	(0.011)	(0.005)	(0.002)		

Between Regressions

α	γ	$(\mu - 1)$	R^2	MSE
0.324	—	—	0.522	0.079
(0.027)				
0.317	0.072	—	0.538	0.077
(0.027)	(0.034)			
0.341	0.053	−0.033	0.551	0.075
(0.029)	(0.035)	(0.017)		

Within Regressions

α	γ	$(\mu - 1)$	λ	R^2	MSE
0.232	—	—	0.017	0.402	0.0211
(0.017)			(0.001)		
0.160	0.150	—	0.018	0.422	0.0204
(0.020)	(0.020)		(0.001)		
0.150	0.080	−0.126	0.025	0.437	0.0199
(0.019)	(0.022)	(0.019)	(0.002)		

Within Regressions with Year Dummies

α	γ	$(\mu - 1)$	R^2	MSE
0.250	—	—	0.420	0.0206
(0.017)				
0.176	0.158	—	0.442	0.0198
(0.019)	(0.020)			
0.163	0.091	−0.121	0.455	0.0194
(0.019)	(0.022)	(0.020)		

the appendix). The time-series within estimates are, however, rather different: α being about .15 and γ about .15 or .08 depending on whether constant returns to scale are imposed or not. It is also clear that using separate year dummy variables instead of a linear trend makes little difference.

Unfortunately, these first results did not improve with further analysis; on the contrary. While the measurement of variables (within the range of our experimentation) does not really matter, trying to allow for sectoral and period differences and cleaning the sample of observations contaminated by mergers sharply degraded our within estimates of the R & D capital elasticity γ. The pattern of results already evident in table 17.2 is much amplified, especially in the time dimension: a tendency of the estimated γ's to be substantial, whenever the estimated α's seem too low; and a tendency for them to diminish or even to collapse when constant returns to scale are not imposed. We shall now document these different problems in detail before considering their possible causes and solutions.

17.2.2 Alternative Variable Definitions and Sectoral Differences

One of the original aims of this study was to experiment with various ways of defining and measuring physical and R & D capital. Using all the information available to us, we tried a number of different ways of measuring these variables but to little effect. The resulting differences in our estimates, even when they were "statistically significant," were nonetheless quite small and not very meaningful. In particular, they did not alter the order of magnitude of our two parameters of interest, α and γ. The various measures we tried turned out to be very good substitutes for each other and the choice between them had little practical import. Our final choices were based, therefore, primarily on a priori considerations, external evidence, and convenience. The appendix describes these choices and some of our experiments.

Since our sample consisted of R & D performing firms in rather diverse industries, it was also of interest to investigate the influence of sectoral (industrial) differences. Table 17.3 gives our main estimates separately for firms in research-intensive industries (so-called scientific firms) and the rest of our sample.

Dividing the sample into two allows for much of the heterogeneity, bringing down the sum of square of errors (SSE) by about 20 percent for the total regressions and 10 percent for the within regressions (with the division corresponding to very high F ratios of about 100 and 70, respectively). The two groups are indeed a priori very distinct: as a matter of fact, the average rate of productivity growth is about four times higher for the scientific firms, while the average R & D to sales ratio is about twice as high (see table 17.1).

In spite of this sharp contrast, the differences in our estimates are not

Table 17.3 Production Function Estimates Separately for the Scientific and Other Firms (complete sample, 77 and 56 firms, respectively, 1966–77)

	Total Regressions						Within Regressions					
	α	γ	$(\mu-1)$	λ	R^2	MSE	α	γ	$(\mu-1)$	λ	R^2	MSE
Scientific firms	0.243	—	—	0.030	0.423	0.088	0.194	—	—	0.033	0.607	0.0170
	(0.012)			(0.003)			(0.020)			(0.002)		
	0.203	0.223	—	0.025	0.570	0.066	0.150	0.111	—	0.032	0.615	0.0167
	(0.011)	(0.013)		(0.003)			(0.022)	(0.026)		(0.007)		
	0.250	0.185	-0.051	0.026	0.604	0.061	0.140	0.021	-0.200	0.044	0.653	0.0151
	(0.011)	(0.013)	(0.006)	(0.002)			(0.021)	(0.026)	(0.020)	(0.002)		
Other firms	0.364	—	—	-0.008	0.609	0.093	0.243	—	—	-0.001	0.172	0.0202
	(0.011)			(0.003)			(0.028)			(0.002)		
	0.365	-0.007	—	-0.008	0.609	0.093	0.169	0.124	—	0.001	0.196	0.0196
	(0.012)	(0.018)		(0.004)			(0.032)	(0.028)		(0.002)		
	0.351	0.010	0.025	-0.008	0.614	0.092	0.133	-0.015	-0.207	0.011	0.223	0.0190
	(0.013)	(0.019)	(0.009)	(0.003)			(0.032)	(0.039)	(0.043)	(0.003)		

that large, except for the estimated time-trend coefficients (rates of technical progress λ). The within estimates of α and γ (and also μ) are, in fact, quite comparable, only the fit being much lower in the "other firms" equation. Yet the total estimates of γ are very large in the scientific firms and insignificant for the other firms. Part of this discrepancy can be accounted for by the higher estimates of α in the other firms group.

Disaggregating to the industrial level decreases the total and within sums of square of errors by another 20 percent or so. The main effect is, however, to worsen the collinearity between R & D and physical capital in the within dimension. Some of the within estimates actually fall apart: two extreme cases being the computer industry with an estimated α of -0.06 and an estimated γ of 0.50, and the instruments industry with an estimated α of 0.49 and an estimated γ of -0.32. Without a larger sample, we do not really have the option of working at the detailed industrial level.[7]

17.2.3 Differences between Subperiods

Current discussions of "the productivity slowdown" suggest that some of it may be due not only to "the slowdown in R & D," but also to a significant decrease in the efficiency of recent R & D investments (Griliches 1980b); hence, our interest in whether we could find any evidence of a decrease in the R & D capital elasticity γ over time. Table 17.4 shows what happens (for the scientific firms group) when we divide our data into two six-year subperiods, 1966–71 and 1972–77.[8] Table 17.5 explores the resulting differences further by presenting the within estimates for the two subperiods (as well as the overall period and "between subperiods") and comparing the estimated γ when α and λ are constrained to .25 and .025, respectively. Table 17.5 also lists the rates of growth of the main variables, their within standard deviations, and the decomposition of their within variability for the subperiods (the overall period and "between subperiods").

As might be expected, the total estimates differ only slightly, while the within estimates change a lot. Yet the striking feature is not a decrease in the estimated γ but rather in $\hat{\alpha}$. The decomposition of variance shows, however, that by breaking down our data into two subperiods we keep only about half of the within variability in the overall period (the other half being between subperiods.) Our capital stock variables as well as the time variable itself are slowly changing, trendlike variables, and there is not enough variability in them to allow us to estimate all of their coef-

7. An intermediate step, without going fully to the sectoral level, is to allow for separate sectoral time trends and intercepts. While the total and within estimates change only slightly for the scientific firms, the total estimates of γ and α for the other group move up and down respectively, making them less different from those of the scientific group.

8. We also looked at the preceding six-year subperiod (1960–65) for our longer but smaller subsample of firms. The estimates are very similar to those for 1966–71.

Table 17.4   Production Function Estimates for Two Subperiods: 1966–71 and 1972–77 (scientific firms, complete sample, 77 firms)

Periods	Total Regressions						Within Regressions					
	α	γ	$(\mu-1)$	λ	R^2	MSE	α	γ	$(\mu-1)$	λ	R^2	MSE
1966–71	0.219	—	—	0.013	0.264	0.103	0.250	—	—	0.011	0.307	0.0115
	(0.018)			(0.009)			(0.029)			(0.004)		
	0.169	0.241	—	0.007	0.463	0.076	0.106	0.250	—	0.013	0.380	0.0103
	(0.016)	(0.019)		(0.007)			(0.033)	(0.034)		(0.004)		
	0.235	0.189	−0.068	0.008	0.528	0.067	0.113	0.040	−0.307	0.041	0.501	0.0083
	(0.017)	(0.019)	(0.009)	(0.007)			(0.030)	(0.036)	(0.029)	(0.004)		
1972–77	0.273	—	—	0.033	0.434	0.071	0.083	—	—	0.043	0.459	0.0080
	(0.016)			(0.007)			(0.028)			(0.003)		
	0.242	0.207	—	0.029	0.578	0.053	−0.012	0.225	—	0.041	0.486	0.0076
	(0.014)	(0.017)		(0.006)			(0.034)	(0.047)		(0.003)		
	0.269	0.183	−0.032	0.028	0.594	0.051	−0.023	0.175	−0.076	0.044	0.488	0.0076
	(0.015)	(0.017)	(0.008)	(0.006)			(0.034)	(0.057)	(0.050)	(0.004)		

Table 17.5   Analysis of Subperiod Differences (scientific firms, complete sample, 77 firms)

Periods	Within Degrees of Freedom	Rates of Growth, (within standard deviations), [% within variability]				Within Regressions ($\mu = 1$)					
		$q - \ell$	$c - \ell$	$k - \ell$	ℓ	α	γ	λ	γ ($\alpha = 0.25$)	λ ($\alpha = 0.25$)	γ ($\alpha = 0.25$) ($\lambda = 0.025$)
1966–77	847	4.3 (0.22) [100.0]	6.2 (0.33) [100.0]	3.0 (0.22) [100.0]	4.6 (0.28) [100.0]	0.15	0.11	0.032	0.06	0.027	0.08
1966–71	385	3.3 (0.14) [19.0]	9.1 (0.25) [27.6]	4.5 (0.20) [38.3]	6.3 (0.24) [33.4]	0.11	0.25	0.013	0.17	0.003	0.09
1972–77	385	5.1 (0.13) [17.0]	4.3 (0.19) [15.1]	2.4 (0.13) [15.8]	2.4 (0.13) [10.2]	−0.01	0.23	0.041	0.02	0.034	0.08
Between subperiods	77	4.7 (0.58) [64.0]	6.2 (0.81) [57.3]	2.9 (0.50) [45.9]	4.2 (0.69) [56.4]	0.25	−0.02	0.032	−0.02	0.032	0.07

ficients separately and precisely. What we get are relatively wide gyrations in the estimated coefficients α, γ, and λ, with some of them going down as the others go up. If we impose a reasonable a priori value of $\alpha = .25$, which corresponds to estimating the impact of R & D capital on total factor productivity, (TFP), we do indeed get a large decline in γ, from .17 in the first period to effectively zero in the second. However, this decline is associated with a correspondingly large increase in λ, from .003 to .034. Since such an acceleration in "disembodied" technological change goes against all other pieces of information available to us, we reestimate again, imposing also an a priori $\lambda = .025$. With this new restriction everything falls into place: $\hat{\gamma}$ being estimated at approximately .08 for both subperiods (as well as between subperiods and for the overall period).

This, of course, does not mean that we have strong evidence that γ is about .08, but only that one should not interpret the data as implying a major decline in γ over time. What the data tell us is that one cannot tell and that there is not enough independent variation in the subperiods to estimate the contribution of physical capital, R & D capital, and trend separately. If, however, we are willing to impose a priori, reasonable values on α and λ, then the implied $\hat{\gamma}$ is both reasonable and stable. Moreover, the imposition of such constraints is not inconsistent with the data; while they are not "statistically" accepted given our relatively large sample size, the actual absolute deterioration in fit is rather small, the standard deviation of residuals changing by less than .01.[9]

This may not be all that surprising considering the other major fact that emerges from table 17.5: our "scientific firms" did not actually experience a productivity slowdown in 1972–77 relative to 1966–71 (as against the experience of manufacturing as a whole). There was a slowdown in the growth of both physical and R & D capital, but this was associated with an *acceleration* in labor productivity growth and, hence, also in total factor productivity growth. (The latter rises from about 0.6 percent in the first period to about 3.8 percent in the second.)[10] Given these facts, it is not surprising that correlation of productivity growth with capital input growth tends to vanish, leading to a collapse of the estimated α and γ. These strange events are not limited to the firms in our sample, they also actually happened in the science-based industries as a whole, as can be

---

9. Our estimated regression standard errors are about .1 in the within dimension, implying that we explain annual fluctuations in productivity up to an error whose standard deviation is about 10 percent. Imposing the a priori values of α and λ increases this error by less than one additional percent.

10. This is computed from the average yearly rates of growth given in table 17.5, using .65, .25, and .1 as relative weights for labor, physical capital, and R & D capital, respectively.

seen by examining the aggregate data collected by NSF and the BLS.[11] (Average TFP growth in "scientific" industries increases in these data from about .8 percent in 1966–71 to 3.2 percent in 1972–77.) If anything, the puzzle is why there was so little "exogenous" productivity growth in 1966–71. One possible answer would invoke errors of measurement in the dating of physical and R & D investments (longer lag structures); another might be based on different cyclical positions of the endpoints of these two periods. In any case, since there is no evidence that there has been a significant productivity slowdown in R & D intensive industries, it is unlikely that whatever slowdown did occur could be attributed to the slowdown in R & D growth.[12]

17.2.4 The Problem of Mergers

Starting from our original sample of 157 firms, we first eliminated 24, primarily because of missing observations (in the number of employees generally and in gross plant occasionally) or obvious large errors in the reported numbers. In the case of one or two missing observations we "interpolated" them. In some instances we managed to go back to the original source and obtain the missing figure or correct an error. Fortunately, most firms did not present such difficulties, and the construction of our "complete sample" was straightforward enough. We were still left with the important issue of mergers. About one firm out of five in our "complete" sample (as many as twenty among the seventy-seven "scientific" firms) appeared to be affected (at least for one year over the 1966–77 period) by considerable and generally simultaneous "jumps" (80 percent or more year-to-year increases) in gross plant, number of employees, and sales. We have been able to check and convince ourselves that most of these jumps do, in fact, result from mergers, although some may be the result of very rapid growth. Since the problem was of such a magnitude (as is bound to be the case in a panel of large companies over a number of years), we had to be careful about it.

---

11. The data are taken from sources given in Griliches (1980b). The numbers that correspond to those of table 17.5 are:

Scientific Industries Aggregate: Based on NSF and BLS Statistics
(Average yearly rates of growth)

Subperiods	$q - \ell$	$c - \ell$	$k - \ell$	ℓ
1960–65	4.3	2.0	8.2	2.8
1966–71	3.3	7.4	6.3	0.9
1972–77	3.8	2.0	0.6	2.3

Although the definitions and measures are quite different, and although our firms are much faster growing than the scientific industries as a whole, the growth patterns are very similar.

12. For possible contrary evidence, see Scherer (1981) who emphasizes the impact of R & D on productivity growth in the R & D *using* rather than R & D *doing* industries.

One way of dealing with this problem is simply to drop the offending firms. This results in what we have called the "restricted" sample. An alternative is to create an "intermediate" sample in which a firm before and after a major merger is considered to be two different "firms." If mergers were occurring precisely in a given year, we would have as many observations in the intermediate sample as in the complete one (and more "firms" but some of them over shorter periods), and we would eliminate only the "variability" corresponding to the "jumps." In fact, we lost a few observations because some mergers affect our data for more than one year (primarily because we chose gross plant at the beginning of the year as our measure of capital for the current year) or because they occur in the first or last years of the study period (since we decided not to have "firms" with less than three years of data in the intermediate sample). Estimates for the restricted sample and its complement, the "merger" sample, are given in table 17.6 for the scientific firms group. (Estimates for the other group behave similarly, although there were fewer mergers there.) Table 17.7 provides more detail, showing separately the results for the complete, intermediate, and restricted samples and decomposing the merger group into "jump" and "no-jump" periods. To facilitate interpretation, it also presents estimates of γ based on constraining α to .25 and λ to .025, and it lists the rate of growth, the standard deviations and the variance decomposition of the main variables.[13]

The total estimates (reported in table 17.6) manifest their usual stoutness, remaining practically unchanged whatever the sample. The within estimates are, on the contrary, very sensitive, and the estimated γ collapses, declining from 0.11 to 0.05 and -0.03 in the complete, intermediate, and restricted samples, respectively (even when constant returns to scale are imposed). It is clear from table 17.7 that the merger firms are responsible for the difference. They correspond to a major part of the within variability of our variables (much of it being from the "jumps"). Moreover, they seem to account for the significant, positive within estimates of γ in our complete sample, especially through their "no-jumps"

13. The variance decomposition of a variable y for a firm i going through a merger at the end of year t_0 is identical to its decomposition into the two subperiods before and after the merger, the "jump" component corresponding to the between subperiods component. It can be written

$$\sum_{t=1}^{T} (y_{it} - y_{i\cdot})^2 = \sum_{t=1}^{t_0} (y_{it} - y_{i\cdot}^{(1)})^2 + \sum_{t=t_0+1}^{T} (y_{it} - y_{i\cdot}^{(2)})^2$$
$$+ t_{\delta}(y_{i\cdot}^{(1)} - y_{i\cdot})^2 + (T - t_0)(y_{i\cdot}^{(2)} - y_{i\cdot})^2,$$

where $y_{i\cdot}, y_{i\cdot}^{(1)}$, and $y_{i\cdot}^{(2)}$ are the respective means of y_{it} over the whole period $(1, T)$, the before merger period $(1, t_0)$, and the after merger period $(t_0 + 1, T)$. The practical way to run the regressions corresponding to the jump component is simply to substitute $(y_{i\cdot}^{(1)} - y_{i\cdot})$ and $(y_{i\cdot}^{(2)} - y_{i\cdot})$ for $(y_{it} - y_{i\cdot})$ in the before and after merger years.

Table 17.6   Separate Production Function Estimates for the Restricted and the Merger Samples (57 and 20 firms respectively, scientific firms, 1966–77)

Samples	Total Regressions						Within Regressions					
	α	γ	$(\mu-1)$	λ	R^2	MSE	α	γ	$(\mu-1)$	λ	R^2	MSE
Restricted	0.264	—	—	0.032	0.510	0.075	0.221	—	—	0.035	0.737	0.010
	(0.012)			(0.003)			(0.025)			(0.002)		
	0.230	0.210	—	0.028	0.645	0.054	0.239	−0.034	—	0.035	0.737	0.010
	(0.011)	(0.013)		(0.003)			(0.030)	(0.028)		(0.002)		
	0.278	0.170	−0.048	0.028	0.671	0.050	0.211	−0.062	−0.112	0.041	0.745	0.010
	(0.012)	(0.014)	(0.006)	(0.003)			(0.030)	(0.028)	(0.025)	(0.002)		
Merger	0.204	—	—	0.021	0.235	0.117	0.200	—	—	0.021	0.379	0.034
	(0.032)			(0.007)			(0.036)			(0.004)		
	0.146	0.292	—	0.017	0.462	0.093	0.	0.270	—	0.020	0.437	0.031
	(0.028)	(0.029)		(0.005)			(0.038)	(0.055)		(0.004)		
	0.171	0.265	−0.064	0.020	0.524	0.073	0.114	0.135	−0.229	0.042	0.506	0.027
	(0.027)	(0.028)	(0.011)	(0.006)			(0.036)	(0.057)	(0.040)	(0.005)		

Table 17.7 **Analysis of Merger Differences (scientific firms, 1966–77)**

Samples	Within Degrees of Freedom	Rates of Growth, (within standard deviations), [% within variables]				Within Regressions ($\mu - 1$)					
		$q - \ell$	$c - \ell$	$k - \ell$	ℓ	α	γ	λ	γ ($\alpha = 0.25$)	λ ($\alpha = 0.25$)	γ ($\alpha = 0.25$) ($\lambda = 0.025$)
Complete (1)	847	4.3 (0.22) [100.0]	6.2 (0.33) [100.0]	3.0 (0.22) [100.0]	4.6 (0.28) [100.0]	0.15	0.11	0.032	0.06	0.027	0.08
Intermediate (2)	783	4.7 (0.21) [82.5]	5.8 (0.26) [58.6]	3.7 (0.21) [79.6]	3.2 (0.20) [49.2]	0.79	0.05	0.035	0.01	0.033	0.09
Restricted (3)	627	4.7 (0.21) [67.3]	6.1 (0.27) [50.4]	3.4 (0.22) [63.6]	3.2 (0.21) [42.6]	0.24	−0.03	0.035	−0.04	0.045	0.05
Merger (4) = (1) − (3)	220	3.3 (0.24) [32.7]	6.5 (0.45) [49.6]	2.0 (0.26) [36.4]	8.6 (0.41) [57.4]	0.12	0.27	0.020	0.18	0.012	0.11
"Jump" (5) = (1) − (2)	64	0.0 (0.33) [17.5]	11.5 (0.76) [41.4]	−4.8 (0.36) [20.4]	21.9 (0.72) [50.8]	0.20	0.11	0.011	0.07	0.006	0.02
"No-Jump" (6) = (2) − (3)	156	4.7 (0.20) [15.2]	4.5 (0.22) [8.2]	4.9 (0.21) [16.2]	3.1 (0.17) [6.6]	−0.18	0.65	0.019	0.30	0.017	0.23

component. In other words, R & D seems most effective for firms growing rapidly through mergers, and both phenomena (mergers and R & D growth) are apparently related.

Merger firms have higher R & D than physical capital growth rates during their nonmerger ("no-jumps") periods, while the opposite is true for nonmerging ("restricted") firms. The labor productivity growth rates are about equal for both, but they are much more closely related to R & D growth for the merger firms. Actually, not enough variability is left to estimate the separate contributions of the two capital terms and the time trend term precisely. If one imposes $\alpha = .25$ and $\lambda = .025$ a priori, one gets back from the restricted subsample a reasonable though still low estimate of $\hat{\gamma} = .05$. The intermediate sample, however, is the most relevant one from our point of view, yielding a much higher $\hat{\gamma} = .09$, which can be interpreted as a weighted average of about .2 for the merger firms and .05 for the rest.[14]

Such a finding raises questions that deserve additional analysis: Who are these "merger" firms and why would their R & D investment be more successful? What kind of selectivity is at work here? How does one expand this type of analysis to allow for different R & D-related success rates by different firms? A random coefficient model does not, at first thought, appear to be the most appropriate way to go. Unfortunately, given the small size of our sample, we cannot pursue these questions further here.

Our tentative conclusion is that we should *not* exclude the merger firms from our sample entirely. These are firms whose R & D has apparently been very effective. Throwing them out would seriously bias our estimates of the contribution of R & D to productivity downward.

17.3 Misspecification Biases or an Exercise in Rationalization

17.3.1 Three Possible Sources of Bias

Our within estimates of the production function are unsatisfactory in the sense that they attribute unreasonably low coefficients to the physical and research capital variables and imply that most of our firms are handicapped by severely diminishing returns to scale. The simplest explanation is to impute these "bad results" to a major misspecification of our model. The trouble is that when we start thinking about possible misspecifications, many come to mind. The most important appear to be: (1) the omission of labor and capital intensity of utilization variables, such as hours of work per employee and hours of operation per machine; (2) the use of gross output or sales rather than value added or, alterna-

---

14. Here also the imposition of the a priori values of $\alpha = .25$ and $\lambda = .025$ does not result in an economically meaningful deterioration of fit.

tively, the omission of materials from the list of included factors; (3) overlooking the jointness (simultaneity) in the determination of employment and output.[15]

These three misspecifications are similar in the sense that they all imply the failure of the ordinary least-squares assumption of no correlation between the included factors, c, l, k, and the disturbance e in the production function, resulting in biases in our estimates of the elasticities of these factors (and in our estimate of the elasticity of scale). In all three cases the correlation of the disturbance e with the labor variable ℓ is likely to be relatively high in the time dimension, affecting especially our within estimates.

If we consider the "auxiliary" regression connecting e to c, ℓ, k:

$$E(e) = b_{ec \cdot \ell k} c + b_{e\ell \cdot ck\ell} + b_{ek \cdot c\ell} k$$

(where we suppress for simplicity the constant and trend terms by taking deviations of the variables from the appropriate means, i.e., respectively, $[y_{it} - y_{.t}]$ and $[y_{it} - y_{.t} - y_{i.} + y_{..}]$ for the total and within regressions), the specification biases in our estimates can be written in the following general form:

$$E(\hat{\alpha} - \alpha) = \text{bias}\,\hat{\alpha} = b_{ec \cdot \ell k},$$

$$E(\hat{\beta} - \beta) = \text{bias}\,\hat{\beta} = b_{e\ell \cdot ck},$$

$$E(\hat{\gamma} - \gamma) = \text{bias}\,\hat{\gamma} = b_{ek \cdot c\ell}.$$

If we assume more specifically that the physical and research capital variables c and k are predetermined and that only the labor variable is correlated with e, we can go one step further and formulate the biases in α and γ as proportional to the bias in β (see Griliches and Ringstad 1971, appendix C):

$$\text{bias}\,\hat{\alpha} = -(\text{bias}\,\hat{\beta})\,b_{\ell c \cdot k},$$

$$\text{bias}\,\hat{\gamma} = -(\text{bias}\,\hat{\beta})\,b_{\ell k \cdot c}.$$

There is no good reason why the coefficients $b_{\ell c \cdot k}$ and $b_{\ell k \cdot c}$ should be both small, or one much smaller than the other, or very different for the within and total estimates. One will expect them to be positive and less than one, but large enough to result in a significant transmission of an upward bias in $\hat{\beta}$ into downward biases in both $\hat{\alpha}$ and $\hat{\gamma}$. One would also expect the absolute biases in $\hat{\alpha}$ and $\hat{\gamma}$ to be of the same order of magnitude

---

15. Three other possible misspecifications are the following: (4) ignoring the possibility of random errors in our measures of labor and capital; (5) assuming wrongly that firms operate in competitive markets; and (6) ignoring the peculiar selectivity of our sample. We shall allude briefly to (4) and (5) in what follows, but continue to ignore the selectivity issue, postponing the investigation into this topic to a later study based on a much larger post-1972 sample.

and, therefore, to have a much larger relative effect on $\hat{\gamma}$ than on $\hat{\alpha}$ (assuming that the true γ is small relatively to the true α). For example, a bias of -0.1 might reduce $\hat{\alpha}$ from a true .3 to .2 but could wipe out $\hat{\gamma}$ if its true value were .1.

We can actually estimate such bias transmission coefficients in our sample. They are relatively large and of comparable magnitude, on the order of .3 to .4.[16]

To the extent that the correlation between labor and the disturbance in the production function is the main problem, we are left with the evaluation of the bias in labor elasticity and the question of whether we can ascertain the "within" bias to be positive and sizeable in contrast to a small "total" bias. This is much more difficult, and we have to consider specifically our three possible misspecifications. We shall say a few words about the first two and then concentrate on the simultaneity issue. This issue seems most important, and we have been able to progress further toward its solution by considering a simultaneous equations model composed of the production function and a labor demand function, and by estimating what we call the semireduced form equations for this model.

Consider first the omission of the hours worked per worker variable h (or machine hours operated per machine) and let the "true" model be:

$$q = \alpha c + \beta(\ell + h) + \gamma k + \epsilon,$$

where labor is measured by the total number of hours of work.

The disturbance in the estimated model is then $e = \epsilon + \beta h$, and we get for the labor elasticity bias: bias $(\hat{\beta}) = b_{e\ell \cdot ck} = \beta b_{h\ell \cdot ck}$. Cross-sectionally, hours per worker h should be roughly uncorrelated with any of the included variables c, ℓ, and k and, hence, cause no bias in the between regression or in the total regression (which is similar since the between variances of the variables dominate their total variance). In the time dimension, however, short-run fluctuations in demand (say a business expansion) will be met partly by modifying employment (hiring) and partly by changing hours of work (increase in overtime). Hence, $b_{h\ell \cdot ck}$ should be positive and rather large (perhaps .5 or higher), and therefore the within estimate of $\hat{\beta}$ should be biased upward and substantially so (perhaps by $.6 \times .5 = 0.3$). Considering then that the within correlations of h with c and k are likely to be negligible, we have seen that a significant

---

16. The auxilary regression of ℓ on c and k giving these coefficients is precisely what we shall call our semireduced form labor equation; tables 17.8, 17.9, and 17.10 provide their exact values for our various samples. Note that since the order of magnitude of the sum of these coefficients is less than one, we cannot explain the downward biases in $\hat{\alpha}$ and $\hat{\gamma}$ and also in the returns to scale $\hat{\mu}$ solely by the transmission of an upward bias in $\hat{\beta}$. Our second misspecification example, the omission of materials, does not assume that c and k are predetermined and hence that the biases are only caused by the correlation of ℓ and e; it provides, as we shall see below, a rationalization of the decreasing returns to scale estimates in the within dimension.

downward bias should be transmitted to the within estimates of $\hat{\alpha}$ and $\hat{\gamma}$ (about $-.3 \times .4$ or $-.3 \times .3 \approx -.1$).

The same type of analysis applies to the exclusion of materials as a factor in the production function (or to not using value added but gross output or sales to measure production). The total estimates of $\hat{\alpha}$, $\hat{\beta}$, and $\hat{\gamma}$ should all move up roughly in proportion to the elasticity of materials δ [by $1/(1 - \delta)$], while the within estimates $\hat{\alpha}$, $\hat{\beta}$, and $\hat{\gamma}$ will be raised in lesser proportions, with the plausible result of a negligible bias in the total and a large downward bias in the within estimates of the scale elasticity.

This time let the "true" model be:

$$q = \alpha c + \beta \ell + \gamma k + \delta m + \epsilon$$

(i.e., a generalized Cobb-Douglas production function where materials come in as another factor). Estimating a gross output equation ignoring m assumes implicitly that materials are used in fixed proportion to output. This may be a belief about the technical characteristics of the production processes (the form of the production function) or the consequence of assuming that materials are purchased optimally and that their price relative to the price of output remains roughly constant over firms and over time. In any case, omitting m where it should be included means that the error in the estimated model is $e = \epsilon + \delta m$, resulting in the following biases for our estimates:

$$\text{bias}\,\hat{\alpha} = \delta b_{mc \cdot \ell k}, \text{bias}\,(\hat{\beta}) = \delta b_{m\ell \cdot ck}, \text{bias}\,(\hat{\gamma}) = \delta b_{mk \cdot c\ell}.$$

Across firms, in the between dimension, it is quite likely that the sum of the auxiliary regression coefficients b's will not depart far from unity, so that the sum of estimates $\hat{\alpha} + \hat{\beta} + \hat{\gamma}$ will approach the relevant true scale elasticity $\mu = \alpha + \beta + \gamma + \delta$. If the proportionality assumption of q and m holds well enough, then the b's would be more or less proportional to the corresponding elasticities and the relative biases roughly the same:

$$\hat{\alpha} = \alpha/(1 - \delta), \hat{\beta} = \beta/(1 - \delta), \hat{\gamma} = \gamma/(1 - \delta).$$

Over time, however, it is more likely that material usage may change less than proportionally, since it will respond incompletely or with lags to short-run output fluctuations. Hence, the sum of the b's might be much less than one in the within dimension, causing the misleading appearance of decreasing returns to scale. As a plausible example, we can take

$$b_{mc \cdot \ell k} = b_{mk \cdot \ell c} = 0, \text{ and } b_{m\ell \cdot ck} = .5,$$

and if the true coefficients are $\alpha = .15$, $\beta = .3, \gamma = .05$ and $\delta = .5$ ($\mu - 1 = 0$), we get the following within estimates when m is omitted:

$$\hat{\alpha} = .15, \hat{\beta} = .55, \hat{\gamma} = .05, \text{and}\,\hat{\mu} - 1 = -.25.$$

Turning to the problem of simultaneity and assuming that firms try to maximize their profits in the short run, given their stocks of physical and R & D capital, the true model will consist of a production function and a labor demand function:

$$q = \alpha c + \beta \ell + \gamma k + e,$$

$$q = \ell + w + v,$$

where w is the real price of labor, and v is a random optimization error. We can assume that the errors in the two equations (e and v) are independent or, more generally, that they are of the following form: $(e + f)$ and $(v + f)$, where e and f are respectively the parts of the disturbance in the production function transmitted and not transmitted to the labor variable. The OLS bias in $\hat{\beta}$ can be written as

$$E(\beta - \hat{\beta}) = b_{e\ell \cdot ck} = (1 - \beta)R,$$

where

$$R = \sigma_e^2/[\sigma_e^2 + \sigma_w^2(1 - r_{w \cdot ck}^2) + \sigma_v^2]$$

is the ratio of the random transmitted variance in the production function to the sum of itself and the independent variance in the labor equation. Thus, to get some notion about the value of R and the bias in $\hat{\beta}$, we need to discuss the potential sources of variation in e, v, and w.

Schematically, we can think of the disturbance in the production function as consisting of: (1) long-term differences in factor productivity between firms; (2) short-run shifts in demand which are being met (partly) by changes in (unmeasured) utilization of labor and capital; and (3) errors of measurement in the deflators of output, errors arising from the use of gross rather than net output concepts, and errors arising from the use of sales rather than output concepts. Only items (1) and (2) matter as far as the formulas are concerned since (3) (errors of measurement) are not really transmitted to labor. Moreover, only (1) matters in the cross-sectional (between) dimension under the assumption that (2) cancels out over time, while only (2) matters in the time (within) dimension.

Similarly, the independent variation in the labor equation can be partitioned into: (4) the independent variation in real wage and (5) other short-run deviations from the profit-maximizing level of employment because of implicit contracts, shortages, or mistaken expectations. It is probably the case that most of the factor price variation to which firms respond is either permanent and cross-sectional or is common to all firms in the time dimension and hence is captured by the time dummies or trend coefficients. Thus, we anticipate that (4) manifests itself largely in the between dimension while (5) is all that is left in the within dimension.

On the basis of the estimated variances and covariances of the residuals for the semireduced form equations to be discussed below, we can give the following illustrative orders of magnitude (for $\beta \sim .6$):

$$\sigma^2_{(1)} = \sigma^{2(B)}_e = .004, \quad \sigma^2_{(2)} = \sigma^{2(W)}_e = .002,$$

$$\sigma^2_{(3)} = \sigma^{2(B)}_f + \sigma^{2(W)}_f = .04 + .008,$$

$$\sigma^2_{(4)} = \sigma^{2(B)}_{w \cdot ck} = .04, \quad \sigma^2_{(5)} = \sigma^{2W}_v = .002.\text{[17]}$$

The R would equal $(.004/.044) \sim .10$ in the between dimension and $(.002/.004) \sim .50$ in the within dimension. With a true β of .6, the OLS between and within estimates $\hat{\beta}$ would be respectively biased upward by about .04 and .20.

17.3.2 The Semireduced Form Estimates

If one takes the simultaneity story seriously, it is not surprising that the OLS within estimates of the production function are unreasonable. We should be estimating a complete simultaneous equations system instead. We cannot do that, unfortunately, lacking information on factor prices. But we can estimate semireduced form equations (i.e., reduced form equations omitting factor price variables) which may allow us to infer the relative size of our two parameters of interest α and γ.

Let the true production function be (ignoring constants, time trends, or year dummies)

$$q = \alpha c + \beta \ell + \gamma k + \delta m + e,$$

where both c and k are assumed to be predetermined and independent of e, while q, ℓ, and m are endogenous, jointly dependent variables. Short-run profit maximization in competitive markets implies:

$$q - \ell = w + v, q - m = p + \epsilon,$$

where w and p are the real prices of labor and of materials, respectively, and v and ϵ are the associated optimization errors. Solving for q, ℓ, and m yields:

$$q = \frac{1}{1 - \beta - \delta} [\alpha c + \gamma k + e - \beta(w + v) - \delta(p + \epsilon)],$$

---

17. The variances of the residual e' and v' in our semireduced form production and labor equations are respectively:

$$[\sigma^2_e + \beta^2(\sigma^2_w + \sigma^2_v)]/(1 - \beta)^2 + \sigma^2_f, \text{ and } (\sigma^2_e + \sigma^2_w + \sigma^2_v)/(1 - \beta)^2,$$

while the covariance is $[\sigma^2_e + \beta(\sigma^2_w + \sigma^2_v)]/(1 - \beta)^2$. For a given β, we can thus derive estimated values of σ^2_e, $(\sigma^2_w + \sigma^2_v)$, and σ^2_f. However, these values are extremely sensitive to the value of β chosen and to small differences in the variances and covariance of the semireduced form equations residuals.

$$\ell = \frac{1}{1 - \beta - \delta} [\alpha c + \gamma k + e - (1 - \delta)(w + v) - \delta(p + \epsilon)],$$

$$m = \frac{1}{1 - \beta - \delta} [\alpha c + \gamma k + e - \beta(w + v) - (1 - \beta)(p + \epsilon)].$$

Since materials and factor prices are unobserved in our data, we have to drop the last equation and lump w and p with the other error components in these equations. We are thus left with two semireduced form equations for output and labor. Coming back for the sake of coherence to our previous notations of the production function with m solved out $[\alpha = \alpha/(1 - \delta), \ldots, e = e - \delta(p + \epsilon)/(1 - \delta)]$, we can rewrite these two equations more simply:

$$q = \frac{1}{1 - \beta} (\alpha c + \gamma k) + e',$$

$$\ell = \frac{1}{1 - \beta} (\alpha c + \gamma k) + v',$$

(where $e' = [e - \beta(w + v)]/(1 - \beta)[e - \beta(w + v)]/(1 - \beta)$ and $v' = [e - (w + v)]/(1 - \beta)[e - (w + v)]/(1 - \beta)$.

The semireduced form equation should provide unbiased estimates of $\alpha/(1 - \beta)$ and $\gamma/(1 - \beta)$ to the extent that factor prices w and p are more or less uncorrelated with the capital variables c and k. This condition seems quite plausible in the within dimension. There is little independent variance left in w and p in the within dimension after one takes out their common time-series components with time dummies or a trend variable. In the between dimension, however, one would expect that w and p might vary across firms and be positively correlated with c and k, leading to downward biases in $\alpha/(1 - \beta)$ and $\gamma/(1 - \beta)$ in both equations (and more so in the labor equation).

Tables 17.8, 17.9, and 17.10 present estimates of such semireduced form equations comparable to the production function estimates reported in the earlier tables 17.2–17.7: total and within estimates for all firms and for scientific and other firms separately; for the two subperiods 1966–71 and 1972–77 (and between these two subperiods); for the restricted and merger samples (and the merger-no-jump sample). Since the "theory" of the semireduced form equations implies that corresponding coefficients should be the same in the two equations, we also present the constrained system (SUR) estimates.

A first look at the results shows that they are in the right ball park. They are not very strikingly different in the two dimensions, and most remarkably, the within estimates of the research capital coefficient are quite significant and rather large. Also, the corresponding estimates in the two equations are rather close. Given the large number of degrees of freedom, all differences are "statistically" significant, but constraining the

Table 17.8   Semireduced Form Equations Estimates (complete sample, 1966–77)

Different Regressions		Total Regressions			Within Regressions		
		$\alpha/(1-\beta)$	$\gamma/(1-\beta)$	System R^2	$\alpha/(1-\beta)$	$\gamma/(1-\beta)$	System R^2
All firms ($N = 133$)	Output	.574 (.010)	.296 (.014)	.873	.407 (.022)	.265 (.027)	.559
	Labor	.415 (.013)	.416 (.017)		.400 (.021)	.288 (.026)	
	Constrained	.554 (.010)	.311 (.014)	.857	.403 (.019)	.278 (.024)	.558
Scientific firms ($N = 77$)	Output	.488 (.013)	.378 (.017)	.910	.321 (.025)	.291 (.031)	.711
	Labor	.464 (.019)	.375 (.024)		.283 (.025)	.423 (.030)	
	Constrained	.490 (.013)	.378 (.017)	.909	.301 (.023)	.395 (.028)	.706
Other firms ($N = 36$)	Output	.544 (.018)	.380 (.024)	.860	.510 (.037)	.067 (.052)	.340
	Labor	.290 (.021)	.558 (.029)		.559 (.036)	.122 (.051)	
	Constrained	.506 (.018)	.407 (.024)	.802	.536 (.033)	.096 (.041)	.337

Table 17.9 Semireduced Form Equations Estimates for Subperiods: 1966–71 and 1972–77 and between Subperiods (scientific firms, complete sample)

Different Regressions	Total Regressions			Within Regressions		
	$\alpha/(1 - \beta)$	$\gamma/(1 - \beta)$	System R^2	$\alpha/(1 - \beta)$	$\gamma/(1 - \beta)$	System R^2
Subperiod 1966–71						
Output	0.480 (0.019)	0.363 (0.025)	0.902	0.350 (0.036)	0.164 (0.047)	0.582
Labor	0.482 (0.027)	0.341 (0.035)		0.437 (0.043)	0.230 (0.057)	
Constrained	0.480 (0.019)	0.363 (0.025)	0.902	0.371 (0.035)	0.180 (0.046)	0.571
Subperiod 1972–77						
Output	0.500 (0.018)	0.394 (0.023)	0.917	0.060 (0.046)	0.622 (0.071)	0.418
Labor	0.447 (0.026)	0.408 (0.033)		0.107 (0.040)	0.579 (0.062)	
Constrained	0.506 (0.018)	0.392 (0.022)	0.915	0.093 (0.039)	0.592 (0.059)	0.417
Between subperiods						
Output				0.413 (0.022)	0.264 (0.024)	
Labor				0.259 (0.020)	0.464 (0.022)	0.830
Constrained				0.320 (0.019)	0.385 (0.027)	0.822

Table 17.10 Semireduced Form Equations Estimates for the Restricted, Merger, and Merger No Jump Samples (scientific firms, 1966–77)

Different Regressions		Total Regressions			Within Regressions		
		$\alpha/(1-\beta)$	$\gamma/(1-\beta)$	System R^2	$\alpha/(1-\beta)$	$\gamma/(1-\beta)$	System R^2
Restricted Sample	Output	0.521 (0.014)	0.343 (0.019)		0.500 (0.038)	0.146 (0.037)	.730
	Labor	0.481 (0.022)	0.343 (0.029)	.923	0.392 (0.035)	0.281 (0.034)	
	Constrained	0.527 (0.013)	0.343 (0.017)	.921	0.433 (0.033)	0.230 (0.032)	.725
Merger Firms	Output	0.402 (0.028)	0.484 (0.031)	.896	0.208 (0.042)	0.434 (0.059)	.714
	Labor	0.461 (0.038)	0.438 (0.042)		0.179 (0.045)	0.572 (0.063)	
	Constrained	0.407 (0.028)	0.480 (0.031)	.895	0.196 (0.038)	0.492 (0.053)	.709
Merger: no-jumps sample	Output	0.460 (0.028)	0.414 (0.032)	.925	−0.117 (0.083)	0.652 (0.106)	.519
	Labor	0.521 (0.039)	0.355 (0.045)		0.178 (0.077)	0.372 (0.098)	
	Constrained	0.468 (0.027)	0.405 (0.031)	.924	0.049 (0.066)	0.495 (0.085)	.504

coefficients to be equal in the two equations results in a negligible loss of fit, changing the systemwide R^2 only in the third (or second) decimal place.

A more careful examination confirms, more or less, our previous production function findings. The estimates for the two, scientific and other firms, are close, given the collinearity between c and k, which causes the much lower within estimate of $\gamma/(1 - \beta)$ for the other firms group to be largely counterbalanced by the higher estimates of $\alpha/(1 - \beta)$. The estimates for the two subperiods are also quite comparable, since the higher within estimates of $\gamma(1 - \beta)$ for 1972–77 can be explained, similarly, by the lower estimate of $\alpha/(1 - \beta)$. Also, the merger firms do not seem to behave as differently as it appeared earlier. The within estimates of $\gamma/(1 - \beta)$ for the nonmerger firms are significant, and the discrepancy between the estimates for the two types of firms may also be a result of the collinearity between c and k.

The remaining difficulty with our semireduced form estimates is their absolute size. It is different from our a priori expectations. If the true coefficients of the production function were $\alpha = .15$, $\beta = .3$, $\gamma = .05$, and $\delta = .5$, or in value-added terms $\alpha = .3$, $\beta = .6$, and $\gamma = .1$, the semireduced form coefficients should be about .75 and .25, respectively. The estimated physical capital coefficients should be about .75 and .25, respectively. The estimated physical capital coefficient is much smaller, being about .5 at best, while the estimated R & D coefficient is of the expected order of magnitude but often higher. Although the total and within estimates do not differ too strikingly, it should be noted that the estimated sum $(\alpha + \gamma)/(1 - \beta)$ is about .8 or .9 cross-sectionally and about .5 to .7 in the time dimension. This is quite similar to what happened to our production function returns to scale estimates.

We can think of two possible explanations for these shortfalls: (1) errors in variables, and (2) failure of the perfect competition assumption.

To the extent that errors in measurement are random over time (which is a difficult position to maintain for *stock* variables), their effects can be mitigated by averaging and by trying to increase the signal-to-noise ratio in the affected variables. The between subperiods estimates given in table 17.9 represent an attempt to accomplish this by using differences between two six-year subperiod averages. It is clear from this attempt (and from others not reported here) that averaging does not solve the problem of the absolute magnitude of our estimates. Either our solution for the errors of measurement is not effective (because the errors are correlated over time) or the problem is caused by something else entirely.

The perfect competition assumption is especially dubious for our large firms and short-run context. To explore the consequences of such a misspecification, we have to expand our model by adding a demand equation:

$$q_{it} = \alpha_i + z_t + \eta \, p_{it} + \phi k_{it} + \epsilon,$$

where α_i is a permanent firm demand level variable, z_t is a common industry demand shifter, η is the relative price elasticity of demand (where the price of the firm's products p_{it} is measured relative to the overall price level in the industry), and ϕ is the direct effect of R & D capital on the demand for the firm's products.

Given this model, we reinterpret our output variable as sales (which it really is), make price endogenous, and use the demand equation to solve it out of the system. This yields comparable semireduced form equations, but the coefficients are now

$$\frac{\alpha\left(1 + \dfrac{1}{\eta}\right)}{1 - \beta\left(1 + \dfrac{1}{\eta}\right)} \quad \text{and} \quad \frac{\gamma\left(1 + \dfrac{1}{\eta} - \dfrac{\phi}{\eta}\right)}{1 - \beta\left(1 + \dfrac{1}{\eta}\right)}$$

for physical and research capital, respectively. With $\eta < 0$, the research capital coefficient is seen to be a combination of both its production and demand function shifting effects.

The introduction of the $(1 + 1/\eta)$ terms into these coefficients provides an explanation for the "shortfall" in our estimates. Assuming $\eta = -4$ (i.e., if a firm lowers the relative price of its product by 25 percent, it would double its market share) and $\alpha = .3$, $\beta = .6$, $\gamma = .1$, and $\phi = .1$, implies .4 and .18 as the respective coefficients in the semireduced forms. That is not too far off and the assumptions are plausible enough, but that is about all that we can say. We shall need more data and more evidence from other implications of such a model before we can put much faith in this interpretation of our results.

17.4 Summary and Conclusions

We have analyzed the relationship between output, employment, and physical and R & D capital for a sample of 133 large U.S. firms covering the years 1966 through 1977. In the cross-sectional dimension, there is a strong relationship between firm productivity and the level of its R & D investments. In the time dimension, using deviations from firm means as observations and unconstrained estimation, this relationship comes close to vanishing. This may be due, in part to the increase in collinearity between the trend, physical capital, and R & D capital in the within dimension. There is little independent variability left there. When the coefficients of the first two variables are constrained to reasonable values, the R & D coefficient is both sizeable and significant. Another reason for these difficulties may be the simultaneity of output and employment decisions in the short run. Allowing for such a simultaneity yields rather high estimates of the importance of R & D capital relative to physical capital. Our data do not enable us, however, to answer any detailed questions about the lag structure of the effects of R & D on productivity. These effects are apparently highly variable, both in timing and magnitude.

Appendix
Variables and Additional Results

In this appendix we present more information on our sample and summarize the results of various additional computational experiments.

Table 17.A.1 lists means, standard deviations, and growth rates for our major variables, and indicates that most of the observed variance in the data (90 + percent) is between firms, rather than within firms and across time. It also underscores the fact that these firms are rather large, with an average of more than 10,000 employees per firm.

Table 17.A.2 compares our main measure of physical capital stock C to four alternatives: C', CA, CN, and CD. C is gross plant adjusted for inflation, which we assume to be proportional to a proper capital service flow measure. Since our adjustment for inflation is based on a rough first-order approximation, assuming a fixed service life, a linear depreciation pattern, and an estimate of the age of capital (AA) from reported depreciation levels, we also tried different variants of it.[18] C' is one of them in which we assume the same average service life for plant and equipment of sixteen years for all our firms. The fit is somewhat improved, but the changes in the estimates are only minor. Actually, using the reported gross plant figure without any adjustment does not make that much difference either. CA is our C measure taken at the end of the year instead of the beginning of the year. The fit is slightly improved, and the within estimates of α are increased a little. This could indicate that end of the year measures are appropriate but may also reflect a simultaneity bias arising from the contemporaneous feedback of changes in production on investment. CN and CD are net plant and depreciation adjusted for inflation, respectively. CN can be advocated on the grounds that in some sense it allows for obsolescence and embodied technical progress, and CD on the grounds that it is nearer in principle to a service

18. To be precise C_t is computed as reported gross plant $\times P(72)/P(t - AA_t)$, where P is the GNP price deflator for fixed investment and AA_t (the average age of gross plant) is computed as reported gross plant minus reported net plant (i.e., accumulated depreciation) divided by an estimate of the average service life LL_t. LL_t itself is computed as the five-year moving average of reported gross plant/reported depreciation.

Table 17.A.1   Characteristics of Variables, Complete Sample (133 firms)[a]

Main Variables[b]	Scientific Firms (77)		Percent Variability			Other Firms (56)		Percent Variability		
	Geometric Mean	Standard Deviation	Between	Within	Rate of Growth (%)	Geometric Mean	Standard Deviation	Between	Within	Rate of Growth (%)
Q Deflated sales	297.0	1.66	95.1	4.9	8.9	442.8	1.74	97.9	2.1	3.9
L Number of employees	10.4	1.63	97.4	2.6	4.6	12.5	1.52	97.6	2.4	2.9
C Gross plant adjusted for inflation	188.4	2.12	95.3	4.7	10.8	295.7	2.11	97.3	2.7	8.4
K R & D capital stock computed using a 0.15 rate of obsolescence	58.1	1.64	95.7	4.3	7.6	39.6	1.53	82.3	17.7	4.4
Q/L Deflated sales per employee	28.7	0.39	71.6	28.4	4.3	35.3	0.49	89.8	10.2	0.9
C/L Gross plant adjusted per employee	18.1	0.85	86.6	13.4	6.2	23.6	1.05	93.2	6.8	5.4
K/L R & D capital stock measure per employee	5.6	0.70	90.6	9.4	3.0	3.2	0.67	87.5	12.5	1.5

[a]Standard deviations and the decomposition of the variance are given for the logarithms of the variables.

[b]Deflated sales, gross plant adjusted, and R & D capital stock are in 10^6 and constant 1972 prices. Number of employees is in 10^3 persons.

Table 17.A.2   **Production Function Estimates for Different Measures of Physical Capital Stock and Output, All Sectors, Complete Sample (133 firms), 1966–77 (annual and three-year averages)**

Different Regressions[a]	Total Regressions			Within Regressions		
	α	γ	MSE	α	γ	MSE
C	0.310	0.073	0.097	0.160	0.150	0.0204
	0.332	0.054	0.095	0.150	0.080	0.0199
C'	0.323	0.070	0.095	0.180	0.142	0.0202
	0.350	0.048	0.092	0.173	0.069	0.0197
CA	0.322	0.074	0.095	0.201	0.156	0.0201
	0.344	0.054	0.092	0.186	0.101	0.0197
CN	0.304	0.076	0.096	0.124	0.184	0.0204
	0.325	0.050	0.094	0.114	0.115	0.0199
CD	0.361	0.062	0.099	0.194	0.163	0.0202
	0.383	0.044	0.097	0.189	0.086	0.0196
QC	0.305	0.073	0.100	0.102	0.127	0.0229
	0.325	0.055	0.098	0.093	0.060	0.0224
Three-year averages	0.313	0.074	0.091	0.195	0.154	0.0153
	0.336	0.055	0.090	0.187	0.092	0.0149

[a]Constant returns to scale are imposed for estimates reported in the first line of each cell but not in the second.

flow measure. CN results in a small decrease of the within estimate of α and a corresponding increase in γ, while CD results in an increase in both total and within estimates of α with no noticeable effect on γ. We have also run regressions including an age of capital variable, AA. While our estimates of α and γ are not affected by its inclusion, this variable in conjunction with our gross capital measure C (but not so in conjunction with the net capital measure CN) is clearly significant both in the cross-sectional and time dimensions, tending to indicate a rate of embodied technical progress of 5.5 percent per year (see Mairesse 1978).

Table 17.A.2 also gives the estimates obtained with an alternative measure of deflated sales, QC, tentatively corrected for inventory change. The correction, however, is problematic since it is based on all inventories and not just finished products. In any case, QC performs much worse both in terms of fit and in terms of the order of magnitude of the within estimates. Finally, we also list estimates based on three-year averages of the observations. While errors of measurement appear to be a priori an important issue (if they were random and uncorrelated, going to averages should reduce the resulting biases), the changes are not striking and the discrepancy between total and within estimates remains. Yet there is a sizeable increase (about 20 percent) in the within estimate of α, which might reflect an error in the capital-labor ratio accounting for about 30 percent of the observed "within" variance in this ratio.

Because we did not want to give up hope of gaining some evidence on the lag structure of R & D effects, we experimented with a large number of R & D capital stock measures, but mostly in vain. Table 17.A.3 compares K, the measure we finally settled on based on a 15 percent depreciation rate, to six rather different alternatives. $K00$ and $K30$ are computed similarly to K but assuming 0 and 30 percent per year obsolescence rates instead. K' and $K'00$ differ from K and $K00$ respectively in assuming that R & D vintages older than eight years are completely obsolete. Since information on R & D is available only from 1958 (i.e., for eight years before 1966), this is also a way to test our initial condition assumption. In the K and $K00$ measures, the 1958 R & D capital levels are based on extrapolating R & D expenditures back to 1948, using the 1958–63 individual firm R & D growth rate shrunk toward the overall industry rate. KP is also a summation of past R & D expenditures over eight years but with a very different peaked lag structure: $w_{-1} = w_{-8} = 0.05$, $w_{-2} = w_{-7} = 0.10$, $w_{-3} = w_{-6} = 0.15$, and $w_{-4} = w_{-5} = 0.20$. Finally, K, P_{-34}, P_{-56}, P_{-78}, P_{-9+} is one of the free-lag version experiments we have attempted. The P variables are the following proportion of past R & D expenditures (over two years plus the tail) to total cumulated expenditures (with a .15 rate of obsolescence):

$$(R_{-3} + R_{-4})/K, \ (R_{-5} + R_{-6})/K, \ (R_{-7} + R_{-8})/K,$$

$$(R_{-9} + R_{-10} + \ldots)/K.$$

Table 17.A.3   **Production Function Estimates Based on Different Measures of R & D Capital, Complete Sample (133 firms), 1966–77**

Alternative R & D Capital Measures[a]	Total Regressions			Within Regressions		
	α	γ	MSE	α	γ	MSE
K	0.310	0.073	0.097	0.160	0.150	0.0204
	0.332	0.054	0.095	0.150	0.080	0.0199
K'	0.311	0.075	0.096	0.173	0.119	0.0206
	0.333	0.057	0.094	0.153	0.064	0.0199
$K00$	0.309	0.059	0.098	0.152	0.172	0.0202
	0.334	0.040	0.095	0.154	0.081	0.0199
$K'00$	0.311	0.070	0.097	0.178	0.106	0.0207
	0.333	0.051	0.095	0.158	0.050	0.0200
$K30$	0.311	0.079	0.096	0.167	0.137	0.0204
	0.332	0.061	0.094	0.147	0.084	0.0198
KP	0.311	0.065	0.097	0.195	0.070	0.0209
	0.334	0.046	0.095	0.165	0.027	0.0200
K and	0.318	0.070	0.094	0.149	0.205	0.0197
P_{-34}, P_{-56}, P_{-78}, P_{-9+}	0.340	0.051	0.092	0.152	0.120	0.0196

[a]First line regressions assume constant returns to scale, second line regressions do not.

Hence, the coefficients of the P's should give an indication of how far the respective true weights are from the assumed declining weights in K: 1, .85, .72, .61, .52, . . . , etc.

As was the case for the different physical capital measures, the total estimates are almost unaffected by all this experimentation, while the within estimates are more sensitive. The initial conditions seem to matter very slightly, showing some influence of a truncation remainder or tail effect. The within regressions with the K and $K00$ measures perform a little better in terms of fit than those with the corresponding K' and $K'00$ measures (which assume no effective R & D before 1958), and the estimated γ is a bit higher. The assumption about the order of magnitude of the rate of obsolescence δ is even less important. Still, there is some tenuous evidence here for a rather rapidly declining lag structure. The KP measure (which assumes a peaked lag structure) has the lowest fit and the lowest within γ, while the "free lag" version in the neighborhood of the K measure performs best on both grounds. The estimated P coefficients (within) are:

$$P_{-34}: -0.35, \quad P_{-56}: -0.17, \quad P_{-78}: -0.10, \quad P_{-9+}: 0.05,$$
$$(0.09) \qquad\qquad (0.07) \qquad\qquad (0.07) \qquad\qquad (0.02)$$

implying that around lag 3 and 4 the weight of past R & D is about .22 rather than .57, around lag 5 and 6 it is .24 rather than .41, around lag 7 and 8 it is .20 rather than .30, and around lag 11 it is .22 rather than .17. That is, there is a reasonably strong immediate effect in the first two years which then drops sharply and stays constant through most of the rest of the observable range.

References

Griliches, Z. 1973. Research expenditures and growth accounting. In *Science and technology in economic growth*, ed. B. R. Williams, 59–83. London: Macmillan.

———. 1979. Issues in assessing the contribution of research and development to productivity growth. *Bell Journal of Economics* 10, no. 1:92–116.

———. 1980a. Returns to research and development expenditures in the private sector. In *New developments in productivity measurement and analysis*, ed. J. Kendrick and B. Vaccara, 419–61. *Conference on Research in Income and Wealth: Studies in Income and Wealth*, vol. 44. Chicago: University of Chicago Press for the National Bureau of Economic Research.

————. 1980b. R & D and the productivity slowdown. *American Economic Review, Proceedings Issue* 70, no. 2:343–48.

Griliches, Z., and V. Ringstad. 1971. *Economics of scale and the form of the production function.* Amsterdam: North-Holland.

Hoch, I. 1962. Estimation of production function parameters combining time-series and cross-section data. *Econometrica* 30:34–53.

Mairesse, J. 1978. New estimates of embodied and disembodied technical progress. *Annales de l'INSEE* 30–31:681–719.

Mundlak, Y. 1961. Empirical production function free of management bias. *Journal of Farm Economics* 43:44–56.

Nadiri, M. I., and G. C. Bitros. 1980. Research and development expenditures and labor productivity at the firm level: A dynamic model. In *New developments in productivity measurement and analysis*, ed. J. Kendrick and B. Vaccara, 387–412. *Conference on Research in Income and Wealth: Studies in Income and Wealth*, vol. 44. Chicago: University of Chicago Press for the National Bureau of Economic Research.

Scherer, F. M. 1981. Interindustry technology flows and productivity growth. Working paper, Northwestern University.

Terleckyj, N. E. 1974. *Effects of R & D on the productivity growth of industries: An exploratory study.* Washington, D.C.: National Planning Association.

18   Productivity and R & D at the Firm Level in French Manufacturing

Philippe Cuneo and Jacques Mairesse

18.1 Introduction

Following Griliches and Mairesse's study for the United States (this volume), we use a similar analysis to assess whether a significant relationship exists between R & D expenditures and productivity performance at the firm level in French manufacturing. Our purpose is twofold: to check the results obtained by Griliches and Mairesse on their U.S. sample of firms doing R & D against a comparable sample of French firms and to set the stage for a careful comparison of industrial productivity growth in the two countries (Griliches and Mairesse 1983). The framework and data used are basically the same as in the U.S. study. We have, however, the advantage of being able to use value added which may be a more appropriate measure of production than sales. Moreover, having detailed information on R & D expenditures permits us to correct the measures of physical capital, labor, and production for the double counting or expensing out of R & D labor, capital, or materials. One important drawback of our study is the shorter period, 1972–77 as compared to 1966–77 in the U.S. study.

On the whole, our main findings are quite close to the results obtained by Griliches and Mairesse. We come up with similar discrepancies be-

Philippe Cuneo is with the Institut National de la Statistique et des Etudes Economiques. Jacques Mairesse is a professor at Ecole des Hautes Etudes en Sciences Sociales and Ecole de la Statistique et de l'Administration Economique, and a research affiliate of the National Bureau of Economic Research.

The authors are grateful to Zvi Griliches for his interest and advice. Thanks are due also to Alan Siu for his help, and to the Ministry of Industry and Research, and INSEE for providing the various data bases used. Financial support from the Naional Science Foundation (PRA79–13740) and the Centre National de la Recherche Scientifique (ATP–070199) is gratefully acknowledged.

375

tween the total and within-firm estimates of the two parameters of main interest: the elasticities of physical and R & D capital stocks, α and γ, based on differences across firms and changes over time, respectively. However, because of better measures of the variables, the problem is much less serious than it could have been, and on the whole our estimates are statistically significant and of a likely order of magnitude.

We describe our framework and data and present our main results in section 18.2. We document and discuss the changes in our estimates because of our improved measures of variables in section 18.3. In both sections we systematically refer to Griliches and Mairesse and stress the various comparative aspects of the two studies.

18.2 Framework, Data, and Main Results

18.2.1 Framework

Our basic model, as in Griliches and Mairesse, is the simple extended Cobb-Douglas production function, which can be written in logarithmic form as

$$v_{it} = a + \lambda t + \alpha c_{it} + \beta \ell_{it} + \gamma k_{it} + e_{it},$$

or

$$(v_{it} - \ell_{it}) = a + \lambda t + \alpha(c_{it} - \ell_{it}) + \gamma(k_{it} - \ell_{it})$$
$$+ (\mu - 1)\ell_{it} + e_{it}.$$

The subscripts i, t refer to the firm i and the current year t; e is the error term in the equation; v, c, ℓ, and k stand for production (value added), physical capital, labor, and R & D capital, respectively; α, β, and γ are the parameters (elasticities) of interest; $\mu = \alpha + \beta + \gamma$ is the coefficient of returns to scale; and λ is the rate of disembodied technical change.

We follow the common practice in analyses of panel data by assuming that the error term e_{it} is composed of two components: a permanent effect specific to the firm u_i and a transitory effect w_{it}. Such a decomposition generates two types of estimates, which can be viewed as providing cross-sectional and time-series estimates, respectively: the between-firm estimates based on the firm means $y_{i\cdot}$ and the within-firm estimates based on the deviations of the observations from the firm means $(y_{it} - y_{i\cdot})$. The between-firm estimates are not affected by the biases from possible correlations between the explanatory variables with the w_{it}'s (at least in a long enough sample), while the within-firm estimates are not affected by correlations with the u_i's. Both estimates should be consistent under the assumption of uncorrelated errors, while significant differences between them imply some sort of model misspecification. The least-squares estimates based on the original observations y_{it}, the total estimates, differ very little from the between-firm estimates, since most of the variability in our data comes from the between-firm dimension rather than from the within-firm dimension. Therefore, as in the U.S. paper, we shall report only on the total estimates and the within-firm estimates, and not on the between-firm ones.

18.2.2 Data

Our sample is based primarily on the match of two different data sources: INSEE provided us with balance sheet and current account figures (from the SUSE files), while the Ministry of Industry and Research (DGRST and STISI) provided the R & D information (from the annual survey on company R & D expenditures). The size of the sample is larger than that of the U.S. sample: 182 firms against 133 for the complete U.S. sample, or 103 for the U.S. sample restricted to nonmerger firms. The study period, however, is much shorter: 1972–77, compared with 1966–77 for the U.S. samples.

Like the U.S. sample, ours is very heterogeneous. This led us to divide it into two subsamples: so-called *scientific firms* belonging to the R & D intensive industries (chemicals, drugs, electronics, and electrical equipment) and *other firms* belonging to the other manufacturing industries.

Our variables are defined and measured on a basis similar to Griliches and Mairesse; however, we have taken advantage of the additional information we had on materials and on the components of R & D expenditures. We measure production by deflated value added (V) rather than by deflated sales. We also correct our value-added variable by adding back the materials consumption component of R & D expenditures, which is normally expensed out in current accounts. Labor (L) is measured by the number of employees, physical capital stock (C) by gross plant adjusted for inflation, and R & D capital stock (K) by the weighted sum of past R & D expenditures using a constant rate of obsolescence of 15 percent per year. Both our labor and physical capital stock variables are corrected for the double-counting of R & D already included in the R & D capital stock variable. The available number of R & D employees is thus simply subtracted from the total number of employees, while the part of physical capital stock used in R & D is computed on the basis of the average ratio of the physical investment component of R & D expenditures to total R & D expenditures and is likewise subtracted.

Detailed information on the sample and the variables is given in appendix A (see in particular appendix tables 18.A.1 and 18.A.2 which are comparable to the corresponding tables in the U.S. study). The much more rapid productivity growth and higher R & D intensiveness of the scientific firms subsample (than for the other firms subsample) are remarkable in both countries. Since our study period is shorter, the within-firm variability is even a smaller proportion of the total variability (about 1 percent for levels, and 5–10 percent for ratios) than is the case in the U.S. sample. Note also that the French firms are much smaller in size than their U.S. counterparts: an average of 1500 employees in French firms as against 10,000 employees in U.S. firms.

18.2.3 Main Results

Our main results are presented in table 18.1, again in a format comparable to the U.S. study: total and within-firm estimates of the production function with and without R & D capital stock, assuming or not assuming constant returns to scale, for all firms and for the scientific and other firms separately.

The total estimates are quite satisfactory on the whole. The elasticity of physical capital (α) is perhaps somewhat too low but still of a likely order of magnitude: about .20. In contrast, the elasticity of R & D capital (γ) may be too high: about .20 for the scientific firms and .10 for the other firms. The returns to scale are not significantly different from unity. As could be expected from the average rates of productivity growth over our study period, the rate of disembodied technical change is quite high (3 percent) for the scientific firms, while it is actually negative (minus 2 percent) for the other firms.

The within-firm estimates tend to differ from the total ones, although not as much as in the U.S. study. When assuming constant returns to scale, both types of estimates are actually quite close, the only significant discrepancy being the higher within-firm estimate of α for the other firms. However, when we relax this assumption, just as in the U.S. case, we obtain lower estimates of α and γ with rather implausible decreasing returns to scale estimates.

18.3 Further Results

18.3.1 Value Added versus Sales

The use of gross output or sales (S) instead of value added, or, alternatively, the omission of materials (M) among the factors in the production function is one of the possible misspecifications and sources of bias stressed by Griliches and Mairesse. With our data we are able to check whether this makes a real difference. Table 18.2 gives the results of such comparisons for the scientific and other firms separately. The estimates on the first three lines are comparable to those in table 18.1, except that we use sales instead of value added to measure output. In the estimates on the fourth line, materials are included as another factor of the Cobb-Douglas production function (with an elasticity δ).

The Griliches and Mairesse conjectures are verified by and large. The

Table 18.1 Production Function Estimates for All Firms, and Scientific and Other Firms Separately

	Total Regressions						Within-Firm Regressions					
	α	γ	$\mu-1$	λ	R^2	MSE	α	γ	$\mu-1$	λ	R^2	MSE
All firms (182)	.182 (.018)	—	—	.019 (.007)	.105	.1326	.392 (.042)	—	—	.007 (.004)	.165	.0219
	.213 (.013)	.209 (.007)	—	.005 (.005)	.512	.0724	.333 (.046)	.114 (.035)	—	.003 (.004)	.172	.0217
	.220 (.013)	.203 (.007)	−.018 (.006)	.005 (.005)	.516	.0719	.249 (.050)	.050 (.039)	−.228 (.060)	.015 (.005)	.183	.0215
Scientific firms (98)	.273 (.020)	—	—	.039 (.008)	.287	.1106	.351 (.050)	—	—	.034 (.005)	.348	.0201
	.237 (.017)	.206 (.014)	—	.030 (.007)	.489	.0794	.232 (.055)	.229 (.048)	—	.029 (.005)	.372	.0194
	.237 (.017)	.206 (.014)	−.001 (.009)	.030 (.007)	.489	.0794	.152 (.060)	.144 (.054)	−.243 (.074)	.042 (.006)	.383	.0190
Other firms (84)	.134 (.020)	—	—	−.008 (.006)	.086	.0568	.412 (.067)	—	—	−.023 (.005)	.071	.0196
	.175 (.018)	.116 (.010)	—	−.019 (.006)	.288	.0443	.371 (.071)	.079 (.047)	—	−.027 (.005)	.076	.0195
	.189 (.019)	.114 (.010)	−.018 (.007)	−.018 (.006)	.297	.0439	.268 (.080)	.027 (.050)	−.245 (.090)	−.015 (.007)	.090	.0193

Table 18.2 Production Function Estimates Using Sales instead of Value Added and Excluding or Including Materials, Scientific and Other Firms Separately

	Total Regressions							Within-Firm Regressions						
	α	γ	δ	$\mu-1$	λ	R^2	MSE	α	γ	δ	$\mu-1$	λ	R^2	MSE
Scientific firms (98)	.391 (.026)	—	—	—	.015 (.011)	.293	.1854	.446 (.041)	—	—	—	.011 (.004)	.342	.0137
	.355 (.024)	.199 (.019)	—	—	.006 (.010)	.404	.1565	.341 (.045)	.200 (.040)	—	—	.007 (.004)	.369	.0131
	.368 (.024)	.176 (.019)	—	−.061 (.012)	.007 (.009)	.430	.1501	.181 (.047)	.028 (.043)	—	−.491 (.058)	−.032 (.005)	.437	.0117
	.123 (.012)	.143 (.009)	.535 (.012)	−.021 (.006)	.014 (.004)	.877	.0324	.128 (.034)	.083 (.031)	.404 (.017)	−.160 (.044)	.021 (.003)	.710	.0061
Other firms (84)	.263 (.030)	—	—	—	−.012 (.009)	.131	.1257	.341 (.060)	—	—	—	−.016 (.005)	.062	.0157
	.291 (.030)	.093 (.016)	—	—	−.019 (.009)	.187	.1179	.248 (.063)	.177 (.041)	—	—	−.022 (.005)	.096	.0152
	.289 (.031)	.093 (.016)	—	.002 (.012)	−.019 (.009)	.187	.1181	.038 (.068)	.071 (.043)	—	−.497 (.077)	.000 (.006)	.166	.0140
	.137 (.016)	.058 (.008)	.486 (.013)	−.036 (.006)	−.006 (.005)	.797	.0296	.133 (.042)	.052 (.027)	.474 (.017)	−.158 (.049)	−.005 (.004)	.682	.0050

total estimates using sales and omitting materials do not differ much from those obtained with value added: the elasticity of R & D capital γ is practically unaffected, and returns to scale remain constant; however, the elasticity of physical capital α tends to be significantly higher. When materials are included, we find a plausible total estimate of the elasticity of materials δ of .5, while the estimate of the elasticities of physical and R & D capital α and γ are multiplied approximately by a factor of $(1 - \delta)$ $\sim$ .5 as expected. The within-firm estimates with sales instead of value added also are similar when we impose constant returns to scale. However, if we do not, large discrepancies occur; we get even more sharply decreasing returns to scale (.5 instead of .75), while the estimate of γ collapses for the scientific firms (.03) and also that of α for the other firms (.04). When materials are taken into account, the within-firm estimates are much improved; they become coherent again with the within-firm estimates obtained using value added (granted the multiplicative factor $1 - \delta$) as well as closer to the total estimates.

Our results confirm that the omission of materials in the sales specification affects especially the within-firm estimates, which is related to the fact that, in the short run, materials usage varies much less than proportionally with changes in output and other inputs. The value-added specification has the advantage of being largely immune to such problems (implying in a sense that output and materials vary proportionally). It is clear, however, that the sales specifications duly including materials and the value-added specification both suffer from other problems since they still give rise to estimates of large decreasing returns to scale in the within-firm dimension. One possible explanation is the disregard for the simultaneity in the determination both of output and labor, and of materials. Griliches and Mairesse have investigated this second possibility by estimating what they call the semireduced form model, and we consider it too.

18.3.2 Semireduced Form Estimates versus Production Function Estimates

If we assume with Griliches and Mairesse that firms maximize their short-run profits and are price-takers on competitive markets, and if we lump together the unobserved factor price variables with the errors in the equations, we derive a "semireduced form model" expressing the relationship between the endogenous output and labor variables only in terms of the predetermined physical and R & D capital stocks. Using value added and omitting materials (and ignoring also constants, time trends, or year dummies), we get:

(1)
$$v_{it} = \alpha' c_{it} + \gamma' k_{it} + e'_{it},$$
$$\ell_{it} = \alpha' c_{it} + \gamma' k_{it} + e''_{it},$$

where $\alpha' = \alpha/(1 - \beta)$ and $\gamma' = \gamma/(1 - \beta)$.

If we use sales rather than value added and include materials as another variable factor, we have the same two equations for output and labor (with sales instead of value added) and a parallel third equation for materials:

(2)
$$s_{it} = \alpha' c_{it} + \gamma' k_{it} + e'_{it},$$
$$\ell_{it} = \alpha' c_{it} + \gamma' k_{it} + e''_{it},$$
$$m_{it} = \alpha' c_{it} + \gamma' k_{it} + e'''_{it},$$

where $\alpha' = \alpha/(1 - \beta - \delta)$ and $\gamma' = \delta/(1 - \beta - \delta)$. Note that if this last system of equations holds, it implies that materials vary proportionally to sales, and hence that the value-added equation of the first system and the first system itself will also be verified; that is, the elasticities for value added, α, β, and γ, will be equal to the corresponding ones for sales multiplied by $1/(1 - \delta)$, and the reduced form coefficients in the first system, $\alpha/(1 - \beta)$ and $\gamma/(1 - \beta)$, will be equal to the ones in the second system, $\alpha/(1 - \beta - \delta)$ and $\gamma/(1 - \beta - \delta)$.

Unconstrained and constrained total and within-firm estimates of the two semireduced form models using value added or sales are given in table 18.3 for the scientific firms and the other firms separately. The corresponding estimates, α' and γ', in the various equations are rather close. Although most differences appear statistically significant given the large number of observations, constraining the coefficients to be equal entails only a very small loss of fit. The within-firm estimates of the materials equation are the most out of line, and also the poorest looking ones. All other estimates (i.e., the within-firm estimates of the other equations and the total estimates of all the equations) are coherent enough with the direct estimates of the production function (given in tables 18.1 and 18.2).

The total estimates of the research capital coefficient γ' are all very significant and large; compared to the estimates of the physical capital coefficient α', they indicate that the relative magnitude of the two capital elasticities, γ/α ($= \gamma'/\alpha'$), is about two-thirds for the scientific firms and one-third for the other firms. This is somewhat small but also more reasonable than what we get from the direct estimates. Taking, for example, the true elasticity of labor β to be .6 in terms of value added, we obtain indeed very sensible numbers for α and γ: respectively, .22 and .13 for the scientific firms, and .26 and .09 for the other firms.

The within-firm estimates of the research and physical capital coefficients are much smaller than the corresponding total estimates, and they also indicate a smaller relative magnitude for the research capital elasticity: about 30–40 percent for the scientific firms and 15–20 percent for the other firms. Thus, the absolute size of the within-firm estimates is also a problem for the semireduced form model, and the discrepancy between

Table 18.3   **Semireduced Form Equations Estimates, Scientific and Other Firms Separately**[a]

		Total Regressions			Within-Firm Regressions		
		α'	β'	System R^2	α'	β'	System R^2
Scientific firms (98)	Sales	.601 (.018)	.284 (.020)	.796	.286 (.046)	.068 (.044)	.252
	Materials	.697 (.028)	.173 (.030)	—	.196 (.076)	−.111 (.073)	—
	Value Added	.565 (.017)	.358 (.018)	.844	.314 (.059)	.204 (.056)	.304
	Labor	.588 (.022)	.273 (.024)	—	.350 (.040)	.132 (.038)	—
	Constrained I	.566 (.017)	.353 (.018)	.839	.341 (.037)	.149 (.035)	.303
	Constrained II	.552 (.016)	.336 (.017)	.788	.325 (.034)	.107 (.032)	.244
Other firms (84)	Sales	.712 (.017)	.204 (.017)	.848	.210 (.064)	.098 (.044)	.105
	Materials	.786 (.027)	.196 (.027)	—	−.006 (.105)	.069 (.072)	—
	Value Added	.653 (.013)	.236 (.014)	.898	.469 (.075)	.058 (.052)	.134
	Labor	.683 (.015)	.179 (.015)	—	.437 (.042)	.067 (.029)	—
	Constrained I	.660 (.013)	.223 (.013)	.894	.442 (.040)	.065 (.028)	.133
	Constrained II	.674 (.013)	.195 (.014)	.843	.396 (.039)	.077 (.027)	.093

[a]Constrained I estimates assume equal coefficients in the value-added and labor equations. Constrained II estimates assume equal coefficients in the sales, materials, and labor equations. The systemwide R^2 given are those of the unconstrained and constrained systems of equations.

these estimates and the total estimates remains. The fact that the estimated sum $(\alpha + \gamma)/(1 - \beta)$ is only about .4 to .5 in the within-firm dimension, while it is about .9 in the total dimension, is equivalent to finding decreasing returns to scale for the within-firm estimates of the production function, while finding nearly constant returns to scale in the total estimates. The same pattern is also observed in the U.S. study, but to a lesser extent: the semireduced form, within-firm estimates are much better looking than the production function, within-firm estimates.

On the whole, the semireduced form estimates do confirm the direct production function estimates, but, contrary to what could be hoped, they do not constitute a major improvement. Clearly the simultaneity between output and labor is only one source of trouble. Other problems may affect both types of estimates. The omission of labor and capital intensity of utilization variables (such as hours of work per employee) in the production function considered by Griliches and Mairesse is presumably a very important one. The failure of the assumption of competitive markets and errors in the variables are two other possibilities also suggested by them. In what follows we are able to show that the measurement problem of the double counting of R & D matters a lot.

18.3.3   Correcting for the Double Counting of R & D

The Griliches and Mairesse study, as well as the other studies of the contribution of R & D to productivity, suffers from the fact that R & D labor and physical capital are normally counted twice, once in the available measures of labor and physical capital and again in the measure of R & D capital stock. When a value-added measure is used for output, it also suffers from the fact that R & D expenditures (because of special fiscal rules in favor of R & D spending) are treated as intermediate inputs and are expensed out. This is true for materials used in R & D activities in France and for all R & D expenditures in the United States. These problems are generally overlooked for lack of information to make the necessary adjustments. At best it is considered that the marginal product or rate of return ρ, which derives from the estimated elasticity γ of R & D in the production function, should be interpreted as the "net rate of return to R & D above and beyond its normal remuneration" (Griliches 1979). For our sample of firms, we can illustrate the importance of correcting the different variables for the double counting of R & D, and we can verify the excess return interpretation. We find that such interpretation is roughly valid for the total (or between-firm) estimates in the cross-sectional dimension of the data but not for the within-firm estimates in the time dimension of the data. Both types of estimates are biased downward in the absence of correction, but in a rather untypical fashion the total estimates are much more affected than the within-firm estimates. We document these findings in table 18.4; we also attempt to rationalize them in appendix B.

Following Schankerman (1981), the biases from R & D double counting and expensing out can be analyzed in terms of the following omitted variables in the production function: $(v' - v)$, $- \alpha(c' - c)$, and $- \beta(\ell' - \ell)$, where $(v' - v)$, $(c' - c)$, and $(\ell' - \ell)$ are the log differences of the uncorrected and corrected measures of value added, physical capital, and labor. These three corrections are approximately $- 3, 5$, and 10 percent, respectively, in our sample of scientific firms and $- 1, 1$, and 3 percent for the other firms. Using the appropriate auxiliary regressions, the overall biases (i.e., the differences between the estimates based on the uncorrected and corrected measures) can be decomposed into three components corresponding to the three corrections for R & D materials, capital, and labor. Table 18.4 gives the overall biases and their components

Table 18.4   Production Function Estimates with Uncorrected Measures and Approximate Decomposition of the Overall Biases due to Expensing Out R & D Materials and Double Counting R & D Capital and Labor, Scientific and Other Firms Separately[a]

	Total Regressions					Within-Firm Regressions				
	α	γ	λ	R^2	MSE	α	γ	λ	R^2	MSE
Scientific firms (98)										
Estimates with uncorrected measures	.267 (.017)	.107 (.014)	.036 (.007)	.414	.0743	.217 (.060)	.170 (.052)	.035 (.005)	.362	.0199
Overall biases	.030	−.099	.006	—	—	−.015	−.059	.006	—	—
due to: R & D materials	.013	−.030	.002	.343	.0012	−.008	−.026	.003	.028	.0006
R & D capital	.010	−.012	.000	.398	.0034	.007	−.008	.000	.160	.0001
R & D labor	.013	−.048	.003	.498	.0048	−.004	−.023	.002	.029	.0013
Other firms (84)										
Estimates with uncorrected measures	.182 (.018)	.093 (.010)	−.018 (.006)	.240	.0440	.384 (.072)	.061 (.048)	−.027 (.006)	.072	.0197
Overall biases	.006	−.023	.000	—	—	.013	−.018	.000	—	—
due to: R & D materials	.002	−.006	−.000	.302	.0001	.003	−.000	.001	.048	.0000
R & D capital	.001	−.001	.000	.372	.0001	.004	−.004	.000	.330	.0000
R & D labor	.004	−.015	.001	.506	.0005	.003	−.015	.001	.116	.0001

[a]The estimates of the biases due to R & D materials, capital, and labor are the estimates of the auxiliary regressions on $(v' - v)$, $(c' - c)$ and $(\ell' - \ell)$, respectively, multiplied by 1, $-\alpha$ and $-\beta$, where α and β are taken as the unbiased estimates from the production function using the corrected measures (i.e., the estimates in table 18.1). The corresponding R^2 and mean square errors (MSE) are those of the auxiliary regressions. These numbers are given in the case of the production function with constant returns to scale; they are practically unchanged, however, if we do not impose this restriction, and the biases in the estimated returns to scale themselves are negligible.

for the scientific and other firms separately. These numbers correspond to the estimates we get when we impose constant returns to scale, but they are practically unchanged if we do not.

In spite of the limited magnitude of our corrections for R & D double counting, the overall biases in the estimated elasticity of R & D capital γ are quite sizeable. On the other hand, the biases in the estimates of the elasticity of physical capital α (and also of λ and β, or $\mu - 1$) are relatively small. The total estimates of γ are increased from about .10 to .20 (a doubling) and from .09 to .12 for the scientific firms and other firms, respectively, while the within-firm estimates rise from .17 to .23 and from .06 to .08, respectively. The discrepancy between the within-firm and total estimates for the scientific firms thus nearly vanishes. It is interesting to note that all three γ-bias components are always negative and that they tend to be larger when the corresponding corrections are more substantial, that is, for the scientific firms compared to the other firms and for the R & D labor correction compared to the other two corrections.

18.4 Summary and Conclusion

In a companion study to that of Griliches and Mairesse for the United States, we have investigated the relationship between output, labor, and physical and R & D capital during the 1972–1977 period for a sample of 182 R & D performing firms in the French manufacturing industries. Our results are quite comparable to those obtained for the United States. The relationship between firm productivity and R & D appears both strong and robust in the cross-sectional dimension of the data; it is less so in the time dimension. However, the within-firm estimates are still significant and of a likely order of magnitude. In this respect, they are more satisfactory than the U.S. ones. We show that this is largely the result of better measurement of the variables: (1) we can use a value-added measure of output instead of sales (or equivalently we include materials among the factors of the production function); (2) we can correct the measures of labor, physical capital, and output for the double counting or expensing out of the labor, capital, and materials components of R & D expenditures. As in the U.S. study, the semireduced form estimates which allow for simultaneity in the determination of output, labor, and materials agree with the production function direct estimates and confirm the importance of R & D capital relative to physical capital. However, both specifications yield rather implausible decreasing returns to scale estimates in the within-firm dimension. This is a pervasive problem in this type of work that needs to be solved before we shall be able to reconcile our cross-sectional and time-series results completely.

Appendix A

Additional Information on the Sample, the Variables, and Various Experiments

The construction of our sample is quite similar to that of the U.S. sample by Griliches and Mairesse. Based on the two-digit French NAP and U.S. SIC classification, the definition of the group of scientific firms is the same in the two countries; however, we do not have firms in the computer and instruments industries in the French sample. We preferred to exclude from our sample the firms belonging to the aircrafts, boats, and space vehicles industry (ten of them); this is an extremely R & D intensive industry (with an average R & D to value added ratio of 35 percent), but most of it is publicly financed (about 80 percent) contrary to the other R & D intensive industries. The group of other firms in the United States include some nonmanufacturing companies, such as petroleum refining or food processing companies, which we have not considered as part of manufacturing in constructing our sample. As it is, the French sample accounts for nearly one-half of the total R & D expenditures performed by French firms, while the similar ratio is about one-third for the U.S. sample.

Actually, our sample is more comparable to the U.S. restricted sample, since we removed about twenty-five "merger firms" (or firms we assumed to be such because they showed large jumps of more than 100 percent increase or 50 percent decrease in gross plant, sales, and/or number of employees). Since our study period covered only six years, it was not possible for us to deal with such firms by distinguishing "premerger" and "postmerger" firms.

Our value-added measure is at "factor costs," that is, after deduction of the value-added tax and after it is adjusted for inventory changes. Materials are taken simply as total purchases. Computing a proxy for value added as sales minus purchases changed our within estimates slightly. We have deflated value added, sales, and materials by the relevant national account industry price indices (at the two-digit classification level). Using the gross output price indices (rather than the value-added ones) to deflate value added did not change our estimates.

In our data, the numbers of employees are generally given at the end of the year and are not computed as yearly averages, which is the case for the U.S. data. We used, therefore, the beginning of the year numbers (i.e., the lagged numbers), as is also done for the capital stock measures. Taking the end of year number of employees tended to deteriorate our within estimates. This is another indication that simultaneity between employment and output is one of the sources of discrepancy between the total and within estimates.

The adjustment for inflation of the gross plant book value is made on the basis of an estimated average age of capital and an assumed average service life of sixteen years. The average age of capital is derived from the ratio of net plant to gross plant, this ratio being itself corrected to take into account that the fiscal lengths of life used to compute depreciation in French are much shorter than the actual service lives. Experiments using gross plant adjusted for inflation in various ways, or even without any adjustment, made only very little differences in our estimates, as was also the case in the U.S. study.

We have been able to obtain the (internal) R & D expenditures before 1972 and back to 1963 for most of the firms in our sample by consulting original listings of the first R & D surveys. Our R & D capital stock measures are thus constructed from the past R & D flows for a long enough presample period (at least nine years). Again as in the U.S. study, alternative measures assuming 0 or 30 percent rate of obsolescence per year instead of 15 percent, or using quite different initial conditions in 1963, had only minor effects on the estimates and the quality of the fit.

In addition to information on the materials, wages, and physical investment components of total R & D expenditures (and the number of R & D employees), which we used to correct our measures of value added, physical capital, and labor for R & D expensing out and double counting, different definitions and measures of R & D are available: total expenditures (whether they are financed by the firm or not), expenditures financed by the firm itself (this is the sole measure available in the U.S. study), and internal expenditures spent inside the firm (this is the measure we have preferred, since we could obtain it before 1972). We also have the distinction between development, applied, and basic research expenditures. Experiments with R & D capital stock constructed from these various measures yielded basically the same results. Further detailed attempts to investigate differences in the efficiency of company-financed and public-financed R & D or development, applied, and basic R & D did not prove very successful. At best there is some indication of positive composition and interaction effects of the sort found by Mansfield (this volume). Public-financed R & D appears to be less productive per se than company-financed R & D, but it appears also to enhance the productivity of the latter significantly. Similarly, basic research, though it may not be as directly productive, interacts positively with applied research and development.

Finally, and following the example of the first studies by Terleckyj (1974), we have considered the number of R & D employees as a proxy for the R & D capital stock in the production function. The total estimates are practically unaffected, but the within estimates became much poorer: the estimated $\hat{\gamma}$ is about halved for the scientific firms and is not anymore significant for the other firms.

Table 18.A.1   **Sample Composition and Size, Labor Productivity Growth Rate, R & D to Value Added Ratio[a]**

NAP Industry Classification 'Niveau 40'	Number of Firms	Productivity Growth Rate (%)	R & D Value Added Ratio (%)
Scientific firms:			
11—Chemicals	19	4.2	7.1
12—Drugs	33	7.2	11.5
15—Electronic and			
electronic and electrical equipment	46	6.6	11.6
Total scientific firms	98	6.4	10.7
Other firms:			
7–8 Primary metal industries	8	−.3	2.4
9–10 Stone, clay and glass			
products	7	3.7	2.6
13 Fabricated metal products	8	.0	2.9
14 Machinery and instruments	26	.6	4.8
16 Automobile and ground			
transportation equipment	21	1.2	5.3
18 Textiles and apparel	3	2.7	3.0
21 Paper and allied products	6	−.1	1.7
23 Rubber, miscellaneous			
plastic products	5	−1.3	3.7
Total other firms	84	.8	4.0
Total all firms	182	3.8	7.6

[a]Firm and year average over the study period 1972–77.

For details on all these different experiments, see Cuneo (1982). Table 18.A.1 indicates our sample composition and size at the two-digit industry level; it also gives the average labor productivity growth rate and the average R & D to value added ratio over our study period, 1972–77. Table 18.A.2 lists the (geometric) means, (logarithmic) standard deviations, (logarithmic) between and within-firm decomposition of variance, and the average rates of growth of our major variables, separately for the scientific and other firms.

Table 18.A.2   Characteristics of Variables[a]

Main Variables[b]	Scientific Firms (98)					Other Firms (84)				
	Geo-metric Mean	Stan-dard[a] Devi-ation	Percent Variability Between	Within	Rate of Growth	Geo-metric Mean	Stan-dard[a] Devi-ation	Percent Variability Between	Within	Rate of Growth
VA Deflated value added including R & D materials	59.6	1.41	98.2	1.8	7.3	91.2	1.40	98.9	1.1	1.6
L Number of employees excluding R & D employees	0.86	1.38	99.4	0.16	1.0	1.93	1.38	99.6	0.4	0.8
C Gross plant adjusted for inflation and excluding research capital	49.2	1.59	98.9	1.1	7.2	131.2	1.63	99.3	0.7	6.1
K R & D capital stock measure computed using a .15 rate of obsolescence	24.0	1.47	98.8	1.2	6.6	12.4	1.59	98.8	1.2	7.5
S Deflated sales	127.6	1.38	98.9	1.1	5.3	216.6	1.48	99.3	0.7	1.7
M Deflated materials excluding R & D materials	44.0	1.48	98.3	1.7	3.7	92.5	1.66	98.6	1.4	1.6
VA/L Deflated value added per employee	69.4	0.39	80.2	19.8	6.4	47.4	0.25	66.1	33.9	0.8
C/L Gross plant adjusted per employee	57.2	0.70	94.9	5.1	6.2	68.1	0.53	93.7	6.3	5.3
K/L R & D capital stock measure per employee	28.0	0.87	96.5	3.5	5.7	6.4	0.98	96.6	3.4	6.7
S/L Deflated sales per employee	148.5	0.51	95.9	4.1	4.3	112.4	0.38	88.4	11.6	0.9
M/L Deflated materials per employee	51.2	0.72	91.8	8.2	2.7	48.0	0.65	90.1	9.9	0.8

[a]Standard deviation and the decomposition of variances are given for the logarithms of the variables.

[b]All values are in 10^6 francs and constant 1972 prices. The number of employees is in 10^3 persons. Rates of growth are yearly averages over 1972–77.

Appendix B
R & D Double Counting and the
Excess Return Interpretation

In a recently published article, Schankerman (1981) pointed out forcefully and analyzed explicitly the importance of R & D double counting and expensing in measuring the returns to R & D. Using a large cross-section sample of firms (already investigated by Griliches 1980), he was able to show that the resulting biases could indeed be quite large. He also made the point that the excess return interpretation, even though it happened to be roughly verified in his particular sample, should be considered as "conceptually incorrect." Using our sample we can provide another striking illustration of the importance of such R & D double-counting biases, particularly in the cross-sectional dimension (between or total estimates) and less so in the time dimension (within-firm estimates). We find also that the excess return interpretation is not too far off, at least for our total estimates. If ρ_k and ρ_c are the marginal products or (gross) rates of return to R & D capital and physical capital, respectively, we should verify that $\rho_k \sim \hat{\gamma}(V/K) + \rho_c$, or, restated in terms of elasticities: $\gamma \sim \hat{\gamma} + \alpha(K/C)$. For the scientific firms, we can take α and γ to be .25 and .20 (total estimates with corrected measures), and $\hat{\gamma}$ to be .10 (total estimate with uncorrected measures), implying that K/C should be around .4, which is about the actual order of magnitude. The same is also roughly true for the other firms.

It is not by mere chance that the excess return interpretion is, in fact, roughly valid, and Schankerman's analysis must be qualified in this respect. It is easy to see intuitively why such interpretation might apply to a certain degree of approximation. Schankerman's analysis in terms of biases from omitted corrections, although quite right, tends to obscure the matter. The question is one of functional form, log-linear rather than linear, as much as one of mismeasurement. If we consider only the issue of double counting R & D labor and capital (and ignore that of expensing out R & D materials), and if we assume a linear production function (instead of the Cobb-Doublas function), the excess return interpretation becomes quite intuitive. Assuming a linear formulation, we must be more careful about the "units" of measurement of our variables. Define C, L, and K as the true service flows of physical capital, labor, and R & D capital in value units, and suppose K is made of R & D labor L_r and R & D physical capital C_r, that is, $K = L_r + C_r$, then the true equation and the estimated one are, respectively:

$$V = \rho_c C + \rho_\ell L + \rho_k K + e, \text{ and } V = \rho_c C' + \rho_\ell L' + \rho_k^e K + e,$$

where $C' = C + C_r = C(1 + C_r/C)$ and $L' = L + L_r = L(1 + L_r/L)$, and where $\rho_k^e = \rho_k - [(C_r/K)\rho_c + (L_r/K)\rho_\ell]$ is the rate of return of R & D capital in excess of the "normal remuneration" of its labor and physical capital components. One will actually estimate the excess rate of return ρ_k^e, if the variation in (C_r/K) and (L_r/K) is small relative to that of K. This seems reasonable enough across firms of widely different sizes, that is, in the cross-sectional dimension, for the total estimates. However, for a given firm over time the relative stability of (C_r/C) and (L_r/L) may seem as plausible as that of (C_r/K) and (L_r/K). If this is really so, whether one used the corrected or uncorrected measures of the variables, one would estimate the rate of return ρ_k itself in the time dimension, that is, for the within-firm estimates.

References

Cuneo, P. 1982. Recherche developpement et productivité: Une étude économetrique sur donńees de panel. *Thèse de 3eme cycle, Ecole des Hautes Etudes en Sciences Sociales*. Paris.

Griliches, Z. 1979. Issues in assessing the contribution of research and development to productivity growth. *Bell Journal of Economics* 10, no. 1:92–116.

———. 1980. Returns to research and development expenditures in the private sector. In *New developments in productivity measurement and analysis*, ed. J. Kendrick and B. Vaccara, 419–61. Conference on Research in Income and Wealth: Studies in Income and Wealth, vol. 44. Chicago: University of Chicago Press for the National Bureau of Economic Research.

Griliches, Z., and J. Mairesse. 1983. Comparing productivity growth: An exploration of French and U.S. industrial and firm data. *European Economic Review* 21:89–119.

Schankerman, M. 1981. The effects of double-counting and expensing on the measured returns to R & D. *Review of Economics and Statistics* 63, no. 3:454–58.

Terleckyj, N. E. 1974. *Effects of R & D on the productivity growth of industries: An exploratory study*. Washington, D.C.: National Planning Association.

19 Productivity Growth and R & D at the Business Level: Results from the PIMS Data Base

Kim B. Clark and Zvi Griliches

The recent slowdown in productivity growth in the United States and elsewhere has increased interest in understanding its determinants. Among the determinants commanding attention have been expenditures for research and development. R & D investment has attracted attention because a slowdown in its growth seemed to coincide with the productivity slowdown, and because earlier studies of the R & D–productivity connection had found R & D to be an important determinant of productivity growth. Recent work on R & D and productivity growth, however, presents a relatively mixed picture. While studies on 1950s and 1960s data generally found positive effects, productivity equations for the 1970s found the coefficient alternately collapsing (Griliches 1980; Agnew and Wise 1978; Scherer 1981; Terleckyj (1980) and reviving (Griliches and Lichtenberg, this volume; Scherer 1981), depending on the data used and, in particular, on the level of aggregation. Where disaggregated data were explored, a relatively sizeable effect of R & D was found, even in the turbulent 1970s.

This paper presents the results of a study of productivity growth and R & D in the 1970s using data on narrowly defined "business units" within a firm. The principal focus of the analysis is estimation of the productivity of R & D at the margin. Estimates are developed under different assumptions about technology, industry effects, and changes in the return to R & D over time. Our R & D data are classified into process and product expenditures, and we examine the effect of proprietary technology and technological opportunity on R & D productivity.

Kim B. Clark is an associate professor at the Graduate School of Harvard University, and a research associate of the National Bureau of Economic Research. Zvi Griliches is professor of economics at Harvard University, and program director, Productivity and Technical Change, at the National Bureau of Economic Research.

The results reported below suggest a significant relationship between R & D and the growth of productivity; in versions using total factor productivity as the dependent variable, the estimated marginal product or rate of return is about 18 percent. There is no evidence in these data of a deterioration in the productivity of R & D in the 1970s. Irrespective of model specification, trends in the R & D coefficient are substantively and statistically insignificant. We also find some evidence that, all else equal, a shift in the mix toward more product R & D lowers the measured rate of growth of productivity, and that R & D has its biggest effect on productivity in those businesses where major technical changes have occurred within the recent past.

The paper has three parts. We discuss the data used and present summary information about our key variables in section 19.1. Particular attention is paid to the reported price indexes. Estimates of price changes in the PIMS data are compared with estimates based on government surveys. Section 19.2 sets out the analytical framework and presents estimates of the effect of R & D on productivity under several model specifications. The paper concludes in section 19.3 with a brief summary and some suggestions for further work.

19.1 The Data Set

The data we use are drawn from the PIMS project of the Strategic Planning Institute (SPT).[1] The Institute is composed of over 1500 member companies which participate in the project by supplying annual data on individual businesses within the company. Our sample covers 924 U.S. manufacturing businesses over the period 1970–80.

A "business" in the PIMS lexicon is a unit of a firm "selling a distinct set of products to an identifiable set of customers in competition with a well-defined set of competitors." Businesses tend to be synonymous with operating divisions of a company but may be defined in terms of product lines within divisions. In addition to annual income statements and balance sheets, each business provides information on several measures of market structure, technology, previous competitive experience, and competitive strategy. Along with its panel structure and level of detail, the richness of the PIMS data set makes it a potentially valuable source of information on the determinants and impact of R & D.

But richness has its price. Several aspects of the data must be kept in mind when interpreting the evidence presented below. In the first place, we are not dealing here with typical or representative firms. The companies in the project tend to be large, diversified corporations; many are found in the *Fortune* 500; and almost all of them are found in the *Fortune*

1. A description of the PIMS data can be found in Schoeffler (1977). For an analysis of R & D and profitability using the PIMS data, see Ravenscraft and Scherer (1981).

1000. The analysis thus deals with the impact of R & D on productivity among firms that may not be representative of all firms in a given sector, but which probably account for a significant fraction of the assets and people employed.

The unit of observation is a further problem. Although SPI provides guidelines for defining "business units," the choice is left to the company and will depend on the availability of data and the company's assessment of the usefulness of the definition.[2] In a related fashion, much of the structural data are subject to the company's assessments and perceptions. Of course, a good deal of the information requested by SPI is available through accounting systems and is subject to uniformity of definition and guidelines developed and imposed by SPI. But variables, like the number of competitors or the relative quality of the business's products, depend to some extent on the respondent's perceptions.

Finally, the self-reported character of the data and their use in comparative modeling raise questions about their quality and integrity. Two considerations suggest that the quality of the data is reasonably high. First, the information requested is of value to the business itself (e.g., its market share), and it seems reasonable to suppose that the firm is in a position to know and has expended effort to acquire accurate data. Second, a firm's participation in the project is motivated by a desire to use the data in the strategic planning models developed by SPI. Considerable effort is made to preserve confidentiality and ensure quality: only the firms themselves have access to their own data; sensitive variables (e.g., profits) are only reported in disguised or ratio form; analysts at SPI run the data through an elaborate procedure to check for consistency, and gross errors are followed up with the company.

19.1.1 Major Variables

The annual income statement and balance sheet provided by each firm can be used to construct measures of productivity, R & D, and capital. We use sales, deflated by an index of product prices, as the basic measure of output. Although available information permits calculation of value added, we found that treating materials as a separate factor of production fit the data much better. The output price index and an index of materials prices are provided by the business under guidelines set forth by SPI. The guidelines define the relevant concept of output price as a weighted average of the business's selling prices, holding the mix of products constant. Since the quality of the output and productivity series depends on the quality of the output price indexes, they are examined below in more detail.

2. Definition of a business as developed in the PIMS guidelines is based on the concept of a "strategic business unit." This concept is spelled out in more detail in Abell and Hammond (1979).

Information on labor input is limited. The only variable available to us is the number of employees, and that is only available on a disguised basis and thus can only be used in ratio form. There are no data on hours per employee, nor are the data broken down by occupation or type of employment. Output per employee and capital-labor ratios are defined for all employees, including sales and managerial personnel, as well as those engaged in R & D activities and production. These variables are not adjusted for differences in quality, since no wage data or data on education or other characteristics are available.

Estimates of the real stock of capital are derived from information on the firm's balance sheet and annual investment. The value of plant and equipment in the firm is reported at historical cost, but each firm provides an estimate of the replacement value of gross plant and equipment in the initial year of its participation in the survey. This gives an initial capital stock value in current prices. Since firms may enter the sample in different years, we restate the initial value in current prices into constant (1972) dollars using the deflator for business fixed investment (BFI) from the National Income and Product Accounts. Subsequent investment in plant and equipment is deflated by the BFI price index and added to the initial year stock. The investment series we use is net of retirements, but we have not subtracted out reported depreciation.[3] To provide a comparative perspective, we shall estimate the models using gross book value of capital as well as the stock of capital adjusted for inflation as described above.

As with most data sets, information on R & D comes in the form of current spending. Expenditures on research and development are treated as an expense in the PIMS accounting system and are, therefore, reported in the income statement. Businesses are asked to include in this category all expenses (material, labor, etc.) incurred to improve existing products or to develop new products, and all expenses to improve the efficiency of the manufacturing process. Total R & D expenditures are thus classified into product and process categories. How that split is implemented, however, is left to the business to decide. All R & D expenses are specific to the business and exclude charges for research and development done in a central corporate facility. They may, however, include expenses shared with other businesses but conducted below the corporate level.

3. The nominal investment series is calculated as the difference in the gross book value of plant and equipment. It thus reflects both gross investment and retirements. Estimates of real capital can be obtained in other ways. One possibility is to estimate the age of capital using the ratio of accumulated depreciation to annual depreciation, and then to adjust current book values based on changes in the BFI deflator since the year the average piece of capital (determined by the age calculation) was purchased. For an example of this approach, see Griliches and Mairesse in this volume. Their results, as well as our own estimates reported below, suggest that the R & D estimates are relatively insensitive to adjustments of this sort.

Table 19.1 presents definitions, means, and standard deviations of the basic variables used in the analysis. The sample covers 924 businesses, with a total of 4,146 observations; not all firms are present in each year, so the design of the sample is unbalanced. Data on real sales, materials, and capital per employee show a substantial amount of variability around relatively high average rates of growth. In real terms, sales per employee grew at an annual rate of 4 percent in these data, while capital and materials per person grew at rates between 3.5 and 4.0 percent. The data on newness of the capital stock (ratio of net to gross book value) suggest that, on average, productivity growth occurred during a period in which the capital stock was aging.

Variables measuring R & D intensity and mix are listed in part 2 of table 19.1. These data are of a reasonable order of magnitude and imply that the businesses in the sample cover a wide range of R & D intensities. As in data collected at other levels of aggegation, the majority of R & D (65 percent) is devoted to improving old or developing new products. Although the sample covers most of the two-digit industries, almost half of the observations are from businesses in chemicals, electrical and nonelectrical machinery, and instruments.

We have used the PIMS data to calculate R & D intensity for these two-digit industries, as well as for primary and fabricated metal products, and compared them to data published by the National Science Foundation (NSF). This comparison, presented in table 19.2, shows the same ranking of industries by R & D intensity in the two data sets. Since the NSF is a company-based data set, and since the mix of subindustries within the two-digit industries may not be identical, differences in the R & D-to-sales ratio in the two series are to be expected. But the two sources yield intensity estimates that are quite similar. Only in machinery (SIC 35) does a sizeable discrepancy emerge.

We make no attempt to estimate the stock of R & D capital, but rather use R & D intensity to capture the effects of R & D on productivity. To allow for lagged effects and to break any spurious correlation induced by the presence of lagged output as an independent variable, we define R & D intensity as:

$$RQ(-1) = \frac{R_{-1}}{\frac{1}{2}(S + S_{-1})},$$

where R_{-1} is R & D expenditure in the previous period, and S indicates total sales. Other measures, including R & D intensity lagged one and two periods, and an instrumental variable procedure, had no effect on the results. We shall report only the estimates with $RQ(-1)$.

Part 3 of table 19.1 provides information on three variables that we use as indicators of previous technical activity. The first two indicate whether the business "derives significant benefit" from proprietary products or processes, either through patents or what the SPI guidelines call "trade

Table 19.1   **Means and Standard Deviations U.S. Manufacturing Businesses, PIMS Data Base, 1971–80**

Variable	Definition	Mean	Std. Dev.
(1) Rates of Growth (in percent):			
$(s - \ell)$	real sales per employee	3.95	17.33
p	output price index	7.40	9.10
$(m - \ell)$	real purchases per employee	3.93	22.31
p_m	materials price index	9.17	12.42
$(g - \ell)$	gross book value of plant and equipment per employee	7.32	19.22
$(c - \ell)$	gross plant and equipment per employee in 1972 $	3.55	17.00
util	rate of capacity utilization	2.71	16.62
new	ratio of net to gross book value of plant and equipment	−1.15	12.54
(2) R & D Variables (in percent):			
$RQ(-1)$	ratio of total R & D expenses to average of current sales and sales lagged one period	2.21	3.76
RMIX	ratio of product R & D expenses to total R & D expenses	65.49	29.94
(3) Proprietary Technology and Technological Opportunity:			
DPROD	= 1 if business derives significant benefit from proprietary products (patents etc.) = 0 otherwise	0.21	—
DPROC	= 1 if business derives significant benefit from proprietary processes (patents etc.) = 0 otherwise	0.21	—
DTECH	= 1 if there has been major technological changes in product or process of the business or its major competitors in last eight years = 0 otherwise	0.28	—

secrets." The last variable indicates whether "major" technological change (either product or process) had occurred in the business or in its major competitors in the last eight years. These questions are asked only once (when the business enters the PIMS project) so that the dummy variables are constant over time. The data suggest that a sizeable fraction of the businesses have carried out R & D projects that have led to patents or some other form of proprietary products or processes. An issue we examine below is whether R & D capability defined in this way affects the current connection between R & D investments and productivity.

The mean growth rates of the basic variables are of a reasonable order of magnitude, but a somewhat more detailed look at the data, particu-

Table 19.2   **R & D Expenditures as a Percent of Sales in PIMS and NSF Data[a]**
for Selected Two-Digit Industries, 1974

Industries (SIC)	PIMS	NSF
Chemicals (28)	2.8	3.0
Primary metals (33)	0.5	0.5
Fabricated metal prod. (34)	1.3	1.1
Machinery (35)	2.0	3.8
Electrical equipment (36)	3.5	3.5[b]
Instruments (38)	4.8	5.2

Source:
  NSF = National Science Foundation.
  PIMS = Calculated from PIMS data base.

[a]NSF data pertain to company expenditures on R & D; the PIMS data pertain to business level R & D, excluding R & D performed in corporate research laboratories.

[b]The NSF data for electrical equipment include data on communication (SIC 48).

larly at the output price series, seems in order. Although our focus is productivity, the measures of output underlying the analysis are only as good as the price indexes used to deflate nominal sales. A full-scale analysis of the data is beyond the scope of this paper, but we can provide some perspective by comparing rates of change of prices in the PIMS data with those found in the statistics published by the government. To do that we have focused on price changes in a group of industries where the number of observations available in the PIMS data set is sufficient to justify comparison with the published figures.

Table 19.3 presents annual rates of price change for nine two-digit SIC industries over the period 1971–79. Each cell in the table contains three entries. The first is the percentage change in the two-digit industry deflator calculated by the Bureau of Economic Analysis as part of the National Income and Product Accounts. The second entry is the average percentage change in the price indexes of PIMS firms in the corresponding two-digit industry. The last number is the number of PIMS firms in the industry in that year. The comparisons in table 19.2 are necessarily rough. Because the mix of four-digit industries underlying the PIMS two-digit calculations is different than the mix used in the BEA calculations, it is not reasonable to expect the two sources to yield identical estimates. However, to the extent that similar economic forces affect the constituent four-digit industries in similar ways, a two-digit level comparison should give us some idea of comparability.

Perhaps the most noticeable aspect of the BEA/PIMS comparison in table 19.3 is the similar pattern of change over time. Both data sets generally show small changes in prices in the first three years, followed by an explosion in 1974–75, with rates of price increases running as high as

Table 19.3 Comparison of Rates of Price Change in the PIMS Data Set and the National Income Accounts for Selected Two-Digit Manufacturing Industries

Industry (SIC)	Data Set	1971	1972	1973	1974	1975	1976	1977	1978	1979
(1) Food (20)	BEA	1.5	−4.4	−7.4	15.4	22.7	−2.5	6.6	5.4	4.0
	PIMS	4.0	4.3	9.5	18.5	13.9	2.6	4.5	6.6	11.6
	N	29	35	41	49	40	33	27	17	11
(2) Chemicals (28)	BEA	1.1	−.2	−.8	11.0	12.4	4.2	2.8	4.5	2.7
	PIMS	−0.5	0.2	5.3	23.2	16.0	4.7	5.2	6.0	11.0
	N	75	89	108	95	94	91	55	36	15
(3) Rubber and plastics (30)	BEA	3.2	1.4	−1.0	6.9	9.5	4.9	5.3	4.4	4.3
	PIM	−0.7	−0.6	1.5	18.3	8.2	3.5	4.4	5.6	4.6
	N	22	29	37	46	43	32	21	17	12
(4) Stone, clay and glass (32)	BEA	9.1	3.2	2.0	6.8	13.5	7.0	8.5	10.0	5.8
	PIMS	3.6	3.9	3.8	15.2	14.4	7.8	7.7	8.1	4.7
	N	15	23	30	36	36	38	34	22	7
(5) Primary metals (33)	BEA	3.1	8.6	−1.8	24.5	20.2	2.8	8.5	9.1	10.9
	PIM	0.5	2.3	10.3	29.2	12.4	2.0	5.2	9.7	9.4
	N	13	16	29	28	32	31	31	26	7

(6) Fabricated	BEA	7.3	3.3	3.1	15.9	19.2	1.4	4.8	6.9	5.5
metals (34)	PIMS	5.5	4.8	6.2	17.1	9.6	6.0	6.3	7.6	9.1
	N	12	25	42	56	63	57	49	36	34
(7) Nonelectrical	BEA	3.7	0.9	1.1	5.6	17.8	3.2	7.3	7.0	6.2
machinery (35)	PIM	4.9	3.5	5.5	13.7	10.3	6.9	7.2	7.0	7.4
	N	42	60	84	95	100	91	71	45	23
(8) Electrical	BEA	3.0	−0.2	−0.3	4.8	12.5	3.0	5.1	3.8	6.7
equipment (36)	PIMS	0.7	1.4	2.3	12.7	8.7	5.4	5.2	6.2	8.5
	N	51	67	78	62	62	61	53	34	26
(9) Instruments	BEA	1.4	−0.2	−0.6	−.08	9.0	6.2	1.7	6.5	4.0
(38)	PIMS	1.1	2.0	3.2	9.4	8.3	4.9	5.4	4.1	5.3
	N	21	27	31	33	41	41	30	15	7

Source:
BEA = Bureau of Economic Analysis, unpublished data, National Income and Product Accounts.
PIMS = SPI/PIMS data set.

25–30 percent in some industries. In the latter part of the period, the rate of change is once again much smaller, although higher than the rates found at the beginning of the decade.

Amidst this broad pattern of similarity there are clear differences between the published data and the data from PIMS. In most of the industries, for example, the 1974–75 explosion in prices shows up earlier in the PIMS data, but lasts longer in the BEA estimates.[4] A comparison of the sums of the rates of change in the two years (1974–75) yields values much closer together than comparisons of the years taken individually. Even before the oil shock and the expiration of controls, the two data sets show different patterns in some years in several industries. In fact, the comparisons before the oil shock are much more diverse than those made in the 1976–1979 period. Although differences are present in the latter period, the large discrepancies found in the 1971–74 period are less frequent. This pattern may reflect the influence of wage-price controls on reporting practices or the different sources of inflationary pressure in the two periods.

19.2 Empirical Analysis

The connection between R & D and productivity growth is studied in the context of a fairly conventional model. In its simplest form, output (Q) of the ith business at time t is assumed to be a function of the stock of capital (C), the number of employees (L), accumulated investment in R & D (K), and a factor accounting for disembodied technical change ($Ae^{\lambda t}$), as in

$$(1) \qquad Q_{it} = Ae^{\lambda t} \, Q \, (K_{it}, L_{it}, C_{it}) \, .$$

It is standard procedure to assume that K_{it} can be represented by a distributed lag of past investments in R & D with the weights presumed to depend on the way in which past activities affect the current state of technical knowledge.

Assuming the production function is Cobb-Douglas and separable in R & D, we can totally differentiate (1) and rearrange terms to derive an expression in terms of rates of growth:

$$(2) \qquad q_i = \lambda + \gamma k_i + \alpha c_i + (1 - \alpha)\ell_i \, ,$$

where γ and α are output elasticities with respect to R & D and capital, and lowercase letters have been used to indicate relative rates of growth of their uppercase counterparts (e.g., $k = (dK/dt)/K$). Note that we have assumed constant returns to scale with respect to the conventional

---

4. The use of these data to deflate industry level output would change the estimated pattern of the productivity slowdown quite a bit. It would imply a much slower rise in the 1971–73 period and much less of a fall in 1975.

measures of capital and labor. Rearranging terms yields a productivity equation:

$$(3) \qquad (q - \ell)_i = \lambda + \gamma k_i + \alpha (c - \ell)_i,$$

Where $(q - \ell)_i$ is the growth rate of labor productivity, and $(c - \ell)_i$ is the rate of growth of the capital-labor ratio.

The effect of R & D is measured by γ; estimation in this context requires data on the growth of the stock of R & D capital. If, however, investments in R & D do not depreciate, then data on R & D intensity can be used to capture the R & D effect. If R_{it} is R & D expenditures in year t, then $k_i = R_{it}/K_{it}$, and $\gamma k_{it} = \rho (R_{it}/Q_{it})$, where ρ is the marginal product of R & D. Under competitive assumptions, ρ can also be interpreted as the rate of return.[5] Because employment and capital employed in R & D have not been segregated explicitly, this is an excess return to R & D expenditures. Further, it is a private return because the data pertain to individual businesses. Returns accruing to other firms and investors are not captured here.

Equation (3) provides a starting point for empirical analysis, but several adjustments seem warranted. In the first place, the model as specified ignores the role of intermediate products in production by implicitly assuming that materials (including purchases of intermediate products and energy) are proportional to output.[6] This problem can be dealt with by using information on purchases to expand the input list. It is, of course, possible to use data on materials to calculate a value-added version of output. But this too makes assumptions about the nature of the production process (e.g., materials are used in fixed proportion) which may not apply across all firms. While we have used materials in both ways, treating them explicitly as an input yields much better statistical results, and we shall focus on such results in the empirical work reported below. The variable we use is total purchases deflated by an index of materials prices.[7]

One of the reasons for adding materials as an input is our view that the technology of production is likely to vary across firms and industries. If that is true, estimation of (3) without adjustment could lead to misleading inferences about R & D. A first cut at this problem is to add a set of industry dummies so that parameter estimates are based on variation in

5. If R & D investments depreciate, as they most likely do, especially as far as private returns are concerned (see Pakes and Schankerman, this volume) then the equation is misspecified by leaving out a term of the $-\delta K/Q$ form. Since K/Q and R/Q are likely to be positively correlated, this omission may bias the estimated R/Q coefficient downward, possibly by a rather large amount (since the R/Q coefficient in the K/Q auxiliary equation is likely to be significantly above unity).

6. As Griliches and Mairesse (this volume) show, failure of the proportionality assumption may induce bias into the estimated R & D effects.

7. The date set contains no breakdown of purchases into energy and other intermediate inputs; use of aggregate purchases implicitly treats materials and energy as interchangeable.

productivity and its determinants within industries, with each industry having its own value of λ. Firm-specific variations in technology can be introduced by casting the estimation problem in a total factor productivity framework. Instead of estimating the output elasticities of capital and materials directly, we can use the observed factor shares for each business as an approximation (the two are identical in competitive equilibrium).

After rewriting the R & D variable in intensity form, adding materials and industry dummies and using factor shares, equation (3) becomes:

$$(4) \qquad f_i = \sum_{j=1}^{N} \lambda_j \, D_j + \rho \, (R_{it}/Q_{it}),$$

where j indexes industries, D is an industry dummy, and f_i is defined as:

$$(5) \qquad f_i \cong q_i - \alpha_i c_i - \delta_i m_i - (1 - \alpha_i - \delta_i)\ell_i.$$

The parameters α_i and δ_i are respectively the shares of capital and materials in the sales of the ith firm. To better approximate equilibrium values, we have averaged each firm's share over the sample period. Material's share can be calculated directly, since it is simply the value of purchases divided by sales. No data are provided on the wage bill, however, hence capital's share was estimated as depreciation plus profits divided by sales.[8] Profits are defined gross of R & D expenditures (we treat R & D as an investment), but net of marketing expenses.[9]

The specification of the basic productivity equation is based on what is essentially a long-term perspective. It is assumed that movements in total factor productivity reflect movements in the production frontier caused by R & D investment and disembodied technical change. In practice, businesses may deviate from the frontier, not only because of errors in optimization, but because of disequilibrium phenomena associated with fluctuations in demand and consequent changes in utilization.

One way to incorporate such factors into the model is to assume that the production function (and thus productivity growth) is composed of a long-term and a short-term component. R & D and disembodied technical change are assumed to affect only the long-term component in the manner specified in (4). The short-term component is specified to be a simple linear function of capacity utilization. Cast in growth rate form,

---

8. The use of total profits in the calculation of the share of physical capital is likely to overstate capital's share, since some of the returns that accrue to R & D will be counted as return to capital. The error thus introduced may lead to a downward bias in the estimate of the rate of return to R & D. If total profits include returns to physical capital and the stock of R & D capital, so that $\Pi = rC + \rho K$, then the estimated share of capital will be equal to the true share plus the elasticity of output with respect to R & D capital (note that $\rho K/Q = \gamma$). Use of the estimated share in a total factor productivity framework introduces $-\gamma_i c_i$ into the error term. If c and $RQ(-1)$ are positively correlated, estimates of ρ will be downward biased.

9. In those cases where profits in a given year were negative for a given firm, the average share for that firm was calculated excluding the negative year.

these assumptions introduce the rate of change of capacity utilization as a variable in the analysis.

19.2.1 The Main Results

Estimates of several versions of the basic productivity model are presented in table 19.4. The dependent variable in columns (1)–(4) is the rate of growth of real sales per employee, while the growth of total factor productivity (TFP) is examined in columns (5) and (6). In addition to R & D intensity, the model includes variables measuring the R & D mix, the growth of capacity utilization, the newness of the capital stock, and the percent of employees unionized. Capital and materials per employee

Table 19.4   **Estimates of Alternative Productivity Model Specifications (standard errors in parentheses)**

Independent Variables	Specification[a]					
	Real Sales (1)	Real Sales (2)	Real Sales (3)	Real Sales (4)	TFP (5)	TFP (6)
CONS	0.49	2.13	0.88	2.34	1.08	2.53
	(0.51)	(1.32)	(0.52)	(1.35)	(0.52)	(1.35)
$RQ(-1)$	0.18	0.18	0.19	0.19	0.20	0.20
	(0.05)	(0.05)	(0.05)	(0.05)	(0.05)	(0.05)
RMIX	−1.42	−1.22	−1.16	−1.11	−1.22	−1.15
	(0.59)	(0.61)	(0.60)	(0.62)	(0.60)	(0.62)
$c - \ell$	0.25	0.25	—	—	—	—
	(0.01)	(0.01)				
$m - \ell$	0.45	0.44	—	—	—	—
	(0.01)	(0.01)				
$(c - \ell)^{*b}$	—	—	1.17	1.17	—	—
			(0.06)	(0.06)		
$(m - \ell)^{*b}$	—	—	1.05	1.05	—	—
			(0.02)	(0.02)		
util	0.32	0.32	0.28	0.28	0.28	0.28
	(0.01)	(0.01)	(0.01)	(0.01)	(0.01)	(0.01)
new	−0.05	−0.05	−0.04	−0.03	−0.03	−0.03
	(0.01)	(0.01)	(0.01)	(0.01)	(0.01)	(0.01)
%UN	0.01	0.01	0.01	0.01	0.01	0.01
	(0.01)	(0.01)	(0.01)	(0.01)	(0.01)	(0.01)
Ind. effect[c]	no	yes	no	yes	no	yes
R^2	0.587	0.591	0.574	0.577	0.148	0.154
SEE	11.1	11.1	11.3	11.3	11.3	11.3
d.f.	4138	4119	4138	4119	4140	4121

[a]The dependent variable in columns (1)–(4) is real sales per employee; in columns (5)–(6) the dependent variable is TFP (total factor productivity), calculated as described in the text.

[b]$(c - \ell)^*$ is $(c - \ell)$ multiplied by capital's share; $(m - \ell)^*$ is $(m - \ell)$ multiplied by material's share.

[c]Industry effects are captured by two-digit SIC dummies.

are included as independent variables in (1)–(4) and are incorporated into the dependent variable in the TFP regressions.

Irrespective of specification, the estimates in table 19.4 show a significant effect of R & D on the growth of productivity. In column (1), the model yields an estimated rate of return to R & D investment of 0.18 with a standard error of 0.05. The utilization rate as well as capital and materials per employee are significantly related to sales per employee. Correcting capital for inflation appears to have little effect on the estimated R & D effect. When the growth of gross book value per employee is substituted for $c - \ell$ in column (1), for example, the estimated return to R & D is still 0.18.

The newness variable has a negative sign, while unionization's impact is statistically insignificant. It is possible that the sign of the newness variable reflects measurement problems as well as the differential effects of newer capital. Although capital has been adjusted for inflation, the procedure relies on estimates of replacement value in the first year of participation in the survey. To the extent that the correction fails to remove the effects of inflation, the rate of increase in the stock of capital will be overstated, a problem likely to be more serious for newer equipment. In fact, when column (1) is estimated with the book value of capital, the newness variable remains negative but increases by 30 percent. It is also possible that the negative sign remaining after the inflation correction is the result of adjustment costs of new capital. The integration of new equipment into existing plants or the start-up of new facilities may require time and effort to bring on line and may be disruptive to existing operations.

Measurement problems may also be a factor in the estimates of product-process mix effects. The coefficients on RMIX indicate that an increase in product R & D's share in total R & D investment is associated with a lower rate of productivity growth. High shares of product R & D may indicate a high rate of new product introduction which may be associated with lower rates of productivity growth for two reasons: First, much like new equipment, new products tend to be disruptive to established production processes. Product introductions generally involve a start-up and debugging phase of varying length in which new equipment or new tasks are specified and learned. Productivity growth is likely to suffer as a result. Second, where new products are an important aspect of competition, the business may adopt a relatively adaptable and flexible process technology. The firm is likely to avoid equipment and processes dedicated to a specific product and thus somewhat rigid. Some sacrifice in productivity is likely in the interests of flexibility. Although some of this should be picked up in the capital-labor ratio, this variable is likely to be too broad and rough to capture the distinctions we have in mind. It is well known, for example, that a highly capital-intensive machine shop can be quite flexible in adapting to new products. The R & D mix effect may, therefore, be an indication of the type of technology and the importance of new products.

While these possibilities are interesting, too much should not be made of the mix effect. The distinction between product and process R & D is likely to involve a good deal of arbitrariness. This arises because the guidelines are vague and because the distinction may not be meaningful at this level. Not only are process and product efforts jointly pursued on a project basis, and thus difficult to disentangle, but even pure product development can change the efficiency of the process. A new product design, for example, may lead to a reduction in the number of operations required or in a simplification of tasks, so that labor input is reduced even without any capital investment. Furthermore, if higher product R & D is associated with new products, and if firms base their price index on a fixed set of products, the reported rates of inflation may overstate the extent of price change. Output and productivity growth may, therefore, be understated. The fact that the standard errors on RMIX are relatively large, given the number of observations, lends some support to the importance of measurement error.

Finding a significant effect of R & D on productivity is unaffected by the specifications changes introduced in columns (2)–(6). Column (2) adds two-digit industry dummies, which allows each industry to have its own trend term. Estimation within industries has little effect on the results. In column (3), a new version of the capital and materials variables is used. The new variables are the rates of growth of capital and materials per employee multiplied by their average shares in sales. If the technology were Cobb-Douglas and the businesses were fully competitive, then coefficients on the new variables should equal unity. The materials and capital coefficients are significantly different from one in a statistical but not substantive sense, implying that the Cobb-Douglas specification is not too far off the mark. It is clear that the fit of the equation deteriorates only marginally when the average shares are imposed, and these changes, with or without industry effects, have little impact on the estimated return to R & D investment.

The same is true of the TFP equations in columns (5) and (6). We estimate that R & D had a return of 20 percent in the TFP results, slightly higher than the estimate in columns (1) and (2) but essentially similar to the earlier results. The other coefficients are little changed as well, although the newness variable declines from $-.05$ to $-.03$. As before, the industry dummies have no effect on the results.

19.2.2 Proprietary Knowledge, R & D Capability, and Technological Opportunity

Estimates of R & D's effect on productivity in table 19.4 are obtained under the assumption of a common effect across businesses. While differencing has eliminated fixed firm effects from the production function formulation, firms may also differ in their ability to translate R & D effort into actual products or processes. The productivity of R & D investment may depend on the "opportunity" for technical change in the firm's product or process. Some firms participate in industries where the scientific knowledge related to the product or process technology is rich and growing, while others use techniques where the possibility of new understanding is much more limited. Moreover, where the potential for innovation is high, firms may differ in their ability to exploit those opportunities because of differences in organization or management skill.

The likelihood of interfirm differences in technical opportunity and R & D capability suggests that the average effect of R & D in table 19.4 may mask significant variation across firms. A simple way to model the distinction between R & D effort (expenditures on R & D) and R & D output (new products or processes) and consequent gains in productivity is to assume that ρ is a function of the firm's R & D capability (or technical opportunity). If we assume that past R & D success is an indicator of that ability and if we are willing to specify a linear relationship between ρ and past success, we can write

$$(6) \qquad\qquad \rho = b_o + b_1 P,$$

where b_o and b_1 are parameters, and P indicates previous R & D success (e.g., patents). It seems reasonable to allow for the possibility that past R & D success may affect productivity independent of the current R & D effort. The total factor productivity model then becomes

$$(7) \qquad\qquad f_i = \lambda + b_o (R/Q)_{it} + b_1 (R/Q)_{it} P_i + dP_i,$$

where the effects of utilization, unionization, newness, and industry have been suppressed.

Although we have no data on the number of patents the businesses have produced, we have three variables that provide some indication of R & D capability, and technological opportunity. The first two are dummy variables based on answers to the question: Does this business derive significant benefit from (1) proprietary products and/or (2) proprietary processes? Patented products or processes are included in the definition, but firms are also instructed to consider processes (products) regarded as proprietary but not patented. The broader definition seems reasonable, since the decision to seek a patent depends not only on the significance of the invention or development and potential gains, but also on the costs of the legal process. Moreover, the firm may derive significant benefit from R & D results that are not clearly patentable.

The third variable is based on the question: Have there been major technological changes in the products or processes of this business or its major competitors within the last eight years? Inclusion of the firm and its competitors in the definition means that the variable provides information about the potential for change and development in the technology used in the industry, whether or not the firm itself has experienced a major change. The fact that a firm or its competitors have experienced a major change in technology can be interpreted in several ways. To the extent that an affirmative answer refers to the firm, one could infer that the firm has the capability to apply R & D and make use of the results. A similar conclusion would apply to competitors. However, the change in technology could have come through the purchase of equipment or licensing of new techniques rather than the firm's own R & D effort. Whatever the source of change, the fact that it has occurred implies the existence of further opportunities for technical development.

It is important to note, however, that asking a business about the occurrence of technical change may be equivalent to asking it about the productivity of its R & D investments. In that sense, inferences about the effects of technological opportunity based on the technical change variable may have little substantive content, since the estimated coefficient would be little more than a reflection of how accurately the businesses answered the question. While the possible tautology between our measure of technical opportunity and R & D productivity remains in the analysis to follow, it is mitigated to some extent by the fact that R & D investments are measured in the previous period, while changes in technology may have occurred sometime in the previous eight years.

It would clearly be useful to have more information about what firms have in mind when they answer yes to the technical change question. The PIMS guidelines warn respondents only to answer in the affirmative if there is no doubt that a major change has occurred. The meaning of the variable measuring technical change and proprietary products and processes deserves more analysis, but the nature of the data and the confidentiality provisions of the PIMS project make an in-depth analysis difficult and beyond the scope of this paper.[10]

Table 19.5 presents estimates of the TFP model after inclusion of our measures for R & D capability and technical opportunity. Although the results in line (1) with the proprietary product/process dummies show

10. While our ability to be precise about the substantive content of these variables is limited, we have examined them for internal consistency. A comparison of mean R & D intensity in samples selected on the basis of the presence or absence of technical change (DTECH) and proprietary technology (DPROD, DPROC) shows that firms with DTECH = 1 are almost twice as R & D intensive as their DTECH = 0 counterparts. A similar difference exists for firms where DPROD or DPROC equals one. We also found that 45 percent of firms with DPROD = 1 answer yes to the question about major technical change; for firms with DPROD = 0, the number is 23 percent. The results for DPROC are almost identical. This kind of consistency also shows up in analysis by industry. Not only are changes in technology correlated with proprietary products and processes within industries, but the industrial focus of major technical change is consistent with other information. The industries with high mean values of DTECH—paper, chemical, plastics, transportation equipment (including aerospace), instruments, and electrical equipment—are industries where major changes in technology have occurred.

Table 19.5   The R & D Effect, R & D Capability and Technological Opportunity[a] (standard errors in parentheses)

Specification/ Dependent Variable	CONS	RQ(−1)	DPROD	DPROC	DTECH	RQ DPROD	RQ ROPROC	RQ DTECH	R^2	SEE	d.f.
(1) TFP	0.26 (0.33)	0.19 (0.05)	−1.21 (0.48)	1.11 (0.48)	—	—	—	—	0.149	11.3	4139
(2) TFP	0.38 (0.36)	0.13 (0.07)	−1.06 (0.56)	0.69 (0.55)	—	−.05 (.10)	.15 (.10)	—	0.149	11.3	4137
(3) TFP	0.24 (0.34)	0.19 (0.05)	—	—	0.05 (0.40)	—	—	—	0.147	11.4	4140
(4) TFP	0.58 (0.37)	0.02 (0.08)	—	—	−0.51 (0.47)	—	—	0.24 (0.10)	0.148	11.3	4139

[a]Each equation includes util, new, and %UN, in addition to the variables listed.

little change in the R & D effect, the new dummy variables are statistically and substantively significant. Furthermore, the sign pattern—negative on product; positive on process—is reminiscent of the R & D mix effect noted above. When the dummy variables are interacted with R & D intensity in line (2), however, we find little evidence of a significant relationship between R & D productivity and proprietary technology. Each of the interaction terms has the same sign as its dummy variable counterpart, but the coefficients are not statistically significant.

Lines (3) and (4) present TFP estimates with the technological change variable. While there appears to be no relationship between TFP growth and DTECH, there is a strong connection between DTECH and R & D intensity; the coefficient on RQDTECH is 0.24 and statistically significant. Moreover, the coefficient on $RQ(-1)$ in line (4) (which measures the R & D effect in businesses where DTECH = 0) is close to zero. If interpreted literally, the results imply that R & D has no effect on productivity in businesses where technical opportunities are apparently low. The connection between DTECH and R & D intensity links these finding with results reported by Griliches and Mairesse (this volume), where R & D's largest effect on productivity was in R & D intensive firms. While interesting and worthy of further analysis, the statistical evidence in line (4) can be overinterpreted. It is useful to note that the addition of DTECH and its interaction with $RQ(-1)$ has little effect on the explanatory power of the equation.

19.2.3 Time Effects

Attention has been focused in recent years on possible changes in the productivity of R & D over time. Using aggregated industry data (two-digit SIC) from the 1970s, a number of researchers have documented the collapse of what had been a relatively strong R & D effect. Griliches (1979), Terleckyj (1980), Scherer (1981), and Kendrick and Grossman (1980) all find little evidence in two-digit level data that R & D affected productivity in the post-1973 period. Once the data are disaggregated, however, some R & D effect emerges. Griliches and Lichtenberg (this volume), for example, find that the strong relationships found in the 1960s persisted into the later period.

Figure 19.1 presents a profile of the growth rates of TFP in the PIMS data and in published data on manufacturing. The published TFP estimates where prepared by Kendrick and Grossman (1980). Their output measure is based on real value added, and labor input is total hours worked. The TFP series from the PIMS data shows a downward trend over the 1970s, accompanied by sharp fluctuations associated with the business cycle. A similar pattern is apparent in the published data, although the timing and magnitude of cyclical swings in the 1974–76 period are somwhat different. These differences likely reflect differences

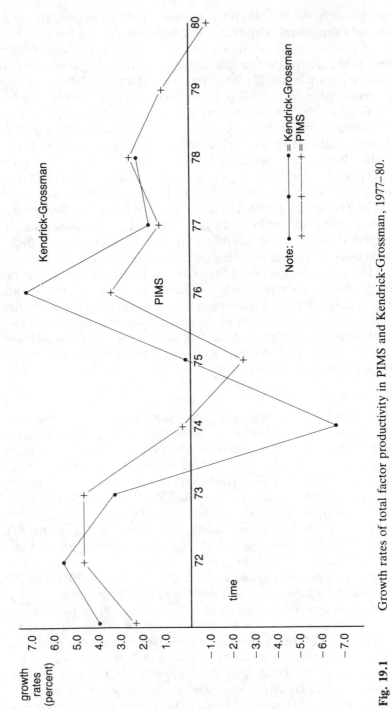

Fig. 19.1   Growth rates of total factor productivity in PIMS and Kendrick-Grossman, 1977–80.

in price indexes as noted earlier and differences in output and input definitions.

We examine the question of a decay in the potency of R & D in table 19.6, where estimates of the TFP model with a time trend and time–R & D interaction are presented. The specification also includes the variables measuring proprietary technology. Line (1) provides a base case, with the time trend entered separately without interaction with R & D intensity. It is evident that TFP growth slowed over the period covered by the data. The coefficient on TIME, negative and statistically significant, implies an average decline of .2 percent per year. The productivity of R & D, however, shows no tendency to decline. In line (2), the TIME–R & D interaction term is negative, but its standard error is quite large, and its actual value is quite small. The estimate of -0.171, for example, implies a decline of 1.7 percentage points in the rate of return over the decade of the 1970s. Evaluated at the midpoint of the time period, the implied rate of return in line (2) is 0.18, quite close to the estimate in line (1).

Lines (3) and (4) present estimates of the TFP model in the sample of firms where DTECH = 0 and in the sample where DTECH = 1. Looking first at line (4), there is some indication of a sizeable drop in R & D productivity, but the evidence is quite weak. The interaction term shows a decline of 4.8 percentage points per year in the return to R & D, but the standard error is relatively large. At the midpoint of the time period, the estimated return to R & D is -5 percent. When line (3) is reestimated without the time trend or the interaction term, the return to R & D is 1.3 percent with a standard error of 8.2.

In line (4) a very different picture emerges. As the estimates in table 19.5 indicated, R & D investment has a substantial impact on TFP growth in businesses where a major change in technology has occurred. In 1975, for example, the estimated return to R & D in line (5) is 26 percent. The interaction term implies a small increase of 0.3 percentage points per year in the return to R & D, but, once again, the standard error is enormous.

The evidence thus suggests that if one looks at businesses where technological opportunity apparently is high and where most of the R & D–productivity effect occurs, there is little statistical support for the notion that the return to R & D declined in the 1970s. In the rest of the sample, where the average return to R & D is very small, there is stronger support for a decline in R & D productivity, but the data do not provide us with a very precise estimate. Further analysis and data may help to clarify trends in the return to R & D in businesses where technological opportunity is low, but for now the evidence is inconclusive.

Table 19.6   Trends in the Productivity of R & D (standard errors in parentheses)

Specification[a] and Sample	CONS	RQ(−1)	TIME	RQTIME (×10²)	DPROD	DPROC	Rate of Return on R & D at Midpoint[b]	R^2	SEE	d.f.
Total Sample:										
(1) TFP	401.0 (157.8)	0.17 (0.05)	−0.20 (0.08)	—	−1.31 (0.49)	1.06 (0.48)	0.17	0.150	11.3	4138
(2) TFP	393.3 (177.8)	3.56 (36.0)	−0.20 (0.09)	−0.171 (1.83)	−1.30 (0.49)	1.06 (1.48)	0.18	0.150	11.3	4137
Tech Change Samples:[c]										
(3) TFP; DTECH = 0	386.2 (220.6)	9.471 (76.03)	−0.20 (0.11)	−4.798 (3.85)	−0.54 (0.59)	0.52 (0.59)	−0.05	0.156	10.7	2995
(4) TFP; DTECH = 1	79.9 (380.2)	−6.00 (46.79)	−0.04 (0.19)	0.317 (2.37)	−2.42 (0.91)	2.25 (0.91)	0.26	0.148	12.8	1133

[a]Lines (3) and (4) are based on observations for firms with DTECH = 0 and DTECH = 1, respectively.

[b]The midpoint of the time period was 1975; the rate of return in that year is equal to the coefficient on $RQ(−1)$ plus the quantity 1975 times the coefficient on RQTIME.

[c]All equations include new, util, and %UN.

19.3 Conclusions and Implications

The estimates presented in tables 19.4–19.6 suggest that R & D investment has a significant positive effect on the growth rate of total factor productivity. All of the specifications examined yielded estimates of an 18–20 percent rate of return to R & D investment. We also found an important connection between the potency of R & D and technical opportunity. And while use of proprietary process technology appears to increase TFP growth, there is only weak statistical evidence of a relationship between the returns to R & D and the use of proprietary processes. Finally the notion that the potency of R & D declined in the 1970s finds little support in these data. Irrespective of model specification or sample used, the coefficient of the time and R & D intensity interaction is both small and statistically insignificant.

The fact that R & D investment continued to have a strong positive effect on productivity growth in the 1970s means that R & D may have played a role in the slowdown of productivity growth. From the early 1970s to the late 1970s, for example, the mean R & D-to-sales ratio fell from 2.7 to 1.9 percent in the PIMS data. With a rate of return to R & D of 20 percent, this would imply a decline of TFP growth of 0.16 percentage points, or about 10 percent of the decline observed over the period. We have found, however, that most of the effect of R & D comes in businesses where technological opportunity is high. Among those firms, a somewhat different perspective emerges. In that group, R & D intensity fell from 3.9 to 3.0 percent, while at the same time TFP growth fell from 4.1 to 3.0 percent. With a return to R & D of about 24 percent, the fall in R & D intensity could explain close to 20 percent of the decline in productivity growth in the high technical opportunity sector.

19.3.1 Further Work

Our analysis has uncovered some interesting relationships and left a number of issues open for further research. One of these issues is the mix between product and process R & D. Both the R & D mix variable and the variable indicating the use of proprietary products had negative effects on productivity growth. This suggests the possibility of some interesting connections between the product development process, choice of technology, and growth of productivity. Analysis of these questions in the PIMS data (and probably in other data sets as well) will have to confront serious measurement problems, especially difficulties in the measurement of prices and output.

There is also the possibility of improving the statistical methodology. All of the estimates presented here are based on ordinary least squares. Except for the use of growth rates, which sweeps out fixed effects, we have ignored the panel structure of the data. Using growth rates does eliminate an important source of autocorrelation, but other forms of covariation in the residuals of a given business may be present and could affect our estimates. If the sample were balanced, there would be little difficulty in applying some form of generalized least squares. An unbalanced design, however, calls for an approach accounting for the differences in numbers of observations within a business over time in calculating the relevant covariance matrix.

Finally, we have not examined explicitly the effect of R & D on costs, prices, and profits. It is well known that under competition the production function and TFP have a dual representation in the cost function as the difference between the sum of share-weighted input price growth rates and the growth of the output price. Although we have no data on the "price" of R & D, its effect in a price-side version of the TFP equation can be estimated using R & D intensity.

References

Abell, D. F., and J. S. Hammond. 1979. *Strategic market planning: Problems and analytical approaches*. Englewood Cliffs, N.J.: Prentice-Hall.

Agnew, C. E., and D. E. Wise. 1978. The impact of R & D on productivity: A preliminary report. Paper presented at the Southern Economic Association Meeetings. Princeton, N.J.: Mathtech, Inc.

Griliches, Z. 1979. Issues in assessing the contribution of research and development to productivity growth. *Bell Journal of Economics* 10, no. 1:92–116.

———. 1980. R & D and the productivity slowdown. *American Economic Review* 70, no. 2:343–48.

Kendrick, J. W., and E. Grossman. 1980. *Productivity in the United States: Trends and cycles*. Baltimore: The Johns Hopkins University Press.

Ravenscraft, David, and F. M. Scherer. 1981. The lag structure of returns to R & D. *Applied Economics* 14, no. 6:603–20.

Scherer, F. M. 1981. Research and development, patenting, and the microstructure of productivity growth. Final report, National Science Foundation, grant no. PRA–7826526.

Schoeffler, Sidney. 1977. Cross-sectional study of strategy, structure, and performance. In Hans B. Thorelli, ed., *Strategy + structure = performance*, ed. Hans B. Thorelli, 108–21. Bloomington: Indiana University Press.

Terleckyj, Nestor E. 1980. *R & D and the U.S. industrial productivity in the 1970s*. Paper presented at the International Institute of Management Conference on Technology Transfer, Berlin.

20 Using Linked Patent and R & D Data to Measure Interindustry Technology Flows

Frederic M. Scherer

20.1 Introduction

This paper discusses in some detail a methodology for linking patent and R & D data to construct a matrix of interindustry technology flows through the U. S. economy. A somewhat aggregated (41 × 53) version of the matrix is presented, as are more detailed disaggregations of the row and column sums.

The motivation for developing these new data was straightforward. During the 1970s the United States experienced a pronounced slump in the rate of productivity growth. One of many possible suggested causes was a slowdown in the emergence or absorption of new technology. New technology comes in significant measure from the research and development (R & D) activities of industrial enterprises. Beginning in the late 1960s, there was a decleration in the growth of company-financed, real (i.e., GNP deflator-adjusted) industrial R & D sufficiently large that, had growth trends continued, real 1979 outlays would have been roughly 40 percent higher than their measured values. The key questions remain: What quantitative links exist between R & D and productivity growth? Did the parameters of any such relationships shift between the 1960s and 1970s?

F. M. Scherer is a professor of economics at Swarthmore College. The underlying research was conducted under National Science Foundation grant PRA–7826526. This paper was drafted at the Max-Planck-Institut für ausländisches und internationales Patent-, Urheber- und Wettbewerbsrecht under a stipendium from the Max-Planck Gesellschaft. The author is also grateful to numerous research assistants, and especially to Chun-Yue Lai, Mary Gianos, Brett Spencer, and Pin Tai, who did most of the patent coding, and to Joe Cholka of the Federal Trade Commission, who provided indispensable computer systems assistance. Use is made of line of business data collected by the Federal Trade Commission. A review by FTC staff has determined that the figures presented here do not disclose individual company line of business data. The conclusions are the author's and not necessarily those of the FTC.

Economists ought to know a considerable amount about this subject. Data on industrial R & D outlays have been collected under National Science Foundation (NSF) auspices since 1953. However, serious obstacles have blocked the path to understanding.

For one, the NSF data leave a good deal to be desired. NSF's industry breakdowns are at a high level of aggregation. With one exception, R & D data are assigned to primary industries by the "whole company" method, which for multi-industry enterprises often leads to substantial misclassification of R & D in companies' secondary lines. NSF's newer and slightly more disaggregated "product field" statistics depart from the "whole company" approach, but the departures are unsystematic. The reporting instructions are confusing and virtually impossible to implement in decentralized companies, and it is evident from a 1975 survey that companies responded to the instructions inconsistently.[1]

Even more important is a fundamental conceptual problem. With the partial exception of the NSF product field data, all research and development spending surveys link R & D to an *industry of origin*—usually the principal industry in which a surveyed enterprise operates. However, it has long been known (e.g., from McGraw-Hill surveys) that the bulk of industrial enterprises' R & D is oriented toward the creation and improvement of *new products* sold to others, as distinguished from R & D on new or improved production processes used internally within the performing company. The latter should in principle lead directly to productivity gains within the industry of R & D performance, assuming that the industry classification is correct. For product R & D, however, the linkage is much less clear. Both behavioral and measurement considerations lead us to believe that performing industries will secure at best only a modest fraction of the productivity benefits from their product R & D.[2]

On the behavioral side, an innovator will capture all the benefits from a productivity-enhancing new product only if it can engage in first-degree price discrimination. Under simple monopoly pricing, some of the benefits will necessarily be passed on to users. And when new product competition is vigorous, price competition may also break out, permitting innovators to retain only a small share of the superiority rents associated with their products.

There are also practical measurement problems. The first step in the compilation of productivity indices is estimating real output, usually by

1. U.S. Bureau of the Census (1975), question 30, reveals that 57 percent of the surveyed R & D expenditures were reported on an end product (i.e., line of business) basis, presumably contrary to instructions. Twenty-nine percent were reported (consistent with the instructions) in technological fields different from the end product fields. For the remaining 14 percent, the technological and end product fields were said to be identical.
2. For surveys of the problems, see Griliches (1979) and Scherer (1979, pp. 200–204).

dividing some dollar measure of output by price deflators. If the price deflators were perfect hedonic price indices, innovators would be found to capture all or most of the productivity benefits resulting from their new products. Few price deflators meet this standard, however. More commonly, the actual prices of new products are linked into a price index at parity with the prices of older products, and the linking often occurs only after the new product has been on the market for a considerable time. One consequence is that price deflators severely underestimate product quality improvements, which in turn means that the measured output of product innovators is lower than it would be if hedonic price indices were used, so that measured productivity gains are not observed at the originating industry stage. (An exception may occur when, because of enhanced monopoly power, profit margins in the product-originating industry rise.) A further implication is that the productivity impact of new products is observed "downstream" at the buying and using industry stage, both because the prices *measured* for inputs used by buying industries do not reflect their superiority value and because (thanks to competition) the prices *actually paid* do not reflect that superiority value. Thus, to ascertain the productivity effects of new product R & D, one must trace the flow of technology from the industry in which a new product originated to the industry(ies) using the product.

The first to propose a solution to this set of problems was Jacob Schmookler (1966, chap. 8). He postulated a kind of input-output matrix of invention flows in which the rows represented industries making an invention, the columns the sectors using inventions, and the diagonal elements process inventions. Row sums correspond more or less closely to the R & D data collected by NSF according to industries of origin. Column sums give the total amount of technology *used* by an industry. With patent data, Schmookler was able to estimate column sums for a small sample of capital-goods-invention-using industries, but his untimely death prevented him from progressing further toward the realization of a complete technology flow matrix.

Since then, Nestor Terleckyj (1974, 1980) combined NSF survey data with conventional input-output statistics to estimate something like Schmooklerian matrix column elements as well as source data row sums. What is described in this paper is an effort to apply methods like those pioneered by Schmookler in estimating more disaggregated matrices at a higher level of precision.

Some substantive results, presented fully in other papers (Scherer 1982a, 1983), are summarized later. However, one finding deserves immediate attention. As expected, process R & D—that is, R & D devoted to improving a firm's own internal production processes—was found to comprise only a small fraction of all company-financed industrial R & D: 26.2 percent when measured by a count of patents and 24.6

percent when measured by linked R & D expenditures data with adjustment for sample coverage. Most industrial R & D is indeed product oriented. A fuller breakdown of the patent count by user category is as follows:

Process inventions		26.2%
Consumer goods products only		7.4
Industrial capital goods products		44.8
Subset used both as producer and consumer goods	7.8%	
Industrial material products		21.6
Subset used both as producer and consumer goods	8.7%	

20.2 The R & D Data

It seems clear that a full understanding of the impact of R & D on productivity growth requires one to go beyond mere industry of origin classifications and find out where the fruits of R & D are actually used. The starting point for such a venture should be R & D or other technological input data of good quality. "Good quality" here means at least three things: reasonable accuracy (recognizing the difficulties of measuring what is and what is not R & D); considerable disaggregation (especially since analyses of R & D–productivity links have proved sensitive to the degree of aggregation); and a correct matching of expenditures with industries. The third criterion, though obvious, deserves further attention. If, for example, as is standard NSF and McGraw-Hill survey practice, all R & D performed by Exxon is assigned to Exxon's primary industry category, substantial amounts of R & D occurring in the organic chemicals and resins, agricultural chemicals, synthetic rubber, office equipment, and communications equipment industries will be wrongly assigned to petroleum extraction and refining. This problem has grown increasingly severe as U.S. corporations have become more diversified. Already by 1972, before the most recent conglomerate merger wave, the average manufacturing corporation had 33 percent of its employment outside its primary (roughly three-digit) field (Scherer 1980a, pp. 76–77).

The data source that best satisfies these three criteria is the Federal Trade Commission's Line of Business survey. The first full survey, covering 443 large corporations,[3] was for 1974. It required reporting companies to break down their privately supported and contract applied R & D outlays, among other financial items, into 262 manufacturing lines of business (LBs), usually defined at the three- or four-digit SIC level, and

3. The count of corporations and lines of business reported here does not agree exactly with official FTC figures because of slight differences in how both corporations and lines of business were consolidated.

14 nonmanufacturing categories. These 1974 line of business data were a principal basis for the work reported here.

They are not without problems. Perhaps most important, 1974 was the first year for which the survey was fully implemented. No survey can achieve perfect reporting, especially on the first iteration for an activity as difficult to measure as research and development. The data were therefore subjected to an extensive verification and correction process before use. Reported company R & D totals were compared to 10–K report R & D figures; individual 1974 line of business figures were compared against 1975 and 1976 reports; and a general check for significant omissions or peculiarities was made. Several classes of difficulties were discovered. First, R & D expenditure reporting was in some instances incomplete. The standard correction was to replace the 1974 figure with the comparable 1975 (or if need be, 1976) figure deflated by the ratio of 1974 to 1975 10–K R & D outlays. Second, some companies failed to distribute their R & D outlays over all relevant lines of business, instead lumping them together in a single (e.g., the largest) line or a few lines. Such problems were normally remedied by applying 1975 or 1976 distributional weights, although, in a few cases, breakdowns were made on the basis of sales or (where some LBs were known to be more R & D intensive than others) patents obtained by the various company LBs. Third, companies were asked to report basic research outlays in a single separate category for the whole corporation. There were quite clearly enormous qualitative differences among companies in what was categorized as basic research. Moreover, a few companies reported *all* their R & D outlays as basic research, and others reported all of their central corporate laboratory outlays, basic or applied, in the basic research category or in some other category such as "services." All outlays reported under basic research were spread by the author over the various lines of business in proportion to private applied R & D outlays. Special problems were resolved through other allocation methods, usually after consultation with company accountants. Fourth, companies were allowed to assign the costs and assets of LBs with sales of less than $10 million to a catchall (99.99) reporting category. This option was exercised fairly frequently in connection with new endeavors that had high R & D outlays but low sales. The clearest cases were reclassified to their proper home industries, but less obvious or more complex cases had to be left as they were, so the 99.99 category (included in "miscellaneous manufactures") entails misclassification and is unusually research intensive. Fifth, companies were permitted to have a limited amount of "contamination" in their reports; that is, activities that should ideally have been accounted for in a different LB, but whose segregation would have imposed appreciable accounting costs. The average level of contamination was on the order of 4 percent (U.S. Federal Trade Commission 1981, pp. 50–53).

For my analysis this poses problems whose solution will be described later. Finally, companies were to report only their domestic, unregulated business activities; foreign operations were excluded. However, when domestic R & D expenditures supported manufacturing operations abroad, the R & D could be prorated between domestic and foreign branches, leading to some understatement of R & D relative to NSF definitions. Nothing could be done about this except to test its effect on the average number of patents received per million dollars of reported R & D expenditures. The elasticity of patenting with respect to the percentage of total corporate sales occurring domestically was found to be -0.14 with a t-ratio of 2.34 (Scherer 1983).

The total amount of company-financed R & D reported by the 443 sample corporations after corrections was $10.64 billion, or 73 percent of the universe total in NSF's 1974 R & D survey. Sample contract R & D outlays (mostly under federal government prime or subcontracts) amounted to $5.97 billion, also 73 percent of the NSF survey estimate.

These percentages are relatively high compared to the FTC sample's coverage of other financial variables such as sales (estimated to be roughly 54 percent of the total manufacturing sector universe) or assets, for which the coverage ratio was approximately 67 percent (U.S. Federal Trade Commission 1981, pp. 69–76). Apparently, the FTC sampled relatively more heavily in R & D–intensive industries. Estimating exact coverage ratios is difficult because the FTC survey emphasized financial accounting variables whereas other universe figures are either heavily contaminated by the mixing of manufacturing and nonmanufacturing activities (e.g., the FTC's *Quarterly Financial Report* series and the Internal Revenue Service's *Statistics of Income* series) or have sales and assets (i.e., under Census reporting rules) defined on quite different bases.

Since individual industry coverage ratios were needed to implement my technology flow matrix concept, a different and rather unorthodox estimation technique was adopted. The basis of comparison was the set of four-digit industry value of shipments concentration ratios published in connection with the 1972 Census of Manufactures. For each industry, concentration curves were interpolated (or sometimes extrapolated) on lognormal probability paper. Aggregations to the FTC line of business category level were carried out following the guidelines in Stigler (1963, pp. 206–11). One could then locate on the relevant concentration curve the maximum fraction of industry sales accounted for by the number of companies reporting in a given FTC line of business. This LB coverage estimate is biased upward to the extent that the company units reporting under the LB program are not uniformly the largest sellers in their industries. Downward biases intrude to the extent that the companies report as "contamination" with some other line of business sales that are reported to the Census Bureau in the correct industry category. The coverage ratios estimated in this way for manufacturing industries (which originated 95.1 percent of total sample R & D) ranged from .06 to .99. The value-added weighted-mean coverage ratio was 0.61—somewhat higher than the FTC's most closely comparable value-of-shipments coverage ratio estimate of 0.54.

The coverage ratios derived in this manner were used to inflate sample line of business R & D outlays and obtain whole-industry estimates. For company-financed R & D, the sum of the inflated values across all lines of business is $14.72 billion, which agrees quite closely with the 1974 NSF survey figure of $14.65 billion. This suggests that measurement errors, sampling ratio estimation errors, sampling errors, and their intersection had on average no serious systematic bias. For contract R & D, there is an evident bias: the sum of the inflated estimates is $6.77 billion, or 18 percent less than the NSF survey universe figure of $8.22 billion.

20.3 Estimating Technology Flow Matrices

What has been described thus far is a procedure for getting R & D expenditure data organized by industry of origin. This is in principle what has been done in other surveys. The improvements consist mainly of considerably greater disaggregation and a more accurate match of expenditures to true origin industries.

Much more difficult and important steps were required to flow those originating industry outlays out to industries of use. The information needed to do so was obtained through a detailed analysis of invention patents. To begin with an overview, a sample of patents was drawn that matched as closely as possible the sample of companies on which R & D data by line of business were available. Each patent was inspected and coded as to industry (LB) of origin and industry(ies) in which use of the invention was anticipated. The industry of origin classifications were employed to link the patents to the lines of business in which corresponding R & D expenditures had been recorded. Each patent then became, in effect, a carrier of the average R & D expenditure per patent in its origin LB, transmitting by a fairly complicated algorithm those expenditures out to the coded, using industries. Summed R & D outlays could then be collected for cells and columns of appropriately aggregated technology flow matrices.

20.3.1 The Patent Sample

The R & D data employed are for companies' 1974 fiscal year, which is centered on the 1974 calendar year. U.S. and West German surveys suggest that the average lag between conception of an invention and filing a patent application is about nine months (Sanders 1962a; 1962b, p. 71;

Grefermann et al. 1974, pp. 34–37). During the mid-1970s, the average period of patent pendency—that is, the lag between application and issue of a U.S. patent—was about nineteen months. Thus, the total lag between invention, which is assumed to accompany R & D expenditure, and the issuance of a patent is estimated to be twenty-eight months. The time span of the patent sample was therefore set for the ten-month period from June 1976 through March 1977, whose midpoint is lagged twenty-eight months from 30 June 1974. Some timing error is inescapable here, since the distribution of patent application to issuance lags is skewed, with a few patents in the sample having been applied for as early as the 1940s. However, 92 percent of the sample patents had applications dated in the years 1974–76.

There is no simple, consistent practice with respect to the names to which corporate patents are assigned. Some patents resulting from corporate R & D go only to the inventor, but this is now extremely rare in large corporations. Some patents issued to corporations are in fact acquired during their pendency from outside or spare-time inventors, but this too, our analysis suggested, also appears to be unusual. The principal company name matching problems come from mergers and the fact that many industrial patents are assigned not to the parent corporation but to some subsidiary. An extensive effort was made to identify patent-receiving subsidiaries. Mergers were identified through the FTC's annual *Statistical Report on Mergers and Acquisitions*. Several protocols were adopted to ensure that patents were in fact linked to the correct 1974 parent companies (see Scherer 1980b, p. 6). In cases where mergers following a parent company's 1974 fiscal year led to an undesirable scrambling of patents, company patent counsel were helpful in providing the needed unscrambling. Failure to have attended to these subsidiary and merger-timing problems would have led to matching error rates on the order of 20–25 percent.

Some seventy-five companies were found to have obtained no patents during the ten-month sample period. For these a more extensive three-year sweep was made, yielding sixty-nine additional patents accorded a weight of 10/36 each. This procedure imparts sampling bias, but a minor bias was considered acceptable in exchange for better coverage of low-patent industries. Unfortunately, there was no feasible means of identifying a universe and weights on the basis of which more efficient stratified sampling techniques could be applied.

Because the R & D expenditure data gathered were for U.S. operations only, patents whose inventor had a foreign address (or in the case of multiple inventors, all or most of whose addresses were foreign) were excluded from the sample.

Altogether, the final patent sample consisted of 15,112 patents counted with unit weights, or 15,062 patents when oversampled company patents are fractionally weighted. After adjustment for foreign inventor exclusions, this was roughly 61 percent of all patents issued during the sample period to U.S. *industrial* corporations (i.e., excluding universities, nonprofit research institutes, patent management firms, retailers, public utility corporations, and the like). Of the 443 sample corporations, 397 were patent recipients. The most prolific assignee, General Electric, received 706 patents originally classified to 51 distinct lines of business.

20.3.2 Patent Classification

Once the patent sample was drawn, the printed specification of each patent was inspected individually by members of a team including an electrical engineering student, an organic chemistry major, a graduate management student with undergraduate honors in chemical engineering, and a "utility infielder" with a joint chemistry-economics major and a farming background. Mirroring the team's specialities, patents were presorted into four groups: electrical inventions, organic chemical inventions, other chemical inventions, and everything else. The primary objective was to classify each patent according to industry of origin and industry(ies) of use. On the latter, up to three specific industries of use (including final consumption) could be identified, or the invention could be coded to either of two "general use" categories: (1) use proportional to the origin industry's normal customer sales distribution (e.g., a machine tool invention); or (2) ubiquitous use throughout the industrial economy (e.g., a corporate jet aircraft invention).

Coding industries of use was for the most part the more straightforward and simpler task. U.S. law requires that inventions be useful to be patentable. Applicants therefore take some pains to point out in patent specifications what the actual or prospective uses of their inventions are. Instructions to classification team members emphasized the importance of coding uses to match as closely as possible the industrial locus where productivity impacts were most likely to occur. In cases of doubt, category (1) general use classification was favored. Of the 15,112 unit value patents in the final sample, 42 percent were classified to one specific nonconsumer industry of use, 11 percent to two specific using industries, 6 percent to three specific using industries, 29 percent to category (1) general use, and 5 percent to category (2) ubiquitous use.

In coding industries of origin, it is not enough to say, for example, "This is a petroleum refining invention, so that must be the relevant industry" (see Scherer 1982b). A catalyst might come from an inorganic chemicals maker, an antiknock additive from the organic chemicals industry, or a process design from a company like UOP, whose home base is engineering services. Origin depends at least as much on how the R & D–performing company is organized as on function. Each classification team member was provided with a set of industry codes in which, according to published information, sample companies purported to operate, along with a qualitative description of the companies' product offerings. Even this, however, was not enough. The objective of the classification effort was not to identify industries of origin that were "correct" in some absolute sense, but to classify the patents in such a way that the origin industry codes corresponded with the LB codes in which enterprises chose to report the R & D expenditures that gave rise to the patents. Because of confidentiality restrictions, however, the structure of companies' LB reporting codes was not, and could not be, known in advance. This required in difficult cases a target-bracketing approach. As many as three industries of origin could be coded. In the original coding, 15.6 percent of the patents had two origin industries and 2.8 percent had three industries of origin. Uncertainty about company account organization was not, however, the only reason for multiple origin codes. Some inventions are genuinely joint: for example, an aerospace company's metal fatigue testing system that can be used in either aircraft or missile assembly operations, or a fuel injection system microcircuit installed in either cars or trucks. Therefore, an additional set of codes was created to guide the ultimate patent–R & D-dollar matching process. Inventions could be coded to be matched with a single preferred industry, and only if that match failed, with others; or a spread over multiple industries of origin could be specified to occur in equal parts; or the spread could be effected in proportion to matched LBs' total R & D expenditures. Additional options existed to deal with problems of vertical integration, for example, when it was expected that an electronic systems producer would report R & D concerning a semiconductor component production process under its systems LB code, even though the production (and hence the productivity impact) was likely to occur in a separate semiconductor plant.

Even after the classification team had acquired considerable expertise from on-the-job-training, 20–30 percent of the patents proved "hard" to classify. An important breakthrough in reducing that fraction was the discovery that, from sources such as telephone books, annual reports, and a rich data base developed by Roger Schmenner at the Harvard-MIT Urban Studies Center, one could tell what specific divisions or industry activities a company had at a geographic location. The company unit location in turn could often be inferred from the residence of the inventor, especially when there were multiple inventors with a similar patent specification address. All industry codings were double-checked by the author against abstracts in the Patent Office's *Official Gazette*. In questionable cases, the entire specification was reviewed. Problem cases resistant to solution by these methods were resolved through telephone calls to company officials or the relevant inventors. In these and other ways, an attempt was made to enforce high standards of accuracy.

In addition to industries of origin and use, the individual patents were coded according to complexity (number of pages and claims), economic characterization (process vs. material vs. capital goods vs. consumer goods), technological characterization (system vs. device vs. circuit vs. composition of matter vs. chemical process), whether the invention originated under a federal government contract, and various other pieces of information. The federal contract invention coding proved to be unexpectedly difficult because, it was learned, contractors did not uniformly comply with the federal requirement that they include a notice of contract support in their patent specification, and the larger contracting agencies lacked complete records of their government-supported, contractor-owned inventions. Through an extensive effort, 325 contract inventions were indentified, but it is believed that another 75 or so eluded the search. Since all were military-related, later adjustments could be made to minimize biases in estimating technology flow matrices.

A tape containing the original coding of the individual patents is available from the author on a cost reimbursement basis.

20.3.3 The Patent–R & D Link

With the main patent coding task completed, the patent tape was brought to the FTC's Line of Business program office in Washington, where the link to R & D data broken down into individual company lines of business commenced. At this point, the original list of 276 LB categories was condensed to 263, partly because it had proved impossible to make distinctions between certain origin industry categories (e.g., ethical and proprietary drugs, electric motors and motor controls, and storage vs. primary batteries) and partly to mirror industry consolidations made by the FTC for disclosure avoidance reasons. Following these consolidations, the number of individual company LBs to which a patent might be linked totalled 4,274. The average company broke its operations down into 9.65 LB categories.

After certain origin industry recodings were made to correct anticipated matching problems, the first link was executed. Among the 15,112 sample patents, there were matching problems on approximately 18 percent, including 1,101 patents on which no match at all was achieved and roughly 1,570 on which multiple origins had been coded, some but not all of which matched. Each patent with a partial or total matching problem was analyzed against company LB program submissions to determine the reason for the problem and to effect, if appropriate, a correct recoding. Extremely valuable in this effort were Schedule II of the FTC's LB reporting form, which broke down reporting LB sales to the five-digit product level of detail, and an appendix that gave the geographic location of every major establishment covered by a reporting LB. The principal reason for matching problems was that companies had not organized their LB reports according to our expectations. "Contaminated" reporting was one subreason. Another was that our salvo approach to questionable classifications had indeed both hit and missed the target. When all recodings were completed, there were 306 three-industry matches (as compared to 429 initial three-industry codings) and 1,619 two-industry matches (as compared to 2,359 codings originally). The remainder were single-industry matches. Altogether, 1,851 of the 4,274 reporting LBs had at least a finite fraction of a matched patent. Of the 3,003 individual company LBs that reported nonzero company-financed R & D outlays, 1,691 had matched patents.

Because at low R & D expenditure levels (i.e., less than $1 million per year) the probability of patenting is finite but well below 1.0, the $732 million of private R & D in individual company LBs with no patents was spread proportionately over LBs *with* patents in the same industry before computing the average amount of private R & D associated with a patent. The average value of this inflation factor was 7.3 percent, although it ranged from zero to as much as 30 times the R & D of patent-receiving LBs within an industry.

For each individual company LB with patents, the average private R & D cost per patent, that is, the quotient of inflated R & D divided by the weighted sum of matched patents, was computed. For each patent, the average cost of the patent was then tallied. When the patent had more than one matched industry of origin, the cost was a weighted average of the originating LBs' average costs, with the weights having either been prespecified to be equal for truly joint inventions or proportional to the matched originating LBs' total R & D outlays. Government contract invention patents were handled differently because it was known that not all such inventions had been identified. For them, the average contract R & D cost input was an industrywide average, not an average within individual company LBs.

The final output of this matching effort consisted of two computer tapes, one organized by individual company LB and one by individual patent. The patent tape contains for each patent all original input data plus matched LB codes, the weights assigned each matched LB, the average company-financed R & D expenditure underlying the patent (hereafter, ACP), and (when relevant) the average federal contract R & D expenditure underlying the patent (FACP). The company LB tape contains R & D expenditure totals, patent counts, average R & D costs per patent, and twelve weighted average values of characteristics (e.g., proportion of process patents, proportion of consumer goods patents, average patent length, etc.) of patents in that LB's portfolio. Since these tapes include individual company line of business information, they can only be accessed within the FTC Line of Business program office.

20.3.4 Technology Flow Matrix Estimation

The completed patent tape became a primary input into the computer programs creating technology flow matrices for the U.S. industrial economy. The essence of the problem was to take the R & D dollars (ACP or FACP) associated with a patent, inflate them by the reciprocal of the origin industry's sample coverage ratio, and then flow them through from industry(ies) of origin to industry(ies) of use, accumulating sums for each relevant cell.

The first substantive step was to retag patents by industry of origin. When the original coding procedure specified a preference for some single industry of origin, that preferred industry code was adopted, whether or not a match to LB reports had been achieved. In the absence of such a preference, multiple origin patents were divided among industries of origin in proportion to the weights determined through the earlier matching procedure. Patents originally coded as having probable vertical integration characteristics received special treatment. If the invention was a process and the vertical integration industry code differed from the industry code under which companies were expected to report their financial data, the invention was assigned to the industry of origin in which its actual use as a process was anticipated, whether or not an LB code match had been achieved. To have done otherwise would have generated process invention data inaccurate for purposes of analyzing the relationship between R & D and productivity growth.

After inflation to correct for differing origin industry coverage ratios, the inventions and their accompanying R & D dollars were flowed out to industries of use. For inventions coded as having a single industry of use or process inventions, this was quite simple. The R & D dollars went fully to the specific industry of use or, in the case of process inventions with multiple surviving origins, were divided among using industries in proportion to origin industry weights.

For inventions with multiple or general uses, the problem was more complex. Plainly, some using industries will use an invention more intensively than others. The question is: How does one determine the relative weights? The basic solution chosen was that inventions and their R & D would be flowed out to multiple using industries in proportion to the using industries' purchases from the origin industry. The natural basis for the needed "carrier matrix" $\underset{\sim}{A}$ was the 1972 input-output tables for the U.S. economy. However, substantial modifications had to be made before the input-output carrier matrices were consistent with our objective of tracing technology flows in such a way as to analyze their productivity growth impact.

The starting point was the 496-order 1972 current transaction matrix recording the use of commodities by industries (U.S. Department of Commerce 1979). This had to be aggregated down to the 263×286 industry level at which the most detailed technology flow estimates were to be prepared. Certain disaggregations also had to be made, usually on the basis of simple relative size weights. However, for the industrial gas, glass, and electron tube industries, new row vectors were estimated from primary data. Also, many input-output industries have large diagonal elements associated with interplant, intraindustry transfers. These might be viewed as surrogates for process inventions, but the correspondence is at best strained, and internal process inventions were in any event separately accounted for in our analysis. Therefore, the diagonals were "cleaned out" so that they did not exceed the row industry's proportionate share of aggregate output, except in a few cases (such as organic chemicals) where productivity might plausibly have been affected by substantial intraindustry technologically advanced intermediate materials transfers. Corrections based on primary data were also made to defense-oriented rows to reflect the fact that the sales pattern for products emerging from private R & D is different from the contract R & D pattern.

A more serious problem was posed by the input-output transactions matrix's handling of capital-goods-producing industries, which tend to be especially important R & D performers. Most of those industries' output is reported as sales to the "gross domestic fixed investment" column of final demand. This is obviously wrong in terms of identifying the industries in which capital goods technology is actually used. Basing technology use estimates on the small fraction of total output spread over using industries in the intermediate commodity output sector could be quite inaccurate. Therefore, the separately available capital flows input-output table for 1972[4] was integrated with the current transactions matrix—something that, to the best of my knowledge, has not been done previously. The most detailed version of that table is available only in an 80 column (i.e., using sector) version, so the columns had to be disaggregated to 286 industries. For any capital flow matrix column spanning two or more of our industries, cell entries were split in proportion to the disaggregated industry's 1972–74 new capital investment as a fraction of the capital investment by all LB industries encompassed by the input-output column sector. Row aggregations to our level of detail were routine. Once a properly dimensioned capital flow matrix was available, the transaction and capital flow matrices were integrated. If T_{ic} is the capital formation element of the ith row in the current transaction matrix and I_{ij} is a representative element of the capital flow matrix, a representative element T_{ij}^* of the revised transaction matrix is formed as $T_{ij}^* = T_{ij} +$

---

4. For a summary, see Coughlin et al. (1980). The detailed capital flow data were available on tape BEA IED 80–001. The detailed current transactions matrix is on tape BEA IED 79–005.

T_{ic} $(I_{ij}/\Sigma_{i=1}^{285} I_{ij})$. This was done for seventy capital-goods-producing industries with positive general use (category [1] or [2]) inventive activity.

Input-output conventions concerning the construction industry(ies) as a using industry posed similar problems. Substantial fractions of the output of the heating equipment, fabricated structural metal, office partitions, valves and pipe fittings, bathtub, and other industries are shown as used in the construciton sector. It is true that construciton is a large purchaser of such items, but it purchases them to install them for use by others. Inventions whose main utility lay in greater ease of installation by the building trades were specifically coded as having a construction industry use. Allowing the received input-output table structure to stand for general-use inventions would have inaccurately measured productivity-affecting technology flows. Consequently, output to construction industry subsectors was rerouted to "downstream" using sectors to the extent that the input-output table detail permitted. Where it did not, all or part of the remaining output originating from twenty-nine industries and reported as used in construction was spread over all using industries in proportion to the industries' purchases of capital goods from the construction sector.

This problem has still another analog. Consider the output of a technologically important component-producing industry, such as semiconductors. According to the input-output tables, that industry's output flows to using industries like computer manufacturing, radio and television set production, and communications equipment. Yet who actually realizes the productivity-enhancing benefits of a more efficient large-scale integrated circuit: the computer maker who installs it in his newly designed computer, or the university or bank or manufacturer who purchases (or leases) and uses the faster, higher capacity computer? Some sharing of benefits may occur, but if the forces of competition are working with reasonable vigor or if deflators for new component-embodying systems are less than perfect hedonic price indices, one would expect much of the productivity-enhancing benefit from component product inventions to be passed on from the industry that assembles the components to the industries that buy and use the products embodying the improved components. To implement this notion, twenty-two industries that specialized in supplying components to some set of first-order using industries (usually assembly-type industries) were identified.[5] Relevant parts of component industries' output were subjected to a second-order (or for synthetic

---

5. The industries were weaving mills, fabric knitting mills, organic fibers, tires and tubes, rubber hose and belting, flat glass, pressed and blown glass, internal combustion engines, pumps, antifriction bearings, compressors, speed changers and industrial drives, mechanical power transmission equipment, automotive carburetors etc., vehicular lighting equipment, electron tubes, cathode-ray tubes, semiconductors, other electronic components, starter and traction batteries, aircraft engines, and buttons, zippers, etc. Not all elements in these industries' rows were subjected to second-order flows. Only those elements that were preponderantly of a "component sale to further assemblers" nature were so handled.

fibers, third-order) flow correction. Thus, let T_{ij}^* be the integrated first-order matrix sales of component origin industry i to assembly industry j. Then for any element k in industry j's output use row, the adjusted value is

$$T'_{jk} = T_{jk}^* + T_{ij}^* (T_{jk}^* / \sum_{k=1}^{285} T_{jk}^*),$$

with T_{ij}^* set equal to zero before second-order flows to row i $(i \neq j)$ are computed. Because it was not clear a priori whether the benefits of component product inventions would in fact be passed on as measured productivity gains to second-order buyers, complete carrier matrices were calculated both with and without these component flow adjustments, and corresponding pairs of technology flow matrices were estimated. In fact, regression analyses of productivity growth revealed that technology flow variables *without* second-order component flow adjustments consistently had slightly greater explanatory power (see Scherer 1982a).

The result of these modifications was a set of input-output tables unlike any previously available, but suited as well as possible to performing the carrier matrix role in the estimation of technology flow matrices. All row elements were converted to ratios whose sum over all using sectors, except end consumption, equalled unity. After unneeded rows were purged and some further aggregation, a set of four carrier matrices $\underset{\sim}{A}$—one each with and without second-order component flows at the 263 × 285 and 263 × 56 levels of aggregation—was taken to the Federal Trade Commission to be linked with the patent data tape for the final technology flow matrix estimation stage.

For general-use inventions of category (1), the R & D cost of a patent ACP flowed through to using industry j (excluding the final consumption sector) from single origin industry i with coverage ratio c_i and carrier matrix coefficient a_{ij} was as a first approximation a_{ij} (ACP/c_i). For general-use inventions of category (2) (i.e., ubiquitous industrial use), a_{ij} ratios relating using industry value added to value added in all industries were applied. When there were three or fewer (e.g., M) specific industries of use, the coefficient for the kth designated using industry was $a_{ik}/\sum_{k=1}^{M} a_{ik}$, except that when this value was less than 0.15, the coefficient was set equal to 0.15 and all the specifically designated using industry coefficients were renormalized to sum to unity. Although arbitrary, this convention ensured that specific-use industries received some of an invention's value even when input-output tables showed no relevant transactions between the origin and using industry pair.

20.3.5 The Public Goods Problem

Under the procedures described thus far, R & D dollars (or patents) were flowed out to using industries in such a way that the sum of the flows equalled the sum of the origin industry's R & D. An exception was made for final consumption goods uses, for which no productivity analysis could have been contemplated. For any patent covering consumer goods, the final consumption sector column received the full R & D cost of that patent, whether or not there were also industrial uses. In effect, the consumer goods applications of such inventions were treated as public goods not reducing the amount of R & D available for transmission to industrial sectors.

It can plausibly be argued that multiuse industrial inventions should also be handled as public goods, with use by industry k not reducing potential use by industry j. There are, however, both theoretical and practical difficulties in implementing such a public goods approach. It can be shown (Scherer 1983) that as the number of using industries (i.e., the scope of the market) increases, firms will do more R & D and receive more patents, all else (such as the size of the average using industry) held equal. This increase in inventive activity may be channeled in either or both of two directions: perfecting a given narrow array of products, or increasing the variety of products geared to specialized demands of the diverse using markets. When product variety increases with rising market scope, particular inventions may be applicable in only a subset of the relevant using industries. This goes against the spirit of the public goods hypothesis. When R & D emphasizes perfecting a narrow array of products, other problems arise. If the same product is sold in many markets, it may pay to carry the product's development to a high state of refinement. For any single using market, considerable progression into the stage of diminishing marginal benefits is implied. In contrast, for single industry of use inventions, development is more apt to cease where the marginal benefit is high. This difference in marginal benefits per using industry is difficult to capture under a public goods approach.

If increased market scope led mainly to the perfection of a fixed range of products rather than increased product differentiation, one might also expect (because of increasing marginal invention costs) the R & D cost per patented invention to be greater, the broader the scope of the market. Crude tests of this increasing cost hypothesis failed to provide support (see Scherer 1983). There was no significant evidence of systematically rising R & D cost per patent as the number of using industries increased from one to three and then from category (1) general use to category (2) ubiquitous use, all else equal.

Despite the possibility of increasing product differentiation and diminishing single-market marginal benefits as the number of using industries rises, an attempt was made to construct technology flow matrices under the assumption that multiple industry-of-use inventions were public goods. The conceptual problems were substantial. The very nature of public goods makes a certain amount of arbitrariness unavoidable. A basic guiding principle was that even though use by one industry should not detract from use by another, industries purchasing large amounts of an origin industry's output should enjoy a larger technology flow than relatively small purchasers. One alternative considered and ultimately rejected was to multiply each ACP or FACP value by the numbers equivalent of the Herfindahl-Hirschman index for origin industry carrier matrix $\underset{\sim}{A}$ row elements before flowing out general-use invention values in proportion to the carrier matrix a_{ij} coefficients.[6] Instead, a suggestion made at an NBER workshop by Richard Levin was adopted. For any multiple industry of use invention, the using industry with the largest a_{ij} value was assigned a unit value and all other industries' a_{ij} coefficients were normalized to this value. That is, if the maximum a_{ij} is a_{im}, the coefficient for industry k would be a_{ik}/a_{im}, so the R & D dollars distributed to that industry would be (a_{ik}/a_{im}) (ACP/c_i).

This convention, like the numbers equivalent approach, has the property of assigning greater weight to individual inventions, the larger the number of industries using the invention is and the more equal in size the using industries are. Ubiquitous-use inventions in particular (with a numbers equivalent value of nearly 24) received far more weight in total than specific using industry inventions. Whether such weighting is appropriate cannot be determined on a priori grounds; the question is essentially an empirical one.

Another problem with the public goods approach is that, because R & D dollars are in effect double counted, estimated R & D coefficients in regressions explaining productivity growth cannot be interpreted as steady-state returns on R & D investment. This is a significant disadvantage relative to the private goods approach, under which such rate of return inferences can (with appropriate caveats) be drawn.

Given the conceptual and practical difficulties faced in implementing a public goods approach to technology flow estimates, the question of

6. Thus, with 285 industries of use, the numbers equivalent for origin industry i is $1/\Sigma_{j=1}^{285}$, a_{ij}^2, where a_{ij} is an element from the carrier matrix $\underset{\sim}{A}$.

This numbers equivalent is in effect a purchasing industry dispersion index and, because there is a tradition of using such indices in industrial organization studies of pricing behavior, it has interest in its own right. For the 172 (out of 263) LB categories on which complete capital flows, construction, and other corrections were made, the median numbers equivalent value without adjustment for second order component flows was 8.8. The mean was 13.4. The highest values were for miscellaneous plastic products (71.3), paperboard containers (61.7), conveyors (48.2), industrial trucks (47.4), and metal-cutting machine tools (43.8). Twenty-one industries had values in the 1.00–1.99 range. The numbers equivalent for the total value added of all industries (i.e., the ubiquitous-use carrier matrix row) was 23.9.

It should be noted that this analysis calls attention to what may be a serious problem in prior studies using purchasing industry dispersion indices. It is not clear what those studies do about the gross capital formation element in input-output transaction matrix rows. If it is included in the computation, there will usually be serious understatement of buyer dispersion for capital goods industries relative to what one would obtain integrating the transactions and capital flow matrices, as should be done. If it is excluded, actual sales patterns may be badly measured from intermediate output data alone.

which approach—public or private—to use in productivity analyses was left open. Some evidence will be presented in a later section.

20.4 The Output

A principal end product of the effort described here is a set of technology flow matrices and vectors. Full matrices were constructed only at the 48 row by 57 column level of aggregation. These were estimated both for patents and company-financed R & D dollars under both the private and public goods assumptions, with and without adjustment for second-order component invention flows. For federal government contract R & D outlays, similar matrices were constructed only under the public goods assumption. Table 20.1 provides an example of a technology flow matrix for company-financed R & D expenditures. It is aggregated further to the 41 × 53 level, mainly to minimize confidential data problems. It is defined under the private goods assumption (except for final consumption) *with* adjustment for second-order component invention flows. The rows are industries of origin; the columns are industries of use; and the diagonal elements approximate internal process inventions (except for a few sectors like organic chemicals with extensive intraindustry intermediate product invention flows). All entries are in millions of dollars. Blank cells denote R & D flows of less than $50,000. Entries marked "d" had to be suppressed to comply with the FTC requirement that no underlying R & D data be disclosed for any group of fewer than four companies.

Examining row 3,4, we see that a majority of food and tobacco products industry R & D is internal process oriented, with most of the remainder flowing, not surprisingly, into final consumption or trade (i.e., restaurants and food stores). Reading down column (3), we see that the food products sector used appreciable amounts of R & D embodied in products purchased from the paper, miscellaneous chemicals (16), plastic products, fabricated metal products (e.g., containers), other machinery, office equipment, motor vehicles, and instruments industries. For food and tobacco products, the balance between R & D originated ($444.9 million) and R & D used ($493.4 + 29.8 = $523.2 million) is fairly even. This is not true for all sectors. At one extreme among manufacturing industries is the printing and publishing sector, which originated $67.4 million of R & D but used $147.7 million. At the other extreme is farm machinery, which originated $199.3 million but used only $19.2 million. Nonmanufacturing industries, as has been well known, originate very little R & D, but they use roughly half of the R & D originating in the manufacturing sector.

The appendix presents more disaggregated industry R & D sums classified in three ways: by industry of origin, by industry of use with second-order component flows under the private goods assumption, and

Table 20.1   Technology Flow Matrix

	Origin R & D	Agriculture (1)	Mining (2)	Food products (3)	Tobacco products (4)	Textile products (5)	Apparel products (6)	Lumber & wood (7)	Furniture (8)	Paper products (9)	Publishing (10)	Inorganic chemicals (11)
Agriculture & forestry	128.1	d	—	—	—	—	—	—	—	—	—	—
Mining, exc. petroleum	60.3	—	45.2	—	—	—	—	—	—	—	d	—
Food and tobacco products	444.9	7.9	—	257.8	20.4	0.2	—	—	—	d	d	—
Textile mill products	179.3	0.8	—	0.2	—	128.4	8.1	0.3	0.8	0.5	0.2	—
Apparel & leather products	55.5	d	d	d	d	d	16.5	d	d	0.1	d	—
Lumber & wood products	72.6	0.4	—	0.2	—	0.2	—	64.2	0.6	0.2	0.2	—
Furniture	51.1	0.1	0.1	0.3	—	0.1	0.1	0.3	7.8	0.2	0.2	—
Paper mill products	202.3	6.0	0.1	25.8	0.1	1.2	0.3	0.4	0.4	86.4	3.1	0.1
Printing & publishing	67.4	0.1	—	0.4	—	0.1	—	—	—	0.1	32.7	—
Industrial inorganic chemicals	159.2	1.2	0.1	d	0.8	1.7	0.1	0.1	0.1	4.6	0.3	90.8
Industrial organic chemicals	297.2	3.0	0.2	2.7	d	11.8	0.4	1.3	0.1	5.9	d	1.1
Synthetic resins, fibers, rubber	601.6	1.6	0.5	2.8	d	32.0	11.8	1.1	0.8	13.9	0.5	d
Pharmaceuticals	557.3	32.0	—	0.2	—	—	—	—	—	d	—	—
Agricultural chemicals	186.7	142.8	—	d	—	—	—	—	—	d	—	—
Paints, toiletries, explosives & other chemical products	485.7	3.9	4.2	11.1	0.2	26.3	0.2	7.8	4.5	20.5	13.4	0.7
Petroleum extraction & refining	380.3	4.0	0.7	1.1	0.1	0.3	0.4	0.9	0.1	1.0	0.3	0.2
Rubber & plastic products	419.8	13.7	2.8	18.1	0.7	3.0	1.1	d	3.5	5.0	2.2	0.7
Stone, clay & glass products	265.0	0.9	0.1	0.1	0.1	d	0.1	2.4	1.0	d	0.3	0.1
Ferrous metals	189.3	d	0.1	0.1	—	—	—	0.1	0.4	—	—	—
Nonferrous metals	156.9	d	0.1	d	—	0.1	0.1	0.3	0.2	0.1	0.4	0.2

Fabricated metal products	552.7	5.3	3.0	28.0	0.2	0.7	0.7	4.5	2.7	7.4	0.9	1.0
Engines & turbines	282.2	10.4	3.2	2.2	0.3	0.6	0.6	1.1	0.3	2.5	0.9	0.4
Farm machinery	199.3	165.4	—	—	—	—	—	—	—	—	—	—
Construction, mining & materials handling equipment	351.2	2.0	65.9	4.6	0.3	1.2	0.3	7.3	0.3	1.3	0.6	0.3
Metalworking machinery	121.5	5.4	0.2	0.4	—	0.1	—	2.8	0.5	d	0.1	—
Other machinery	691.0	17.6	8.0	56.9	1.5	18.9	1.9	7.8	1.1	20.9	40.1	5.0
Computers & office equipment	1153.0	2.9	2.4	15.9	0.6	9.4	3.4	1.6	1.7	10.1	13.4	1.8
Industrial electrical equipment	205.9	0.3	1.6	1.3	—	1.1	0.2	0.5	0.3	2.5	0.3	0.7
Household appliances	102.7	0.1	—	0.1	—	0.4	2.2	0.6	—	—	—	—
Lamps, batteries, ignition, X-ray & other electrical equipment	233.3	8.1	0.6	2.5	0.1	0.5	0.3	1.3	0.4	0.7	0.7	0.2
Radio & communication equipment	1227.7	5.1	1.6	5.3	0.7	1.5	1.5	1.5	0.7	1.7	2.4	0.5
Electronic components	594.9	0.5	0.2	0.7	—	0.6	0.2	0.2	0.1	0.4	0.7	0.1
Motor vehicles & equipment	1518.0	78.0	10.6	25.4	1.1	5.0	3.6	17.1	3.9	5.4	10.0	0.9
Aircraft	659.4	0.3	—	0.2	—	—	—	0.1	—	0.2	0.1	—
Missiles, spacecraft & ordnance	122.7	—	—	d	—	—	—	—	—	—	—	—
Other transportation equipment	140.1	—	d	—	—	—	—	—	—	—	—	—
Measuring & medical instruments, photo equipment & timepieces	1036.4	6.3	2.1	8.1	0.8	3.3	1.5	2.0	0.9	7.2	20.7	2.4
Miscellaneous manufactures	211.6	d	0.1	0.6	0.1	0.9	1.3	0.6	0.3	0.2	0.5	0.1
Trade & finance & real estate	39.7	—	—	—	—	—	—	—	—	—	—	—
Transportation & public utilities	47.2	—	—	—	—	—	—	—	—	—	—	—
Construction & services, including R & D services	266.0	0.7	d	0.7	0.1	0.2	0.2	0.2	0.1	0.3	0.3	d
Total R & D Dollars Used		561.8	157.3	493.4	29.8	250.8	57.2	131.1	33.7	206.0	147.7	108.8

Table 20.1 (cont.)

	Origin R & D	Organic chemicals (12)	Synthetic resins etc. (13)	Pharmaceuticals (14)	Agri. chemicals (15)	Other chem. products (16)	Petroleum (17)	Rubber & plastics (18)	Leather products (19)	Stone, clay & glass (20)	Ferrous metals (21)	Non-ferrous metals (22)
Agriculture & forestry	128.1	—	—	—	—	—	—	d	—	—	—	—
Mining, exc. petroleum	60.3	—	d	—	—	d	—	d	—	—	—	—
Food & tobacco products	444.9	0.1	0.1	d	—	d	0.1	0.1	0.9	0.1	0.1	0.1
Textile mill products	179.3	d	—	—	—	—	0.1	21.6	0.9	0.1	d	d
Apparel & leather products	55.5	d	—	—	—	d	d	d	—	d	d	d
Lumber & wood products	72.6	—	—	—	—	—	0.1	0.1	—	0.3	0.2	0.2
Furniture	51.1	—	—	—	—	0.1	0.1	0.1	—	0.1	0.1	0.1
Paper mill products	202.3	0.1	1.0	0.7	0.1	1.0	1.5	1.2	0.2	0.9	0.1	0.1
Printing & publishing	67.4	—	0.1	0.1	—	0.1	0.1	0.1	—	0.1	0.1	d
Industrial inorganic chemicals	159.2	3.3	1.8	0.2	0.3	6.1	9.3	0.4	—	0.9	2.5	d
Industrial organic chemicals	297.2	163.3	24.1	1.2	2.3	19.0	7.1	13.8	d	2.7	1.6	2.2
Synthetic resins, fibers, rubber	601.6	0.8	70.8	d	—	23.9	2.3	169.1	1.7	6.4	0.9	10.0
Pharmaceuticals	557.3	—	—	71.0	—	d	—	d	—	—	—	—
Agricultural chemicals	186.7	0.1	—	d	34.2	d	—	d	—	—	—	—
Paints, toiletries, explosives & other chemical products	485.7	1.4	9.6	1.0	—	85.4	13.8	16.2	2.5	5.4	3.5	3.8
Petroleum extraction & refining	380.3	0.3	0.3	0.1	0.1	0.8	312.7	0.7	0.1	0.6	0.6	0.4
Rubber & plastic products	419.8	0.8	1.3	1.3	0.1	4.3	3.8	203.0	1.9	2.5	1.0	5.1
Stone, clay & glass products	265.0	0.2	0.1	d	—	1.3	d	d	—	155.3	11.3	0.3
Ferrous metals	189.3	d	—	d	d	—	d	0.1	—	0.1	164.2	0.3
Nonferrous metals	156.9	0.4	—	0.1	—	0.2	d	0.1	0.1	0.3	2.8	102.9

Fabricated metal products	552.7	2.7	1.5	0.7	1.4	d	7.4	1.1	1.3	12.3	2.3
Engines & turbines	282.2	1.0	0.5	0.2	0.3	0.5	6.5	0.6	1.1	2.0	1.1
Farm machinery	199.3	—	—	—	—	—	—	—	—	—	—
Construction, mining & materials handling equipment	351.2	0.8	0.7	0.2	0.2	0.3	17.7	0.4	5.8	3.4	1.3
Metalworking machinery	121.5	—	0.1	—	1.6	d	0.1	1.4	0.9	14.5	7.5
Other machinery	691.0	10.1	5.4	2.1	1.3	3.4	16.2	9.7	13.6	35.4	9.1
Computers & office equipment	1153.0	5.0	4.8	6.2	0.5	8.6	20.6	8.1	8.5	19.0	5.3
Industrial electrical equipment	205.9	1.6	0.6	—	—	0.3	4.0	0.5	d	8.2	4.9
Household appliances	102.7	—	—	—	—	—	0.1	—	—	—	—
Lamps, batteries, ignition, X-ray & other electrical equipment	233.3	0.4	0.3	0.2	0.1	0.3	1.2	2.0	0.8	4.6	0.5
Radio & communication equipment	1227.7	0.9	0.6	0.6	0.2	1.0	7.6	1.6	1.9	2.4	1.0
Electronic components	594.9	0.3	0.2	0.2	0.1	0.3	0.9	0.4	0.6	0.7	0.2
Motor vehicles & equipment	1518.0	2.2	0.9	1.7	0.6	2.3	16.7	2.6	11.4	4.2	2.6
Aircraft	659.4	0.1	—	—	—	—	0.5	0.1	0.1	0.1	0.1
Missiles, spacecraft & ordnance	122.7	d	—	d	—	—	d	—	—	—	—
Other transportation equipment	140.1	d	—	—	d	—	d	—	—	0.1	—
Measuring & medical instruments, photo equipment & timepieces	1036.4	5.3	2.8	4.2	1.3	4.3	13.5	4.0	5.1	6.6	3.1
Miscellaneous manufactures	211.6	0.1	0.1	0.1	—	0.2	1.4	0.3	0.2	0.4	0.1
Trade & finance & real estate	39.7	—	—	—	—	—	—	—	—	—	—
Transportation & public utilities	47.2	—	—	—	—	—	—	—	—	—	—
Construction & services, including R & D services	266.0	5.5	d	0.1	d	—	24.4	0.3	0.3	5.0	0.7
Total R & D Dollars Used	207.6	332.1	95.3	45.7	180.3	496.5	470.0	16.6	232.0	307.8	166.1

Table 20.1 (cont.)

	Origin R & D	Fab. metal products (23)	Engines & turbines (24)	Farm machinery (25)	Construction equip. (26)	Metalworking mach'y (27)	Other machinery (28)	Office equipment (29)	Indust. electrical (30)	Appliances (31)	Other electrical (32)	Communications equip. (33)
Agriculture & forestry	128.1	—	—	—	—	—	—	—	—	—	—	—
Mining, exc. petroleum	60.3	—	—	—	—	—	—	—	—	—	—	—
Food & tobacco products	444.9	d	—	—	—	—	—	—	—	—	—	—
Textile mill products	179.3	d	—	—	d	—	0.2	—	—	0.1	d	d
Apparel & leather products	55.5	d	—	—	d	—	d	—	—	—	d	0.1
Lumber & wood products	72.6	0.3	—	—	—	—	0.1	—	—	0.1	0.1	0.5
Furniture	51.1	0.2	—	—	0.1	—	0.2	0.1	0.1	0.1	0.1	0.2
Paper mill products	202.3	0.6	—	—	—	—	0.3	0.1	—	0.2	d	0.1
Printing & publishing	67.4	0.1	—	—	—	—	0.1	—	—	—	—	0.1
Industrial inorganic chemicals	159.2	1.6	d	0.1	0.1	0.1	0.2	d	—	—	d	0.2
Industrial organic chemicals	297.2	0.8	—	—	—	—	0.4	d	—	0.1	0.6	d
Synthetic resins, fibers, rubber	601.6	2.7	—	—	—	—	0.5	0.1	0.8	1.9	1.9	1.1
Pharmaceuticals	557.3	—	—	—	—	—	—	—	—	—	—	—
Agricultural chemicals	186.7	—	—	—	—	—	—	—	—	—	—	—
Paints, toiletries, explosives & other chemical products	485.7	31.8	0.1	0.6	0.8	0.3	2.0	2.4	2.0	5.7	0.9	d
Petroleum extraction & refining	380.3	1.5	0.1	d	0.1	0.2	0.7	0.1	0.2	—	0.1	0.1
Rubber & plastic products	419.8	4.6	0.2	0.4	0.9	0.3	1.6	1.1	1.0	1.5	2.6	1.0
Stone, clay & glass products	265.0	2.7	1.1	0.1	0.4	0.7	d	0.1	d	0.3	0.7	1.0
Ferrous metals	189.3	9.7	d	0.3	0.6	0.3	1.2	d	1.7	0.4	0.3	0.3
Nonferrous metals	156.9	7.0	0.8	0.2	0.3	d	1.3	0.2	1.3	0.3	2.3	0.6

Fabricated metal products	552.7	127.7	3.0	1.0	2.0	1.5	7.2	0.8	1.6	1.4	1.6	1.8
Engines & turbines	282.2	1.9	38.9	0.3	1.7	0.3	3.9	0.2	0.9	0.2	0.3	0.5
Farm machinery	199.3	0.1	—	d	—	—	—	—	—	—	—	—
Construction, mining & materials handling equipment	351.2	1.9	1.0	0.5	16.0	0.4	2.0	0.7	0.4	0.4	0.5	1.0
Metalworking machinery	121.5	15.1	1.0	1.3	2.2	d	7.4	0.8	1.8	0.7	2.0	1.3
Other machinery	691.0	16.0	1.1	0.8	1.6	2.7	64.9	1.5	1.3	8.0	2.5	1.8
Computers & office equipment	1153.0	10.2	5.5	3.5	7.3	2.9	12.9	110.5	4.2	1.4	3.3	16.5
Industrial electrical equipment	205.9	5.1	0.7	0.6	2.9	2.2	15.6	2.6	34.4	3.8	1.7	0.7
Household appliances	102.7	0.1	—	—	—	—	—	—	—	6.8	d	0.1
Lamps, batteries, ignition, X-ray & other electrical equipment	233.3	1.3	0.4	1.6	1.4	0.3	1.3	0.6	0.2	0.4	27.8	0.6
Radio & communication equipment	1227.7	3.4	0.5	0.4	0.9	0.7	2.2	1.2	1.1	0.5	1.2	106.3
Electronic components	594.9	0.6	0.2	0.1	0.3	0.1	0.8	3.9	0.6	0.4	0.6	6.5
Motor vehicles & equipment	1518.0	14.2	0.8	2.4	2.9	1.2	5.1	1.5	1.0	0.8	1.1	2.0
Aircraft	659.4	0.2	—	—	—	—	0.1	—	—	—	—	—
Missiles, spacecraft & ordnance	122.7	—	—	—	—	—	—	—	—	—	—	—
Other transportation equipment	140.1	—	—	—	—	—	—	—	—	—	—	—
Measuring & medical instruments, photo equipment & timepieces	1036.4	5.7	0.7	0.5	1.0	0.9	5.5	2.3	3.0	5.9	2.1	12.4
Miscellaneous manufactures	211.6	0.7	—	0.1	0.1	0.1	0.3	0.1	—	0.1	0.1	0.1
Trade & finance & real estate	39.7	—	—	—	—	—	—	—	—	—	—	—
Transportation & public utilities	47.2	—	—	—	—	—	—	—	—	—	—	—
Construction & services, including R & D services	266.0	0.4	—	—	0.1	0.1	0.3	0.1	—	—	d	d
Total R & D Dollars Used		270.3	56.9	19.2	43.8	22.4	141.1	131.9	59.0	41.7	72.2	185.6

Table 20.1 (cont.)

	Origin R & D	Electronic components (34)	Motor vehicles (35)	Aircraft (36)	Missiles & ordnance (37)	Other trans. equip. (38)	Instruments (39)	Misc. manufactures (40)	Trade (41)	Finance, insurance & real estate (42)	Ground transportation (43A)	Air transportation (43B)
Agriculture & forestry	128.1	—	—	—	—	—	—	—	—	—	—	—
Mining, exc. petroleum	60.3	—	d	—	—	—	—	—	16.0	0.4	d	0.3
Food & tobacco products	444.9	—	0.6	—	—	d	0.2	0.3	0.8	1.0	0.2	0.3
Textile mill products	179.3	—	d	—	—	d	d	d	d	d	d	d
Apparel & leather products	55.5	—	0.1	0.1	0.1	0.2	—	0.3	0.6	0.4	0.2	—
Lumber & wood products	72.6	0.1	0.5	0.1	—	0.2	—	0.3	7.0	2.0	0.3	d
Furniture	51.1	0.1	0.3	0.1	—	—	0.9	0.1	14.7	1.0	0.5	0.3
Paper mill products	202.3	—	0.3	0.3	—	—	0.1	0.6	2.5	1.1	0.2	0.1
Printing & publishing	67.4	—	0.1	0.1	—	—	0.1	—	1.7	0.9	0.3	0.1
Industrial inorganic chemicals	159.2	1.1	2.7	0.3	0.1	0.4	d	0.1	1.6	0.7	0.3	0.1
Industrial organic chemicals	297.2	2.0	0.3	0.1	—	0.1	d	0.3	1.2	1.0	0.3	0.1
Synthetic resins, fibers, rubber	601.6	1.8	12.0	1.2	0.2	0.7	3.7	4.8	1.2	1.0	0.5	0.2
Pharmaceuticals	557.3	—	—	—	—	—	—	—	—	—	—	—
Agricultural chemicals	186.7	—	—	—	—	—	—	—	0.2	0.1	0.2	—
Paints, toiletries, explosives & other chemical products	485.7	4.7	19.3	1.3	0.3	3.6	1.3	3.4	9.5	2.7	2.5	0.3
Petroleum extraction & refining	380.3	0.1	0.5	0.3	0.1	0.1	0.2	0.3	13.3	2.9	5.1	2.9
Rubber & plastic products	419.8	1.2	7.3	0.8	0.3	0.7	1.9	2.9	22.2	9.5	12.4	0.9
Stone, clay & glass products	265.0	d	5.6	2.1	0.2	0.3	d	0.7	8.3	2.0	0.9	0.3
Ferrous metals	189.3	0.1	2.8	0.1	0.2	0.5	0.2	0.3	d	—	d	—
Nonferrous metals	156.9	1.2	1.4	1.3	0.5	0.3	0.5	0.4	0.8	0.5	0.7	—

Fabricated metal products	552.7	0.9	11.0	16.3	0.7	2.9	1.7	1.0	11.7	9.4	8.1	3.3
Engines & turbines	282.2	0.3	1.8	0.9	0.3	8.5	0.6	0.3	12.0	8.9	7.8	1.0
Farm machinery	199.3	—	0.1	—	—	0.1	—	—	d	5.9	—	—
Construction, mining & materials handling equipment	351.2	0.9	5.7	0.2	0.1	2.4	0.4	0.3	10.2	1.6	3.9	2.3
Metalworking machinery	121.5	1.1	12.2	3.2	0.8	0.7	3.9	1.1	2.3	0.1	0.3	—
Other machinery	691.0	3.3	16.2	1.5	0.9	2.7	2.4	2.1	64.0	11.5	8.4	3.5
Computers & office equipment	1153.0	6.6	17.1	20.5	3.8	3.5	16.8	3.7	97.5	176.4	20.9	4.3
Industrial electrical equipment	205.9	0.8	1.4	0.4	0.2	2.4	1.6	0.8	2.1	1.8	2.5	0.3
Household appliances	102.7	—	0.1	—	—	0.2	—	—	4.1	2.8	0.2	—
Lamps, batteries, ignition, X-ray & other electrical equipment	233.3	0.3	12.2	0.4	0.1	1.1	2.4	0.5	13.0	6.1	11.4	19.7
Radio & communication equipment	1227.7	2.0	6.2	17.6	3.6	2.2	1.7	0.9	31.8	20.9	11.7	75.6
Electronic components	594.9	386.4	3.1	4.6	1.5	0.5	2.2	0.8	4.7	5.7	2.2	3.9
Motor vehicles & equipment	1518.0	1.5	158.8	0.8	0.8	1.1	7.0	3.8	288.8	99.0	185.3	5.9
Aircraft	659.4	—	0.4	160.5	0.9	0.2	—	—	0.7	0.4	0.5	360.4
Missiles, spacecraft & ordnance	122.7	—	—	d	22.0	—	—	—	—	—	—	0.3
Other transportation equipment	140.1	d	d	—	—	d	—	—	d	d	59.9	d
Measuring & medical instruments, photo equipment & timepieces	1036.4	11.4	6.8	9.7	2.5	2.1	88.8	1.2	33.1	26.8	14.7	35.4
Miscellaneous manufactures	211.6	0.1	0.5	0.1	—	1.2	0.4	77.5	3.9	2.3	0.7	d
Trade & finance & real estate	39.7	—	—	—	—	—	—	—	(	d	—	—
Transportation & public utilities	47.2	—	—	—	—	—	—	—	)	)	—	—
Construction & services, including R & D services	266.0	d	0.5	0.1	—	0.2	0.2	0.2	4.3	2.5	0.9	d
Total R & D Dollars Used		446.4	308.1	245.0	40.8	44.8	147.0	108.8	727.9	409.7	364.9	524.1

Table 20.1 (cont.)

	Origin R & D	Water transportation (43C)	Telecommunications (43D)	Electric, gas & sanitary utilities (43E)	Construction (44)	Medical services (45A)	All other services (45B)	General government (46)	Defense & space (47)	Final consumption (48)	Row No.
Agriculture & forestry	128.1	—	—	—	—	—	—	—	—	d	1
Mining, exc. petroleum	60.3	—	—	d	—	—	2.5	0.4	—	0.0	2
Food & tobacco products	444.9	—	d	d	0.3	0.9	d	0.4	d	143.1	3, 4
Textile mill products	179.3	0.1	0.1	0.1	0.8	0.7	d	0.4	0.2	18.2	5
Apparel & leather products	55.5	—	d	d	d	d	d	d	d	37.7	6, 19
Lumber & wood products	72.6	—	0.1	0.4	0.3	0.2	1.5	0.4	0.2	2.8	7
Furniture	51.1	—	0.2	0.2	0.6	d	7.1	1.3	1.1	22.4	8
Paper mill products	202.3	—	0.2	0.3	1.4	2.4	4.0	1.6	0.4	73.3	9
Printing & publishing	67.4	—	0.2	0.2	0.4	0.2	d	0.8	1.3	13.9	10
Industrial inorganic chemicals	159.2	—	0.2	d	1.0	0.8	1.6	1.7	1.9	8.8	11
Industrial organic chemicals	297.2	d	0.2	0.8	0.6	2.9	2.3	1.5	9.9	14.1	12
Synthetic resins, fibers, rubber	601.6	0.1	0.3	0.4	1.1	2.0	4.6	1.5	2.9	10.0	13
Pharmaceuticals	557.3	—	—	—	—	321.4	d	0.8	0.1	462.0	14
Agricultural chemicals	186.7	—	—	0.1	0.1	d	1.0	0.5	0.1	80.2	15
Paints, toiletries, explosives & other chemical products	485.7	0.5	2.1	7.5	16.3	4.7	23.7	9.1	3.9	161.9	16
Petroleum extraction & refining	380.3	1.0	0.3	5.7	6.1	1.0	5.5	3.1	3.2	53.2	17
Rubber & plastic products	419.8	d	1.3	5.8	9.7	4.8	18.1	7.2	3.9	111.9	18
Stone, clay & glass products	265.0	d	1.6	5.3	9.0	1.7	3.7	4.6	2.1	36.8	20
Ferrous metals	189.3	d	—	d	1.6	—	d	—	0.1	0.0	21
Nonferrous metals	156.9	—	d	1.8	0.4	1.2	0.7	0.5	3.0	9.6	22

Fabricated metal products	552.7	0.4	5.1	151.2	17.1	3.2	19.8	8.4	29.7	106.6	23
Engines & turbines	282.2	6.4	1.7	75.8	10.2	2.4	21.1	15.6	15.3	59.6	24
Farm machinery	199.3	—	—	0.1	2.4	0.3	4.6	0.5	0.1	33.1	25
Construction, mining & materials handling equipment	351.2	0.2	0.4	1.2	154.4	4.7	3.5	8.1	10.5	4.0	26
Metalworking machinery	121.5	—	—	d	6.6	—	6.6	1.6	0.9	12.2	27
Other machinery	691.0	1.2	2.3	33.0	20.5	7.8	64.3	12.1	26.8	111.6	28
Computers & office equipment	1153.0	1.5	25.8	26.1	2.3	19.8	224.2	75.6	75.2	45.0	29
Industrial electrical equipment	205.9	0.4	0.5	52.0	1.8	13.6	7.9	5.2	4.7	25.4	30
Household appliances	102.7	—	0.1	0.1	0.2	0.2	10.1	0.4	d	94.1	31
Lamps, batteries, ignition, X-ray & other electrical equipment	233.3	0.6	8.8	11.5	5.7	26.0	13.5	7.2	20.9	120.0	32
Radio & communication equipment	1227.7	12.2	417.7	11.0	16.1	8.8	38.1	39.4	220.6	223.2	33
Electronic components	594.9	0.2	43.7	2.3	1.7	2.6	28.2	11.7	50.0	46.2	34
Motor vehicles & equipment	1518.0	2.5	9.5	43.9	125.1	16.9	204.7	79.8	27.7	1345.9	35
Aircraft	659.4	—	0.1	4.1	0.5	0.2	0.4	6.5	120.1	d	36
Missiles, spacecraft & ordnance	122.7	—	d	d	d	d	d	d	93.2	d	37
Other transportation equipment	140.1	6.4	d	—	—	d	d	d	d	51.7	38
Measuring & medical instruments, photo equipment & timepieces	1036.4	2.7	9.3	35.0	13.9	207.0	142.0	35.6	60.2	342.5	39
Miscellaneous manufactures	211.6	d	d	0.6	1.6	1.7	29.0	4.3	1.8	86.9	40
Trade & finance & real estate	39.7	—	—	—	—	—	—	—	—	0.0	41, 42
Transportation & public utilities	47.2	d	d	—	—	—	—	—	—	0.0	43
Construction & services, including R & D services	266.0	d	0.6	4.7	2.2	d	72.6	d	d	14.0	44, 45
Total R & D Dollars Used		84.1	542.8	493.9	432.9	684.6	1000.3	378.1	842.2	4111.4	

Source: This table originally appeared on pages 232–41 of F. M. Scherer, "Inter-Industry Technology Flows in the United States," *Research Policy* 11 (1982): 227–45. Reprinted here with permission of North-Holland Publishing Company.

by industry of use with second-order component flows under the public goods assumption. The industry categories have been consolidated somewhat relative to the original source computations to avoid possible disclosure problems. Because nonmanufacturing industries perform so little R & D but are heavy users, a more detailed level of disaggregation is implemented on the use side of certain nonmanufacturing sectors.

Table 20.2 provides a matrix of the zero-order correlation coefficients between industry totals for some of the principal technology flow variables. Because of the asymmetry of origin versus use disaggregation detail among nonmanufacturing industries, the correlations are for 247 manufacturing industries only. Note that the variables with and without second-order component flows are highly correlated: between USERD1 and USERD2, $r = 0.996$. There is more difference between the private and public goods measures; for example, with component flows, $r = 0.877$.

Also included in the appendix is a variable for each industry with origin data giving internal production process patents as a percent of total coverage ratio-inflated patents. Patents are the focus rather than R & D dollars because of disclosure limitations. The two, however, are fairly closely related. If PRD measures process R & D spending as a fraction (not percentage) of total origin industry spending and PP measures process patents as a fraction of total origin industry patents, the simple regression equation is:

(R1) $PDR = .02 + .956 \ PP; \ r^2 = .855, \ SEE = .128.$
                  (.026)

Examining the individual data in the appendix, one finds wide interindustry differences in the degree of process patent orientation. But there are consistent and plausible similarities within like groups of industries. Thus, complex capital goods producers tend to be very product invention oriented (i.e., with low process invention ratios), while producers of basic raw materials are process oriented. It should be noted, however, that some of the process percentage values in the appendix are computed from rather small numbers of patents, and so possibly substantial sampling errors may exist for the individual industry estimates.

Another potential hazard in the process invention percentage estimates is that they stem, as stated before, from detailed examination of 15,112 individual patents. It is generally believed that process inventions (used largely within the originating firm) are easier to keep secret than product inventions, and from this may follow a propensity for firms to patent relatively fewer process than product inventions, all else (such as the economic significance of the invention) held equal (see Scherer et al. 1959, pp. 153–154). If so, our process patent ratio estimates could have a systematic downward bias. When patents are linked to the privately

Table 20.2   Correlation Matrix: R & D Sums, Manufacturing Industries Only ($N = 247$)

	ORGPAT	ORGRD	USERD1	USERD2	USEPUB1	USEPUB2	USEPAT
ORGPAT	1.000						.658
ORGRD	—	1.000					.422
USERD1	—	—	1.000				.871
USERD2	—	—	—	1.000			.862
USEPUB1	—	—	—	—	1.000		.754
USEPUB2	—	—	—	—	—	1.000	.715
	.724	.673	.690	.565	.591		
		.608	.649	.681	.742		
			.996	.877	.847		
				.889	.871		
					.988		

Definitions:
  ORGPAT: Coverage ratio-inflated count of patents by industry of origin.
  ORGRD: Coverage ratio-inflated company-financed R & D outlays by industry of origin.
  USERD1: R & D by industry of use, private goods assumption, with second-order component flows.
  USERD2: R & D by industry of use, private goods assumption, without second-order component flows.
  USEPUB1: R & D by industry of use, public goods assumption, with second-order component flows.
  USEPUB2: R & D by industry of use, public goods assumption, without second-order component flows.
  USEPAT: Coverage ratio-inflated patent count flowed to industries of use, private goods assumption, without second-order component flows.

financed 1974 R & D dollars of the company LBs in which they originated, one finds that process inventions accounted for 24.6 percent of
total coverage ratio-inflated sample R & D expenditures. There are two
benchmarks against which this figure can be compared. Recent McGraw-
Hill research and development expenditure surveys (1978, 1979) have
asked inter alia what fraction of corporate respondents' R & D outlays
involved process development and improvement. The universe estimates
appear to be sensitive to survey response, varying since 1974 in the range
of 17–24 percent. Second, the Strategic Planning Institute's PIMS data
base contains among other things a breakdown of applied R & D expenditures between product and process categories. These estimates are made
at the level of finely subdivided "businesses" within companies, and are
therefore likely to be more accurate than the corporate aggregates estimated for McGraw-Hill surveys. The simple average process R & D
share for some 948 businesses reporting in PIMS during the mid-1970s
was 25.5 percent.[7] Thus, from comparison with available alternative
benchmark data, there is no reason to believe that our process R & D
share estimates are seriously biased downward.

20.5 Productivity Relationships

Although the technology flow data described in this paper also provide
new insight into a facet of American industry structure, the principal
reason for compiling them was to permit a better-specified analysis of the
links between R & D and productivity growth. The detailed results of that
analysis are described elsewhere (Scherer 1982a). Here a brief overview
must suffice.

Of three productivity data sets analyzed, we focus here on one following input-output industry definitions and published by the U.S. Department of Labor, Bureau of Labor Statistics (March 1979) and supplemented by unpublished computer printouts. With 1974 R & D
expenditures as the independent variable of central interest, the principal
regression analyses examined annual labor productivity growth ΔLP (in
percentage terms) over the peak-to-peak business cycle interval 1973–78.
Productivity indices and corresponding gross capital stock change indices
ΔK were available for a total of eighty-seven industry groups, including
nearly all of manufacturing plus agriculture, crude oil and gas, railroads,
air transport, communications, and the electric-gas-sanitary utilities. Following a formulation developed by Terleckyj (1974, pp. 4–5), the industry R & D flow sums are divided by 1974 industry sales S.

7. Because industries performing relatively little R & D tend to have relatively high
process R & D ratios, the simple average of ratios for the 210 industries covered by the
appendix is 31.4. Relative to a weighted average, as our 24.6 percent figure is, the PIMS
simple average could conceivably be similarly upward biased.

As noted earlier, R & D outlays USERD2 linked to industries of use without second-order component flows had slightly greater explanatory power than the variable USERD1 *with* second-order flows. The simple correlation coefficients with ΔLP were 0.249 and 0.233, respectively. R & D flowed to industries of use under the public goods assumption had appreciably less explanatory power than under the private goods assumption; for example, the zero-order productivity growth correlations were 0.160 for USEPUB1/*S* as compared to 0.233 for USERD1/*S*. A similar but even more pronounced disparity was found with other quite differently measured industry productivity growth data sets. This implies either a lack of support for the public goods approach to technology flows measurement generally or deficiencies in the specific (and necessarily arbitrary) assumptions made to implement that approach.

A strong a priori hypothesis underlying this research was that R & D flowed through to industries of use would better "explain" productivity growth than R & D measured by industry of origin. Product R & D was expected to have especially little explanatory power. The support for this hypothesis with the BLS input-output data set was surprisingly equivocal. Where USERD1/*S* is the used R & D variable and PRODRD/*S* measures product R & D classified by industry of origin, the relevant full-sample multiple regression was:

(R2) $\Delta LP = -.14 + .35 \ \Delta K + .289 \ \text{PRODRD}/S$
                    (.11)   (.144)

      $+ .742 \ \text{USERD1}/S;$
      (.393)

      $R^2 = .193; \ N = 87;$

with standard errors given in parentheses. Both R & D variables are significant at the .05 level, but product R & D has a slightly higher *t*-ratio (2.01 vs. 1.89).

The results were quite different when the industry sample was split into two mutually exclusive subsets, one for which the price deflators underlying the productivity indices were reasonably comprehensive in their industry product line coverage and another for which deflator coverage was skimpy. For the more comprehensive deflator subset, the hypothesis favoring used R & D is clearly supported:

(R3) $\Delta LP = -.16 + .40 \ \Delta K - .182 \ \text{PRODRD}/S$
                    (.14)   (.337)

      $+ 1.039 \ \text{USERD1}/S;$
      (.411)

      $R^2 = .241; \ N = 51.$

Used R & D is highly significant; product R & D negative but insignificant. For the subset based on meager price deflators, the opposite pattern is observed:

(R4) $\Delta LP = .08 + .31 \ \Delta K + .431 \ \text{PRODRD}/S$
                    $(.17)$ $(.205)$

        $+ .096 \ \text{USERD1}/S;$
        $(.96)$

        $R^2 = .197; \ N = 36.$

Since used R & D was also significant and product R & D insignificant in another quite different sample with well-measured productivity indices (see Scherer 1982a), it would appear that the superior performance of the product R & D variable in equations (R4) and (R2) is somehow associated with especially severe problems in measuring productivity growth. With somewhat less compelling support, one is inclined to conclude that the difficult task of tracing R & D flows to industries of use was indeed worthwhile.

20.6 Conclusions

I have described in some detail a methodology for estimating a technology flow matrix for the U.S. industrial economy. Many problems had to be overcome; there are undoubtedly appreciable errors of measurement; and the matrix is incomplete because it has no foreign, university, government laboratory, and individual inventor technology origin sectors. Yet from the standpoint of investigating the relations between R & D and productivity growth, the data developed are surely much closer to what the relevant theory demands than anything previously available.

From regression equations (R2) and (R3) plus additional information, it can be ascertained that a two standard deviation increase in an industry's use of R & D was associated during the 1970s with an annual increase in labor productivity of 1.1 to 1.5 percentage points. Rates of return on investment in used R & D of from 74 to 104 percent are suggested. The magnitudes involved are important economically. I do not know how we can progress further toward understanding the impact of R & D on productivity growth without obtaining additional data similar to, but more accurate and comprehensive than, the R & D use data described here. Yet the thought of linking on an even larger scale patent to R & D data by the extremely labor-intensive methods used in my project is daunting, to say the least. A simpler and more accurate approach would be to have patent applicants provide the necessary information by filling out a form similar to the one used by my patent classification team. The marginal costs would be small, and the rewards in terms of improved information about the structure of technology flows and productivity growth could be substantial.

Appendix

Detailed Industry R & D Data (R & D figures in $millions)

SIC Codes	Description	Origin R & D	Percent Process Patents	Used R & D: Private Goods	Used R & D: Public Goods
01–09	Agriculture, forestry, fisheries	128.2	25	556.2	1939.1
10	Metal mining	11.6	97	35.1	107.1
11, 12	Coal mining	21.1	97	35.1	107.1
13	Oil and gas extraction	179.6	99	275.4	596.1
14	Nonmetallic mineral mining	27.6	77	49.0	177.4
15–17	Construction	28.0	0	435.8	2635.2
2011, 13	Meat packing	31.5	86	57.8	173.4
2016–17	Poultry and egg processing	4.2	67	8.6	33.2
2026	Fluid milk	3.7	0	7.5	68.4
2021–24	Other dairy products	26.2	23	32.3	99.3
2032	Canned specialties	9.1	84	19.9	71.0
2037	Frozen fruits and juices	8.5	100	12.7	43.7
2038	Frozen specialties	9.7	67	13.6	43.9
2033–35	Canned and dehydrated foods	40.6	83	51.3	135.1
2043	Cereal breakfast foods	12.5	62	11.8	26.0
2047	Pet foods	17.4	0	2.4	22.0
2048	Prepared feeds	39.3	60	33.9	63.7
2041, 44, 45	Flour and grain mill products	10.9	7	7.3	35.2
2046	Wet corn milling	16.6	54	13.3	30.2
2051	Bread and cakes	8.5	20	19.3	99.4
2052	Cookies and crackers	6.3	100	9.5	32.2
2061–63	Sugar	5.5	70	10.6	39.7
2065	Confectionary products	5.9	75	5.5	34.9

Appendix (cont.)

SIC Codes	Description	Origin R & D	Percent Process Patents	Used R & D: Private Goods	Used R & D: Public Goods
2066	Chocolate and cocoa products	2.0	87	2.0	10.4
2076	Chewing gum	6.8	13	0.8	5.9
2074–79	Fats and oils	20.0	65	15.8	58.9
2082–83	Beer and malt	12.1	83	36.1	128.3
2084–85	Wines and liquors	5.9	81	13.5	66.1
2086–87	Flavorings, syrups, and soft drinks	25.1	30	41.3	140.9
2095	Coffee	13.1	62	12.1	28.0
2091, 92, 97, 98, 99	Miscellaneous food products	70.3	41	51.8	138.0
21	Tobacco products	31.5	43	29.8	91.4
221, 222, 223, 226	Textile weaving and finishing				
2251, 52	Hosiery				
2253–59	Knitting mills [a]	179.3	65	250.8	457.7
227	Carpets and rugs				
228	Yarn and thread mills				
229	Miscellaneous textiles				
231, 232	Men's clothing	15.4	90	16.4	74.5
233–38	Women's, children's, other clothing	13.0	50	17.3	95.2
239	Miscellaneous fabricated textiles	27.1	26	24.0	84.3
241, 242	Logging and sawmills	56.8	100	92.0	296.4
243	Millwork and plywood	7.7	44	18.2	141.2
245	Wood buildings	0.0	0	12.5	80.3
244, 249	Miscellaneous wood products	8.1	39	13.4	87.2
251	Household furniture	21.5	35	22.6	185.2
252	Office furniture	6.2	22	4.4	27.7
253, 259	Miscellaneous furniture	12.3	6	3.3	30.0

SIC					
254	Partitions and fixtures	11.0	0	3.3	37.8
261–63	Pulp, paper, and paperboard	45.8	73	108.7	340.8
2641–43	Bags, envelopes, and paper coating	39.5	24	30.9	120.1
2647	Sanitary paper products	74.1	26	35.6	65.0
2648	Stationery and tablets	0.7	100	1.3	8.5
2645, 46, 49	Converted paper products	5.1	39	4.6	24.4
265	Containers and boxes	32.6	18	18.9	109.0
266	Building paper and paperboard	4.5	48	4.5	16.7
271–73	Newspapers, periodicals, and books	27.4	19	56.8	260.3
274	Miscellaneous publishing	0.2	0	2.3	12.4
275	Commercial publishing	9.6	68	68.3	260.4
276	Business forms	4.5	39	5.7	24.3
277–79	Other printing and printing services[b]	—	—	—	—
2813	Industrial gases	17.8	49	11.2	25.0
2816	Inorganic pigments	35.1	69	29.2	46.4
2812, 19	Industrial inorganic chemicals	106.3	49	68.4	144.9
2821	Plastic materials and resins	289.6	35	132.7	258.0
2822	Synthetic rubber	97.8	50	50.2	78.9
2823–24	Organic fibers	215.9	59	151.2	255.7
283	Drugs	557.3	12	96.6	244.0
2844	Toiletries	53.7	4	16.3	90.6
2841–43	Soap, detergents, polishes	96.5	13	51.0	181.6
2851	Paints	124.3	13	67.0	144.7
2861, 65, 69	Industrial organic chemicals	297.2	58	211.2	425.0
2873–75	Fertilizers	22.6	57	18.7	60.7
2879	Other agricultural chemicals	164.1	11	27.0	47.0
2892	Explosives	7.0	19	2.2	8.4
2891, 93, 95, 99	Other miscellaneous chemicals	204.3	17	39.4	100.2
29	Petroleum products	201.3	64	202.8	489.2
301	Rubber tires and tubes	117.5	45	135.4	291.3
302–06	Other rubber products	33.7	35	47.2	132.1
3079	Miscellaneous plastic products	268.7	51	289.0	480.0

Appendix (cont.)

SIC Codes	Description	Origin R & D	Percent Process Patents	Used R & D: Private Goods	Used R & D: Public Goods
31	Leather goods	0.0	0	17.6	107.2
321	Flat glass	16.9	71	18.8	60.0
3221	Glass containers	24.9	87	28.7	70.4
3229, 31	Other glass products	58.7	57	44.8	105.2
324	Hydraulic cement	7.3	33	16.9	69.2
325	Structural clay products	7.5	100	16.2	40.5
326	Porcelain and pottery products	42.3	59	28.7	53.9
3271–74, 3281	Concrete and stone products	7.9	33	19.0	135.1
3275	Gypsum products	5.6	15	3.6	14.7
3291	Abrasive products	22.6	34	11.7	26.7
3292	Asbestos products	5.0	28	3.5	20.0
3296	Mineral wool	17.6	54	14.0	41.8
3293, 95, 97, 99	Other mineral products	48.7	31	20.9	46.0
331	Steel mills	123.6	74	223.1	640.8
332	Iron and steel foundries	65.6	100	92.1	239.7
3331	Primary copper	14.4	95	17.6	34.6
3332	Primary lead	2.1	38	1.1	3.3
3333	Primary zinc	1.8	0	0.5	2.8
3334, 53, 54, 55	Aluminum and aluminum products	56.9	64	51.3	137.5
3339	Other primary nonferrous metals	18.3	61	15.7	27.6
3341	Secondary nonferrous metals	6.1	70	8.3	29.6
3351, 56	Nonferrous rolling and drawing	9.1	53	11.0	54.7
3357	Wire drawing and insulating	27.1	67	36.4	96.2
3361, 62, 69	Nonferrous foundries	0.3	67	8.9	54.8
3398–99	Miscellaneous primary metal products	20.8	65	14.8	33.3

341	Metal cans and barrels	54.9	39	46.1	158.7
3421	Cutlery	4.6	20	1.9	10.1
3423, 25	Hand and edge tools	23.3	8	7.0	32.1
3429	Other hardware	13.5	0	7.1	69.7
343	Plumbing and heating ware	23.8	0	7.1	49.5
3441	Fabricated structural metal	2.7	33	8.8	74.4
3442	Metal doors, windows, etc.	3.7	0	3.8	36.0
3443	Nuclear reactors and fabricated plate	225.5	19	71.1	133.1
3444, 46, 48, 49	Miscellaneous fab. metal work	52.1	11	21.3	148.2
345	Screw machine products	33.6	6	12.9	66.0
3462–63	Metal forgings	4.0	100	6.7	26.3
3465	Automotive stampings	8.2	12	13.8	101.3
3466	Crowns and closures	4.2	12	2.8	11.7
3469	Other metal stampings	2.7	17	8.5	66.5
3471, 79	Metal plating and coating	44.4	24	24.2	117.6
348	Ordnance	22.6	10	6.6	44.1
3494	Valves and pipe fittings	40.2	0	7.8	66.8
349 × 3494	Other fabricated metal products	11.3	5	11.6	100.4
3511	Turbines and turbogenerators	117.2	6	15.2	56.9
3519	Internal combustion engines	165.0	10	41.1	106.1
3523	Farm machinery	172.8	2	15.3	97.0
3524	Lawn and garden equipment	26.4	0	3.9	31.2
3531	Construction machinery	242.4	0	26.3	129.2
3532	Mining machinery	22.9	0	1.9	14.7
3533	Oil field equipment	15.6	2	4.1	32.7
3534	Elevators and escalators	13.5	0	1.3	11.7
3535	Conveyors	12.3	0	2.1	16.6
3536	Hoists and cranes	4.5	0	1.5	10.4
3537	Industrial trucks	40.2	7	6.7	36.2
3541	Metal-cutting machine tools	30.9	0	2.6	22.1
3545	Machine tool accessories	8.2	0	2.9	21.6
3546	Power driven machine tools	20.6	0	2.2	18.5

Appendix (cont.)

SIC Codes	Description	Origin R & D	Percent Process Patents	Used R & D: Private Goods	Used R & D: Public Goods
3542, 44, 47, 49	Other metalworking machinery	61.8	3	14.6	123.3
3551	Food products machinery	56.8	2	3.5	18.9
3552	Textile machinery	11.2	0	1.8	17.0
3553	Woodworking machinery	1.1	0	2.4	23.4
3554	Paper industries machinery	10.8	0	1.2	9.2
3555	Printing trades machinery	40.9	4	2.3	15.0
3559	Special industry machinery	40.1	1	7.4	56.5
3561	Pumps	38.3	0	7.4	54.0
3562	Antifriction bearings	30.7	30	13.1	42.1
3563	Compressors	29.4	0	2.5	20.0
3564	Blowers and industrial fans	12.6	0	2.1	16.6
3566	Speed changers and drives	12.5	7	1.6	14.2
3567	Industrial furnaces and ovens	27.5	0	1.2	7.4
3568	Mechanical power transmission equipment	24.4	0	1.9	17.6
3565, 69	General industrial machinery	70.4	10	9.9	34.7
3573	Computers and peripheral equipment	1027.3	6	115.9	307.6
3574	Calculators and accounting machines	51.1	4	11.8	50.4
3576	Scales and balances	5.3	0	0.5	5.3
3572, 79	Other office machines (incl. typewriters)	69.3	0	5.1	39.5
3585	Refrigeration and heating equipment	110.1	10	31.9	188.2
3581, 82, 86, 89	Service industry machinery	65.1	0	5.8	41.3
3592, 99	Miscellaneous machinery	110.0	4	44.5	134.8
3612	Electrical transformers	23.6	11	7.8	34.8
3613	Switchgear	42.5	3	7.3	46.0
3621–22	Motors, generators, controls	64.4	15	21.7	101.6

3623	Electric welding apparatus	9.1	10	5.3	22.6
3624	Carbon and graphite products	13.5	48	7.7	19.3
3629	Other electrical industrial equip.	52.9	20	9.6	19.1
3631	Household cooking equipment	18.1	3	6.4	33.7
3632	Household refrigerators and freezers	16.8	11	13.1	69.7
3633	Household laundry equipment	13.1	5	8.7	48.1
3634, 35, 36, 39	Other household appliances	54.8	3	16.1	82.9
364	Light bulbs and fixtures	67.4	12	23.6	118.7
3651	Radio, TV, and high fidelity sets	141.3	3	13.9	59.9
3652	Records and tapes	4.5	67	3.9	11.8
366	Telephone and communications equipment	1082.0	7	192.8	415.1
3671, 73	Receiving and transmitting tubes	33.6	19	20.3	30.1
3672	Cathode-ray picture tubes	34.7	47	19.6	29.3
3674	Semiconductors	436.3	50	329.2	410.6
3675–79	Other electronic components	90.3	43	71.0	159.4
3691, 92	Batteries	86.5	22	13.9	35.6
3694	Engine electrical equipment	34.5	10	18.3	47.3
3693, 99	Miscellaneous electrical equipment	44.9	2	2.1	14.1
3711	Passenger cars	1263.8	5	157.9	823.0
3713, 95	Trucks, buses, and combat vehicles	139.1	2	49.1	377.1
3714	Motor vehicle parts	109.4	10	99.7	403.0
3715	Truck trailers	5.7	21	4.3	22.6
3721, 28	Aircraft	306.6	22	137.1	361.1
3724	Aircraft engines	352.8	23	107.8	171.6
373	Ship and boat building	22.2	0	26.6	168.2
374	Railroad equipment	66.3	1	11.0	60.0
376	Guided missiles and spacecraft	100.1	16	34.2	97.5
3792	Travel trailers and campers	31.6	0	4.3	28.5
3751, 99	Motorcycles, bicycles, and miscellaneous transportation equipment	20.1	0	3.1	27.7
3811	Scientific and engineering instruments	135.1	2	5.1	27.9
382	Measuring and controlling devices	159.6	3	19.2	111.8

Appendix (cont.)

SIC Codes	Description	Origin R & D	Percent Process Patents	Used R & D: Private Goods	Used R & D Public Goods
383	Optical instruments	66.0	16	12.4	23.7
3843	Dental equipment	25.6	0	1.6	13.3
3841–42	Surgical equipment and supplies	117.5	7	16.8	78.2
3851	Ophthalmic goods	14.4	22	6.4	17.1
3861 (part)	Photocopying equipment	177.9	9	26.9	79.9
3861 (part)	Other photo equipment and supplies	319.0	8	53.2	152.1
3873	Watches and clocks	21.4	2	3.4	17.3
3931	Musical instruments	3.7	0	1.4	10.2
3949	Sporting and athletic goods	18.0	19	14.9	52.1
3942, 44	Dolls, games, and toys	45.5	8	9.9	69.8
395	Pens, pencils, office supplies	10.4	55	10.5	32.2
3911, 14, 15, 396	Jewelry and silverware	4.8	22	4.6	40.6
399, catchall	Miscellaneous manufactures (incl. FTC code 99.99)	129.2	35	67.6	157.0
40–49	Transportation and public utilities	47.2	97	—	—
40	Railroads	—	—	101.5	579.3
41	Suburban transit	—	—	70.0	351.8

Code	Industry				
42	Trucking	—	—	177.8	1684.2
44	Water transportation	—	—	38.8	136.9
45	Air transportation	—	—	524.3	1253.6
46	Petroleum pipelines	—	—	9.4	44.8
49	Electric, gas, sanitary utilities	—	—	497.7	1638.5
43, 47	Other transportation and utilities	—	—	15.5	117.7
50–67	Trade, finance, insurance, real estate	39.7	100	—	—
54	Retail food stores	—	—	96.1	692.7
55	New car dealers and gas stations	—	—	102.1	859.8
58	Eating and drinking places	—	—	82.2	361.1
52, 53, 56, 57, 59	Other retail trades	—	—	226.6	1874.1
60–67	Finance, insurance, real estate	—	—	409.8	3006.8
70–89	Services, including R & D services	238.0	4	—	—
70	Hotels and motels	—	—	43.5	303.8
75	Auto repair services	—	—	116.5	1069.5
78	Motion picture production and exhibition	—	—	11.0	56.0
80	Medical, dental, and health services	—	—	686.8	1382.0
82	Educational services	—	—	147.8	693.6
	Other services	—	—	656.0	3557.2
	Government, except postal and defense	—	—	378.7	2244.7
	Defense and space operations	—	—	1206.3	2689.1
	Final consumption	—	—	4111.0	4111.0

[a]Combined because of line of business disclosure limitations.

[b]Suppressed because of line of business data limitations.

References

Coughlin, Peter E., et al. 1980. New structures and equipment by using industries, 1972. *Survey of Current Business* 60:45–54.

Grefermann, Klaus, K. H. Oppenländer, E. Peffgen, K. Röthlingshöfer, and L. Scholz. 1974. *Patentwesen and technischer Fortschritt*. Göttingen: Schwartz.

Griliches, Zvi. 1979. Issues in assessing the contribution of research and development to productivity growth. *Bell Journal of Economics* 10:92–116.

McGraw-Hill Department of Economics. 1978. 23rd annual McGraw-Hill survey of business' plans for research and development expenditures, 1978–81.

———. 1979. 24th annual McGraw-Hill survey of business' plans for research and development expenditures, 1979–82.

Sanders, Barkev. 1962a. Speedy entry of patented inventions into commercial use. *Patent, Trademark, and Copyright Journal* 6:87–116.

———. 1962b. Some difficulties in measuring inventive activity. In *The rate and direction of inventive activity*, ed. R. R. Nelson. Princeton: Princeton University Press for the National Bureau of Economic Research.

Scherer, F. M., et al. 1959. *Patents and the corporation*, rev. ed. Boston: James Galvin and Associates.

Scherer, F. M. 1979. The causes and consequences of rising industrial concentration. *Journal of Law and Economics* 22: 191–208.

———. 1980a. *Industrial market structure and economic performance*. 2d ed. Chicago: Rand-McNally.

———. 1980b. Interim progress report: Research and development, patenting, and the microstructure of productivity growth. Northwestern University.

———. 1982a. Inter-industry technology flows and productivity growth. *Review of Economics and Statistics* 64:627–34.

———. 1982b. The Office of Technology Assessment and Forecast concordance as a means of identifying industry technology origins. *World Patent Information* 4, no. 1:12–17.

———. 1983. The propensity to patent. *International Journal of Industrial Organization* 1:107-28.

Schmookler, Jacob. 1966. *Invention and economic growth*. Cambridge: Harvard University Press.

Stigler, George J. 1963. Capital and rates of return in manufacturing industries. Princeton: Princeton University Press.

Terleckyj, Nestor. 1974. Effects of R & D on the productivity growth of industries: An exploratory study. Washington, D.C.: National Planning Association.

————. 1980. R & D and the U.S. industrial productivity in the 1970s. Paper presented at the International Institute of Management Conference on Technology Transfer, Berlin.

U.S. Bureau of the Census. 1975. Findings on the research and development response analysis project, 1975. Washington, D.C.: Mimeo.

U.S. Department of Commerce, Bureau of Economic Analysis. 1979. *The detailed input-output structure of the U.S. economy, 1972,* vol. 1, *The use and make of commodities by industries.* Washington, D.C.: GPO.

U.S. Department of Labor, Bureau of Labor Statistics. 1979. *Time series data for input-output industries: output, prices, and employment.* Bulletin 2018, March. Washington, D.C.: GPO.

U.S. Federal Trade Commission. 1981. *Statistical report: Annual line of business report, 1974.* September. Washington, D.C.: FTC.

Comment   Edwin Mansfield

Professor Scherer has carried out a very interesting and useful study. As has frequently been pointed out, an industry's rate of productivity increase depends on technological change in other industries, as well as on its own rate of technological change. Unfortunately, many studies (but not all[1]) of the relationship between R & D and productivity increase have ignored such interindustry technology flows, presumably because of the lack of adequate data. In this paper, Scherer links patent and R & D data to estimate interindustry technology flows. As he describes in detail, his results have to be based on a considerable number of arbitrary decisions. Without question, the results are very rough. But in my judgment, they are worth the toil—and the frustration—described in his paper.

Before turning to more specific comments, I should say that Professor Scherer devotes relatively limited attention to the theoretical justification of some of the procedures he carries out and to guidance for potential users about the way in which his results should (and should not) be interpreted. Most of the paper is devoted to a blow-by-blow account of the mechanics involved, which, of course, is appropriate. In my brief comments, I'll discuss a few of the limitations and problems, and make a few suggestions.

First, I think that Professor Scherer might have explained more fully some of the limitations of R & D expenditures in measuring technology

Edwin Mansfield is a professor in the Department of Economics, University of Pennsylvania.

1. For exceptions, see Mansfield (1980) and Terleckyj (1974).

flows. In this regard, it is important to recognize that the role of R & D is broader than economists often assume. In their models, they view R & D as basically an invention-producing activity. Although this certainly is part of what R & D does, it is by no means its only mission. In addition, R & D provides the firm with a window opening on various parts of its environment; it sometimes is a device to recruit and train people who eventually will move on to general management; and it often includes many activities that are essentially technical service for other parts of the firm.

For some purposes, R & D includes too much. For example, research expenditures might be more appropriate in some cases because only research findings are transferred. For other purposes, R & D includes too little. For example, it is well known that salesmen, sales engineers, and other sales and technical representatives of firms play a major role in transferring technology to their customers. Consider, for example, the computer industry, where IBM has transferred a considerable amount of important technology by training potential users, providing software, and servicing computer installations. It seems reasonable to believe that the amount of technology transferred is measured in part by the cost of such sales, educational, and customer support activities. Indeed, to the extent that much of the technology transferred from one industry to another is old not new (which I would regard as likely), these costs may be more relevant than current R & D expenditures.[2] And it is by no means obvious that they are proportional to the R & D expenditure data Scherer uses.

Second, even if we forget about the limitations of R & D expenditures as a measure of the amount of technology transferred from a given industry to others (or itself), there is a question concerning the use of patents to determine how much of the technology transferred by a given industry to others was received by a particular industry. As no one knows better than Professor Scherer, the value and cost of individual patents vary enormously within and across industries. In some industries, like pharmaceuticals, patents are of considerable economic importance; in others, like electronics, they are of much less importance. Many inventions are not patented. And in some industries like electronics, there is considerable speculation that the patent system is being bypassed to a greater extent than in the past.[3] Moreover, as Scherer recognizes, biases would be expected, since some types of technologies are much more likely to be patented than others. For all these reasons, it seems to me that results based on patent statistics are bound to be rough.

Third, I wonder whether it would be possible for Professor Scherer to compare his findings with what would have resulted if he had simply used

2. For some relevant discussion of technology transfer, see Mansfield et al. (1982).
3. For example, see *Science*, 24 November 1978; and Mansfield, Schwartz, and Wagner (1981).

an input-output matrix to allocate R & D expenditures. From the present paper, one gets little feel for how big the differences are. Many technological changes in one industry find uses or prompt changes in other industries without being transferred via some purchased item. For example, the continuous casting of steel benefited from the previous continuous casting of nonferrous metals. And in some cases, firms provide technology to their suppliers, as documented by a recent study of ours.[4] Thus, I suspect that the technology flows differ perceptibly from the results based on the use of an input-output table. It would be helpful if Scherer could compare the two sets of results, with an eye toward analyzing the differences between them and seeing whether these differences seem reasonable when considered in the light of other evidence. Also, it would be interesting to know whether Scherer's "R & D by industry of use" is more highly correlated with productivity change in an industry than "R & D by industry of use" based simply on an input-output table.

Fourth, as Scherer points out at some length, there is another basic problem in trying to determine how much of the technology generated by a particular industry is transferred to other industries: Should technology be treated as a private or public good? I agree with Scherer that most economists would regard the more reasonable of the two options as being its treatment as a public good. However, there are many costs in transferring technology that economists often neglect. But with regard to the sort of relatively straightforward transfer process that Scherer seems to visualize, the public goods treatment appears more reasonable.

Fifth, still another very difficult problem arises because the transfer of technology from industry A to industry B may result in a higher rate of productivity increase in industry C or D. For example, a manufacturer of equipment to produce thread may develop a new type of equipment, which may increase productivity in the clothing industry, because the new equipment may result in cheaper thread which may increase productivity in clothing. Scherer tries to take this into account in some cases, but the treatment seems somewhat ad hoc. (Construction and twenty-two other industries are singled out for special attention.) Also, it appears that there is seldom an attempt to carry out more than a "second-order flow correction." One wonders whether there are not many cases where technology transferred from industry A to B has an impact on productivity in industries several stages downstream in the economy. The history of technology, as I read it, would suggest this to be true.

Sixth, I wonder whether observed relationships between the rate of increase of total factor productivity and "R & D by industry of use" may not reflect some mismeasurement of inputs like computers. If some of these inputs were measured correctly, if there were better price indexes,

4. See Mansfield and Romeo (1980).

and if quality changes were properly taken into account, the rate of productivity increase in some of the industries buying relatively large amounts of R & D-intensive products might be reduced considerably.

Turning to another matter, it is unfortunate that overseas R & D expenditures by U.S. firms are omitted entirely from Scherer's data. In 1974, U.S. firms spent over $1 billion on overseas R & D.

In conclusion, I think that Scherer's paper addresses an important problem which he correctly regards as being central to some of the issues discussed in this volume. We are a long way from having adequate measures of interindustry technology flows, and it would be unrealistic to suppose that any single paper would solve all or most of the problems in this nettlesome area. But based on any sort of reasonable standard, Scherer's paper is a valuable contribution to work along this line. We all are in his debt.

References

Mansfield, Edwin. 1980. Basic research and productivity increase in manufacturing. *American Economic Review* 70:863–73.

Mansfield, Edwin, and Romeo, Anthony, 1980. Technology transfer to overseas subsidiaries by U.S.-based firms. *Quarterly Journal of Economics* 94:737–50.

Mansfield, Edwin, Anthony Romeo, Mark Schwartz, David Teece, Samuel Wagner, and Peter Brach. 1982. *Technology transfer, productivity, and economic policy*. New York: Norton.

Mansfield, Edwin, Mark Schwartz, and Samuel Wagner. 1981. Imitation costs and patents: An empirical study. *Economic Journal* 91:907–18.

Terleckyj, Nestor. 1974. *Effects of R & D on the productivity growth of industries: An exploratory study*. Washington, D.C.: National Planning Association.

21 R & D and Productivity Growth at the Industry Level: Is There Still a Relationship?

Zvi Griliches and Frank Lichtenberg

21.1 Introduction

A previous paper (Griliches 1980) explored the time-series relationship between total factor productivity (TFP) and cumulated past research and development (R & D) expenditures within different "2-1/2 digit" SIC level manufacturing industries. It used the Bureau of Labor Statistics' (BLS) Input-Output (I-O) sector level productivity and capital series and the National Science Foundation's (NSF) applied research and development series by product class as its data base and focused on the potential contribution of the slowdown in the growth of R & D expenditures to the explanation of the recent slowdown in productivity growth in manufacturing. Its main conclusions were: (1) The magnitude of the R & D slowdown together with the size of estimated elasticities of output with respect to R & D stock do not account for more than a small fraction of the observed decline in productivity. (2) When the data are disaggregated by period, almost no significant relationship was found between changes in R & D stock and productivity growth in the more recent 1969–77 period.[1] This led one commentator (Nordhaus 1980) to interpret these results as evidence for the hypothesis of the depletion of scientific

Zvi Griliches is professor of economics at Harvard University, and program director, Productivity and Technical Change, at the National Bureau of Economic Research. Frank Lichtenberg is an assistant professor in the Graduate School of Business, Columbia University, and a faculty research fellow of the National Bureau of Economic Research.

The authors are indebted to the National Science Foundation (grants PRA-79-13740 and SOC-78-04279) and the NBER Capital Formation Program for financial support.

1. These findings were also consistent with the evidence assembled by Agnew and Wise (1978), Scherer (1981), and Terleckyj (1980).

opportunities. The paper itself was more agnostic, pointing to the large unexplained annual fluctuations in TFP and arguing that many of the recent observations were affected by unexpected price developments and large swings in capacity utilization and, hence, could not be interpreted as being on the production possibilities frontier and as providing evidence about changes in the rate of its outward shift.

A variety of problems were raised by the data and methodology used in that paper, some of which we hope to explore and to improve in this paper. There were, roughly speaking, three kinds of problems: (1) those associated with the choice of a particular R & D series; (2) those arising from the use of a particular TFP series; and (3) those associated with the modeling of the relationship between R & D and subsequent productivity growth. We shall address these topics in turn. To foreshadow our conclusions, we find that the relationship between an industry's R & D intensity and its productivity growth did not disappear. An overall decline in productivity growth has also affected the R & D intensive industries, but to a lesser extent. If anything, this relationship was stronger in recent years. What cannot be found in the data is strong evidence of the differential effects of the slowdown in R & D itself. The time series appear to be too noisy and the period too short to detect what the major consequences of the retardation in the growth of R & D expenditures may yet turn out to be.

21.2 The R & D Data

The major and only source of R & D data at the industrial level of detail are the surveys conducted by the Census Bureau for the National Science Foundation (see, e.g., National Science Foundation 1977). These surveys are based, however, on company reports and on the industrial designation of the company by its main line of activity. There are at least two problems with these data: (1) Many of the major R & D performers are conglomerates or reasonably widely diversified firms. Thus, the R & D reported by them is not necessarily "done" in the industry they are attributed to. (2) Many firms perform R & D directed at processes and products used in other industries. There is a significant difference between the industrial locus of a particular R & D activity, its "origin," and the ultimate place of use of the results of such activity, the locus of its productivity effects. In addition, one should also keep in mind the possibility of pure knowledge spillovers, the cross-fertilization of one industry's research program by developments occurring in other industries.[2]

2. Cf. Griliches (1979) for a more detailed discussion of these issues.

There are various ways of dealing with such problems. We chose to use the NSF data on applied research and development expenditures by *product class* as the basis for our series.[3] The product-class classification is closer to the desired notion of R & D by industry of *use* and it is available at a reasonable level of SIC detail (twenty-eight distinct "2-1/2" digit groupings). It does attribute the fertilizer research of a "textile" firm to the fertilizer industry (but not to agriculture) and the work on bulldozers of an "automotive" firm to the construction equipment industry (but not to construction itself). It is thus based on a nation of proximate rather than ultimate use. Nevertheless, it is much better conceptually than the straight NSF industrial origin classification scheme.[4]

Unfortunately, it is based on much more spotty reporting than the overall R & D numbers. Moreover, after using these numbers in the earlier study, we discovered rather arbitrary and abrupt jumps in the historical series as published by NSF. It appears that when the Census drew new samples in 1968 and 1977, it did not carry through the revisions of the published data consistently backward, leaving large incomparabilities in some of the years for some of the industries. We had to go back to the original annual NSF reports and splice together and interpolate between the unrevised and revised numbers to keep them somewhat comparable over time.[5]

The industrial classification of a particular R & D data set determines the possible level of detail of subsequent analysis. Since the two-digit industrial categories are rather broad, we would like to use finer detail where possible, for example we would like to separate drugs from chemicals or computers from all machinery. This, of course, influences our choice of total factor productivity series.

3. Other ways of dealing with this problem include the use of R & D by product class by industry of origin table (Schankerman 1979), input-output and capital flow of purchase table (Terleckyj 1974), and patents class by industry of origin and use table (Scherer 1981) to redistribute the NSF R & D data.

4. NSF (1977, p. 70) instructs respondents to the industrial R & D survey to complete the "applied R & D by product field" item on the questionnaire as follows:

Costs should be entered in the field or product group in which the research and development project was actually carried on regardless of the classification of the field of manufacturing in which the results are to be used. For example, research on an electrical component for a farm machine should be reported as research on electrical machinery. Also, research on refractory bricks to be used by the steel industry should be reported as research on stone, clay, glass, and concrete products rather than primary ferrous metals, whether performed in the steel industry or the stone, clay, glass, and concrete industry. Research and development work on an automotive head lamp would be classified in other electrical equipment and supplies, regardless of whether performed by an automotive or electrical company.

In fact, however, the majority of respondents interprets this question as relating to "industry of use" according to a recent internal audit by the Bureau of the Census.

5. This work was done by Alan Siu and is described in more detail in appendix B.

21.3 The TFP Data

Because we are interested in industrial detail below the usual two-digit level breakdown, we could not use some of the already published and carefully constructed total factor productivity series, such as Gollop and Jorgenson (1980) or Kendrick and Grossman (1980). In the previous paper we used the BLS growth study data based on the input-output classification of 145 sectors (95 of them in manufacturing; see U.S. Department of Labor 1979a) and associated physical capital data series. These data are subject to two major drawbacks: First, the output concept used by BLS is based on the product rather than the establishment classification, which introduces an unknown amount of incomparability between the output measure and the associated labor and capital measures. The latter are based on the industrial classification of establishments rather than products. Second, the only available output concept is gross output (not value added), and there are no consistent official numbers on material or energy use below the two-digit industry level. The use of gross output and the lack of data on materials introduce a bias of an unknown magnitude that could be quite large during the seventies, when materials and energy prices rose sharply relative to the prices of other inputs.

Because of these problems, we turned to another source of data: the four-digit level, Annual Survey of Manufactures based series constructed by Fromm, Klein, Ripley, and Crawford (1979) as part of a joint Bureau of the Census, University of Pennsylvania, and SRI International (formerly Stanford Research Institute) project.[6] These data cover the years 1959–76 and contain information on material use by industry as well as separate information on energy use since 1971. Several problems also arise with this data set: First, it only goes through 1976. Second, the information on labor input available to us covered only production worker manhours, and we had to adjust it to reflect total employment. Third, the construction of these data is rather poorly documented, so one does not know how some of the numbers were derived or interpolated on the basis of the published sources. Nevertheless, they are very rich in detail and we hope to explore them further in subsequent work.

We used these data, after an adjustment of the labor input, to construct Tornqvist-Divisia indexes of total factor productivity at the relevant levels of aggregation (see appendix A for more detail). Table 21.1 presents estimated rates of growth of TFP between subperiod averages for manufacturing industries according to the breakdown given in the NSF

---

6. We are indebted to David Crawford for making these series available to us.

R & D publications. In these data, a clear retardation in the rate of growth for most of the industries is evident already in the late sixties.[7]

Almost all TFP data start with some gross sales or revenues concept adjusted for inventory change and then deflated by some price index to yield a measure of "output in constant prices." Such a measure is no better than the price indexes used to create it. The price indexes are components of the Producer Price Index (PPI) and associated series reprocessed by the U.S. Department of Commerce's Bureau of Economic Analysis (BEA) to yield a set of deflators used in the detailed deflation of the GNP accounts. As is well known, the quality of these deflators is quite variable.[8] Moreover, there is some reason to suspect that it may deteriorate further in periods of rapid price change, such as 1974–75, where there may be a widening of the gap between quoted prices and the average realized prices by sellers, many of whose prices may have been actually set earlier or not changed as fast as some of the more standard and widely traded and hence also collected items.

We tried rather hard to pinpoint such a deterioration in the price data and to find ways of adjusting for it, but without much success. Looking at the detailed data (either the BLS I-O sectors set or the Penn-SRI one), it becomes quite clear that many of the large TFP declines that occurred in 1974 and 1975 are associated with above average increases in the output price indexes used to deflate the corresponding industry revenue data. Table 21.2 illustrates the negative relation between TFP and output price growth for selected industries (based on four-digit detail) and its growth over time. Some of the reported price movements are large and bizarre and raise the suspicion that they may be erroneous. But without some alternative direct price or output measurement, it is difficult to go beyond such suspicions since, given the accounting identities and the assumption of competitive behavior, declines in productivity would produce a rise in the associated output price indexes.[9] We can either not believe in the

---

7. It should be pointed out that, because of the volatility of the annual TFP series, estimates of the timing and severity of the TFP slowdown (measured by the change in the average annual growth rate of TFP between two adjacent subperiods) are quite sensitive to the particular way in which the entire sample period is divided into subperiods. The weighted (by value of shipments) averages of the industries' beginning-, middle-, and end-of-period TFP average annual growth rates shown in table 20.1 are 1.72, 0.86, and 0.10, respectively. If instead of measuring changes between the mean level of TFP over several years, we compute average annual TFP changes between single "peak" years in business activity (as measured by the Federal Reserve Board index of capacity utilization for total manufacturing), the beginning, middle, and end subperiod definitions are 1959–65, 1965–73, and 1973–76, and the corresponding weighted TFP growth rates are 1.67, 1.23, and − 1.94; almost all of the apparent slowdown occurs at the end of the period.

8. E.g., consider the obvious ridiculously low estimate of TFP growth for the computer industry in table 21.1. It is caused by the absence of a decent price index.

9. In fact, given these identities, if factor prices move similarly for different industries and if factor shares do not change much, the correlation between TFP changes and product price changes should be close to − 1.

Table 21.1   **Average Annual Rates of Total Factor Productivity Growth between Subperiod Averages: Industries in NSF Applied R & D by Product Field Classification, in Percent[a]**

Industry	1959–63 to 1964–68	1964–68 to 1969–73	1969–73 to 1974–76
Ordnance	3.9	−0.9	1.4
Guided missiles	3.3	1.2	1.3
Food	0.7	1.2	−0.3
Textiles	1.5	1.6	−0.5
Plastics	2.8	2.6	0.3
Agricultural chemicals	1.6	2.3	1.2
Other chemicals	1.6	1.5	−1.3
Drugs	4.9	3.6	2.4
Petroleum refining	3.5	1.4	−9.8
Rubber	1.8	1.5	−1.1
Stone, clay and glass	1.8	0.4	0.2
Ferrous metals	1.6	−0.4	−0.2
Nonferrous metals	0.6	−0.6	−0.3
Fabricated metals	1.9	0.4	−0.9
Engines and turbines	2.0	0.8	−0.9
Farm machinery	1.9	0.2	2.3
Construction machinery	2.2	0.1	−1.0
Metalworking machinery	1.7	−0.3	0.3
Computers	1.9	1.3	3.8
Other machinery	2.1	0.3	−0.3
Electrical transportation equipment	2.7	1.9	−0.3
Electrical industry apparatus	3.4	−0.2	0.0
Other electrical equipment	2.7	1.2	0.0
Communications equipment	2.3	2.0	1.6
Motor vehicles	1.7	0.8	−1.1
Other transportation equipment	2.8	0.5	0.3
Aircraft	3.4	0.4	2.1
Instruments	2.1	1.5	1.5

[a]Based on Tornqvist-Divisia indexes constructed from the Penn-SRI data base.

reality of some of the reported productivity declines, in which case we also cast doubt on the price indexes that "produced" such declines, or we can accept both of them as a fact. Both views are consistent with the data as we have them. It would take an independent source of price or output data to adjudicate between these two points of view.

Before we turn to the analysis of the relationship of TFP growth and R & D expenditures, which can be looked at only at the same level of industrial detail as is available for R & D data, we can use the available

Table 21.2   **Correlation Coefficients between Rates of Growth or Rates of
Acceleration of Prices and of Total Factor Productivity in
Four-Digit Industries within Two-Digit Industries 35, 36, and 37**

	SIC 35 Machinery except Electrical	SIC 36 Electrical and Communication Equipment	SIC 37 Transportation Equipment
Rates of growth by period:			
1959–65	−.505	−.701	−.212
1965–73	−.717	−.816	−.252
1973–76	−.821	−.747	−.633
Rates of acceleration, period to period:			
1959–65 to 1965–73	−.521	−.532	−.217
1965–73 to 1973–76	−.782	−.519	−.622
Number of four-digit industries	44	39	17

four-digit detail to look at a few additional aspects of these data. An analysis of variance of annual changes in TFP at the four-digit level during the 1959–73 period illustrates the rather high level of noise in these data. Even in this earlier, relatively calm period only 20 percent of the variance is common at the two-digit level. That is, most of the variance in TFP changes as computed is within two-digit industries. Similarly, only 8 percent of the variance is accounted for by common movements over time. The vast majority of the computed TFP movements are not synchronized. If these numbers are to be interpreted on their face value, as reflecting changes in industrial efficiency, these changes are highly idiosyncratic. Alternatively, if one believed that substantive causal changes in technological levels occur together for subindustries within a two-digit classification and follow similar time patterns, then this lack of synchronization would indicate a rather high level of error in these data.

Another issue of potential interest is whether the observed retardation in TFP growth at the two-digit level is also apparent at the four-digit level and is not just an artifact of a faster growth of lower productivity industries. Computations for three two-digit industries (35, 36, and 37) presented in appendix C, table 20.A.1 indicate that this is indeed the case. If one held the four-digit industrial mix constant at the beginning period levels, the recorded TFP growth would have been even lower. When one looks at the computed rates of retardation (in the second part of table 21.A.1), the effects are reversed, but the differences are quite small. The observed retardation is not an artifact, a "mix" effect. It actually happened quite pervasively at the four-digit level of industrial detail.

21.4   Modeling the R & D to Productivity Relationship

Many of the theoretical issues that arise in the attempt to infer the contribution of R & D to productivity growth from usual types of data were discussed at some length in Griliches (1979) and will not be considered explicitly here. But we want to mention and try to deal with three specific topics: (1) TFP measures as indicators of growth in technological potential: (2) the lag structure of R & D effects; and (3) the functional form and the econometric model within which such effects are to be estimated.

We have already discussed briefly the possibility that the TFP measures as computed are subject both to significant measurement error (arising mainly from errors in the level and timing of the output price deflators) and to large, short-run, irrelevant fluctuations. Irrelevant in the sense that though they do indicate changes in the efficiency with which resources are used, these changes occur as the result of unanticipated fluctuations in demand and in relative prices, forcing firms to operate their plants and organizations in a suboptimal fashion (at least from the point of view of their original design). Whatever theory one has of such business cycle and capacity utilization fluctuations, observations that are not on the production possibilities frontier are unlikely to be informative about the factors that are intended to shift this frontier. By and large, R & D expenditures are spent on designing new products, which will provide more consumer or producer value per unit of resources used, or new processes, which would reduce the resource requirements of existing products. TFP fluctuations obscure such effects because the observed efficiencies do not reflect the potential ones and because during business cycle downswings there is a significant slowdown in investment with an associated, slower than normal, introduction and diffusion of new products and processes.

Within the limits imposed by our data, we tried three different ways of coping with such problems. The first was to assume that "true" productivity can only improve (no forgetting) and hence allows the TFP series to only increase or stay constant, but not decline, by resetting every "lower" observation to the previously observed peak level. The second approach tried to rule out large downward shifts in TFP that appeared to be caused by large changes in the price deflator and seemed to be inconsistent with the observed variable input (labor and materials) data. For example, if sales went up by 10 percent, and variable inputs went up by 5 percent, while the output price index went up by 15 percent, we would assume that perhaps up to one-half of the price movement was in error. The actual formula used was more complicated than that (it is described in the notes to table 21.3). The gist of it is that in the four-digit industries whose output per unit of variable input declined by more than 3 percent, and whose output price increases exceeded their respective two-digit industry average price increases by more than 5 percent, output was redefined so as to make "variable input productivity" decline *exactly* 3 percent. This adjustment affected about 24 percent (119 out of 486) of our annual observations.

Because neither of these procedures had a noticeable effect on our final results, we ultimately turned to the third and simpler way of coping with some of these problems: averaging. We picked subperiods, averaged the total factor productivity within each of these subperiods, and then computed rates of growth between such subperiod averages. In particular, the growth rate of TFP at the beginning of the 1959–76 period was defined by the average annual change between the mean level of TFP during 1959–63 and its mean level during 1964–68; the growth rates at the middle and end of the period were defined in terms of the changes in the mean level of TFP from 1964–68 to 1969–73, and from 1969–73 to 1974–76, respectively. We hope, in this way, to mitigate, if not solve, some of the difficulties discussed above.

We have very little to contribute on the issue of R & D lag effects. In the earlier work, only some of which was reported in Griliches (1980), we experimented at length with various lag structures, but largely to no avail. The data did seem to prefer, weakly, the no depreciation to the any depreciation assumption, and there was also some evidence of the possibility of rather long lags. Unfortunately, given the shortness of the series and the overall level of noise in the data, we could not really distinguish between a small, slowly decaying effect of R & D long past and fixed industry differences in their average levels of TFP. Thus, in this paper, we do not focus on this issue, but we hope to come back to it some day with better methods and data.

The common approach to the estimation of such models is to use the generalized Cobb-Douglas function in which a term involving some measure of R & D "stock" is added on, paralleling the role of physical capital. There is a problem, however, in applying such a framework across industries, since it is unlikely that different industries have the same production function coefficients. The TFP approach goes some ways toward solving this problem, by assuming that conventional inputs are used at their competitive equilibrium levels and by using the observed factor share as approximations to the relevant production function elasticities. This allows each industry to have its own (a priori imposed) labor, capital, and materials coefficients. One is left, then, only with the estimation of trend and R & D effects.

The usual procedure (e.g., Griliches 1980) still imposes a common trend rate and a common output–R & D elasticity on all the data. The common trend restriction can be lifted by shifting to an analysis of first differences—the acceleration (or deceleration) in TFP growth—at the cost of magnifying the role of errors and short-term fluctuations in both the dependent and independent variables. The assumption of a common elasticity of output with respect to R & D stock is bothersome when the relationship is estimated *across* industries with well-known and long-term differences in R & D intensity. Unless the difference between the observed R & D "shares" in sales and the estimated overall common R & D elasticity parameter is to be interpreted as reflecting exact differences between the level of social and private R & D returns (which is not very likely), the estimated model is not consistent with any reasonable optimal R & D choice behavior. An alternative approach, used earlier by Griliches (1973) and Terleckyj (1980), is to reparameterize the model in terms of a common *rate of return* (marginal product) of R & D across industries, rather than a common elasticity. Writing the contribution of the change in the stock of R & D to TFP growth as

$$\gamma \dot{K}/K = \frac{\partial Q}{\partial K}\frac{K}{Q}\frac{\dot{K}}{K} = \rho\frac{\dot{K}}{Q} = \rho\frac{R - \delta K}{Q} \simeq \rho\frac{R}{Q},$$

where γ is the elasticity of output (Q) with respect to changes in the stock of R & D capital (K); $\rho = \partial Q/\partial K$ is the rate of return or marginal product of R & D; R is investment in R & D; and δ is the average rate of depreciation of R & D capital, the TFP growth rate can be expressed as a function of the R & D *intensity* of an industry, provided that δ is zero or close to it. This is the form that we will use in much of what follows.

21.5 Models and Main Results

We postulate a Cobb-Douglas production function (which may be viewed as a local, first-order logarithmic approximation to any arbitrary production function) which includes the stock of R & D capital as a distinct factor of production:

$$(1) \qquad Q(t) = A \cdot K(t)^{\gamma} \cdot \prod_{i=1}^{4} X_i(t)^{\alpha^i} \cdot \exp(\beta t),$$

where $Q(t)$ = output; A = a constant; $K(t)$ = stock of R & D capital; $X_1(t)$ = labor input; $X_2(t)$ = stock of physical capital (structures and equipment); $X_3(t)$ = energy input; and $X_4(t)$ = nonenergy intermediate materials input. Define a conventional index of total factor productivity, $T(t)$, as

$$(2) \qquad T(t) = Q(t)/\prod_{i=1}^{4} X_i(t)^{\alpha^i},$$

normalized to 1 in 1972. By the first-order conditions for producer equilibrium, α_i—the elasticity of output with respect to the ith input ($i = 1, \ldots, 4$)—is equal to the share of the ith factor in total cost of production. Under the maintained hypothesis of constant returns to scale, $\Sigma\alpha_i = 1.$[10]

Combining (1) and (2),

$$(3) \qquad\qquad T(t) = A \cdot K(t)^\gamma \cdot \exp(\beta t),$$

$$(4) \qquad\qquad \log T(t) = \log A + \gamma \log K(t) + \beta t.$$

Differentiating (4) with respect to time and writing, for example, $[d\log T(t)]/dt = \dot{T}/T$,

$$(5) \qquad\qquad \frac{\dot{T}}{T} = \gamma\frac{\dot{K}}{K} + \beta.$$

It is apparent from (1) that γ is the elasticity of output with respect to the stock of R & D capital, that is,

$$\gamma = \frac{\partial \ell n Q}{\partial \ell n K} = \frac{\partial Q}{\partial K}\cdot\frac{K}{Q}.$$

Hence, one may rewrite (5) as

$$(6) \qquad\qquad \frac{\dot{T}}{T} = \frac{\partial Q}{\partial K}\cdot\frac{K}{Q}\cdot\frac{\dot{K}}{K} + \beta = \rho\frac{\dot{K}}{Q} + \beta,$$

where $\rho = \partial Q/\partial K$.

We estimated each of the three equations (4), (5), and (6) to measure the contribution of research and development expenditures to productivity. Although the deterministic versions of (4) and (5) are equivalent, they are not stochastically equivalent: in general, OLS estimation of (4) and (5) would yield different estimates of the parameter γ. In (4) and (5), the output elasticity of R & D capital is viewed as a parameter, that is, invariant across observations; in (6) the marginal productivity of R & D capital is a parameter. We argue below that ρ may be loosely interpreted as the social gross excess rate of return to investment in R & D. While there is no reason to expect the *social* rate of return to be equalized across industries, under the hypothesis that the discrepancy between social and private returns is distributed randomly across industries (or is at least uncorrelated with R & D intensity), an estimate of ρ obtained from (6) will be a consistent estimate of the *average* excess of social over private returns.

A variant of equation (4) was estimated on pooled time-series data (1959–76) for twenty-seven industries. Two modifications were made.

10. There is a question about whether the coefficient of the R & D–stock variable should be included in the definition of constant returns to scale or not. Since the actual inputs purchased by the R & D expenditures are not segregated out of the conventional measures of labor and capital input, we avoid double counting by not including R & D in $\Sigma\alpha_i = 1$ and by interpreting its coefficient as representing both social and excess returns to this activity. See also note 13.

First, each industry was specified to have its own intercept term, $\log A$. Rather than including twenty-seven industry dummies in the estimating equation, $\log T(t)$ and $\log K(t)$ were measured as deviations from the respective industry means. Second, the time trend was generalized to a set of time dummies. These time dummies control for all "year effects" common to the included industries. The actual specification of the estimating equation is therefore

$$(4') \qquad \log \tilde{T}(t) = \gamma \log \tilde{K}(t) + \sum_{\tau=1}^{T} \beta_{\tau} D_{\tau},$$

where a tilde above a variable denotes the deviation of that variable from its industry mean, and D_{τ} ($\tau = 1, \ldots, T$) is a set of time dummies.

It is well known that much of the year-to-year variation in total factor productivity is attributable to fluctuations in the level of capacity utilization. It is perhaps useful to view the TFP time series as the sum of a long-run trend and a serially correlated deviation from trend. We postulate that the level of the R & D stock is a determinant of the trend component of TFP, but not of its short-run deviations from trend; the latter are primarily the result of fluctuations in capacity utilization. A complete model of TFP should include variables accounting for both forces. Alternatively, if one is interested only in explaining the long-run behavior of TFP, one can attempt to remove some of the short-run variation from the observed series. We have tried both strategies in estimating equation $(4')$. In several equations we included a variable, average annual hours of work, postulated to be an indicator of the level of capacity utilization. In other equations we attempted to adjust TFP to its full-capacity level or to eliminate observations in which TFP was below capacity.

Table 21.3 presents regression results for variants of the model $(4')$. Line (1) includes no variable other than R & D stock and year dummies. Line (2) includes a measure of the age of the industry's plant ([gross plant − net plant]/gross plant), while line (3) also includes a utilization index, average annual hours of work per employee. In line (4), the dependent variable was defined as the minimum of the current level of TFP and the previous peak level of TFP. Observations in which TFP was below its previous peak were excluded in estimating the equation on line (5). The dependent variable in line (6) is "adjusted" TFP; the adjustment formula is described in the notes below the table. The coefficient on the R & D variable is negative in all cases and insignificantly different from zero in all but one case.

Before turning to a discussion of the results of estimating variants of the constant marginal productivity (or R & D intensity) model (6), we present in table 21.4 descriptive statistics on TFP and private R & D intensity (or R & D per unit of output) by subperiod for the twenty-seven

Table 21.3   **Summary of "Within" Industries' Total Factor Productivity Level on R & D Stock Regression Results: 27 Industries, 1959–76**

Dependent Variable	Coefficient (*t*-stat) on R & D Stock ($\delta = 0$)	Other Variables	R^2	Line Number
1	−.0014 (0.10)		.6317	(1)
1	−.0031 (0.22)	age	.6375	(2)
1	−.0048 (0.34)	age, hours	.6379	(3)
2	−.0387 (2.85)	age, hours	.7125	(4)
3	−.0014 (0.72)	age, hours	.7475	(5)
4	−.0012 (0.08)	age	.6589	(6)

Key to Dependent Variable (all variables defined as deviations from industry means):
1: Unadjusted TFP.
2: MIN (TFP, past peak TFP).
3: Excludes observations in which TFP < past peak TFP.
4: "Adjusted" TFP, based on following rule for adjusting data at the four-digit level: If "variable input productivity" (output per unit of weighted index of labor, energy, and materials) declined by more than 3 percent, *and* the increase in the price of output exceeded the respective two-digit industry average price increase by more than 5 percent, redefine output so that variable input productivity declines exactly 3 percent.

Table 21.4   **Descriptive Statistics: TFP Growth and Privately Financed R & D Investment per Unit of Output, by Subperiod, 1959–76**

	Mean	Std. Dev.	Minimum	Maximum
Average annual percent change in TFP, between periods:				
1969–63 and 1964–68	2.25	0.93	0.64	4.85
1964–68 and 1969–73	0.92	1.05	−0.92	3.60
1969–73 and 1974–76	0.39	1.29	−1.33	3.77
Privately financed R & D investment as percentage of output: average during period:				
1959–63	3.53	4.10	0.10	14.70
1964–68	3.01	3.13	0.20	11.46
1969–73	2.71	2.50	0.20	10.54

Correlation coefficients:	(1)	(2)	(3)	(4)	(5)	(6)
(1) TFP growth, 1969–63 to 1964–68	1.00	—	—	—	—	—
(2) TFP growth, 1964–68 to 1969–74	0.23	1.00	—	—	—	—
(3) TFP growth, 1969–73 to 1974–76	0.42	0.22	1.00	—	—	—
(4) R & D intensity, 1959–63	0.35	0.51	0.62	1.00	—	—
(5) R & D intensity, 1964–68	0.39	0.59	0.65	0.97	1.00	—
(6) R & D intensity, 1969–73	0.41	0.54	0.69	0.92	0.97	1.00

industry sample.[11] Table 21.4 indicates that both the (unweighted) mean growth of TFP and the (unweighted) level of R & D declined throughout the period, and that the larger absolute decline in both variables occurred early. There is also a striking increase in the variability of TFP growth over time; the standard deviation rises by over 40 percent.

Plots of TFP growth against private R & D intensity by subperiod are shown in appendix C, figures 21.A.1, 21.A.2, and 21.A.3. Note that the computer industry (R) is a consistent outlier in these charts. This is an industry whose productivity is clearly underestimated by the conventional measures.

At the bottom of table 21.4 we show correlation coefficients between TFP growth rates and R & D intensities. Note the extremely high, positive correlations between period-specific R & D intensities, indicating the stability of the industries' relative positions with respect to R & D performance. An alternative (nonparametric) way of analyzing the relationships between TFP growth and R & D intensity is to classify industries into groups, according to their rank in the R & D intensity distribution and to compute the mean rate of TFP growth for each group. Mean TFP growth rates between adjacent subperiods by quartile of the R & D intensity distribution of the earlier period are reported in table 21.5. Industries were ranked according to both private R & D intensity and total R & D intensity. With a single exception, average TFP growth of industries in higher quartiles of the R & D intensity distribution is higher than average TFP growth of industries in lower quartiles, and this relationship appears to grow stronger over time.

We now turn to a discussion of estimates of the TFP growth, R & D intensity model. This model was estimated separately by subperiod under alternative assumptions about the rate of depreciation of R & D capital.[12] For each subperiod and depreciation rate assumption, two variants of the model were estimated: one in which R & D intensity is divided into privately financed and government-financed components, and one in which only total R & D is included. The estimates, reported in table 21.6, indicate that substitution of the R & D measures classified by source of financing for the total R & D figure results uniformly in an improvement in the R^2; in the latter two periods this improvement is dramatic. This improvement arises from relaxing the a priori constraint that the coefficients on the two types of R & D be equal. Obviously, the

---

11. We dropped petroleum refining (SIC 29) from our sample because of clearly erroneous TFP numbers for recent years. The unadjusted numbers show TFP declining at the rate of 10 percent per year during 1973–76, mainly because the material price deflators are for some reason not rising as fast as the output deflators.

12. Note that the R & D intensity is as of the beginning of the period. That is, the $\bar{R}$ associated with TFP growth between 1969–73 and 1974–76 is computed as $(K_{73} - K_{69})/5$, where K is the R & D capital stock constructed on the basis of the various depreciation assumptions.

Table 21.5   **Mean Rate of Total Factor Productivity Growth of Industries, by Quartile of (Private or Total) R & D Intensity Distribution**

Period and Source of R & D Financing	Industries Excluded from NSF R & D Classification[a]	Quartile of R & D Intensity Distribution			
		lowest 1	2	3	highest 4
1959–63 to 1964–68					
Private R & D	0.34	1.56	1.96	2.72	2.85
Total R & D		1.56	1.96	2.64	2.94
1964–68 to 1969–73					
Private R & D	0.13	0.43	0.39	1.08	1.92
Total R & D		0.43	0.55	0.99	1.84
1969–73 to 1974–76					
Private R & D	0.07	−0.24	−0.12	0.55	1.44
Total R & D		−0.15	−0.22	0.22	1.93

[a]These industries' investment in R & D is negligible.

unconstrained coefficients differ greatly in magnitude and even in sign in half of the regressions. Since we can reject the hypothesis of equality of coefficients for privately and government-financed R & D, we shall confine our attention to estimates with R & D disaggregated by source of financing.

The equation for each of the three TFP growth rates indicates that both the highest R^2 and the highest t-statistic on private R & D are obtained under the 0 percent depreciation rate assumption, and that both of these statistics decline monotonically as the assumed depreciation rises. In this sense, the data clearly favor the hypothesis of no depreciation of R & D capital in terms of its effects on physical productivity of resources at the industry level.[13]

Although the coefficient on private R & D is only marginally significant in the 1959–63 to 1964–68 equation, the corresponding coefficients in the two later equations are significantly different from zero at the 1 percent level. Both the coefficients and the associated t-statistics grow larger over the period. Recall that the coefficient on R & D intensity in the TFP growth equation may be interpreted loosely as the social gross excess rate of return to investment in R & D. It is a social rate of return because it is based on output in constant prices rather than profit calculations. It is gross because it also includes a possible allowance for depreciation. And it is excess because the conventional inputs of labor and capital

13. Strictly speaking, the data favor the hypothesis of no depreciation, conditional on the maintained hypothesis of a constant geometric (declining balance) depreciation scheme. Earlier experimentation with other depreciation schemes and lag structures indicates that this conclusion is rather robust.

Table 21.6   Estimates of the Relationship between Averaged Total Factor Productivity and R & D and R & D Intensity, under Alternative R & D Depreciation Assumptions, by Subperiod (N = 27)

Period and Depreciation Rate	R^2	C	Total R & D	R^2	C	Private R & D	Federal R & D
1959–63 to 1964–68							
0%	.1461	2.06	2.69	.2138	1.89	9.15	1.51
		(10.7)	(2.07)		(8.4)	(1.96)	(1.00)
10%	.1088	2.11	3.84	.1516	1.98	12.90	2.20
		(10.9)	(1.75)		(9.0)	(1.52)	(0.83)
20%	.0906	2.13	4.88	.1261	2.03	17.07	2.76
		(11.1)	(1.58)		(9.3)	(1.34)	(0.73)
30%	.0793	2.15	5.86	.1109	2.05	21.46	3.23
		(11.3)	(1.47)		(9.5)	(1.24)	(0.66)
1964–68 to 1969–73							
0%	.0303	0.83	1.38	.3120	0.37	20.33	−1.35
		(3.7)	(0.88)		(1.5)	(3.28)	(0.84)
10%	.0295	0.84	3.00	.3044	0.41	42.84	−2.82
		(3.8)	(0.87)		(1.7)	(3.20)	(0.80)
20%	.0303	0.83	5.71	.2941	0.42	71.47	−4.50
		(3.7)	(0.88)		(1.8)	(3.15)	(0.68)
30%	.0325	0.82	10.40	.2785	0.41	102.01	−5.96
		(3.5)	(0.92)		(1.7)	(3.04)	(0.52)
1969–73 to 1974–76							
0%	.1538	0.11	5.19	.4574	−0.54	33.86	0.69
		(0.4)	(2.13)		(1.9)	(4.20)	(0.29)
10%	.1495	0.09	32.14	.2981	−0.18	74.63	−14.14
		(0.3)	(2.10)		(0.6)	(3.16)	(0.57)
20%	.0028	0.39	−2.13	.2196	−0.03	103.15	−22.47
		(1.5)	(0.27)		(0.1)	(2.49)	(2.10)
30%	.0110	0.38	−3.98	.1459	0.11	109.04	−22.18
		(1.5)	(0.53)		(0.4)	(1.86)	(1.88)

already include most of the R & D expenditures once at "normal" factor prices.[14] The estimates imply an average 9.2 percent social excess rate of return to privately financed R & D investment undertaken during 1959–63, a 20.3 percent rate of return to 1964–68 R & D, and a 33.4 percent return to 1969–73 investments.

The coefficient on government-financed R & D is not significant in any of the three equations, and it has the wrong sign in the second one. In contrast to the private R & D coefficient, the government R & D coefficient is largest and most significant in the first period.

The regressions reported in table 21.6 are of the form

$$\log\left(\frac{Q(+1)}{Q}\right) - \log\left(\frac{IN(+1)}{IN}\right) = \alpha_0 + \alpha_1 \frac{NRD}{Q},$$

where Q = output; IN = index of total input; and NRD = net investment in R & D. Note the presence of Q on both sides of the equation. This suggests the possibility that the observed positive correlation between R & D intensity and TFP growth may be partly spurious, arising, for example, from errors in measuring current output. One way of eliminating this potential source of spurious correlation is to estimate the equation using the *lagged* value of R & D intensity. Estimates of equations in which the lagged value of R & D intensity *replaced* the current value, and equations in which *both* lagged and current values were included are presented in table 21.7. For convenience, the zero-depreciation equations for the three subperiods from table 21.6 are reproduced in table 21.7. In view of our earlier results, the assumption of no depreciation of R & D capital was maintained throughout.

Substituting the lagged (i.e., 1959–63) value of total R & D investment per unit of output for the current (i.e., 1964–68) value in the 1964–68 to 1969–73 TFP growth rate equation slightly increases the R^2; when both variables are included, the lagged value dominates, although both are insignificant. When R & D intensity is disaggregated by source of financing, the R^2 of the current value equation is higher than that of the lagged value equation, although private R & D is significant in both cases. When both current and lagged intensity are included, current intensity dominates.

The current value of R & D intensity dominates the lagged value in all of the 1969–73 to 1974–76 TFP growth rate equations, although the lagged values are also generally significant, indicating that while perhaps slightly biased upward, the results reported earlier (in table 21.6) are not entirely spurious.

Although one's impressions about the timing and severity of the slow-down in TFP growth are sensitive to the periodization scheme adopted,

14. This is only approximately correct. See Schankerman (1981) for a more detailed discussion.

Table 21.7   Total Factor Productivity Growth Related to "Current" and "Lagged" R & D Intensity

R^2	C	Current Total R & D	Lagged Total R & D	Current Private R & D	Lagged Private R & D	Current Federal R & D	Lagged Federal R & D
A. 1959–63 to 1964–68							
.1461	2.06 (10.1)	2.69 (2.07)					
.2138	1.89 (8.4)			9.15 (1.96)		1.51 (1.00)	
B. 1964–68 to 1969–73							
.0303	0.83 (3.7)	1.38 (0.88)					
.0333	0.82 (3.5)		1.45 (0.93)				
.0341	0.81 (3.3)	−1.23 (0.14)	2.65 (0.31)				
.3633	0.33 (1.4)			20.33 (3.28)		−1.35 (0.84)	

R^2							
.2756	0.47 (2.0)				13.85 (3.02)		−0.97 (0.60)
.4283	0.28 (1.1)			49.99 (1.66)	−22.16 (0.94)	2.30 (0.13)	−4.30 (0.24)
C. 1969–73 to 1974–76							
.1538	0.11 (0.4)						
.1215	0.17 (0.7)	5.19 (2.13)	3.41 (1.86)				
.2777	−0.19 (0.7)	45.11 (2.28)	−29.67 (2.03)				
.4854	−0.58 (2.1)			33.86 (4.20)		0.69 (0.29)	
.4173	−0.41 (1.5)				26.22 (3.91)		0.35 (0.20)
.5263	−0.68 (2.4)			42.82 (1.24)	−7.19 (0.26)	33.89 (1.14)	−24.21 (1.09)

that is, the particular way in which the entire sample period is divided into subperiods, some experimentation with alternative schemes indicated that the TFP growth/R & D intensity estimation results reported in this paper are not substantially altered by changing the subperiod definitions. Indeed, the finding that the association between productivity growth and R & D activity became *increasingly strong* over the period is even more apparent in results not reported in the paper (i.e., those obtained using the "peak-to-peak" periodization scheme described in note 8) than it is in the evidence presented above.

To summarize the regression results reported above: variants of the constant elasticity version of the TFP/R & D model (equation [4']) estimated on pooled "within" annual data yielded estimates of the coefficient on R & D that were negative and insignificantly different from zero, whereas the constant marginal productivity version of the model (equation [6]) estimated on a cross section of subperiod averages yielded estimates of the R & D coefficient that were generally positive and significant, at least for private R & D when R & D expenditure was disaggregated by source of financing. In principle, this marked difference in results could be an artifact of either (a) difference in functional form; (b) difference in time period of observation (annual *vs.* subperiod average); or (c) both differences. To determine what the source of the difference in results was, we estimated the constant elasticity version of the model on subperiod averages, that is, we estimated equations of the form

$$\log\frac{TFP}{TFP(-1)} = \beta_o + \beta_1\log\frac{K}{K(-1)},$$

where K = average net stock of R & D over the period. As before, the model was estimated under alternative assumptions about R & D capital depreciation. The R & D coefficients obtained from estimating these equations were never significantly different from zero and were negative in the first and third subperiods under all depreciation assumptions. We may conclude that the relatively good R & D intensity results (compared to the R & D stock results) are not due to the averaging of periods, but rather to the difference in functional form, that is, to the assumption of a constant marginal product rather than a constant elasticity across industries.

A different source of data allows us a more disaggregated glimpse at the same problem. Estimates of the fraction of all employees engaged in research and development by three-digit industry ($N = 139$) are available from the 1971 Survey of Occupational Employment and enable us to estimate the TFP growth/R & D intensity model on more detailed data.[15]

15. See Sveikauskas (1981) for details about these data. We are indebted to Leo Sveikauskas for making these data available to us.

Results based on these unpublished BLS data must be interpreted with caution, however, since their reliability is subject to question because of the underrepresentation of central office workers in the survey sample. To render the results of this analysis comparable to our earlier estimates, we multiplied the ratio of R & D employment to total employment by labor's share in total cost of production in 1971. Assuming real wages (adjusted for interindustry differences in labor quality) are equal across industries, the resulting figure is proportional to R & D employment expenditures per unit of output, a proxy for the desired measure—real net R & D investment per unit of output. Unfortunately, we have only a single cross section for the year 1971 and are therefore forced to assume stability with respect to relative R & D intensity (an assumption warranted by the evidence presented earlier).

Estimates of the TFP growth/R & D intensity equation based on the 139 industry sample for different periods of TFP growth are shown in table 21.8. The results indicate a positive and significant coefficient on R & D intensity in all subperiods. Given that the costs of R & D scientists account for about half of total R & D expenditures, the estimated R & D intensity coefficients should be divided by about half to make them roughly comparable to those reported in tables 21.6 and 21.7. The resulting numbers are significantly higher than those reported for total R & D there but lower than the comparable numbers for privately financed R & D alone. Since the employment numbers reflect both privately and federally financed R & D activities, this is approximately as it should be if the earlier results are attenuated because of aggregation. In any case, here too no evidence of a *decline* in the "potency" of R & D is found.

21.6 Tentative Conclusion

The relationship between the growth of total factor productivity and R & D did not disappear in recent years, though it was obscured by the

Table 21.8   **Total Factor Productivity Growth Related to 1971 R & D Intensity, 139 Three-Digit Manufacturing Industries[a]**

	R^2	C	R & D Intensity
TFP growth, 1959–63 to 1964–68	.0323	1.572 (11.9)	48.361 (2.14)
TFP growth, 1964–68 to 1969–73	.0294	0.436 (3.4)	44.207 (2.04)
TFP growth, 1969–73 to 1974–76	.0672	−0.646 (3.2)	107.85 (3.14)

[a]R & D data derived from 1971 BLS Survey of Occupational Employment.

overall decline in the average growth rate of TFP. While fine timing effects cannot be deduced from the available data, when one does not impose a constant elasticity coefficient across different industries, there appears to be a rather strong relationship between the intensity of private (but not federal) R & D expenditures and subsequent growth in productivity.

Appendix A
Total Factor Productivity Data

The present investigation has the advantage of making use of consistent data on intermediate inputs as well as on gross output and primary inputs. The index of total factor productivity used in the empirical analysis is defined as the ratio of real gross output (shipments adjusted for inventory change) to a Tornqvist index (a discrete approximation to the Divisia index) of four inputs: capital, labor, energy, and materials.[16]

The Tornqvist index of total input is constructed as follows:

$$\ell n\left(\frac{I_t}{I_{t-1}}\right) = \sum_i [.5^*(S_{it} + S_{i,t-1})]\ell n\left(\frac{X_{it}}{X_{i,t-1}}\right),$$

where I_t = index of total input; S_{it} = share of factor i in total cost, $i = K, L, E, M$; X_{it} = quantity of factor $i, i = K, L, E, M$. This formula generates a sequence of growth rates of aggregate input; the *level* of the index in any given year is determined by an arbitrary normalization. The level of total factor productivity is defined as the ratio of output to aggregate input; the latter is normalized so that TFP equals unity in 1972.

The data base was developed jointly by the University of Pennsylvania, the U.S. Bureau of the Census, and SRI International as part of a project under the direction of Gary Fromm, Lawrence Klein, and Frank Ripley. It consists of annual time series (1959–76) on the value of output (shipments adjusted for inventory change), capital, labor, energy, and materials, in current and constant (1972) dollars, for 450 SIC four-digit industries in U.S. manufacturing. The source for most of these series is the Annual Survey and Census of Manufactures. Data for years prior to 1972 were reclassified to conform to the 1972 SIC scheme so that the industry classification is consistent throughout the period.

16. Because expenditure on energy was included in materials expenditure in most years prior to 1971, the input index for the years 1959–71 is based on only three inputs: capital, labor, and the energy-materials aggregate. The input index for 1971–76 (the period during which the relative price of energy increased dramatically) treats energy and materials separately. Construction of the input index for the whole period consisted of defining a three-input index for 1959–71; defining a four-input index for 1971–76; normalizing both indexes to unity in 1971; and splicing the two indexes together in that year.

The following is a brief summary of salient characteristics of the data underlying the total factor productivity indexes. For a more detailed discussion of data sources and methodology, see the appendix to Fromm et al. (1979).

Output. Current dollar output is defined as value of industry shipments adjusted for changes in finished goods and work-in-process inventories. Constant dollar output is derived by deflating the current dollar series by deflators developed by the Industry Division of the Bureau of Economic Analysis. These deflators are constructed at the five-digit level and are generally weighted averages of BLS producer price indexes.

Capital. Consistent with the maintained hypothesis of constant returns to scale, the current dollar value of capital services is computed as the difference between the value of output and the sum of expenditures on labor, energy, and materials.[17] The real flow of capital services is assumed to be proportional to the real capital stock; the capital stock concept is the gross fixed reproducible stock of capital, that is, the stock of plant and equipment net of discards (land and working capital are excluded). The stocks are computed from a perpetual inventory algorithm, which takes account of the industry- and year-specific distribution of expenditures on investment goods across one plant and twenty-six equipment categories (based on a series of capital flow matrices extrapolated from a 1967 matrix by a biproportional matrix balancing procedure). This information on the composition of capital purchases enables development of industry- and year-specific weights for the construction of investment deflators and service lives (weighted averages, respectively, of the PPI's and the service life assumptions for the twenty-seven types of investments).

Labor. The current dollar value of labor services is measured as total expenditures by operating manufacturing establishments for employee compensation, including wages, salaries, and both legally required and voluntary supplements to wages and salaries. We adjusted for the compensation of employees in central administrative offices and auxiliaries. In the absence of data on hours of work of nonproduction workers, real labor input is defined as the ratio of total wages and salaries to average hourly earnings of production workers; under the assumption that the relative wages of production and nonproduction workers are equal to their relative marginal productivity, this ratio may be viewed as an index of "production worker equivalent" manhours. No adjustment was made for changes in labor quality from for example, shifts in the age or sex distribution of employment.

Energy and other intermediate materials. Current dollar energy input is

---

17. Because expenditures for business services such as advertising and legal services are not accounted for, the value of capital services and capital's share in total cost of production are probably slightly overstated.

defined as the value of energy consumed in the production process; it includes energy produced and consumed within an establishment as well as purchases of energy from other establishments. Real energy input is obtained by deflating the current dollar series by a fixed-weighted index of three principal energy prices. Current dollar cost of materials is deflated by a fixed-weighted index of 450 four-digit manufacturing output price deflators and 7 one-digit nonmanufacturing price deflators. The weights for both energy and materials reflect the composition of the industry's purchases of intermediate inputs, as shown in the 1967 input-output table.

Appendix B
Smoothing the Applied R & D Series[18]

1972–75 Data Revision

The 1972–75 data were revised in 1976 because a new sample was drawn in 1976 and a response analysis study was conducted in 1975 which helped to improve respondents' interpretation of definitions of the survey. Consequently, the 1976 data may not be directly comparable to earlier ones. Among the twenty-seven product fields (excluding ordnance, guided missiles, and spacecraft) were three kinds of revision:

Revisions	No. of Product Fields
1. 1972–74 figures increased, 1975 figure decreased	17
2. 1972–74 figures unchanged, 1975 figure decreased	7
3. 1972–75 figures increased	3

Obviously, the first and second revisions result in sharp deceleration of the growth rates between 1974 and 1975, relative to the original series. The rationale behind this pattern of adjustment is unknown. As an alternative, the 1971–75 original annual growth rates were scaled by the 1975 adjustment factor,[19] thereby preserving the 1971–75 overall growth rates in smoothed series.

Stone, Clay and Glass Products

The data for 1968–70 are given as 130, 157, and 128. The 1970 figure was originally reported s 159 and then revised to 128 in 1971, resulting in a big spike in 1969. The 1969 figure was set as 126 (157 × 128/159).

---

18. Prepared by Alan Siu.
19. Log (1975 revised/1975 original).

Fabricated Metal Products

Between 1967 and 1968 there is a 134 percent jump in the data. This break is the result of an abrupt increase of applied R & D done by the electrical equipment and communication industry in the fabricated metal product field, from $49 million to $224 million. To smooth out the series, the 1962–68 growth rate was used as a control total to adjust the annual growth rates within this period.

Electrical Equipment

The data for this product field are not broken down into four subfields between 1967 and 1970. The average shares in 1966–67 and 1971–72 were used to disaggregate the total figures.[20]

Appendix C

Table 21.A.1   **Weighted Averages of Four-Digit Rates of Total Factor Productivity Growth and Acceleration, 1959–76, by Selected Two-Digit Industries**

	SIC 35	SIC 36	SIC 37
A. Weighted average of four-digit rate of TFP growth:			
1959 value of shipment weights	0.379	1.558	0.910
1976 value of shipment weights	0.421	1.821	0.925
Correlation coefficient between rate of TFP growth and change in share of two-digit industry value of shipments, 1959–76	.164	.304	.505
B. Weighted rates of acceleration of TFP between 1959–63 to 1964–68 and 1964–68 to 1969–73			
1959 weights	.077	−1.81	−1.29
1967 weights	.022	−1.82	−1.31
C. Weighted rates of acceleration of TFP between 1964–68 to 1969–73 and 1969–73 to 1974–76			
1967 weights	−3.52	−2.09	−2.35
1976 weights	−3.62	−2.39	−2.79
Number of industries	44	39	17

20. The 1967 data are available separately for the four subfields.

Table 21.A.2   Selected TFP and R & D Data, by Industry in NSF Product-Field Classification

SIC Code	TFP Growth 1959–63 to 1964–68	1964–68 to 1969–73	1969–73 to 1974–76	R & D Intensity 1959–63	1964–68	1969–73	Federal Share in R & D 1973	1977
348	3.9	−0.9	1.4	10.6	5.3	5.6	74.8	80.2
376	3.3	1.2	1.3	66.1	69.1	50.6	89.7	90.0
20	0.7	1.2	−0.3	0.2	0.2	0.2	0	0
22	1.5	1.6	−0.5	0.1	0.2	0.3	0	0
282	2.8	2.6	0.3	12.8	9.5	5.7	1.6	2.1
287	1.6	2.3	1.2	1.8	3.0	3.1	1.1	0.8
281, 284–286, 289	1.6	1.5	−1.3	3.5	3.1	2.4	1.1	2.1
283	4.9	3.6	2.4	8.5	8.3	7.0	1.6	1.3
30	1.8	1.5	−1.1	1.2	1.2	1.2	34.3	34.3
32	1.8	0.4	0.2	0.6	0.7	0.7	7.5	7.5
331, 332, 339	1.6	−0.4	−0.2	0.4	0.4	0.4	3.8	1.7
333–336	0.6	−0.6	−0.3	0.6	0.5	0.5	3.8	1.9
34	1.9	0.4	−0.9	0.6	0.7	1.4	44.0	52.5
351	2.0	−0.8	−0.9	6.1	5.9	5.0	7.6	9.6
352	1.9	0.2	2.3	3.1	2.5	1.9	7.6	0
353	2.2	0.1	−1.0	1.2	1.3	1.9	7.6	0
354	1.7	−0.3	0.3	1.3	1.1	1.1	7.6	0
357	1.9	1.3	3.8	15.9	12.4	11.4	13.7	7.5
355, 356, 358, 359	2.1	0.3	−0.3	1.7	1.2	1.0	12.3	7.4
361	2.7	1.9	−0.3	4.3	4.0	5.1	21.4	43.1
362	3.4	−0.2	0.0	3.5	3.0	3.7	21.4	13.1
363, 364, 369	2.7	1.2	0.0	2.4	2.1	2.1	21.4	28.8
365–367	2.3	2.0	1.6	25.0	14.7	11.6	55.0	48.7
371	1.7	0.8	−1.1	2.2	1.8	2.3	3.5	3.5
373–375, 379	2.8	0.5	0.3	0.8	0.9	1.5	55.2	55.2
372	3.4	0.4	2.1	14.9	12.5	14.2	67.8	68.8
38	2.1	1.5	1.5	4.5	5.6	5.6	27.6	21.7

**Key to Symbols Used to Represent Industries in
Appendix C Figures 21.A.1, 21.A.2, and 21.A.3**

Symbol	Industry	SIC Code
A	Ordnance and accessories, N.E.C.	348
B	Guided missiles and spacecraft	376
C	Food and kindred products	20
D	Textile mill products	22
E	Plastics materials and synthetic resins, rubbers and fibers	282
F	Agricultural chemicals	287
G	Other chemicals	281, 284–286, 289
H	Drugs and medicines	283
I	Rubber and miscellaneous plastics products	30
J	Stone, clay, and glass products	32
K	Ferrous metals and products	331, 332, 339
L	Nonferrous metals and products	333–336
M	Fabricated metal products	34
N	Engines and turbines	351
O	Farm machinery and equipment	352
P	Construction, mining, and materials-handling machinery and equipment	353
Q	Metalworking machinery and equipment	354
R	Office, computing, and accounting machines	357
S	Other machinery, except electrical	355, 356, 358, 359
T	Electric transmission and distribution equipment	361
U	Electrical industrial apparatus	362
V	Other electrical equipment and supplies	363, 364, 369
W	Communication equipment and electronic components	365–367
X	Motor vehicles and equipment	371
Y	Other transportation equipment	373–375, 379
Z	Aircraft and parts	372
7	Instruments	38

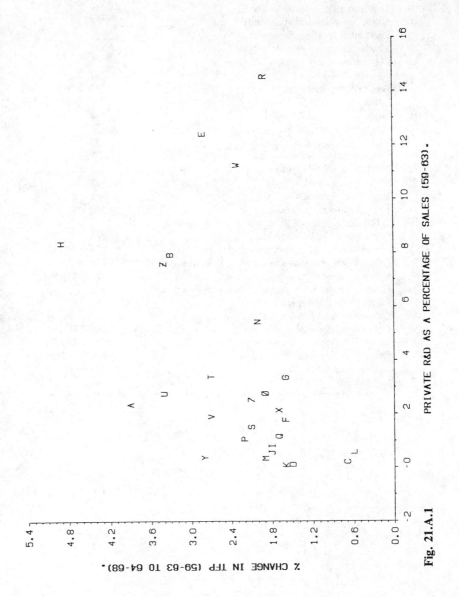

PRIVATE R&D AS A PERCENTAGE OF SALES (59-63).

% CHANGE IN TFP (59-63 TO 64-68).

Fig. 21.A.1

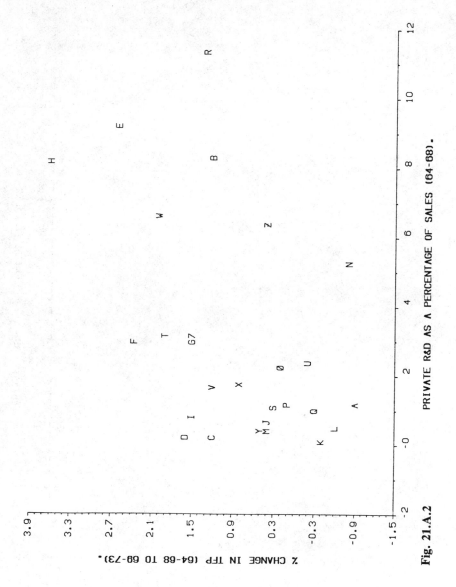

PRIVATE R&D AS A PERCENTAGE OF SALES (64-68).

% CHANGE IN TFP (64-68 TO 69-73).

Fig. 21.A.2

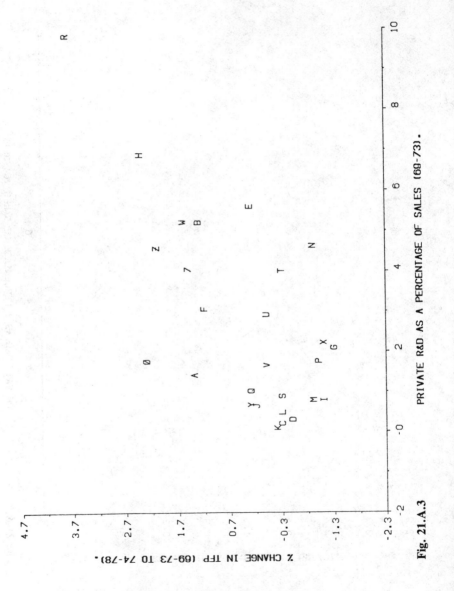

Fig. 21.A.3

References

Agnew, C. E., and D. E. Wise. 1978. The impact of R & D on productivity: A preliminary report. Paper presented at the Southern Economic Association Meetings. Princeton, N.J.: Mathtech, Inc.

Fromm, G., L. R. Klein, F. C. Ripley, and D. Crawford. 1979. Production function estimation of capacity utilization. Manuscript, Univ. of Pennsylvania.

Gollop, F. M., and D. W. Jorgenson. 1980. U.S. productivity growth by industry, 1947–73. In *New developments in productivity measurement*, ed. J. W. Kendrick and B. Vaccara. Conference on Research in Income and Wealth: Studies in Income and Wealth, vol. 44. Chicago: University of Chicago Press for the National Bureau of Economic Research.

Griliches, Z. 1973. Research expenditures and growth accounting. In *Science and technology in economic growth*, ed. B. R. Williams, 59–95. London: Macmillan.

———. 1979. Issues in assessing the contribution of research and development to productivity growth. *Bell Journal of Economics* 10, no. 1:92–116.

———. 1980. R & D and the productivity slowdown. *American Economic Review* 70, no. 2:343–48.

Kendrick, J. W., and E. Grossman. 1980. *Productivity in the United States: Trends and cycles*. Baltimore: The Johns Hopkins University Press.

National Science Foundation. 1977. *Research and development in industry*, serial 1957–77. Washington, D.C.: GPO.

Nordhaus, W. D. 1980. Policy responses to the productivity slowdown. In *The decline in productivity growth*, series no. 22. Federal Reserve Bank of Boston.

Schankerman, M. A. 1979. Essays on the economics of technical change: The determinants, rate of return, and productivity impact of research and development. Ph.D. diss., Harvard University.

———. 1981. The effects of double-counting and expensing on the measured returns to R & D. *Review of Economics and Statistics* 63, no. 3: 454–58.

Scherer, F. M. 1981. Research and development, patenting, and the microstructure of productivity growth. Final report, National Science Foundation, no. PRA-786526.

Sveikauskas, L. 1981. Technological inputs and multifactor productivity growth. *Review of Economics and Statistics* 63, no. 2:275–82.

Terleckyj, N. E. 1974. Effects of R & D on the productivity growth of industries: An exploratory study. Washington, D.C.: National Planning Association.

————. 1980. R & D and U.S. industrial productivity in the 1970s. Paper given at the International Institute of Management Conference on Technology Transfer, Berlin.

U.S. Department of Labor. Bureau of Labor Statistics. 1979a. *Time series data for input-output industries: Output, prices, and employment.* Bulletin 2018. Washington, D.C.: GPO.

————. 1979b. *Capital stock estimates for input-output industries: Methods and data.* Bulletin 2034. Washington, D.C.: GPO.

Comment   Nestor E. Terleckyj

General Comment

In their paper, Griliches and Lichtenberg report on a number of intensive explorations for the relationship at the industry level between R & D expenditures and productivity growth in the 1970s. In past research, relationships between productivity growth and R & D have been found for earlier periods and over longer time intervals, but in many industry-level studies the relationship could not be established for the 1970s.

The authors use a new set of total factor productivity data defined at the industry level approximating the product line categories for which the NSF reports the amounts of applied research and development performance. With these data they explore the relationship between R & D and productivity for three subperiods of the period 1959–76.

Griliches and Lichtenberg reconfirm evidence of a positive relationship between private R & D expenditure and productivity growth of the industries performing this R & D. They conduct a number of careful and sophisticated analyses to control for short-term variation in the data and to test their results. In addition to confirming continued existence of a relationship between R & D expenditures and productivity, their results contribute significantly to our knowledge on a number of subjects ranging from the formulation of the underlying theoretical model to depreciation of R & D. They also throw new light on the recent history of R & D investments and productivity changes which helps us better understand some of the U.S. economic history in the 1960s and the 1970s.

The present paper reports a number of intensive explorations of the data directed to the question of the effect of private R & D expenditures on the productivity of industries conducting this R & D, but it does not attempt to distinguish between the conduct and the utilization of R & D as some of the other recent research did.

Nestor E. Terleckyj is vice-president and director of the Center for Socio-Economic Analysis, National Planning Association.

I have two questions regarding the specification of independent variables by Griliches and Lichtenberg. One deals with omission of non-R & D variables, and the other with the effects of industrial R & D on productivity of industries using the new products developed by the performing industries.

Some Questions

Non–R & D variables. The authors include only the conventional inputs and R & D in their model, implying that other factors influencing productivity represent essentially short-term cyclical, random, or irregular effects. Indeed, the authors do get significant results with this approach. However, apart from the transitory, cyclical, and noise factors, there may be certain long-term structural characteristics of industries that might raise the explanatory power of the relationship and sharpen the estimates of the R & D effects. Such variables representing organizational structure of industries (e.g., unionization of labor) or the long-term structure of the variability of output, as distinct from the short-term effects determining the position of data points within the business cycle, have been related to TFP growth in the past.

Interindustry effects of R & D performance. Griliches and Lichtenberg obtain sufficient explanation with the R & D conducted in industries, and I am not disputing their interpretation of these results. My concern is with the need to recognize the additional effects of R & D on productivity of other industries and sectors. R & D performed in an industry is clearly important to its productivity growth. However, many productivity effects of R & D are transferred to other industries in the form of improved materials and capital goods. The productivity of purchasing industries may be much influenced even if they conduct no or relatively little R & D as in the case of airlines, banking, or the textile industry. Previous work by Link (1982), Scherer (1981), and myself (Terleckyj 1974, 1982) distinguished explicitly between the R & D performed in the industry and the R & D embodied in inputs purchased from the R & D-intensive industries, and further between the product and process R & D. Of course, the authors recognize such transfers and discuss them in section 21.2. But their paper is focused on the direct effects of the R & D performed in an industry and on the continued existence of the relationship between R & D and productivity.

In pursuing my questions, I have reestimated Griliches and Lichtenberg's best-fit equation (equation in table 21.7 for the latest subperiod based on zero rate of depreciation of R & D) by first using their formulation of the R & D variable (total R & D performed in an industry) and then introducing separately two alternative R & D variables, one representing the part of R & D both performed and used in the industry for which the productivity growth is being measured and the other representing the R & D embodied in capital goods imported from other industries.

The allocations of R & D were based on the estimates of technology flows made by Scherer (this volume). I have also included the unionization variable which in the past (Kendrick 1973; Terleckyj 1974; Kendrick and Grossman 1980) was found to be structurally correlated with productivity growth. My table C21.1 contains the results of these changes and additions for the twenty-four industries used by the authors. (Ordnance, guided missiles and petroleum refining were excluded.)

These results actually support the choice of variables made by Griliches and Lichtenberg. Equation (1), which replicates their best-fit equation, has a better statistical fit than the other equations in the table. Also, the regression coefficient for the R & D performed in the industry is statistically stronger than coefficients estimated for other R & D variables. Little is added to the explanation obtained with R & D performed in the industry by introducing other variables in equations (4) and (5). Nevertheless, equation (3), although it has a lower R^2 than equation (1), (.40 instead of .49), suggests that an alternative explanation of productivity

Table C21.1   Estimating Equations for Total Factor Productivity Growth Rates, 1969–73 to 1974–76, with Private R & D Intensity 1973/74 and Percent Unionized 1973–75 (t-ratios)

No.	Constant	R & D in R/Q Ratio			UN	R^2
		Total R & D Performed in Industry	R & D Performed and Used in Industry	R & D Embodied in Capital Goods Purchased from Other Industries		
(1)	.79	36.90	—	—	—	.49
	(2.61)	(4.84)				
(2)	−.46	—	—	209.94	—	.24
	(1.27)			(2.91)		
(3)	1.13	—	—	174.83	−.04	.40
	(1.63)			(2.67)	(2.60)	
(4)	−.03	31.86	—	—	−.02	.50
	(.03)	(3.51)			(1.02)	
(5)	−.04	25.60	—	62.34	−.02	.48
	(.05)	(2.09)		(.78)	(1.16)	
(6)	−.28	—	118.23	—	—	.09
	(.64)		(1.78)			
(7)	−.66	—	57.56	182.75	—	.24
	(1.53)		(.87)	(2.31)		

Source: Griliches and Lichtenberg (this volume); Scherer (1981); Freeman and Medoff (1979).

growth is possible in terms of an R & D variable representing the R & D content of the capital goods used in the industry and the unionization variable. The regression coefficient for unionization is significant at a 1 percent level in equation (3) and is negative in other equations.

The main problem appears to be that the R & D variables are very highly intercorrelated. This is the case here and has been typical of other industries for all types of R & D variables, for example, R & D performed in an industry, R & D performed and used in an industry, R & D attributed to an industry through its purchases from other industries, company-funded and government-funded R & D, product and process R & D, etc. Identifying their separate effects may be extremely difficult. For this reason, I could not add a variable representing the R & D embodied in intermediate inputs to the equations in table C21.1. Moreover, R & D effects usually do not dominate productivity growth over any given period, and the results, in some cases, have been sensitive to the method of measurement of productivity. This complicates further the task of separately identifying the different kinds of R & D effects on productivity, especially with measurement errors present in the output data.

For these reasons I do not believe that the strong results obtained by Griliches and Lichtenberg for the R & D performed in industry should be taken to mean that the productivity impacts of R & D are largely internal to the industries in which the R & D is conducted. The observed results may also reflect some of the various transfer effects. However, the analysis of transfer and diffusion of R & D effects among industries and sectors, and the role they play in generating productivity changes, will require considerable further research in its own right.

Implications of Results and Conclusions

The work that Griliches and Lichtenberg have done in this paper is important and very timely in a number of ways at the present stage of research in this field.

Relationship in the 1970s. Confirmation of the existence of a relationship between R & D and productivity for the 1970s in itself is important. Taken together with the similar results obtained by others (notably by Link and by Scherer), this carefully assembled evidence establishes a continuity for research in the field. Their subperiod analysis may suggest some new explorations into the role of private and possibly government R & D in the productivity slowdown of the mid-1960s.

Depreciation of R & D. At several points their results suggest and the authors indicate that in productivity analysis the depreciation rate of R & D should be zero. I have obtained similar results in my own work (Terleckyj 1982, 1983). I think that we can perhaps clear up the question of depreciation of R & D and suggest the following approach. We should draw sharp distinctions between different economic effects of different types of R & D because the appropriate depreciation rates, even for the same R & D, differ greatly among its different effects. Specifically, for the private expenditures for industrial R & D in the United States, we should distinguish between: (1) R & D as private capital asset, that is, R & D as a source of profit to the investor; (2) R & D as a source of prevailing technology and a determinant of the level of productivity of the U.S. economy; and (3) R & D as a social capital asset determining the rate of growth of the U.S. economy.

As a source of profit, R & D depreciates very rapidly, apparently much more rapidly than has been thought before. As a result of domestic and foreign competition, existing technologies become obsolete regardless of their level of productivity as new, more profitable, and usually more productive, technologies are developed. Estimates reported by Ravenscraft and Scherer (1982) and by Pakes and Schankerman (in this volume) indicate very high rates of depreciation and comparatively short useful lives of R & D capital as a source of profit. This is quite consistent with the research results reported by Mansfield on the cost and time lags of imitations relative to innovations. Mansfield (this volume) reports estimates of imitation cost at two-thirds of the cost of innovation and a 60 percent rate of successful legal imitation within four years. Consequently, the average economically useful life of patent protection appears to be much shorter than its legal life.

On the other hand, as a source of productivity (output per unit of input) R & D does not depreciate at all. A level of total factor productivity reached in the economy as a result of technological improvements based on past R & D can be maintained indefinitely by replacing labor and capital of the same kind without need for any additional R & D conducted to maintain it. As a source of growth in income and output (the degree of utilization of existing technology), the social R & D capital of the United States does depreciate, but less rapidly than the private capital of the R & D investors because it is affected only by foreign competition. However, its rate of depreciation has not yet been analyzed. Clearly, various regional disaggregations and international aggregations of this type of capital may also be defined for specific analytic purposes.

Methodological implications. The results obtained by Griliches and Lichtenberg have two significant methodological implications. First, the marginal product model appears to be rather robust and seems to perform quite well in interindustry productivity analysis. Second, cross-sectional industry analysis, even at this fairly high level of aggregation, apparently continues to be a promising mode of research in this field.

Data quality for high technology industries. Finally, this paper brings out the problem of poor quality of price and output data. Because some of the mesurement biases are most serious for the R & D-intensive products, research on productivity and technological change will continue to be hampered by large distortions in its most important observations. We should explicitly recognize the data problem, but in the near term there is not much we can do about it. In the long-run, however, basic work has to be done by the research community in measuring output and prices in high technology fields before regular maintenance of these indexes is taken over by a government agency and better data become available for research. In my opinion this measurement work is one of the top priority items for the research agenda in the economics of technical change.

References

Freeman, Richard B., and James L. Medoff. 1979. New estimates of private sector unionism in the United States. *Industrial and Labor Relations Review* 32, no. 2:143–74.

Kendrick, J. W. 1973. *Postwar productivity trends in the United States, 1948–1969.* New York: National Bureau of Economic Research.

Kendrick, J. W., and E. S. Grossman. 1980. *Productivity in the United States: Trends and cycles.* Baltimore: The Johns Hopkins University Press.

Link, A. N. 1982. Disaggregated analysis of industrial R & D: Product versus process innovation. In *The transfer and utilization of technical knowledge,* ed. D. Sahal. Lexington, Mass.: D. C. Heath.

Ravenscraft, David, and F. M. Scherer. 1982. The lag structure of Returns to R & D. *Applied Economics* 14 (December): 603–20.

Terleckyj, N. E. 1974. *Effects of R & D on the productivity growth of industries: An exploratory study.* Report no. 140, December. Washington, D.C.: National Planning Association.

Terleckyj, N. E. 1982. R & D and the U.S. industrial productivity in the 1970s. In *The transfer and utilization of technical knowledge,* ed. D. Sahal. Lexington, Mass.: D. C. Heath.

Terleckyj, N. E. 1983. R & D as a source of growth of productivity and of income. In *The science of productivity,* Franke, R. H. and Associates. San Francisco: Jossey-Bass.

List of Contributors

Professor Andrew B. Abel
Department of Economics
Littauer Center 111
Harvard University
Cambridge, MA 02138

Professor John J. Beggs
Dept. of Statistics
Faculty of Economics
Australian National University
Canberra ACT 2601
Australia

Professor Uri Ben-Zion
Faculty of Industry Engineering
  and Management
Technion
Haifa 32000
Israel

Mr. John Bound
National Bureau of Economic
  Research
1050 Massachusetts Avenue
Cambridge, MA 02138

Professor Kim B. Clark
Harvard University
Graduate School of Business
  Administration
Morgan 8, Soldiers Field
Boston, MA 02163

Mr. Clint Cummins
Department of Economics
Littauer Center
Harvard University
Cambridge, MA 02138

Mr. Philippe Cuneo
Ecole Nationale de La Statistique et
  de l'Administration Economique
3, Avenue Pierre Larousse
92240 Malakoff
France

Professor George C. Eads
School of Public Affairs
Suite 1218, Lefrak Hall
University of Maryland
College Park, MD 20742

Professor Robert E. Evenson
Economic Growth Center
Yale University
Box 1987, Yale Station
New Haven, CT 06250

Professor Zvi Griliches
National Bureau of Economic
  Research
1050 Massachusetts Avenue
Cambridge, MA 02138

Ms. Bronwyn H. Hall
National Bureau of Economic
  Research
204 Junipero Serra Boulevard
Stanford, CA 94305

Mr. Adam Jaffe
Department of Economics
Harvard University
Cambridge, MA 02138

Professor Richard C. Levin
Department of Economics
Yale University
37 Hillhouse Avenue
New Haven, CT 06520

Mr. Frank Lichtenberg
Graduate School of Business
Columbia University
Uris Hall
New York, NY 10027

Professor Albert N. Link
School of Business and Economics
University of North Carolina
Greensboro, NC 27412

Professor Jacques Mairesse
Ecole Nationale de La Statistique et
  de l'Administration Economique
3, Avenue Pierre Larousse
92240 Malakoff
France

Professor Edwin Mansfield
Wharton School, Department of
  Economics
University of Pennsylvania
Philadelphia, PA 19104

Professor M. Ishaq Nadiri
Department of Economics
New York University
15–19 West 4th Street
New York, NY 10012

Dr. Ariel Pakes
Department of Economics
Mount Scopus Campus
Hebrew University of Jerusalem
Jerusalem 91905
Israel

Professor Peter C. Reiss
School of Business
Stanford University
Stanford, CA 94305

Professor Mark Schankerman
Department of Economics
New York University
269 Mercer Street
8th floor
New York, NY 10003

Professor Frederic M. Scherer
Department of Economics
Swarthmore College
Swarthmore, PA 19081

Professor John T. Scott
Department of Economics
Dartmouth College
Hanover, NH 03755

Mr. Alan K. Siu
Department of Economics
Chinese University of Hong Kong
Shatin-Hong Kong

Professor Pankaj Tandon
Department of Economics
Boston University
270 Bay State Road
Boston, MA 02215

Dr. Nestor E. Terleckyj, Director
Center for Socio-Economic Analysis
National Planning Association
1606 New Hampshire Avenue, NW
Washington, DC 20009

Author Index

Subject Index